Springer Proceedings in Mathematics & Statistics

Volume 386

This book series features volumes composed of selected contributions from workshops and conferences in all areas of current research in mathematics and statistics, including data science, operations research and optimization. In addition to an overall evaluation of the interest, scientific quality, and timeliness of each proposal at the hands of the publisher, individual contributions are all refereed to the high quality standards of leading journals in the field. Thus, this series provides the research community with well-edited, authoritative reports on developments in the most exciting areas of mathematical and statistical research today.

More information about this series at https://link.springer.com/bookseries/10533

Alexander M. Kasprzyk · Benjamin Nill
Editors

Interactions with Lattice Polytopes

Magdeburg, Germany, September 2017

Springer

Editors
Alexander M. Kasprzyk
School of Mathematical Sciences
University of Nottingham
Nottingham, UK

Benjamin Nill
Fakultät für Mathematik
Otto-von-Guericke-Universität Magdeburg
Magdeburg, Sachsen-Anhalt, Germany

ISSN 2194-1009 ISSN 2194-1017 (electronic)
Springer Proceedings in Mathematics & Statistics
ISBN 978-3-030-98326-0 ISBN 978-3-030-98327-7 (eBook)
https://doi.org/10.1007/978-3-030-98327-7

Mathematics Subject Classification: 14M25, 52B20, 05E10, 06A07, 14J33, 14J45, 14M15, 14M27, 14N15, 52B12, 52B40, 53D05

This Springer imprint is published by the registered company Springer Nature Switzerland AG
The registered company address is: Gewerbestrasse 11, 6330 Cham, Switzerland

Preface

This volume contains original research and survey articles highlighting interdisciplinary connections between a diverse range of topics. The common points of interest are lattice polytopes. Lattice polytopes are fundamental combinatorial objects—convex polytopes whose vertices have integer coordinates—with many beautiful and deep connections across modern mathematics. Topics considered include: algebraic geometry, mirror symmetry, symplectic geometry, discrete geometry, the geometry of numbers, and algebraic combinatorics.

The study of lattice polytopes continues to open up fertile and unforeseen interactions. In order to enhance this exchange of ideas, the workshop *Interactions with Lattice Polytopes* took place 14–16 September, 2017, at the Otto-von-Guericke-Universität Magdeburg, Germany. There were 15 talks given by world-leading experts from several different backgrounds, elaborating upon the theme of applications of lattice polytopes. Many of the presented results can be found in this volume.

Contributions to this volume contain original as well as expository research articles that illustrate some of the varied topical approaches and settings where lattice polytopes play an important role. This volume should be particularly beneficial to researchers and graduate students interested in learning more about the multifaceted use of lattice polytopes across a broad range of active research areas.

This book relies deeply on the enthusiasm and engagement of the diverse and collegial lattice polytope community. We are extremely grateful to the contributors for their high-quality articles, and to the anonymous referees for their careful work. We would like to express our gratitude to everyone involved for their patience and assistance. We are also thankful for logistical and financial support from: the Otto-von-Guericke-Universität Magdeburg; the Research Training Group Mathematical Complexity Reduction, funded by the Deutsche Forschungsgemeinschaft (DFG, German Research Foundation)—314838170, GRK 2297 MathCoRe;

and the Engineering and Physical Sciences Research Council (EPSRC) Fellowship EP/N022513/1.

Nottingham, UK
Magdeburg, Germany
October 2021

Alexander M. Kasprzyk
Benjamin Nill

Contents

Contributors

Gennadiy Averkov Brandenburgische Technische Universität Cottbus-Senftenberg, Fakultät 1, Cottbus, Germany

Victor Batyrev Mathematisches Institut, Universität Tübingen, Tübingen, Germany

Mónica Blanco Department of Mathematics, Statistics and Computation, University of Cantabria, Santander, Cantabria, Spain

Amanda Cameron Max Planck Institute for Mathematics in the Sciences, Leipzig, Germany;
Department of Mathematics and Computer Science, Eindhoven University of Technology, Eindhoven, The Netherlands

Daniel Cavey School of Mathematical Sciences, University of Nottingham, Nottingham, UK

Tom Coates Department of Mathematics, Imperial College London, London, UK

Alessio Corti Department of Mathematics, Imperial College London, London, UK

Rodica Dinu University of Konstanz, Fachbereich Mathematik und Statistik, Konstanz, Germany

Akihiro Higashitani Department of Pure and Applied Mathematics, Graduate School of Information Science and Technology, Osaka University, Osaka, Japan

Johannes Hofscheier School of Mathematical Sciences, University of Nottingham, Nottingham, UK

Katharina Jochemko Department of Mathematics, KTH Royal Institute of Technology, Stockholm, Sweden

Alexander Kasprzyk School of Mathematical Sciences, University of Nottingham, Nottingham, UK

Valentina Kiritchenko Laboratory of Algebraic Geometry and Faculty of Mathematics, National Research University Higher School of Economics, Moscow, Russia; Institute for Information Transmission Problems, Moscow, Russia

Edwin Kutas Mathematics Institute, University of Warwick, Coventry, UK

Anders Lundman Department of Mathematics, KTH Royal Institute of Technology, Stockholm, Sweden

Mateusz Michałek University of Konstanz, Fachbereich Mathematik und Statistik, Konstanz, Germany

Milena Pabiniak Mathematisches Institut, Universität zu Köln, Köln, Germany

Maria Padalko SimilarWeb, Tel Aviv, Israel

Andrea Petracci Dipartimento di Matematica, Università di Bologna, Bologna, Italy

Thomas Prince Mathematical Institute, University of Oxford, Oxford, UK

Sandra Di Rocco Department of Mathematics, KTH Royal Institute of Technology, Stockholm, Sweden

Karin Schaller Mathematisches Institut, Freie Universität Berlin, Berlin, Germany

Tim Seynnaeve Mathematisches Institut, Universität Bern, Bern, Switzerland

Genival da Silva Jr. Department of Mathematics, Imperial College London, London, UK

Hendrik Süß Friedrich-Schiller-Universität Jena, Institut für Mathematik, Jena, Germany

Bach Le Tran School of Mathematics, University of Edinburgh, James Clerk Maxwell Building, Peter Guthrie Tait Road, Edinburgh, UK

Akiyoshi Tsuchiya Graduate School of Mathematical Sciences, University of Tokyo, Tokyo, Japan

Chapter 1
Difference Between Families of Weakly and Strongly Maximal Integral Lattice-Free Polytopes

Gennadiy Averkov

Abstract A d-dimensional closed convex set K in $\mathbb{R}^d$ is said to be lattice-free if the interior of K is disjoint with $\mathbb{Z}^d$. We consider the following two families of lattice-free polytopes: the family $\mathcal{L}^d$ of integral lattice-free polytopes in $\mathbb{R}^d$ that are not properly contained in another integral lattice-free polytope and its subfamily $\mathcal{M}^d$ consisting of integral lattice-free polytopes in $\mathbb{R}^d$ which are not properly contained in another lattice-free set. It is known that $\mathcal{M}^d = \mathcal{L}^d$ holds for $d \le 3$ and, for each $d \ge 4$, $\mathcal{M}^d$ is a proper subfamily of $\mathcal{L}^d$. We derive a super-exponential lower bound on the number of polytopes in $\mathcal{L}^d \setminus \mathcal{M}^d$ (with standard identification of integral polytopes up to affine unimodular transformations).

Keywords Egyptian fraction · Hollow polytope · Lattice-free set · Lattice polytope · Maximality

1.1 Introduction

By $|X|$ we denote the cardinality of a finite set X. Let $\mathbb{N}$ be the set of all positive integers and let $d \in \mathbb{N}$ be the dimension. Elements of $\mathbb{Z}^d$ are called *integral points* or *integral vectors*. We call a polyhedron $P \subseteq \mathbb{R}^d$ *integral* if P is the convex hull of $P \cap \mathbb{Z}^d$. Let $\mathrm{Aff}(\mathbb{Z}^d)$ be the group of affine transformations $A : \mathbb{R}^d \to \mathbb{R}^d$ satisfying $A(\mathbb{Z}^d) = \mathbb{Z}^d$. We call elements of $\mathrm{Aff}(\mathbb{Z}^d)$ *affine unimodular transformations*. For a family $\mathcal{X}$ of subsets of $\mathbb{R}^d$, we consider the family of equivalence classes

$$\mathcal{X}/\,\mathrm{Aff}(\mathbb{Z}^d) := \big\{\{A(X) \ : \ A \in \mathrm{Aff}(\mathbb{Z}^d)\} \ : \ X \in \mathcal{X}\big\}$$

with respect to identification of the elements of $\mathcal{X}$ up to affine unimodular transformations. A subset K of $\mathbb{R}^d$ is called *lattice-free* if K is closed, convex, d-dimensional and

G. Averkov (✉)
Brandenburgische Technische Universität Cottbus-Senftenberg, Fakultät 1, Platz der Deutschen Einheit 1, 03046 Cottbus, Germany
e-mail: averkov@b-tu.de

A. M. Kasprzyk and B. Nill (eds.), *Interactions with Lattice Polytopes*, Springer Proceedings in Mathematics & Statistics 386,
https://doi.org/10.1007/978-3-030-98327-7_1

the interior of K contains no points from $\mathbb{Z}^d$. A set K is called *maximal lattice-free* if K is lattice-free and is not a proper subset of another lattice-free set.

Our objective is to study the relationship between the following two families of integral lattice-free polytopes:

1. The family $\mathcal{L}^d$ of integral lattice-free polytopes P in $\mathbb{R}^d$ such that there exists no integral lattice-free polytope properly containing P. We call elements of $\mathcal{L}^d$ *weakly maximal* integral lattice-free polytopes.
2. The family $\mathcal{M}^d$ of integral lattice-free polytopes P in $\mathbb{R}^d$ such that there exists no lattice-free set properly containing P. We call the elements of $\mathcal{L}^d$ *strongly maximal* integral lattice-free polytopes.

The family $\mathcal{L}^d$ has applications in mixed-integer optimization, algebra and algebraic geometry; see [1, 3, 4, 13], respectively. In [2, 11] it was shown that $\mathcal{L}^d$ is finite up to affine unimodular transformations:

Theorem 1 ([2, Theorem 2.1], [11, Corollary 1.3]) $\mathcal{L}^d / \operatorname{Aff}(\mathbb{Z}^d)$ *is finite.*

Several groups of researchers are interested in enumeration of $\mathcal{L}^d$, up to affine unimodular transformations, in fixed dimensions. This requires understanding geometric properties of $\mathcal{L}^d$. Currently, no explicit description of $\mathcal{L}^d$ is available for dimensions $d \geq 4$ and, moreover, it is even extremely hard to decide if a given polytope belongs to $\mathcal{L}^d$. A brute-force algorithm based on volume bounds for $\mathcal{L}^d$ (provided in [11]) would have doubly exponential running time in d. In contrast to $\mathcal{L}^d$, its subfamily $\mathcal{M}^d$ is easier to deal with. Lovász's characterization [9, Proposition 3.3] of maximal lattice-free sets leads to a straightforward geometric description of polytopes belonging to $\mathcal{M}^d$. This characterization can be used to decide whether a given polytope is an element of $\mathcal{M}^d$ in only exponential time in d. Thus, while enumeration of $\mathcal{M}^d$ in fixed dimensions is a hard task, too, enumeration of $\mathcal{L}^d$ is even more challenging.

For a given dimension d, it is a priori not clear whether or not $\mathcal{M}^d$ is a proper subset of $\mathcal{L}^d$. Recently, it has been shown that the inequality $\mathcal{M}^d = \mathcal{L}^d$ holds if and only if $d \leq 3$. The equality $\mathcal{M}^d = \mathcal{L}^d$ is rather obvious for $d \in \{1, 2\}$, as it is not hard to enumerate $\mathcal{L}^d$ in these very small dimensions and to check that every element of $\mathcal{L}^d$ belongs to $\mathcal{M}^d$. Starting from dimension three, the problem gets very difficult. Results in [1, 2] establish the equality $\mathcal{M}^3 = \mathcal{L}^3$ and enumerate $\mathcal{L}^3$, up to affine unimodular transformations. As a complement, in [11, Theorem 1.4] it was shown that for all $d \geq 4$ there exists a polytope belonging to $\mathcal{L}^d$ but not to $\mathcal{M}^d$.

While Theorem 1.4 in [11] shows that $\mathcal{L}^d$ and $\mathcal{M}^d$ are two different families, it does not provide information on the number of polytopes in $\mathcal{L}^d$ that do not belong to $\mathcal{M}^d$. Relying on a result of Konyagin [6], we will show that, asymptotically, the gap between $\mathcal{L}^d$ and $\mathcal{M}^d$ is very large.

For $a_1, \ldots, a_d > 0$, we introduce

$$\kappa(a) := \kappa(a_1, \ldots, a_d) = \frac{1}{a_1} + \cdots + \frac{1}{a_d}.$$

Reciprocals of positive integers are sometimes called *Egyptian fractions*. Thus, if $a \in \mathbb{N}^d$, then $\kappa(a)$ is a sum of d Egyptian fractions. We consider the set

$$\mathcal{A}_d := \{(a_1, \dots, a_d) \in \mathbb{N}^d : a_1 \le \dots \le a_d,\ \kappa(a_1, \dots, a_d) = 1\}$$

of all different solutions of the Diophantine equation

$$\kappa(x_1, \dots, x_d) = 1$$

in the unknowns $x_1, \dots, x_d \in \mathbb{N}$. The set $\mathcal{A}_d$ represents possible ways to write 1 as a sum of d Egyptian fractions. It is known that $\mathcal{A}_d$ is finite. Our main result allows is a lower bound on the cardinality of $(\mathcal{L}^d \setminus \mathcal{M}^d)/\operatorname{Aff}(\mathbb{Z}^d)$:

Theorem 2 $\left|(\mathcal{L}^{d+5} \setminus \mathcal{M}^{d+5})/\operatorname{Aff}(\mathbb{Z}^{d+5})\right| \ge |\mathcal{A}_d|$.

The proof of Theorem 2 is constructive. This means that, for every $a \in \mathcal{A}_d$, we generate an element in $P_a \in \mathcal{L}^{d+5} \setminus \mathcal{M}^{d+5}$ such that for two different elements a and b of $\mathcal{A}_d$, the respective polytopes P_a and P_b do not coincide up to affine unimodular transformations. The proof of Theorem 2 is inspired by the construction in [11]. Using lower bounds on $|\mathcal{A}_d|$ from [6], we obtain the following asymptotic estimate:

Corollary 3 $\ln \ln \left|(\mathcal{L}^d \setminus \mathcal{M}^d)/\operatorname{Aff}(\mathbb{Z}^d)\right| = \Omega\left(\frac{d}{\ln d}\right)$, *as* $d \to \infty$.

Note 4 We view the elements of $\mathbb{R}^d$ as columns. By o we denote the zero vector and by $e_1, \dots, e_d$ the standard basis of $\mathbb{R}^d$. If $x \in \mathbb{R}^d$ and $i \in \{1, \dots, d\}$, then x_i denotes the i-th component of x. The relation $a \le b$ for $a, b \in \mathbb{R}^d$ means $a_i \le b_i$ for every $i \in \{1, \dots, d\}$. The relations $\ge$, $>$ and $<$ on $\mathbb{R}^d$ are introduced analogously. The abbreviations aff, conv, int and relint stand for the affine hull, convex hull, interior and relative interior, respectively.

1.2 An Approach to Construction of Polytopes in $\mathcal{L}^d \setminus \mathcal{M}^d$

We will present a systematic approach to construction of polytopes in $\mathcal{L}^d \setminus \mathcal{M}^d$, but first we discuss general maximal lattice-free sets.

Definition 5 Let P be a lattice-free polyhedron in $\mathbb{R}^d$. We say that a facet F of P is *blocked* if the relative interior of F contains an integral point.

Maximal lattice-free sets can be characterized as follows:

Proposition 6 ([9, Proposition 3.3]) *Let K be a d-dimensional closed convex subset of $\mathbb{R}^d$. Then the following conditions are equivalent:*

1. *K is maximal lattice-free;*
2. *K is a lattice-free polyhedron such that every facet of K is blocked.*

It can happen that some facets of a maximal lattice-free polyhedron are more than just blocked. We introduce a respective notion. Recall that the *integer hull* K_I of a compact convex set K in $\mathbb{R}^d$ is defined by

$$K_I := \operatorname{conv}(K \cap \mathbb{Z}^d).$$

Definition 7 Let P be a d-dimensional lattice-free polyhedron in $\mathbb{R}^d$. A facet F of P is called *strongly blocked* if F_I is $(d-1)$-dimensional and $\mathbb{Z}^d \cap \operatorname{relint} F_I \neq \emptyset$. The polyhedron P is called *strongly blocked* if all facets of P are strongly blocked.

The following proposition extracts the geometric principle behind the construction from [11, Sect. 3]. (Note that arguments in [11, Sect. 3] use an algebraic language.)

Proposition 8 *Let P be a strongly blocked lattice-free polytope in $\mathbb{R}^d$. Then $P_I \in \mathcal{L}^d$. Furthermore, if P is not integral, then $P_I \notin \mathcal{M}^d$.*

Proof In order to show $P_I \in \mathcal{L}^d$ it suffices to verify that, for every $z \in \mathbb{Z}^d$ such that $\operatorname{conv}(P_I \cup \{z\})$ is lattice-free, one necessarily has $z \in P_I$. If $z \notin P_I$, then $z \notin P$ and so, for some facet F of P, the point z and the polytope P lie on different sides of the hyperplane aff F. Then $\emptyset \neq \mathbb{Z}^d \cap \operatorname{relint} F_I \subseteq \operatorname{int}(\operatorname{conv}(P \cup \{z\}))$, yielding a contradiction to the choice of z. Thus, for every facet F of P, z and P lie on the same side of aff F. It follows $z \in P$. Hence $z \in P \cap \mathbb{Z}^d \subseteq P_I$.

If P is not integral, then $P_I \notin \mathcal{M}^d$ since $P_I \subsetneq P$ and P is lattice-free. □

1.3 Lattice-Free Axis-Aligned Simplices

For $a \in \mathbb{R}^d_{>0}$, the d-dimensional simplex

$$T(a) := \operatorname{conv}\{o, a_1 e_1, \ldots, a_d e_d\}.$$

is called *axis-aligned*. The proof of the following proposition is straightforward.

Proposition 9 *For $a \in \mathbb{R}^d_{>0}$, the following statements hold:*

1. *the simplex $T(a)$ is a lattice-free set if and only if $\kappa(a) \geq 1$;*
2. *the simplex $T(a)$ is a maximal lattice-free set if and only if $\kappa(a) = 1$.*

We introduce transformations which preserve the values of κ. The transformations arise from the following trivial identities for $t > 0$:

$$\frac{1}{t} = \frac{1}{t+1} + \frac{1}{t(t+1)}, \tag{1.1}$$

$$\frac{1}{t} = \frac{1}{t+2} + \frac{1}{t(t+2)} + \frac{1}{t(t+2)}, \tag{1.2}$$

$$\frac{1}{t} = \frac{2}{3t} + \frac{1}{3t}. \tag{1.3}$$

Consider a vector $a \in \mathbb{R}^d_{>0}$. By (1.1), if t is a component of a, we can replace this component with two new components $t+1$ and $t(t+1)$ to generate a vector $b \in \mathbb{R}^{d+1}_{>0}$ satisfying $\kappa(b) = \kappa(a)$. Identities (1.2) and (1.3) can be applied in a similar fashion. For every $d \in \mathbb{N}$, with the help of (1.1)–(1.3), we introduce the following maps:

$$\phi_d : \mathbb{R}^d_{>0} \to \mathbb{R}^{d+1}_{>0}, \qquad \phi_d(a) := \begin{pmatrix} a_1 \\ \vdots \\ a_{d-1} \\ a_d + 1 \\ a_d(a_d+1) \end{pmatrix}, \tag{1.4}$$

$$\psi_d : \mathbb{R}^d_{>0} \to \mathbb{R}^{d+3}_{>0}, \qquad \psi_d(a) := \begin{pmatrix} a_1 \\ \vdots \\ a_{d-1} \\ a_d + 3 \\ a_d(a_d+1) \\ (a_d+1)(a_d+3) \\ (a_d+1)(a_d+3) \end{pmatrix},$$

$$\xi_d : \mathbb{R}^d_{>0} \to \mathbb{R}^{d+1}_{>0} \qquad \xi_d(a) := \begin{pmatrix} a_1 \\ \vdots \\ a_{d-1} \\ \frac{3}{2}a_d \\ 3a_d \end{pmatrix}. \tag{1.5}$$

The map ϕ_d replaces the component a_d by two other components based on (1.1), while ξ_d replaces a_d based on (1.3). The map ψ_d acts by replacing the component a_d based on (1.1) and then replacing the component $a_d + 1$ based on (1.2). Identities (1.1)–(1.3) imply

$$\kappa(\phi_d(a)) = \kappa(\psi_d(a)) = \kappa(\xi_d(a))) = \kappa(a). \tag{1.6}$$

Lemma 10 *Let $P = T(\xi_d(a))$, where $a \in \mathcal{A}_d$ and $d \geq 2$. Then P is a strongly blocked lattice-free $(d+1)$-dimensional polytope. Furthermore, if a_d is odd, P is not integral.*

Proof In this proof, we use the *all-ones vector*

$$\mathbb{1}_d := \begin{pmatrix} 1 \\ \vdots \\ 1 \end{pmatrix} \in \mathbb{R}^d.$$

For the sake of brevity we introduce the notation $t := a_d$. One has $1 = \kappa(a) = \sum_{i=1}^{d} \frac{1}{a_i} \geq \sum_{i=1}^{d} \frac{1}{t} = \frac{d}{t}$, which implies $t \geq d \geq 2$. By (1.6), one has $\kappa(\xi_d(a)) = 1$ and so, by Proposition 9, P is maximal lattice-free.

If t is even, the polytope P is integral and hence every facet of P is integral, too. In view of Proposition 6, integral maximal lattice-free polytopes are strongly blocked, and so we conclude that P is strongly blocked.

Assume that t is odd, then the polytope P has one non-integral vertex. In this case, we need to look at facets of P more closely, to verify that P is strongly blocked. We consider all facets of P.

1. The facet $F = \operatorname{conv}\{o, a_1e_1, \dots, a_{d-1}e_{d-1}, 3te_{d+1}\}$ is a d-dimensional integral integral axis-aligned simplex. Since

$$\kappa(a_1, \dots, a_{d-1}, 3t) < 1,$$

the integral point $e_1 + \cdots + e_{d-1} + e_{d+1}$ is in the relative interior of F. Hence, F is strongly blocked.

2. The facet $F = \operatorname{conv}\left\{o, a_1e_1, \dots, a_{d-1}e_{d-1}, \frac{3}{2}te_d\right\}$ contains the d-dimensional integral axis-aligned simplex

$$G := \operatorname{conv}\left\{o, a_1e_1, \dots, a_{d-1}e_{d-1}, \frac{3t-1}{2}e_d\right\},$$

as a subset. In view of $t \geq 2$, we have

$$\kappa\left(a_1, \dots, a_{d-1}, \frac{3t-1}{2}\right) < 1,$$

which implies that the integral point $e_1 + \cdots + e_d$ is in the relative interior of G. It follows that F is strongly blocked.

3. The facet $F := \operatorname{conv}\left\{a_1e_1, \dots, a_{d-1}e_{d-1}, \frac{3}{2}te_d, 3te_{d+1}\right\}$ contains the integral d-dimensional simplex

$$G := \operatorname{conv}\left\{a_1e_1, \dots, a_{d-1}e_{d-1}, \frac{3t-1}{2}e_d + e_{d+1}, 3te_{d+1}\right\}.$$

as a subset. It turns out that $\mathbb{1}_{d+1}$ is the relative interior of G, because $\mathbb{1}_{d+1}$ is a convex combination of the vertices of relint G, with positive coefficients. Indeed, the equality

$$\mathbb{1}_{d+1} = \sum_{i=1}^{d-1} \frac{1}{a_i}(a_ie_i) + \lambda\left(\frac{3t-1}{2}e_d + e_{d+1}\right) + \mu(3te_{d+1})$$

holds for $\lambda = \frac{2}{3t-1}$ and $\mu = \frac{t-1}{t(3t-1)}$, where

$$\sum_{i=1}^{d-1} \frac{1}{a_i} + \lambda + \mu = 1.$$

4. It remains to consider faces F with the vertex set

$$\left\{o, a_1e_1, \ldots, a_de_d, \frac{3}{2}te_d, 3te_{d+1}\right\} \backslash \{a_ie_i\},$$

where $i \in \{1, \ldots, d+1\}$. Without loss of generality, let $i = 1$ so that

$$F = \operatorname{conv}\left\{o, a_2e_2, \ldots, \frac{3}{2}te_d, 3te_{d+1}\right\}.$$

This facet contains the integral d-dimensional simplex

$$G := \operatorname{conv}\left\{o, a_2a_2, \ldots, a_{d-1}e_{d-1}, \frac{3t-1}{2}e_d + e_{d+1}, 3te_{d+1}\right\}.$$

Similarly to the previous case, one can check that $e_2 + \cdots + e_{d+1}$ is an integral point in the relative interior of G. Consequently, F is strongly blocked. □

1.4 Proof of the Main Result

For $d \geq 4$, Nill and Ziegler [7] construct one vector $a \in \mathbb{R}^d_{>0}$ with $T(a)_I \in \mathcal{L}^d \backslash \mathcal{M}^d$. We generalize this construction and provide many further vectors a with the above properties. We will also need to verify that for different choices of a, we get essentially different polytopes $T(a)_I$.

Lemma 11 *Let P and Q be d-dimensional strongly blocked lattice-free polytopes such that for their integral hulls the equality $Q_I = A(P_I)$ holds for some $A \in \operatorname{Aff}(\mathbb{Z}^d)$. Then $Q = A(P)$.*

Proof Since A is an affine transformation, we have

$$A(P_I) = A(\operatorname{conv}(P \cap \mathbb{Z}^d)) = \operatorname{conv} A(P \cap \mathbb{Z}^d).$$

Using $A \in \operatorname{Aff}(\mathbb{Z}^d)$, it is straightforward to check the equality $A(P \cap \mathbb{Z}^d) = A(P) \cap \mathbb{Z}^d$. We thus conclude that $A(P_I) = A(P)_I$. The assumption $Q_I = A(P_I)$ yields $Q_I = A(P)_I$. Since P is strongly blocked lattice-free, $A(P)$ too is strongly blocked lattice-free. We thus have the equality $Q_I = A(P)_I$ for strongly blocked lattice-free polytopes Q and $A(P)$. To verify the assertion, it suffices to show that a strongly blocked lattice-free polytope Q is uniquely determined by the knowledge of its integer hull Q_I. This is quite easy to see. For every strongly blocked facet G of Q_I, the affine hull of G contains a facet of Q. Conversely, if F is an arbitrary facet of Q, then $G = F_I$ is a strongly blocked facet of Q_I. Thus, the knowledge of Q_I allows to determine affine hulls of all facets of Q. In other words, Q_I uniquely determines a hyperplane description of Q. □

Lemma 12 *Let $a, b \in \mathbb{R}^d_{>0}$ be such that the equality $T(b) = A(T(a))$ holds for some $A \in \mathrm{Aff}(\mathbb{Z}^d)$. Then a and b coincide up to permutation of components.*

Proof We use induction on d. For $d = 1$, the assertion is trivial. Let $d \geq 2$. One of the d facets of $T(a)$ containing o is mapped by A to a facet of $T(b)$ that contains o. Without loss of generality we can assume that the facet $T(a_1, \ldots, a_{d-1}) \times \{0\}$ of $T(a)$ is mapped to the facet $T(b_1, \ldots, b_{d-1}) \times \{0\}$ of $T(b)$. By the inductive assumption, $(a_1, \ldots, a_{d-1})$ and $(b_1, \ldots, b_{d-1})$ coincide up to permutation of components. Since unimodular transformations preserve the volume, $T(a)$ and $T(b)$ have the same volume. This means, $\prod_{i=1}^d a_i = \prod_{i=1}^d b_i$. Consequently, $a_d = b_d$ and we conclude that a and b coincide up to permutation of components.

Proof (Proof of Theorem 2) For every $a \in \mathcal{A}_d$, we introduce the $(d + 5)$-dimensional integral lattice-free polytope

$$P_a := T(\eta(a))_I,$$

where

$$\eta(x) := \xi_{d+4}(\psi_{d+1}(\phi_d(x)))$$

and the functions ξ_{d+4}, ψ_{d+1} and ϕ_d are defined by (1.4)–(1.5).

By (1.6) for each $a \in \mathcal{A}_d$, we have $\kappa(\eta(a)) = 1$. For $a \in \mathcal{A}_d$ the last component of $\phi_d(a)$ is even. This implies that the last component of $\psi_{d+1}(\phi_d(a))$ is odd. Thus, by Lemma 10, $T(\eta(a))$ is strongly blocked lattice-free polytope which is not integral.

Let $a, b \in \mathcal{A}_d$ be such that the polytopes P_a and P_b coincide up to affine unimodular transformations. Then, by Lemma 11, $T(\eta(a))$ and $T(\eta(b))$ coincide up to affine unimodular transformations. But then, by Lemma 12, $\eta(a)$ and $\eta(b)$ coincide up to permutations. Since the components of a and b are sorted in the ascending order, the components of $\eta(a)$ and $\beta(b)$ too are sorted in the ascending order. Thus, we arrive at the equality $\eta(a) = \eta(b)$, which implies $a = b$.

In view of Proposition 8, each P_a with $a \in \mathcal{A}_d$ belongs to $\mathcal{L}^d$ but not to $\mathcal{M}^d$. Thus, the equivalence classes of the polytopes P_a with $a \in \mathcal{A}_d$ with respect to identification up to affine unimodular transformations form a subset of $(\mathcal{L}^{d+5} \setminus \mathcal{M}^{d+5}) / \mathrm{Aff}(\mathbb{Z}^{d+5})$ of cardinality $|\mathcal{A}_d|$. This yields the desired assertion. □

Proof (Proof of Corollary 3) The assertion is a direct consequence of Theorem 2 and the asymptotic estimate

$$\ln \ln |\mathcal{A}_d| = \Omega\left(\frac{d}{\ln d}\right)$$

of Konyagin [6, Theorem 1] (see also [5, Corollary 1.2]). □

Remark 13 In view of the asymptotic upper bound $\ln \ln |\mathcal{A}_d| = O(d)$, determined with different degrees of precision in [8, 10] and [12, Theorem 2], the lower bound of Konyagin is optimal up to the logarithmic factor in the denominator.

Since all known elements of $\mathcal{L}^d$ are of the form P_I, for some strongly blocked lattice-free polytope P, we ask the following:

Question 14 Do there exist polytopes $L \in \mathcal{L}^d$ which cannot be represented as $L = P_I$ for any strongly blocked lattice-free polytope P?

If there is a gap between the families $\mathcal{L}^d$ and the family

$$\left\{P_I \ : \ P \subseteq \mathbb{R}^d \text{ strongly blocked lattice-free polytope}\right\},$$

then it would be interesting to understand how irregular the polytopes from this gap can be. For example, one can ask the following:

Question 15 Do there exist polytopes $L \in \mathcal{L}^d$ with the property that no facet of L is blocked?

Acknowledgements I would like to thank Christian Wagner for valuable comments and Christian Elsholtz for pointing to [5, 8, 10].

References

1. Averkov, G., Krümpelmann, J., Weltge, S.: Notions of maximality for integral lattice-free polyhedra: the case of dimension three. Math. Oper. Res. **42**(4), 1035–1062 (2017)
2. Averkov, G., Wagner, C., Weismantel, R.: Maximal lattice-free polyhedra: finiteness and an explicit description in dimension three. Math. Oper. Res. **36**(4), 721–742 (2011)
3. Blanco, M., Haase, C., Hofmann, J., Santos, F.: The finiteness threshold width of lattice polytopes. Trans. Am. Math. Soc. Ser. B **8**, 399–419 (2021)
4. Del Pia, A., Weismantel, R.: Relaxations of mixed integer sets from lattice-free polyhedra. Ann. Oper. Res. **240**(1), 95–117 (2016)
5. Elsholtz, C.: Egyptian fractions with odd denominators. Q. J. Math. **67**(3), 425–430 (2016)
6. Konyagin, S.V.: Double exponential lower bound for the number of representations of unity by Egyptian fractions. Math. Notes **95**(1–2), 277–281 (2014). Translation of Mat. Zametki **95** (2014), no. 2, 312–316
7. Lagarias, J.C., Ziegler, G.M.: Bounds for lattice polytopes containing a fixed number of interior points in a sublattice. Canad. J. Math. **43**(5), 1022–1035 (1991)
8. Landau, E.: Über die Klassenzahl der binären quadratischen Formen von negativer Discriminante. Math. Ann. **56**(4), 671–676 (1903)
9. Lovász, L.: Geometry of numbers and integer programming. In: Mathematical Programming, Tokyo (1988). Mathematics Applied (Japanese Ser.), vol. 6, pp. 177–201. SCIPRESS, Tokyo (1989)
10. Newman, M.: A bound for the number of conjugacy classes in a group. J. London Math. Soc. **43**, 108–110 (1968)
11. Nill, B., Ziegler, G.M.: Projecting lattice polytopes without interior lattice points. Math. Oper. Res. **36**(3), 462–467 (2011)

12. Sándor, C.: On the number of solutions of the Diophantine equation $\sum_{i=1}^{n} \frac{1}{x_i} = 1$. Period. Math. Hungar. **47**(1–2), 215–219 (2003)
13. Treutlein, J.: 3-dimensional lattice polytopes without interior lattice points. Ph.D. thesis, Eberhart Karls Universität Tübingen (2010)

Chapter 2
On the Fine Interior of Three-Dimensional Canonical Fano Polytopes

Victor Batyrev, Alexander Kasprzyk, and Karin Schaller

Abstract The Fine interior Δ^{FI} of a d-dimensional lattice polytope Δ is a rational subpolytope of Δ which is important for constructing minimal birational models of non-degenerate hypersurfaces defined by Laurent polynomials with Newton polytope Δ. This paper presents some computational results on the Fine interior of all 674,688 three-dimensional canonical Fano polytopes.

Keywords Lattice polytope · Fine interior · Hypersurface

2.1 Introduction

Let $M \cong \mathbb{Z}^d$ be a free abelian group of rank d. We set $M_{\mathbb{Q}} := M \otimes \mathbb{Q}$ and denote by N the dual group $\text{Hom}(M, \mathbb{Z})$ in the dual $\mathbb{Q}$-linear vector space $N_{\mathbb{Q}} := \text{Hom}(M, \mathbb{Q})$. Let $\langle \cdot, \cdot \rangle : M_{\mathbb{Q}} \times N_{\mathbb{Q}} \to \mathbb{Q}$ be the natural pairing.

A convex compact d-dimensional polytope $\Delta \subseteq M_{\mathbb{Q}}$ is called *lattice d-tope* if all vertices of Δ belong to the lattice $M \subseteq M_{\mathbb{Q}}$, i.e., Δ equals the convex hull $\text{conv}(\Delta \cap M)$ of all lattice points in Δ. The usual interior Δ° of Δ is the complement $\Delta \setminus \partial\Delta$, where $\partial\Delta$ is the boundary of Δ. Another interior of a lattice polytope Δ was introduced by Fine [3, 13, 15, 20]:

Definition 1 Let $\Delta \subseteq M_{\mathbb{Q}}$ be a lattice d-tope. Denote by ord_{Δ} the piecewise linear function $N_{\mathbb{Q}} \to \mathbb{Q}$ with

V. Batyrev
Mathematisches Institut, Universität Tübingen, Auf der Morgenstelle 10, 72076 Tübingen, Germany
e-mail: victor.batyrev@uni-tuebingen.de

A. Kasprzyk
School of Mathematical Sciences, University of Nottingham, Nottingham NG7 2RD, UK
e-mail: a.m.kasprzyk@nottingham.ac.uk

K. Schaller (✉)
Mathematisches Institut, Freie Universität Berlin, Arnimallee 3, 14195 Berlin, Germany
e-mail: karin.schaller@fu-berlin.de

A. M. Kasprzyk and B. Nill (eds.), *Interactions with Lattice Polytopes*,
Springer Proceedings in Mathematics & Statistics 386,
https://doi.org/10.1007/978-3-030-98327-7_2

$$\mathrm{ord}_\Delta(y) := \min_{x\in\Delta}\langle x, y\rangle \quad (y \in N_\mathbb{Q}).$$

Then the convex subset

$$\Delta^{\mathrm{FI}} := \bigcap_{n\in N\setminus\{0\}} \{x \in M_\mathbb{Q} \mid \langle x, n\rangle \geq \mathrm{ord}_\Delta(n) + 1\}$$

is called the *Fine interior* of Δ.

One can show that only finitely many linear inequalities $\langle x, n\rangle \geq \mathrm{ord}_\Delta(n) + 1$ are necessary to define Δ^{FI}. Therefore, Δ^{FI} is a convex hull of finitely many rational points $p \in M_\mathbb{Q}$. Moreover, any lattice point $p \in \Delta^\circ \cap M$ in the usual interior of Δ is contained in Δ^{FI}. Therefore, Δ^{FI} contains the convex hull of $\Delta \cap M$, i.e., we get the inclusion $\mathrm{conv}(\Delta^\circ \cap M) \subseteq \Delta^{\mathrm{FI}}$. In particular, Δ^{FI} is non-empty if $\Delta^\circ \cap M$ is non-empty. Moreover, for any lattice polytope Δ of dimension $d \leq 2$ one has the equality $\mathrm{conv}(\Delta^\circ \cap M) = \Delta^{\mathrm{FI}}$ [3]. The Fine interior Δ^{FI} of a lattice polytope Δ of dimension $d \geq 3$ may happen to be strictly larger than the convex hull $\mathrm{conv}(\Delta^\circ \cap M)$. The simplest famous example of such a situation is due to M. Reid. Other similar examples based on hollow 3-topes can be found in Sect. 2.7:

Example 2 ([20, Example 4.15]) Let $M \subseteq \mathbb{Q}^4$ be 3-dimensional affine lattice defined by

$$M := \left\{ (m_1, m_2, m_3, m_4) \in \mathbb{Z}^4 \;\middle|\; \sum_{i=1}^4 m_i = 5,\ \sum_{i=1}^4 i m_i \equiv 0 \pmod 5 \right\}.$$

Consider the M-lattice 3-tope $\Delta \subseteq M_\mathbb{Q}$ defined as the convex hull of 4 lattice points

$$(5, 0, 0, 0),\ (0, 5, 0, 0),\ (0, 0, 5, 0),\ \text{and } (0, 0, 0, 5) \in M.$$

Then $\mathrm{conv}(\Delta^\circ \cap M) = \emptyset$, but Δ^{FI} is the 3-dimensional M-rational simplex

$$\mathrm{conv}\{(2, 1, 1, 1), (1, 2, 1, 1), (1, 1, 2, 1), (1, 1, 1, 2)\}$$

and $\Delta^{\mathrm{FI}} \cap M$ is empty.

In this paper, we are interested in lattice d-topes $\Delta \subseteq M_\mathbb{Q}$ obtained as Newton polytopes of Laurent polynomials f_Δ in d variables $x_1, \ldots, x_d$, i.e.,

$$f_\Delta(\mathbf{x}) = \sum_{m\in\Delta\cap M} a_m \mathbf{x}^m,$$

where $a_m \in \mathbb{C}$ are sufficiently general complex numbers. The importance of the Fine interior is explained by the following theorem [3, 15, 20]:

Theorem 3 *Let $\mathcal{Z}_\Delta \subseteq \mathbb{T}^d$ be a non-degenerate affine hypersurface in the d-dimensional algebraic torus $\mathbb{T}^d$ defined by a Laurent polynomial f_Δ with Newton d-tope Δ. Then the following conditions are equivalent:*

1. *a smooth projective compactification $\mathcal{V}_\Delta$ of $\mathcal{Z}_\Delta$ has non-negative Kodaira dimension, i.e., $\kappa \geq 0$;*
2. *$\mathcal{Z}_\Delta$ is birational to a minimal model S_Δ with abundance;*
3. *the Fine interior Δ^{FI} of Δ is non-empty.*

Remark 4 By well known results of Khovanskii [14], one has vanishing of the cohomology groups

$$h^i(\mathcal{O}_{\mathcal{V}_\Delta}) = 0 \quad (1 \leq i \leq d-2)$$

and the equation $h^{d-1}(\mathcal{O}_{\mathcal{V}_\Delta}) = |\Delta^\circ \cap M|$. The numbers $h^i(\mathcal{O}_{\mathcal{V}_\Delta})$ are birational invariants of $\mathcal{Z}_\Delta$; they do not depend on a smooth projective compactification $\mathcal{V}_\Delta$ of $\mathcal{Z}_\Delta$. In particular, the number $|\Delta^\circ \cap M|$ is the geometric genus p_g of the affine hypersurface $\mathcal{Z}_\Delta \subseteq \mathbb{T}^d$.

Smooth projective compactifications of non-degenerate hypersurfaces in $\mathbb{T}^d$ can be obtained using the theory of toric varieties [14].

Let $\Delta \subseteq M_\mathbb{Q}$ be a lattice d-tope. We consider the *normal fan* Σ^Δ of Δ in the dual space $N_\mathbb{Q}$, i.e., $\Sigma^\Delta := \{\sigma^\theta \mid \theta \preceq \Delta\}$, where σ^θ is the cone generated by all inward-pointing facet normals of facets containing the face $\theta \preceq \Delta$ of Δ. One has $\dim(\sigma^\theta) + \dim(\theta) = d$ for any face $\theta \preceq \Delta$. We denote by X_Δ the normal projective toric variety constructed via the normal fan Σ^Δ. In particular, the above function $\mathrm{ord}_\Delta : N_\mathbb{Q} \to \mathbb{Q}$ is a piecewise linear function with respect to this fan defining an ample Cartier divisor on X_Δ. In particular, the cones $\sigma^\theta \in \Sigma^\Delta$ are defined as

$$\sigma^\theta = \left\{ y \in N_\mathbb{Q} \;\middle|\; \mathrm{ord}_\Delta(y) = \langle x, y \rangle \;\; \text{for all } x \in \theta \right\}.$$

Remark 5 Using the normal fan Σ^Δ, one can compute the fundamental group $\pi_1(\mathcal{V}_\Delta)$ of a smooth projective birational model $\mathcal{V}_\Delta$ of a non-degenerate affine hypersurface $\mathcal{Z}_\Delta$ (given as in Theorem 3). The fundamental group $\pi_1(\mathcal{V}_\Delta)$ does not depend on the choice of the smooth birational model and it is isomorphic to the quotient of the lattice N modulo the sublattice N' generated by all lattice points in $(d-1)$-dimensional cones σ^θ of the normal fan Σ^Δ [4].

Example 6 The minimal model S_Δ of a non-degenerate affine surface $\mathcal{Z}_\Delta$ defined by a Laurent polynomial with the Newton polytope Δ from Example 2 is a *Godeaux surface*. It is a surface of general type with $p_g = q = 0$, $K^2 = 1$, and $\pi_1(S_\Delta) \cong \mathbb{Z}/5\mathbb{Z}$.

Definition 7 A lattice d-tope Δ is called *canonical Fano d-tope* if $|\Delta^\circ \cap M| = 1$. Up to a shift by a lattice vector, we will assume without loss of generality that $0 \in M$ is the single lattice point in the interior Δ° of the canonical Fano d-tope Δ, i.e., $\Delta^\circ \cap M = \{0\}$.

All canonical Fano 3-topes have been classified [16]. There exists exactly 674,688 canonical Fano 3-topes Δ. The aim of this paper is to present computational results of their Fine interiors Δ^{FI} and some related combinatorial invariants. These data are important for computing minimal smooth projective surfaces $\mathcal{S}_\Delta$ with $p_g = 1$ and $q = 0$ which are birational to affine non-degenerate hypersurfaces $\mathcal{Z}_\Delta \subseteq \mathbb{T}^3 \cong (\mathbb{C}^\times)^3$.

The simplest description of the minimal surface $\mathcal{S}_\Delta$ has been obtained when Δ is a reflexive 3-tope [5].

Definition 8 A lattice d-tope $\Delta \subseteq M_{\mathbb{Q}}$ containing the origin $0 \in M$ in its interior is called *reflexive* if the dual polytope

$$\Delta^* := \{y \in N \mid \langle x, y\rangle \geq -1 \ \text{ for all } x \in \Delta\} \subseteq N_{\mathbb{Q}}$$

is a lattice polytope.

There exist 4,319 reflexive 3-topes, classified by Kreuzer and Skarke [17]. They form a small subset in the list of all 674,688 canonical Fano 3-topes [16]. Reflexive 4-topes are also classified by Kreuzer and Skarke [18]. There exist 473,800,776 reflexive 4-topes, but the complete list of all canonical Fano 4-topes is unknown and expected to be much bigger.

If Δ is a reflexive d-tope, then X_Δ is a Gorenstein toric Fano d-fold and the Zariski closure $\overline{Z}_\Delta$ in X_Δ is a Gorenstein Calabi-Yau $(d-1)$-fold. If $d = 3$, then $\overline{Z}_\Delta$ is a $K3$-surface with at worst finitely many Du Val singularities of type A_k. The minimal surface $\mathcal{S}_\Delta$ is a smooth $K3$-surface which is obtained as the minimal (crepant) desingularization of $\overline{Z}_\Delta$ [5].

One motivation for the present paper is due to Corti and Golyshev, who have found 9 interesting examples of canonical Fano 3-simplices Δ such that the affine surfaces $\mathcal{Z}_\Delta$ are birational to elliptic surfaces of Kodaira dimension $\kappa = 1$ [11].

The computation of the Fine interior Δ^{FI} for all canonical Fano 3-topes $\Delta \subseteq M_{\mathbb{Q}}$ has shown that the dimension of the Fine interior Δ^{FI} has only three values: 0, 1, and 3. It is rather surprising that there are no canonical Fano 3-topes Δ with $\dim(\Delta^{\mathrm{FI}}) = 2$.

The condition $\dim(\Delta^{\mathrm{FI}}) = 0$ holds if and only if Δ^{FI} equals the lattice point $0 \in M$. There exist exactly 665,599 canonical Fano 3-topes with $\Delta^{\mathrm{FI}} = \{0\}$, where $0 \in M$ is the only interior lattice point of Δ. These polytopes are characterized in [3, Proposition 3.4] by the condition that $0 \in N$ is an interior lattice point of the lattice 3-tope

$$[\Delta^*] := \mathrm{conv}(\Delta^* \cap N).$$

Remark 9 If Δ is a canonical Fano 3-tope, then $\Delta^{\mathrm{FI}} = \{0\}$ if and only if the non-degenerate affine surface $\mathcal{Z}_\Delta$ is birational to a $K3$-surface [3, Theorem 2.26].

The case $\dim(\Delta^{\mathrm{FI}}) = 1$ splits in two subcases. There exists exactly 20 canonical Fano 3-topes Δ such that $0 \in M$ is the midpoint of the Fine interior Δ^{FI}. Therefore, we call this Fine interior *symmetric*. Canonical Fano 3-topes with 1-dimensional symmetric Fine interior are characterized by the condition that $[\Delta^*]$ is a reflexive 2-tope.

The Fine interior of the remaining 9,020 canonical Fano 3-topes with $\dim(\Delta^{\mathrm{FI}}) = 1$ contains $0 \in M$ as a vertex. Therefore, we call this Fine interior *asymmetric*. Canonical Fano 3-topes with 1-dimensional asymmetric Fine interior are combinatorially characterized by the condition that $0 \in N$ is contained in the relative interior of a facet $\Theta \preceq [\Delta^*]$ of the lattice 3-tope $[\Delta^*]$. The minimal surfaces $\mathcal{S}_\Delta$ corresponding to canonical Fano 3-topes with 1-dimensional Fine interior (symmetric and asymmetric) are elliptic surfaces of Kodaira dimension $\kappa = 1$.

There exist exactly 49 canonical Fano 3-topes with $\dim(\Delta^{\mathrm{FI}}) = 3$. These polytopes are characterized by the condition that $0 \in N$ is a vertex of the lattice 3-tope $[\Delta^*]$. The surfaces $\mathcal{S}_\Delta$ corresponding to canonical Fano 3-topes Δ with 3-dimensional Fine interior Δ^{FI} are of general type (i.e., $\mathcal{S}_\Delta$ has maximal Kodaira dimension $\kappa = \dim(\mathcal{S}_\Delta) = 2$).

Remark 10 The Fine interior computations were done using

$$\Delta^{\mathrm{FI}} = \bigcap_{\theta \preceq \Delta} \bigcap_{n \in \mathcal{H}(\sigma^\theta)} \left\{ x \in M_{\mathbb{Q}} \,\middle|\, \langle x, n \rangle \geq \mathrm{ord}_\Delta(n) + 1 \right\},$$

where $\mathcal{H}(\sigma^\theta)$ denotes the set of all irreducible elements in the monoid $\sigma^\theta \cap N$. It is the minimal generating set of the monoid $\sigma^\theta \cap N$ and is called *Hilbert basis* of $\sigma^\theta \cap N$.

In the next sections we consider examples and discuss additional properties of canonical Fano 3-topes Δ in dependence of their Fine interiors Δ^{FI}. All computations were done using the Graded Ring Database [8], including the data of all 674,688 canonical Fano 3-topes, and MAGMA [7]. Therefore, all statements have been checked by computer calculations. The canonical Fano 3-topes used as examples in this paper appear with an ID that is the example's ID in the Graded Ring Database.[1]

2.2 Almost Reflexive Polytopes of Dimension 3 and 4

Definition 11 A canonical Fano d-tope $\Delta \subseteq M_{\mathbb{Q}}$ is called *almost reflexive* if the convex hull of all N-lattice points in the dual polytope Δ^* is reflexive.

It is easy to show the following statement:

Proposition 12 *If a canonical Fano d-tope Δ is almost reflexive, then*

$$\Delta^{FI} = \{0\}.$$

Proof If $[\Delta^*]$ is reflexive, then $\Delta = (\Delta^*)^*$ is contained in the dual reflexive polytope $[\Delta^*]^*$. Therefore, the Fine interior of Δ is contained in the Fine interior of the reflexive polytope $[\Delta^*]^*$ and $([\Delta^*]^*)^{\mathrm{FI}} = \{0\}$. Thus, $\Delta^{\mathrm{FI}} = \{0\}$. □

---

[1] http://www.grdb.co.uk/forms/toricf3c.

The converse statement is not true in general for $d \geq 5$, but there exist many equivalent characterizations of reflexive and almost reflexive d-topes among canonical Fano d-topes if $d = 3$ or $d = 4$.

Let us recall some combinatorial invariants of arbitrary lattice d-topes.

Definition 13 The *Ehrhart power series* of an arbitrary lattice d-tope $\Delta \subseteq M_{\mathbb{Q}}$ is defined as

$$P_\Delta(t) := \sum_{k\geq 0} |k\Delta \cap M|\, t^k,$$

where $|k\Delta \cap M|$ denotes the number of lattice points in the k-th dilate $k\Delta$ of Δ.

This Ehrhart series is a rational function of the form

$$P_\Delta(t) = \frac{\psi_d(\Delta)t^d + \cdots + \psi_1(\Delta)t + \psi_0(\Delta)}{(1-t)^{d+1}},$$

where $\psi_i(\Delta)$ are non-negative integers for all $0 \leq i \leq d$ [22] such that $\psi_0(\Delta) = 1$ and $\psi_1(\Delta) = |\Delta \cap M| - d - 1$. Moreover, $\sum_{i=0}^{d} \psi_i(\Delta) = v(\Delta)$, where $v(\Delta) := d! \cdot \mathrm{vol}(\Delta)$ denotes the *normalized volume* of Δ.

One has the following characterization of reflexive d-topes:

Proposition 14 ([6, Theorem 4.6]) *A canonical Fano d-tope Δ is reflexive if and only if*

$$\psi_i(\Delta) = \psi_{d-i}(\Delta) \quad (0 \leq i \leq d).$$

The Ehrhart reciprocity implies that the power series

$$Q_\Delta(t) := \sum_{k\geq 1} |(k\Delta)^\circ \cap M|\, t^k$$

is a rational function

$$Q_\Delta(t) = \frac{\varphi_{d+1}(\Delta)t^{d+1} + \cdots + \varphi_2(\Delta)t + \varphi_1(\Delta)t + \varphi_0(\Delta)}{(1-t)^{d+1}},$$

where $\varphi_0(\Delta) = 0$ and $\varphi_1(\Delta) = |\Delta^\circ \cap M|$. Using Serre duality, one obtains [12, Sect. 4, 5.11]

$$\varphi_i(\Delta) = \psi_{d+1-i}(\Delta) \quad (1 \leq i \leq d+1),$$

i.e., in particular

$$\psi_d(\Delta) = \varphi_1(\Delta) = |\Delta^\circ \cap M|$$

and

$$\psi_{d-1}(\Delta) = \varphi_2(\Delta) = |2\Delta^\circ \cap M| - (d+1)|\Delta^\circ \cap M|.$$

Therefore, the lattice d-tope Δ is a canonical Fano d-tope if and only if $\psi_d(\Delta) = 1$. Moreover,

$$\psi_{d-1}(\Delta) = |(2\Delta)^\circ \cap M| - (d+1)$$

if Δ is a canonical Fano d-tope.

Applying the above equations, one immediately obtains the following criterion for reflexivity of canonical Fano d-topes in the case $d = 3, 4$:

Proposition 15 *Let $\Delta \subseteq M_{\mathbb{Q}}$ be a canonical Fano d-tope with $d \in \{3, 4\}$. Then for $d = 3$, one has*

$$P_\Delta(t) = \frac{t^3 + (|(2\Delta)^\circ \cap M| - 4)t^2 + (|\Delta \cap M| - 4)t + 1}{(1-t)^4}$$

and for $d = 4$, one obtains

$$P_\Delta(t) = \frac{t^4 + (|(2\Delta)^\circ \cap M| - 5)t^3 + \psi_2(\Delta)t^2 + (|\Delta \cap M| - 5)t + 1}{(1-t)^5}.$$

In particular, Δ is reflexive if and only if

$$|\Delta \cap M| = |(2\Delta)^\circ \cap M|.$$

Proposition 16 *Let $\Delta \subseteq M_{\mathbb{Q}}$ be a canonical Fano d-tope with $d \in \{3, 4\}$ such that $0 \in N$ is an interior lattice point of $[\Delta^*]$. Then $[\Delta^*]$ is reflexive, i.e., Δ is almost reflexive.*

Proof Let $n \in N$ be an interior lattice point of $[\Delta^*]$. Then $\langle x, n\rangle \geq 0$ for all $x \in \Delta \cap M$ because

$$\Delta^* = \{y \in N_{\mathbb{Q}} \mid \langle x, y\rangle \geq -1 \quad \text{for all } x \in \Delta\}$$

and $\langle x, n\rangle$ is an integer. Since $0 \in \Delta^\circ \cap M$, $M_{\mathbb{Q}}$ is the set of all non-negative $\mathbb{Q}$-linear combinations of all lattice points in $\Delta \cap M$. This implies $\langle x', n\rangle \geq 0$ for all $x' \in M_{\mathbb{Q}}$, i.e., $n = 0$. Therefore, $[\Delta^*]$ has only one interior lattice point $0 \in N$, i.e., $[\Delta^*]$ is a canonical Fano d-tope.

It is clear that $[\Delta^*]$ is contained in the interior of $2[\Delta^*]$. Therefore, we have $[\Delta^*] \cap N \subseteq (2[\Delta^*])^\circ \cap N$. On the other hand, for any lattice point $n \in (2[\Delta^*])^\circ$, $\langle x, n\rangle > -2$ for all $x \in \Delta \cap M$. Since $\langle x, n\rangle$ is an integer, $n \in \Delta^* \cap N$, i.e.,

$$[\Delta^*] \cap N = (2[\Delta^*])^\circ \cap N.$$

Using Proposition 15, $[\Delta^*]$ is reflexive. □

Corollary 17 *Let $\Delta \subseteq M_{\mathbb{Q}}$ be a canonical Fano d-tope with $d \in \{3, 4\}$ such that $0 \in N$ is an interior lattice point of $[\Delta^*]$. Then $[\Delta^*]^*$ is the smallest (referring to inclusion) reflexive polytope containing Δ.*

Proof Let $\Delta' \subseteq M_{\mathbb{Q}}$ be a reflexive d-tope such that $\Delta \subseteq \Delta'$. Then $(\Delta')^* \subseteq \Delta^*$. Since $(\Delta')^*$ is a lattice polytope, it is contained in $[\Delta^*]$. Thus, $[\Delta^*]^*$ is contained in $((\Delta')^*)^* = \Delta'$. □

Remark 18 If Δ is a reflexive d-tope, then $[2\Delta^\circ] = \Delta$. If Δ is a canonical Fano d-tope with $d \in \{3, 4\}$ such that $\Delta^{\text{FI}} = \{0\}$ and Δ is contained in a reflexive d-tope Δ', then $[2\Delta^\circ]$ is contained in $[(2\Delta')^\circ] = \Delta'$. Therefore, $[2\Delta^\circ]$ is contained in the smallest reflexive polytope $[\Delta^*]^*$ containing Δ, i.e.,

$$[2\Delta^\circ] \subseteq [\Delta^*]^*.$$

Computations showed that among all 665,599 canonical Fano 3-topes Δ with $\Delta^{\text{FI}} = \{0\}$ there exist exactly 211,941 canonical Fano 3-tops such that $[2\Delta^\circ]$ is reflexive. For the remaining canonical Fano 3-topes Δ the lattice 3-topes $[2\Delta^\circ]$ are larger than Δ, but are not equal to the reflexive hull $[\Delta^*]^*$.

Remark 19 Let Δ be an almost reflexive 3-tope. We denote by $\tau(\Delta)$ the lattice d-tope $[2\Delta^\circ]$. If $\tau(\Delta)$ is not reflexive, then it is almost reflexive and we can consider the larger lattice d-tope $\tau^2(\Delta) := \tau(\tau(\Delta)) \subseteq [\Delta^*]^*$. After at most five steps, $\tau^k(\Delta)$ is equal to the reflexive hull $[\Delta^*]^*$ of Δ.

In dimension 4, the situation is comparable:

Example 20 Let $\Delta \subseteq \mathbb{R}^4$ be the almost reflexive 4-tope defined by the inequalities $x_i \geq -1$ $(1 \leq i \leq 4)$, $x_1 \leq 2$, and $x_1 + x_2 + x_3 + x_4 \leq 1$. Then $\Delta^{FI} = \{0\}$ and the smallest reflexive 4-tope containing Δ is the 4-simplex $[\Delta^*]^*$ defined by the inequalities $x_i \geq -1$ $(1 \leq i \leq 4)$ and $x_1 + x_2 + x_3 + x_4 \leq 1$. It is easy to see that $\tau(\Delta)$ is not the reflexive 4-tope $[\Delta^*]^*$ because the vertex $(4, -1, -1, -1) \in \text{vert}([\Delta^*]^*)$ is not in $2\Delta^\circ$. However, $\tau^2(\Delta) = [\Delta^*]^*$.

2.3 Canonical Fano 3-Topes with $\Delta^{\text{FI}} = \{0\}$

We note that the set of all reflexive 3-topes forms a rather small part of the set of all canonical Fano 3-topes. The majority of the canonical Fano 3-topes belong to the subset of almost reflexive 3-topes. The proof of the following statement is based on the result of Skarke [21] and the explanations in the previous section.

Proposition 21 *A canonical Fano* 3*-tope* Δ *is almost reflexive if one of the following equivalent conditions is satisfied:*

1. $\Delta^{\text{FI}} = \{0\}$;
2. $0 \in N$ *is an interior lattice point of* $[\Delta^*]$;
3. Δ *is contained in some reflexive* 3*-tope;*
4. $\tau^k(\Delta)$ *is the reflexive* 3*-tope* $[\Delta^*]^*$ *for some sufficiently large* k $(1 \leq k \leq 5)$;
5. *the lattice* 3*-tope* $[2\Delta^\circ]$ *has exactly one interior lattice point;*
6. *the non-degenerate affine hypersurface* $\mathcal{Z}_\Delta$ *defined by a Laurent polynomial with Newton polytope* Δ *is birational to a smooth* $K3$*-surface.*

Computations show that there exist exactly 665,599 almost reflexive canonical Fano 3-topes. The set of almost reflexive 3-topes includes all 4,319 reflexive 3-topes. We have shown that for any almost reflexive 3-tope Δ, the reflexive polytope $\Delta^{\text{ref}} := [\Delta^*]^*$ is the smallest reflexive 3-tope containing Δ. We call Δ^{ref} the *reflexive hull* of Δ. Thus we obtain a natural surjective map $\Delta \mapsto \Delta^{\text{ref}}$ from the set of almost reflexive 3-topes to the set of reflexive 3-topes, which is the identity on the set of reflexive 3-topes. The minimal surface S_Δ is a $K3$-surface if and only if Δ is an almost reflexive 3-tope. If Δ is an almost reflexive 3-tope, but not reflexive, then the minimal surface S_Δ is a crepant desingularization of the Zariski closure of $\mathcal{Z}_\Delta$ in the Gorenstein toric Fano threefold $X_{\Delta^{\text{ref}}}$ defined by the reflexive hull of Δ.

A generalization of the reflexive hull of almost reflexive 3-topes for arbitrary lattice d-topes with non-empty Fine interior can be obtained using the notion of the support of the Fine interior Δ^{FI}.

Definition 22 Let $\Delta \subseteq M_{\mathbb{Q}}$ be a lattice d-tope with $\Delta^{\text{FI}} \neq \emptyset$. Then the set

$$\text{supp}(\Delta^{\text{FI}}) := \{n \in N \mid \text{there exists } x \in \Delta^{\text{FI}} \text{ with } \langle x, n\rangle = \text{ord}_\Delta(n) + 1\}$$

is called *support of the Fine interior* of Δ.

Example 23 If Δ is a reflexive d-tope, then the support of the Fine interior of Δ is the set of all non-zero lattice points in $\Delta^* \cap N$.

Remark 24 It is easy to show that one always has

$$\Delta^{\text{FI}} = \bigcap_{n \in \text{supp}(\Delta^{\text{FI}})} \{x \in M_{\mathbb{Q}} \mid \langle x, n\rangle \geq \text{ord}_\Delta(n) + 1\}.$$

Definition 25 Let $\Delta \subseteq M_{\mathbb{Q}}$ be a lattice d-tope with $\Delta^{\text{FI}} \neq \emptyset$. Then the rational polytope

$$\Delta^{\text{can}} := \bigcap_{n \in \text{supp}(\Delta^{\text{FI}})} \{x \in M_{\mathbb{Q}} \mid \langle x, n\rangle \geq \text{ord}_\Delta(n)\}$$

contains Δ and is called the *canonical hull* of Δ.

Example 26 If Δ is an almost reflexive 3-tope, then $\text{supp}(\Delta^{\text{FI}})$ is the set $(\Delta^* \cap N) \setminus \{0\}$ of boundary lattice points in the reflexive 3-tope $[\Delta^*]$ and the canonical

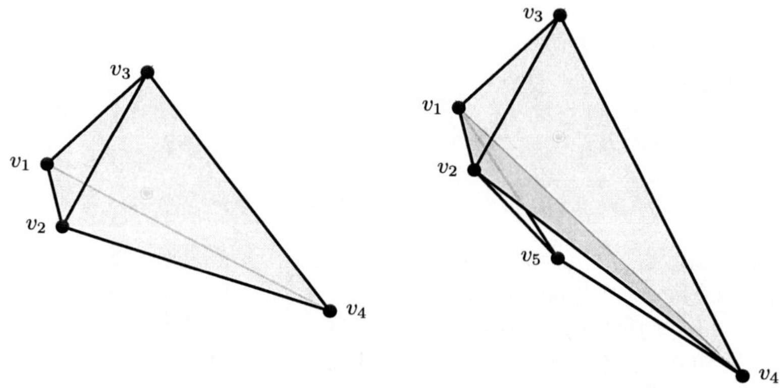

(a) ID 547386. (b) ID 547385.

Fig. 2.1 Canonical Fano 3-topes $\boldsymbol{\Delta}$ with $\boldsymbol{\Delta^{\mathrm{FI}}=\{0\}}$. Shaded faces are occluded and the Fine interior $\{0\}$ is shown in grey with a double border. The whole polytope is the canonical hull Δ^{can} as well as the reflexive hull Δ^{ref} and the grey coloured polytope is Δ. **a** Reflexive polytope $\Delta = \mathrm{conv}\{v_1, v_2, v_3, v_4\}$ with $v_1 = (1, 0, 0), v_2 = (0, 1, 0), v_3 = (0, 0, 1), v_4 = (-1, -1, -1)$, and $\Delta^{\mathrm{ref}} = \Delta^{\mathrm{can}} = \Delta$. All facets of Δ have lattice distance 1 to the origin. **b** Almost reflexive polytope $\Delta = \mathrm{conv}\{v_1, v_2, v_3, v_4\}$ with $v_1 = (1, 0, 0), v_2 = (0, 1, 0), v_3 = (0, 0, 1), v_4 = (-1, -1, -2)$, and $\Delta^{\mathrm{ref}} = \Delta^{\mathrm{can}} = \mathrm{conv}\{v_1, v_2, v_3, v_4, v_5\}$ with $v_5 = (0, 0, -1)$ reflexive. The dark grey coloured facet of Δ has lattice distance 2 and all other facets have lattice distance 1 to the origin

hull Δ^{can} equals the reflexive hull Δ^{ref} of the polytope Δ, i.e., $\Delta^{\mathrm{can}} = \Delta^{\mathrm{ref}} = [\Delta^*]^*$. In particular, in this case Δ^{can} is always a lattice 3-tope.

There exists a smooth projective toric variety X_Σ defined by a fan Σ whose 1-dimensional cones are generated by all lattice vectors from the finite set $\mathrm{supp}(\Delta^{\mathrm{FI}})$. Then the minimal surface S_Δ is a $K3$-surface which is the Zariski closure of Z_Δ in X_Σ [3].

Example 27 Let us consider the (almost) reflexive canonical Fano 3-tope $\Delta = \mathrm{conv}\{v_1, v_2, v_3, v_4\} \subseteq M_{\mathbb{Q}}$ (ID 547386, Fig. 2.1a) with vertices

$$v_1 := (1, 0, 0),\ v_2 := (0, 1, 0),\ v_3 := (0, 0, 1),\ \text{and } v_4 := (-1, -1, -1)$$

and $\Delta^{\mathrm{FI}} = \{0\}$. Moreover,

$$\Delta^{\mathrm{ref}} = \mathrm{conv}(2\Delta^\circ \cap M) = \mathrm{conv}(\Delta \cap M) = \Delta$$

and

$$\Delta^{\mathrm{can}} = [\Delta^*]^* = (\Delta^*)^* = \Delta$$

because Δ is reflexive, i.e., $\Delta^{\mathrm{ref}} = \Delta^{\mathrm{can}} = \Delta$ reflexive (Fig. 2.1a).

Example 28 Consider the almost reflexive canonical Fano 3-tope $\Delta \subseteq M_{\mathbb{Q}}$ (ID 547385, Fig. 2.1b) with vertices

$$v_1 := (1, 0, 0),\ v_2 := (0, 1, 0),\ v_3 := (0, 0, 1),\ \text{and}\ v_4 := (-1, -1, -2)$$

and $\Delta^{\mathrm{FI}} = \{0\}$. Moreover,

$$\Delta^{\mathrm{ref}} = \mathrm{conv}((\Delta \cap M) \cup \{v_5\}) = \mathrm{conv}\{v_1, v_2, v_3, v_4, v_5\}$$

and

$$\Delta^{\mathrm{can}} = [\Delta^*]^* = \mathrm{conv}\{v_1, v_2, v_3, v_4, v_5\}$$

with $v_5 := (0, 0, -1)$ because Δ is almost reflexive, i.e., $\Delta^{\mathrm{ref}} = \Delta^{\mathrm{can}} = \Delta$ reflexive (Fig. 2.1b).

2.4 Asymmetric Fine Interior of Dimension 1

There exist exactly 9,020 canonical Fano 3-topes Δ with 1-dimensional Fine interior such that $0 \in N$ belongs to a facet $\Theta \preceq [\Delta^*]$ of the lattice 3-tope $[\Delta^*]$. This class of canonical Fano 3-topes is characterized by the property that the lattice 3-tope $[2\Delta^\circ]$ has exactly 2 interior lattice points.

The corresponding minimal surfaces S_Δ are *simply connected* (i.e., have trivial fundamental group $\pi_1(S_\Delta)$) elliptic surfaces of Kodaira dimension $\kappa = 1$. We observed that the facet $\Theta \preceq [\Delta^*]$ is a reflexive 2-tope corresponding to one of the three types pictured in Fig. 2.2. All N-lattice points on the boundary of Θ belong to $\mathrm{supp}(\Delta^{\mathrm{FI}})$. It was checked that for all these 3-topes Δ the canonical hull Δ^{can} is again a lattice 3-tope. Moreover, the Fine interior Δ^{FI} is contained in the ray generated by the primitive lattice vector $v_\Delta \in M$ which is the primitive inward-pointing facet normal of Θ, i.e., $\langle x, y \rangle = 0$ for all $x \in \Delta^{\mathrm{FI}}$, $y \in \Theta$. The lattice point $0 \in M$ is a vertex of Δ^{FI}. More precisely, one has

$$\Delta^{\mathrm{FI}} = \mathrm{conv}\{0, \lambda v_\Delta\},$$

where $\lambda \in \{1/2, 2/3\}$. The primitive lattice vector v_Δ is the unique interior lattice point on a reflexive facet $\theta_+ \preceq \Delta$ of Δ of one of the three possible types pictured in Fig. 2.2. These three reflexive polygons θ_+ are characterized by the condition that the dual reflexive polygons θ_+^* are obtained from θ_+ (Fig. 2.3) by enlarging the lattice $\mathbb{Z}^2$ in the following ways: $\mathbb{Z}^2 + \mathbb{Z}(1/3, 2/3)$ (Fig. 2.3a), $\mathbb{Z}^2 + \mathbb{Z}(1/2, 0)$ (Fig. 2.3b), and $\mathbb{Z}^2 + \mathbb{Z}(1/2, 1/2)$ (Fig. 2.3c). Moreover, the reflexive facet θ_+ of Δ is isomorphic to the facet Θ of $[\Delta^*]$. The projection $M \to M/\mathbb{Z}v_\Delta$ of Δ or of θ_+ along v_Δ is a reflexive polygon of one of the three types pictured in Fig. 2.3, which is dual to θ_+ and Θ. The lattice vector v_Δ defines a character of the 3-dimensional torus $\chi : \mathbb{T}^3 \to \mathbb{C}^\times$. For almost all $\alpha \in \mathbb{C}^\times$, the fiber $\chi^{-1}(\alpha)$ is an affine elliptic curve defined by a Laurent

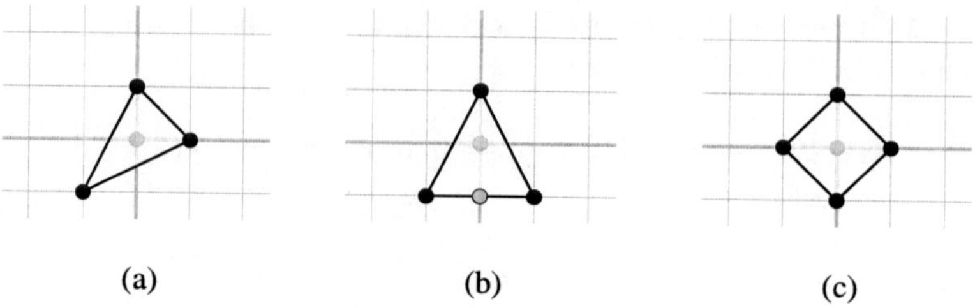

Fig. 2.2 Reflexive Facets of Δ Containing $\pm v_\Delta$. Three types of reflexive facets $\theta_\pm \preceq \Delta$ of Δ containing $\pm v_\Delta$ for all 9,020 + 20 canonical Fano 3-topes Δ with $\dim(\Delta^{\text{FI}}) = 1$. Vertices are coloured black, boundary points that are not vertices grey, and the origin light grey. **a** $\text{conv}\{(1,0),(0,1),(-1,-1)\}$. **b** $\text{conv}\{(1,0),(-1,1),(-1,-1)\}$. **c** $\text{conv}\{(\pm1,0),(0,\pm1)\}$

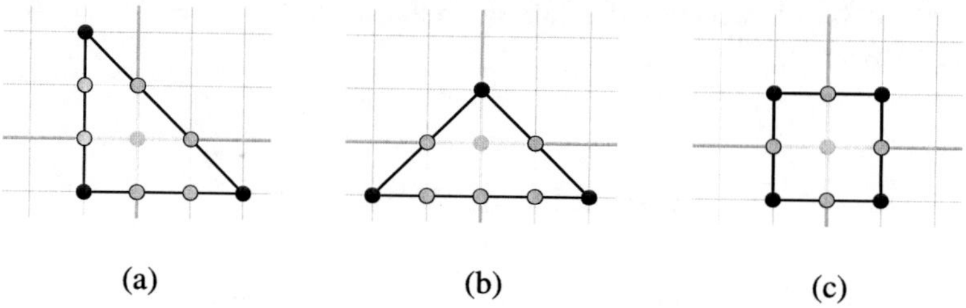

Fig. 2.3 Reflexive Projection Polytopes. Three types of reflexive polytopes obtained via a projection of Δ along $\pm v_\Delta$ for all 9,020 + 20 canonical Fano 3-topes Δ with $\dim(\Delta^{\text{FI}}) = 1$. Vertices are coloured black, boundary points that are not vertices grey, and the origin light grey. **a** $\text{conv}\{(-1,2),(-1,-1),(2,-1)\}$. **b** $\text{conv}\{(-2,-1),(0,1),(2,-1)\}$. **c** $\text{conv}\{(\pm1,\pm1)\}$

polynomial with the reflexive Newton polytope $\Theta^* \cong \theta_+^*$ of one of the three types pictured in Fig. 2.3 with the distribution shown in Table 2.1. So χ defines birationally an elliptic fibration.

Table 2.1 Distribution of the Reflexive Facets of Δ Containing $\pm v_\Delta$. Table contains: Type of the reflexive facet $\theta_\pm$ containing $\pm v_\Delta$, type of the dual reflexive facet $\theta_\pm^*$, the enlarged lattice used to obtain $\theta_\pm^*$ from $\theta_\pm$, the number of canonical Fano 3-topes $\Delta_{\text{asym}} := \{\Delta \mid \text{1-dim. asym. } \Delta^{\text{FI}}\}$, and the number of canonical Fano 3-topes $\Delta_{\text{sym}} := \{\Delta \mid \text{1-dim. sym. } \Delta^{\text{FI}}\}$ with respect to the facet type of $\theta_\pm$ pictured in Fig. 2.2

$\theta_\pm$	$\theta_\pm^*$	Enlarged lattice	$\#\Delta_{\text{asym}}$	$\#\Delta_{\text{sym}}$
Figure 2.2a	Figure 2.3a	$\mathbb{Z}^2 + \mathbb{Z}(1/3, 2/3)$	3,038	7
Figure 2.2b	Figure 2.3b	$\mathbb{Z}^2 + \mathbb{Z}(1/2, 0)$	4,663	9
Figure 2.2c	Figure 2.3c	$\mathbb{Z}^2 + \mathbb{Z}(1/2, 1/2)$	1,319	4

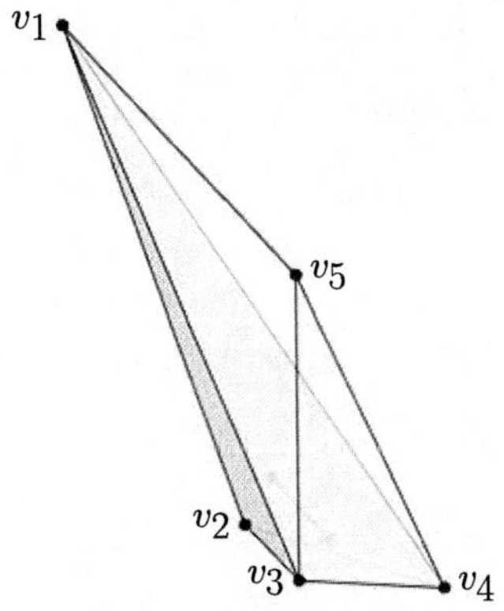

(a) ID 547324

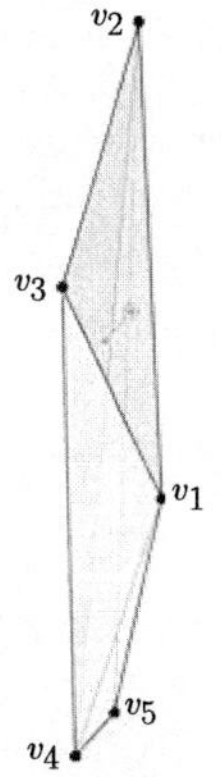

(b) ID 547323

Fig. 2.4 Canonical Fano 3-topes with Asymmetric Fine Interior of Dimension 1 Shaded faces are occluded. The Fine interior and the origin are shown in grey, with a double border around the origin. The facet θ_+ is grey dotted. **a** The whole polytope is $\Delta = \text{conv}\{v_1, v_2, v_3, v_4\}$ with $v_1 = (2, 3, 8)$, $v_2 = (1, 0, 0)$, $v_3 = (0, 1, 0)$, and $v_4 = (-1, -1, -1)$. Moreover, $\Delta^{\text{FI}} = \text{conv}\{(0, 0, 0), (1/2, 1/2, 1)\}$, $\theta_+ = \text{conv}\{v_1, v_2, v_3\}$, and $\Delta^{\text{can}} = \text{conv}\{v_1, v_2, v_3, v_4, v_5\}$ with $v_5 = (0, 1, 4)$. **b** The whole polytope is $\Delta = \text{conv}\{v_1, v_2, v_3, v_4\}$ with $v_1 = (-1, 1, -2)$, $v_2 = (1, -2, 3)$, $v_3 = (1, 0, 0)$, and $v_4 = (-2, 5, -3)$. Moreover, $\Delta^{\text{FI}} = \text{conv}\{(0, 0, 0), (0, 2/3, 0)\}$ and $\theta_+ = \text{conv}\{v_2, v_3, v_4\}$, and $\Delta^{\text{can}} = \text{conv}\{v_1, v_2, v_3, v_4, v_5\}$ with $v_5 = (-2, 4, -3)$

Example 29 Let $\Delta \subseteq M_{\mathbb{Q}}$ be a canonical Fano 3-tope given as the convex hull of

$$v_1 := (2, 3, 8),\ v_2 := (1, 0, 0),\ v_3 := (0, 1, 0),\ \text{and } v_4 := (-1, -1, -1)$$

(ID 547324, Fig. 2.4a, Tables 2.2 and 2.4). Then

$$\Delta^{\text{FI}} = \text{conv}\{(0, 0, 0), (1/2, 1/2, 1)\} = \text{conv}\{0, 1/2 \cdot v_\Delta\},$$

where $v_\Delta = (1, 1, 2)$. One has $v_1 + 2v_2 + v_3 = 4v_\Delta$. Therefore, v_Δ is the interior lattice point of the reflexive facet θ_+ of Δ with vertices v_1, v_2, v_3 and the images $\overline{v}_1, \overline{v}_2, \overline{v}_3$ of v_1, v_2, v_3 in $M/\mathbb{Z}v_\Delta$ are vertices of the dual reflexive triangle θ_+^* (Fig. 2.3b) satisfying the relation

$$\overline{v}_1 + 2\overline{v}_2 + \overline{v}_3 = 0.$$

To compute the canonical hull Δ^{can} of Δ, we obtain $\text{supp}(\Delta^{\text{FI}}) = \{s_i \mid 1 \le i \le 18\}$ with $s_1 := (-1, -1, 1)$, $s_2 := (-1, -1, 2)$, $s_3 := (-1, -1, 3)$, $s_4 := (-1, 0, 1)$, $s_5 := (-1, 0, 2)$, $s_6 := (-1, 1, 0)$, $s_7 := (-1, 1, 1)$, $s_8 := (-1, 2, 0)$, $s_9 := (-1, 3, -1)$, $s_{10} := (0, -1, 1)$, …, $s_{18} := (-2, -2, 1)$, which leads to

$$\Delta^{\text{can}} = \text{conv}\{v_1, v_2, v_3, v_4, v_5\}$$

with $v_5 := (0, 1, 4)$ (Fig. 2.4a).

Table 2.2 9 Canonical Fano 3-topes with Asymmetric Fine Interior of Dimension 1. Table contains: vertices vert(Δ) of Δ, vertices vert(Δ^{FI}) of the Fine interior Δ^{FI}, unique primitive lattice point $v_\Delta \in \theta_+$ in the reflexive facet $\theta_+ \preceq \Delta$, and weights $(w_i)_{0\le i\le 3}$ of the weighted projective 3-space $\mathbb{P}(w_0, \ldots, w_3)$ appearing in [11]

ID	vert(Δ)	vert(Δ^{FI})	v_Δ	$(w_i)_{0\le i\le 3}$
547324	(2, 3, 8), (1, 0, 0), (0, 1, 0), (−1, −1, −1)	$0, 1/2 \cdot v_\Delta$	(1, 1, 2)	(1, 5, 6, 8)
547323	(−1, 1, −2), (1, −2, 3), (1, 0, 0), (−2, 5, −3)	$0, 2/3 \cdot v_\Delta$	(0, 1, 0)	(1, 4, 7, 9)
547311	(−1, 4, 2), (−1, −1, 0), (0, 0, −1), (2, 0, 1)	$0, 2/3 \cdot v_\Delta$	(0, 1, 1)	(2, 5, 8, 9)
547490	(1, 2, 4), (1, 0, 0), (1, −2, 3), (−1, 1, −2)	$0, 1/2 \cdot v_\Delta$	(0, 1, 0)	(1, 5, 8, 14)
547321	(1, −2, 3), (0, 1, 0), (1, 0, 0), (−6, 3, −8)	$0, 1/2 \cdot v_\Delta$	(−1, 1, −2)	(3, 7, 8, 10)
547305	(0, 1, 0), (1, 0, 0), (1, 2, 4), (−4, −6, −7)	$0, 2/3 \cdot v_\Delta$	(−1, −1, −1)	(4, 7, 9, 10)
547526	(1, 0, 0), (0, 1, 0), (−2, 1, 5), (2, −4, −9)	$0, 2/3 \cdot v_\Delta$	(1, −1, −3)	(5, 9, 8, 11)
547454	(2, 1, 7), (1, 0, 0), (0, 1, 0), (−2, −3, −3)	$0, 1/2 \cdot v_\Delta$	(0, 0, 1)	(3, 7, 8, 18)
547446	(0, 1, 1), (−6, 7, −15), (1, −2, 3), (1, 0, 0)	$0, 1/2 \cdot v_\Delta$	(−1, 1, −2)	(5, 8, 9, 22)

Table 2.3 9 Canonical Fano 3-topes with Asymmetric Fine Interior of Dimension 1. Table contains: primitive inward-pointing facet normals $(n_i)_{1\le i\le 4}$ of Δ, vertices vert(θ_+) of the reflexive facet $\theta_+ \preceq \Delta$, and primitive inward-pointing facet normal n_{θ_+} of the reflexive facet $\theta_+ \preceq \Delta$

ID	$(n_i)_{1\le i\le 4}$	vert(θ_+)	n_{θ_+}
547324	(−2, −2, 1), (−1, −1, 3), (−1, 3, −1), (7, −3, −1)	(2, 3, 8), (1, 0, 0), (0, 1, 0)	(−2, −2, 1)
547323	(−3, −3, −2), (−1, 0, 1), (−1, 6, 4), (17, 3, −5)	(1, −2, 3), (1, 0, 0), (−2, 5, −3)	(−3, −3, −2)
547311	(−1, −1, 1), (−1, 2, 1), (1, 2, −5), (7, −2, 5)	(−1, 4, 2), (−1, −1, 0), (2, 0, 1)	(1, 2, −5)
547490	(−2, −2, 1), (−1, 0, 0), (−1, 6, 4), (23, 2, −8)	(1, 2, 4), (1, 0, 0), (−1, 1, −2)	(−2, −2, 1)
547321	(−3, −3, −2), (−2, −2, 1), (−1, 3, 2), (9, −5, −8)	(0, 1, 0), (1, 0, 0), (−6, 3, −8)	(−2, −2, 1)
547305	(−7, −7, 11), (−2, −2, 1), (−1, 2, −1), (7, −3, −1)	(0, 1, 0), (1, 2, 4), (−4, −6, −7)	(7, −3, −1)
547526	(−5, −5, −2), (−3, −3, 1), (−1, 2, −1), (25, −8, 10)	(1, 0, 0), (0, 1, 0), (2, −4, −9)	(−3, −3, 1)
547454	(−7, −7, 2), (−1, −1, 2), (−1, 1, 0), (7, −2, −2)	(2, 1, 7), (0, 1, 0), (−2, −3, −3)	(7, −2, −2)
547446	(−9, 21, 14), (−5, −3, −2), (−1, −1, 0), (9, 1, −3)	(0, 1, 1), (−6, 7, −15), (1, −2, 3)	(9, 1, −3)

Example 30 Let $\Delta \subseteq M_{\mathbb{Q}}$ be a canonical Fano 3-tope given as the convex hull of

$$v_1 := (-1, 1, -2),\ v_2 := (1, -2, 3),\ v_3 := (1, 0, 0),\ \text{and } v_4 := (-2, 5, -3)$$

(ID 547323, Fig. 2.4b, Tables 2.2 and 2.4). Then (Table 2.3)

$$\Delta^{FI} = \mathrm{conv}\{(0, 0, 0), (0, 2/3, 0)\} = \mathrm{conv}\{0, 2/3 \cdot v_\Delta\},$$

where $v_\Delta = (0, 1, 0)$. One has $v_2 + v_3 + v_4 = 3v_\Delta$. Therefore, v_Δ is the interior lattice point of the reflexive facet θ_+ of Δ with vertices v_2, v_3, v_4 and the images $\overline{v}_2, \overline{v}_3, \overline{v}_4$ of v_2, v_3, v_4 in $M/\mathbb{Z}v_\Delta$ are vertices of the dual reflexive triangle θ_+^* (Fig. 2.3a) satisfying the relation

$$\overline{v}_2 + \overline{v}_3 + \overline{v}_4 = 0.$$

Table 2.4 9 Canonical Fano 3-topes with Asymmetric Fine Interior of Dimension 1. Table contains: vertices $\mathrm{vert}(\Theta)$ of the reflexive facet $\Theta \preceq [\Delta^*]$, support $\mathrm{supp}(\Delta^{FI})$ of the Fine interior Δ^{FI}, and vertices $\mathrm{vert}(\Delta^{can})$ of the canonical hull Δ^{can}

ID	$\mathrm{vert}(\Theta)$	$\mathrm{supp}(\Delta^{FI})$	$\mathrm{vert}(\Delta^{can})$
547324	$(-1,3,-1), (-1,-1,1), (1,-1,0)$	$(-2,-2,1), (-1,-1,1), S_1$	$\mathrm{vert}(\Delta), (0,1,4)$
547323	$(-1,0,1), (-1,0,0), (2,0,-1)$	$(-3,-3,-2), (-1,0,0), S_2$	$\mathrm{vert}(\Delta), (-2,4,-3)$
547311	$(-1,-1,1), (0,1,-1), (1,0,0)$	$(-1,-1,1), (-1,0,1), S_3$	$\mathrm{vert}(\Delta), (-1,2,0)$
547490	$(-1,0,0), (-1,0,1), (3,0,-1)$	$(-2,-2,1), (-1,0,0), S_4$	$\mathrm{vert}(\Delta), (1,-1,4)$
547321	$(-1,-1,0), (-1,3,2), (1,-1,-1)$	$(-2,-2,1), (-1,-1,0), (-1,0,0), (-1,1,1), (-1,3,2), (0,-1,-1), (1,-1,-1)$	$\mathrm{vert}(\Delta), (1,0,1), (0,-3,4)$
547305	$(-1,2,-1), (1,-1,0), (0,-1,1)$	$(-1,-1,1), (-1,0,0), (-1,2,-1), (0,-1,1), (1,-1,0), (7,-3,-1)$	$\mathrm{vert}(\Delta), (0,-2,-3), (1,2,2)$
547526	$(-1,-1,0), (-1,2,-1), (2,-1,1)$	$(-3,-3,1), (-1,-1,0), (-1,2,-1), (0,-1,0), (2,-1,1)$	$\mathrm{vert}(\Delta)\setminus\{(-2,1,5)\}, (0,1,3), (-3,1,6)$
547454	$(-1,1,0), (0,-1,0), (2,-1,0)$	$(-1,-1,1), (-1,-1,2), S_5$	$\mathrm{vert}(\Delta), (2,1,2)$
547446	$(-1,-1,0), (0,2,1), (2,0,-1)$	$(-1,-1,0), (-1,0,0), (0,2,1), (1,1,0), (2,0,-1), (9,1,-3)$	$\mathrm{vert}(\Delta), (1,0,-1), (1,0,3)$

where $S_1 := (-1,-1,2), (-1,-1,3), (-1,0,1), (-1,0,2), (-1,1,0), (-1,1,1), (-1,2,0), (-1,3,-1), (0,-1,1), (0,-1,2), (0,0,1), (0,1,0), (1,-1,0), (1,-1,1), (1,0,0), (2,-1,0)$

$S_2 := (-1,0,1), (-1,1,1), (-1,2,2), (-1,3,2), (-1,4,3), (-1,6,4), (0,1,1), (0,2,1), (0,3,2), (0,5,3), (1,1,0), (1,2,1), (1,4,2), (2,0,-1), (2,1,0), (2,3,1), (3,2,0), (4,1,-1)$

$S_3 := (-1,1,1), (-1,2,1), (0,0,1), (0,1,-1), (0,1,0), (0,1,1), (1,0,0), (1,0,1), (1,2,-5)$

$S_4 := (-1,0,1), (-1,1,1), (-1,2,2), (-1,3,2), (-1,4,3), (-1,6,4), (0,1,1), (0,2,1), (0,3,2), (0,5,3), (1,0,0), (1,1,0), (1,2,1), (1,4,2), (2,1,0), (2,3,1), (3,0,-1), (3,2,0), (4,1,-1)$

$S_5 := (-1,0,1), (-1,1,0), (0,-1,0), (0,-1,1), (1,-1,0), (2,-1,0), (7,-2,-2)$

To compute the canonical hull Δ^{can} of Δ, we obtain $\text{supp}(\Delta^{\text{FI}}) = \{s_i \mid 1 \le i \le 20\}$ with $s_1 := (-3, -3, -2)$, $s_2 := (-1, 0, 0)$, $s_3 := (-1, 0, 1)$, $s_4 := (-1, 1, 1)$, $s_5 := (-1, 2, 2)$, $s_6 := (-1, 3, 2)$, $s_7 := (-1, 4, 3)$, $s_8 := (-1, 6, 4)$, $s_9 := (0, 1, 1)$, $s_{10} := (0, 2, 1), \ldots, s_{20} := (4, 1, -1)$, which leads to

$$\Delta^{\text{can}} = \text{conv}\{v_1, v_2, v_3, v_4, v_5\}$$

with $v_5 := (-2, 4, -3)$ (Fig. 2.4b).

Remark 31 The detailed information about a small selection of the 9,020 canonical Fano 3-topes with $\dim(\Delta^{\text{FI}}) = 1$ and $0 \in \text{vert}(\Delta^{\text{FI}})$ can be found in Tables 2.2, 2.3, and 2.4.

2.5 Symmetric Fine Interior of Dimension 1

There exist exactly 20 canonical Fano 3-topes Δ such that 0 is the center of 1-dimensional Fine interior Δ^{FI}. In this case, $\mathcal{S}_\Delta$ is an elliptic surface of Kodaira dimension $\kappa = 1$ with non-trivial fundamental group $\pi_1(\mathcal{S}_\Delta)$ of order 2 or 3. Computations show that one always has $\Delta = \Delta^{\text{can}}$ and

$$\Delta^{\text{FI}} = \text{conv}\{-\lambda v_\Delta, \lambda v_\Delta\}$$

with $\lambda = \frac{1}{2}$ if and only if $|\pi_1(\mathcal{S}_\Delta)| = 2$ and

$$\Delta^{\text{FI}} = \text{conv}\{-\mu v_\Delta, \mu v_\Delta\}$$

with $\mu = \frac{2}{3}$ if and only if $|\pi_1(\mathcal{S}_\Delta)| = 3$. The primitive lattice vectors $\pm v_\Delta$ are the two unique interior lattice points in two reflexive facets $\theta_\pm \preceq \Delta$ of one of the three possible types pictured in Fig. 2.2. The reflexive facets $\theta_\pm$ of Δ are isomorphic to the facet Θ of $[\Delta^*]$. The projections $M \to M/\mathbb{Z}(\pm v_\Delta)$ of Δ or of $\theta_\pm$ along $\pm v_\Delta$ reveal a reflexive polygon of one of the three types pictured in Fig. 2.3, which is dual to $\theta_\pm$ and Θ. The lattice vector v_Δ defines a character of the 3-dimensional torus $\chi : \mathbb{T}^3 \to \mathbb{C}^\times$. For almost all $\alpha \in \mathbb{C}^\times$, the fiber $\chi^{-1}(\alpha)$ is an affine elliptic curve defined by a Laurent polynomial with the reflexive Newton polytope $\Theta^* \cong \theta_\pm^*$ of one of the three types pictured in Fig. 2.3 with the distribution shown in Table 2.1. So χ defines birationally an elliptic fibration. The vertex sets of Δ and these reflexive facets are related via $\text{vert}(\Delta) = \text{vert}(\theta_+) \cup \text{vert}(\theta_-)$. Moreover, every edge of Δ is either an edge of θ_+ or θ_- of these two facets or it is parallel to v_Δ.

Example 32 Let $\Delta \subseteq M_\mathbb{Q}$ be a canonical Fano 3-tope given as the convex hull

$$v_1 := (0, 1, 0),\ v_2 := (2, 1, 1),\ v_3 := (-2, -3, -5),\ \text{and}\ v_4 := (2, 1, 9)$$

(ID 547393, Fig. 2.5b, Tables 2.5 and 2.6). Then

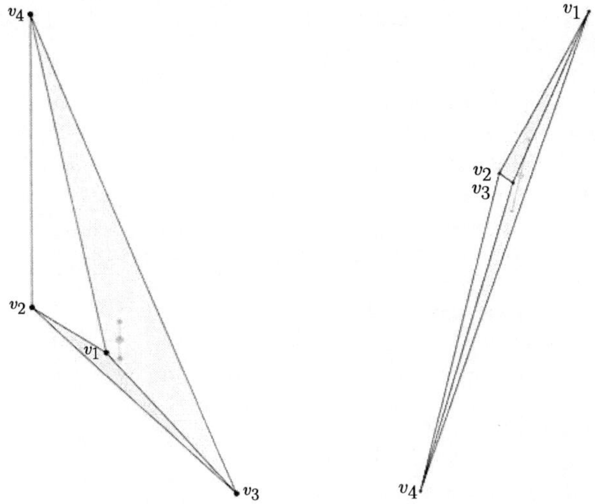

Fig. 2.5 Canonical Fano 3-topes with Symmetric Fine Interior of Dimension 1. Shaded faces are occluded. The Fine interior and the origin are shown in grey with a double border around the origin. The facets $\theta_\pm$ are grey dotted. **a** The whole polytope is $\Delta = \text{conv}\{v_1, v_2, v_3, v_4\}$ with $v_1 = (1, 0, 0)$, $v_2 = (2, 1, 1)$, $v_3 = (-2, -3, -5)$, and $v_4 = (2, 1, 9)$. Moreover, $\Delta^{\text{FI}} = \text{conv}\{(0, 0, -1/2), (0, 0, 1/2)\}$, $\theta_+ = \text{conv}\{v_1, v_3, v_4\}$, $\theta_- = \text{conv}\{v_1, v_2, v_3\}$, and $\Delta^{\text{can}} = \Delta$. **b** The whole polytope is $\Delta = \text{conv}\{v_1, v_2, v_3, v_4\}$ with $v_1 = (-4, 2, 9)$, $v_2 = (1, 0, 0)$, $v_3 = (0, 1, 0)$, and $v_4 = (7, -6, -18)$. Moreover, $\Delta^{\text{FI}} = \text{conv}\{(-2/3, 2/3, 2), (2/3, -2/3, -2)\}$, $\theta_+ = \text{conv}\{v_1, v_2, v_3\}$, $\theta_- = \text{conv}\{v_1, v_3, v_4\}$, and $\Delta^{\text{can}} = \Delta$

$$\Delta^{\text{FI}} = \text{conv}\{(0, 0, -1/2), (0, 0, 1/2)\} = \text{conv}\{-\lambda v_\Delta, \lambda v_\Delta\}$$

with $\lambda = \frac{1}{2}$, where $v_\Delta = (0, 0, 1)$. One has $2v_1 + v_3 + v_4 = 4v_\Delta$ and $2v_1 + v_2 + v_3 = 4(-v_\Delta)$. Therefore, v_Δ is the interior lattice point of the reflexive facet $\theta_+ = \theta_{134}$ of Δ and $-v_\Delta$ is the interior lattice point of the reflexive facet $\theta_- = \theta_{123}$ of Δ (Fig. 2.2b). The images $\overline{v}_1, \overline{v}_3, \overline{v}_4$ of v_1, v_3, v_4 in $M/\mathbb{Z}v_\Delta$ and the images $\overline{v}_1, \overline{v}_2, \overline{v}_3$ of v_1, v_2, v_3 in $M/\mathbb{Z}(-v_\Delta)$ are vertices of the dual reflexive triangle $\theta_\pm^*$ (Fig. 2.3b) satisfying the relation

$$2\overline{v}_1 + \overline{v}_3 + \overline{v}_4 = 0$$

and

$$2\overline{v}_1 + \overline{v}_2 + \overline{v}_3 = 0,$$

respectively.

To compute the canonical hull Δ^{can} of Δ, we obtain $\text{supp}(\Delta^{\text{FI}}) = \{s_i \mid 1 \leq i \leq 6\}$ with $s_1 := (-1, -2, 2)$, $s_2 := (-1, 1, 0)$, $s_3 := (0, -1, 0)$, $s_4 := (1, -1, 0)$, $s_5 := (2, -1, 0)$, and $s_6 := (9, -2, -2)$, which leads to $\Delta^{\text{can}} = \Delta$.

Table 2.5 20 Canonical Fano 3-topes with Symmetric Fine Interior of Dimension 1. Table contains: vertices vert(Δ) of Δ

ID	vert(Δ)
547393	(0, 1, 0), (2, 1, 1), (−2, −3, −5), (2, 1, 9)
547409	(−4, 2, 9), (1, 0, 0), (0, 1, 0), (7, −6, −18)
547461	(0, 1, 0), (2, 1, 1), (−2, −3, −5), (0, 1, 4)
544442	(1, 0, 0), (0, 1, 0), (3, −6, 8), (1, −4, 4), (−5, 6, −12)
544443	(−1, −2, 0), (3, −6, 8), (0, 1, 0), (1, 0, 0), (−3, 4, −8)
544651	(−4, 1, −3), (4, −2, 3), (0, 1, 0), (1, −2, 3), (−1, 1, −3)
544696	(5, −4, −15), (1, 0, 0), (0, 1, 0), (−4, 2, 9), (−3, 1, 6)
544700	(−2, −3, −3), (0, 1, 0), (1, 0, 0), (−1, −4, −6), (2, 5, 9)
544749	(−6, −5, −8), (0, 1, 0), (1, 0, 0), (−2, −1, 0), (3, 2, 4)
520925	(0, 1, 0), (0, 0, 1), (−2, −1, 0), (−2, 0, −1), (8, 2, 3), (−2, −3, −2)
520935	(3, 4, 6), (2, 1, 2), (−3, −2, −2), (1, 0, 0), (0, 1, 0), (−6, −5, −8)
522056	(−1, −1, 0), (0, 1, 0), (1, 0, 0), (−1, −1, −3), (−5, −3, −6), (6, 4, 9)
522059	(2, 5, 6), (−2, −3, −3), (0, 1, 0), (1, 0, 0), (−1, −4, −6), (0, 1, 3)
522087	(1, 0, −3), (1, 0, 0), (0, 1, 0), (−4, 2, 9), (−3, 1, 6), (5, −4, −12)
522682	(2, 1, 4), (−3, −2, −4), (−2, −3, −4), (1, 2, 4), (1, 0, 0), (0, 1, 0)
522684	(−2, −1, −4), (3, 2, 4), (−2, −1, 0), (1, 0, 0), (0, 1, 0), (−4, −3, −4)
526886	(−3, 4, −6), (1, 0, 0), (0, 1, 0), (3, −6, 8), (0, 1, −2), (2, −5, 6)
439403	(1, 2, 2), (−1, 0, 0), (−1, 1, −1), (1, 0, 0), (−1, −2, −2), (1, 1, 3), (1, −3, −1)
275525	(4, 1, 2), (0, 1, 0), (−2, −1, 0), (1, 1, 2), (−3, −1, −2), (−2, −1, −2), (1, 1, 0), (1, −1, 0)
275528	(−1, 0, −1), (−3, −2, 1), (−2, −1, 2), (0, −1, 0), (0, 1, 0), (1, 0, 1), (2, 1, −2), (3, 2, −1)

Example 33 Let $\Delta \subseteq M_{\mathbb{Q}}$ be a canonical Fano 3-tope given as the convex hull

$$v_1 := (-4, 2, 9),\ v_2 := (1, 0, 0),\ v_3 := (0, 1, 0),\ \text{and } v_4 := (7, -6, -18)$$

(ID 547409, Fig. 2.5b, Tables 2.5 and 2.6). Then

$$\Delta^{\text{FI}} = \text{conv}\{(-2/3, 2/3, 2), (2/3, -2/3, -2)\} = \text{conv}\{-\mu v_\Delta, \mu v_\Delta\}$$

with $\mu = \frac{2}{3}$, where $v_\Delta = (1, -1, -3)$. One has $v_1 + v_2 + v_3 = -3v_\Delta$ and $v_1 + v_3 + v_4 = -3(-v_\Delta)$. Therefore, v_Δ is the interior lattice point of the reflexive facet $\theta_+ = \theta_{123}$ of Δ and $-v_\Delta$ is the interior lattice point of the reflexive facet $\theta_- = \theta_{134}$ of Δ (Fig. 2.2b). The images $\overline{v}_1, \overline{v}_2, \overline{v}_3$ of v_1, v_2, v_3 in $M/\mathbb{Z}v_\Delta$ and the images $\overline{v}_1, \overline{v}_3, \overline{v}_4$ of v_1, v_3, v_4 in $M/\mathbb{Z}(-v_\Delta)$ are vertices of the dual reflexive triangle $\theta_\pm^*$ (Fig. 2.3b) satisfying the relation

$$\overline{v}_1 + \overline{v}_2 + \overline{v}_3 = 0,$$

and

$$\overline{v}_1 + \overline{v}_3 + \overline{v}_4 = 0,$$

respectively.

Table 2.6 20 Canonical Fano 3-topes with Symmetric Fine Interior of Dimension 1. Table contains: vertices vert(Δ^{FI}) of the Fine interior Δ^{FI}, unique primitive lattice points $\pm\nu_\Delta \in \theta_\pm$ in the reflexive facets $\theta_\pm \preceq \Delta$, vertices vert($\theta_\pm$) of the reflexive facets $\theta_\pm \preceq \Delta$, and support supp($\Delta^{FI}$) of the Fine interior Δ^{FI} (here: $\Delta^{can} = \Delta$)

ID	vert(Δ^{FI})	$\pm\nu_\Delta$	vert($\theta_\pm$)	supp(Δ^{FI})
547393	$\pm 1/2 \cdot \nu_\Delta$	±(0, 1)	(0, 1, 0), (2, 1, 1), (−2, −3, −5) (0, 1, 0), (−2, −3, −5), (2, 1, 9)	(−1, −2, 2), (−1, 1, 0), (0, −1, 0), (1, −1, 0), (2, −1, 0), (9, −2, −2)
547409	$\pm 2/3 \cdot \nu_\Delta$	±(1, −1, −3)	(−4, 2, 9), (1, 0, 0), (0, 1, 0) (−4, 2, 9), (0, 1, 0), (7, −6, −18)	(−3, −3, −1), (−1, −1, 0), (−1, 2, −1), (2, −1, 1), (15, −3, 7)
547461	$\pm 1/2 \cdot \nu_\Delta$	±(0, 0, 1)	(0, 1, 0), (2, 1, 1), (−2, −3, −5) (2, 1, 1), (−2, −3, −5), (0, 1, 4)	(−3, 6, −2), (−1, −2, 2), (−1, 1, 0), (0, −1, 0), (1, −1, 0), (2, −1, 0)
544442	$\pm 1/2 \cdot \nu_\Delta$	±(1, −1, 2)	(0, 1, 0), (1, −4, 4), (−5, 6, −12) (1, 0, 0), (0, 1, 0), (3, −6, 8)	(−2, −2, −1), (−1, −1, 0), (−1, 1, 1), (1, −1, −1), (3, −1, −2), (10, −2, −5)
544443	$\pm 1/2 \cdot \nu_\Delta$	±(1, −1, 2)	(−1, −2, 0), (0, 1, 0), (−3, 4, −8) (3, −6, 8), (0, 1, 0), (1, 0, 0)	(−2, −2, −1), (−1, −1, 0), (−1, 1, 1), (1, −1, −1), (3, −1, −2), (6, −2, −3)
544651	$\pm 2/3 \cdot \nu_\Delta$	±(1, 0, 0)	(−4, 1, −3), (0, 1, 0), (1, −2, 3) (4, −2, 3), (0, 1, 0), (−1, 1, −3)	(−3, −3, 1), (0, −1, −1), (0, −1, 0), (0, 2, 1), (3, −3, −4)
544696	$\pm 2/3 \cdot \nu_\Delta$	±(1, −1, −3)	(1, 0, 0), (0, 1, 0), (−4, 2, 9) (5, −4, −15), (1, 0, 0), (−3, 1, 6)	(−3, −3, −1), (−3, 12, −4), (−1, −1, 0), (−1, 2, −1), (2, −1, 1)
544700	$\pm 2/3 \cdot \nu_\Delta$	±(1, 2, 3)	(−2, −3, −3), (0, 1, 0), (−1, −4, −6) (0, 1, 0), (1, 0, 0), (2, 5, 9)	(−3, −3, 2), (−1, −1, 1), (−1, 2, −1), (2, −1, 0), (3, −3, 2)
544749	$\pm 1/2 \cdot \nu_\Delta$	±(1, 1, 2)	(−6, −5, −8), (0, 1, 0), (1, 0, 0) (0, 1, 0), (−2, −1, 0), (3, 2, 4)	(−2, −2, 3), (−1, −1, 1), (−1, 1, 0), (−1, 3, −1), (1, −1, 0), (2, −2, −1)

(continued)

Table 2.6 (continued)

ID	vert(Δ^{FI})	$\pm v_\Delta$	vert($\theta_\pm$)	supp(Δ^{FI})
520925	$\pm 1/2 \cdot v_\Delta$	$\pm(2, 1, 1)$	$(-2, -1, 0), (-2, 0, -1), (-2, -3, -2)$ $(0, 1, 0), (0, 0, 1), (8, 2, 3)$	$(-1, -1, 3), (0, -1, 1), (0, 1, -1), (1, -2, -2), (1, -1, -1), (1, 0, 0)$
520935	$\pm 1/2 \cdot v_\Delta$	$\pm(1, 1, 2)$	$(1, 0, 0), (0, 1, 0), (-6, -5, -8)$ $(3, 4, 6), (2, 1, 2), (-3, -2, -2)$	$(-2, -2, 3), (-1, -1, 1), (-1, 1, 0), (-1, 3, -1), (0, 4, -3), (1, -1, 0)$
522056	$\pm 2/3 \cdot v_\Delta$	$\pm(2, 1, 3)$	$(0, 1, 0), (-1, -1, -3), (-5, -3, -6)$ $(-1, -1, 0), (1, 0, 0), (6, 4, 9)$	$(-3, 6, -1), (-1, -1, 1), (-1, 2, 0), (0, -3, 2), (2, -1, -1)$
522059	$\pm 2/3 \cdot v_\Delta$	$\pm(1, 2, 3)$	$(-2, -3, -3), (0, 1, 0), (-1, -4, -6)$ $(2, 5, 6), (1, 0, 0), (0, 1, 3)$	$(-3, 3, -2), (-1, -1, 1), (-1, 2, -1), (2, -1, 0), (3, -3, 2)$
522087	$\pm 2/3 \cdot v_\Delta$	$\pm(1, -1, -3)$	$(1, 0, 0), (0, 1, 0), (-4, 2, 9)$ $(1, 0, -3), (-3, 1, 6), (5, -4, -12)$	$(-3, -3, -1), (-1, -1, 0), (-1, 2, -1), (2, -1, 1), (9, 0, 4)$
522682	$\pm 1/2 \cdot v_\Delta$	$\pm(1, 1, 2)$	$(-3, -2, -4), (-2, -3, -4), (1, 0, 0), (0, 1, 0)$ $(2, 1, 4), (1, 2, 4), (1, 0, 0), (0, 1, 0)$	$(-2, -2, 1), (-2, -2, 3), (-1, -1, 1), (-1, 1, 0), (1, -1, 0), (1, 1, -1)$
522684	$\pm 1/2 \cdot v_\Delta$	$\pm(1, 1, 2)$	$(-2, -1, -4), (1, 0, 0), (-4, -3, -4)$ $(3, 2, 4), (-2, -1, 0), (0, 1, 0)$	$(-2, 2, 1), (-1, -1, 1), (-1, 1, 0), (-1, 3, -1), (1, -1, 0), (2, -2, -1)$
526886	$\pm 1/2 \cdot v_\Delta$	$\pm(1, -1, 2)$	$(-3, 4, -6), (0, 1, -2), (2, -5, 6)$ $(1, 0, 0), (0, 1, 0), (3, -6, 8)$	$(-2, -2, -1), (-1, -1, 0), (-1, 1, 1), (0, 4, 3), (1, -1, -1), (3, -1, -2)$
439403	$\pm 1/2 \cdot v_\Delta$	$\pm(0, 1, 1)$	$(-1, 1, -1), (1, 0, 0), (-1, -2, -2), (1, -3, -1)$ $(1, 2, 2), (-1, 0, 0), (-1, 1, -1), (1, 1, 3)$	$(-2, -1, 3), (-1, -1, 1), (-1, 0, 0), (1, 0, 0), (1, 1, -1), (2, -1, -1)$
275525	$\pm 1/2 \cdot v_\Delta$	$\pm(1, 0, 0)$	$(0, 1, 0), (-2, -1, 0), (1, 1, 2), (-3, -1, -2)$ $(4, 1, 2), (-2, -1, -2), (1, 1, 0), (1, -1, 0)$	$(-2, 0, 3), (0, -1, 0), (0, -1, 1), \pm(0, 1, -1), (0, 1, 0), (2, -2, -1)$
275528	$\pm 1/2 \cdot v_\Delta$	$\pm(1, 1, -1)$	$(-3, -2, 1), (-2, -1, 2), (0, -1, 0), (1, 0, 1)$ $(-1, 0, -1), (0, 1, 0), (2, 1, -2), (3, 2, -1)$	$(-1, 1, 0), (-1, 2, -1), (0, -1, -1), (0, 1, 1), (1, -2, 1), (1, -1, 0)$

To compute the canonical hull Δ^{can} of Δ, we obtain $\text{supp}(\Delta^{\text{FI}}) = \{s_i \mid 1 \leq i \leq 5\}$ with $s_1 := (-3, -3, -1), s_2 := (-1, -1, 0), s_3 := (-1, 2, -1), s_4 := (2, -1, 1)$, and $s_5 := (15, -3, 7)$, which leads to $\Delta^{\text{can}} = \Delta$.

Remark 34 Information about all 20 canonical Fano 3-topes with $\dim(\Delta^{\text{FI}}) = 1$ and $0 \in (\Delta^{\text{FI}})^\circ$ can be found in Tables 2.5 and 2.6.

2.6 Fine Interior of Dimension 3

There exist 49 canonical Fano 3-topes Δ such that $\dim(\Delta^{\text{FI}}) = 3$. Exactly 3 of these polytopes Δ define minimal surface $\mathcal{S}_\Delta$ with non-trivial fundamental group of order 2 and $K^2 = 2$. For these 3 polytopes one has $\Delta = \Delta^{\text{can}}$. The surfaces $\mathcal{S}_\Delta$ were investigated by Todorov [24] as well as Catanese and Debarre [10].

The remaining 46 canonical Fano 3-topes Δ define simply connected minimal surfaces $\mathcal{S}_\Delta$ with $K^2 = 1$. These surfaces were investigated by Kanev [19], Catanese [9], and Todorov [23]. Among these 46 canonical Fano 3-topes there exist exactly 26 polytopes Δ such that $\Delta = \Delta^{\text{can}}$.

Example 35 ([19]) Let $M \subseteq \mathbb{Q}^4$ be the 3-dimensional affine lattice defined by

$$M := \{(m_1, m_2, m_3, m_4) \in \mathbb{Z}^4 \mid m_1 + m_2 + m_3 + 2m_4 = 6,\ m_2 + 2m_3 \equiv 0 \pmod 3)\}$$

and $\Delta' \subseteq M_\mathbb{Q}$ be the convex hull of 4 lattice points

$$(6, 0, 0, 0),\ (0, 6, 0, 0),\ (0, 0, 6, 0),\ \text{and } (0, 0, 0, 3) \in M.$$

Then $(\Delta')^{\text{FI}}$ is the 3-dimensional rational simplex

$$\text{conv}\{(2, 1, 1, 1), (1, 2, 1, 1), (1, 1, 2, 1), (1, 1, 1, 3/2)\}$$

and $(\Delta')^{\text{FI}} \cap M = \{(2, 1, 1, 1)\}$.

The canonical Fano 3-tope Δ' is the Newton polytope of the μ_3-cyclic quotient $\overline{Z}_{\Delta'}$ of the projective surface of degree 6 defined by the polynomial $z_1^6 + z_2^6 + z_3^6 + z_4^3 = 0$ in the weighted projective space $\mathbb{P}(1, 1, 1, 2)$, where the cyclic group μ_3 acts via $(z_1 : z_2 : z_3 : z_4) \mapsto (z_1 : \varepsilon z_2 : \varepsilon^2 z_3 : z_4)$. The single interior lattice point in Δ' corresponds to the monomial $z_1^2 z_2 z_3 z_4$. The surface $\overline{Z}_{\Delta'}$ has 3 cyclic quotient singularities of type A_2. The minimal desingularization $\mathcal{S}_{\Delta'}$ of $\overline{Z}_{\Delta'}$ is a simply connected surface of general type with $K^2 = 1$.

One can identify Δ' with the canonical Fano 3-simplex Δ given as the convex hull of

$$v_1 := (1, 0, 0),\ v_2 := (-2, -4, -5),\ v_3 := (1, 2, 4),\ \text{and } v_4 := (1, 4, 2)$$

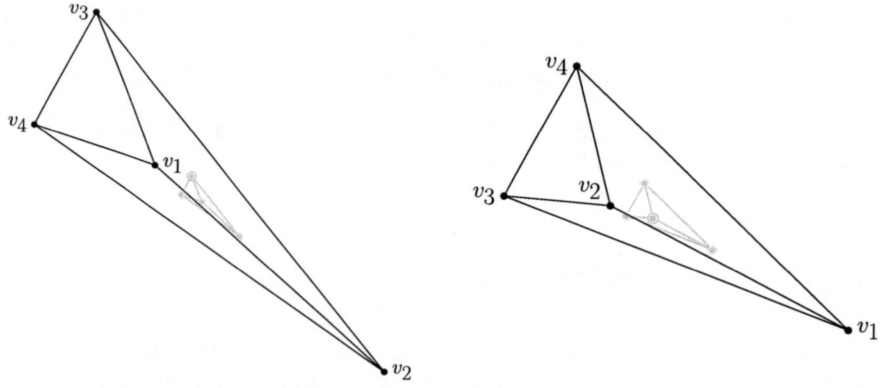

(a) ID 547444 (b) ID 547465

Fig. 2.6 Canonical Fano 3-topes with Fine Interior of Dimension 3. Shaded faces are occluded. The Fine interior and the origin are shown in grey with a double border around the origin. **a** The whole polytope is $\Delta = \operatorname{conv}\{v_1, v_2, v_3, v_4\}$ with $v_1 = (1, 0, 0)$, $v_2 = (-2, -4, -5)$, $v_3 = (1, 2, 4)$, and $v_4 = (1, 4, 2)$. Moreover, $\Delta^{\mathrm{FI}} = \operatorname{conv}\{(0, 0, 0), (-1/2, -1, -3/2), (0, -1/3, -2/3), (0, 1/3, -1/3)\}$ and $\Delta^{\mathrm{can}} = \Delta$. **b** The whole polytope is $\Delta = \operatorname{conv}\{v_1, v_2, v_3, v_4\}$ with $v_1 = (-3, -2, -2)$, $v_2 = (1, 0, 0)$, $v_3 = (1, 3, 1)$, and $v_4 = (1, 1, 3)$. Moreover, $\Delta^{\mathrm{FI}} = \operatorname{conv}\{(0, 0, 0), (-1, -1/2, -1/2), (0, 3/4, 1/4), (0, 1/4, 3/4)\}$ and $\Delta^{\mathrm{can}} = \Delta$

(ID 547444, Fig. 2.6a, Tables 2.7, 2.8, and 2.9). The primitive inward-pointing facet normals of the facets θ_{124}, θ_{234}, θ_{123}, and $\theta_{134} \preceq \Delta$ of this simplex Δ are

$$n_1 := (-2, -1, 2),\ n_2 := (5, -1, -1),\ n_3 := (-1, 2, -1),\ \text{and } n_4 := (-1, 0, 0),$$

respectively. They satisfy the relation

$$n_1 + n_2 + n_3 + 2n_4 = 0.$$

To compute the canonical hull Δ^{can} of Δ, we obtain $\operatorname{supp}(\Delta^{\mathrm{FI}}) = \{s_i \mid 1 \le i \le 6\}$ with $s_1 := (-2, -1, 2)$, $s_2 := (-1, 0, 0)$, $s_3 := (-1, 2, -1)$, $s_4 := (1, 1, -1)$, $s_5 := (3, 0, -1)$, and $s_6 := (5, -1, -1)$, which leads to $\Delta^{\mathrm{can}} = \Delta$.

Example 36 ([24]) Let $M \subseteq \mathbb{Q}^4$ be the 3-dimensional affine lattice defined by

$$M := \{(m_1, m_2, m_3, m_4) \in \mathbb{Z}^4 \mid m_1 + m_2 + 2m_3 + 2m_4 = 8,\ 3m_2 + m_3 + 3m_4 \equiv 0 \pmod 4\}$$

and $\Delta' \subseteq M_{\mathbb{Q}}$ be the convex hull of 4 lattice points

$$(8, 0, 0, 0),\ (0, 8, 0, 0),\ (0, 0, 4, 0),\ \text{and } (0, 0, 0, 4) \in M.$$

Table 2.7 49 Canonical Fano 3-topes with Fine Interior of Dimension 3 Table contains: vertices vert(Δ) of Δ

ID	vert(Δ)
547444	(1, 0, 0), (−2, −4, −5), (1, 2, 4), (1, 4, 2)
547465	(−3, −2, −2), (1, 0, 0), (1, 3, 1), (1, 1, 3)
547524	(0, 2, 1), (−2, −3, −5), (2, 1, 1), (0, 0, 1)
547525	(0, 0, 1), (0, 1, 0), (2, 1, 1), (−2, −5, −7)
545317	(−3, 4, −6), (0, 1, 0), (1, 0, 0), (1, −2, 4), (3, −5, 6)
545932	(0, −1, −1), (1, −1, −3), (−2, 1, 5), (1, 0, 0), (1, 2, −2)
546013	(3, −5, 6), (1, −2, 4), (1, 0, 0), (−1, 1, −2), (−1, 3, −2)
546062	(0, 1, 3), (−2, 1, −1), (0, 1, 0), (1, 0, 0), (−1, −2, −2)
546070	(0, −2, −3), (0, 2, 1), (−2, −3, −5), (2, 1, 1), (0, 0, 1)
546205	(1, 2, −2), (−1, 0, 2), (1, 0, 0), (−2, 1, 5), (1, −1, −3)
546219	(1, 1, 1), (−3, −2, −2), (1, 0, 0), (1, 3, 1), (−1, −1, 1)
546663	(2, −3, −1), (1, 0, 0), (0, 1, 0), (0, 0, 1), (−2, −3, −3)
546862	(1, 0, 0), (0, 1, 0), (−2, 1, 5), (1, −1, −3), (1, 2, −2)
546863	(−1, −1, 1), (1, 3, 1), (0, 0, 1), (1, 0, 0), (−3, −2, −2)
547240	(−1, 1, −2), (0, 1, 0), (1, 0, 0), (1, −2, 4), (3, −5, 6)
547246	(0, −2, −3), (−2, −3, −5), (2, 1, 1), (0, 1, 0), (0, 0, 1)
532384	(1, −1, −3), (−2, 1, 5), (1, 0, 0), (1, −1, −2), (0, −1, −1), (1, 2, −2)
532606	(0, −1, 2), (−1, −1, 0), (0, 1, 0), (1, 0, 0), (2, 2, −3), (−2, 0, −3)
533513	(−1, 1, 2), (1, 0, 0), (0, 1, 0), (1, 1, 2), (−1, −2, −4), (−2, −3, −4)
534667	(1, 0, 3), (−1, −1, −1), (0, 1, 0), (1, 0, 0), (−1, −1, 0), (5, 2, 3)
534669	(1, 3, 0), (5, 3, 2), (−1, −1, −1), (0, 0, 1), (1, 0, 0), (−1, −1, 0)
534866	(−1, −1, −3), (1, 0, 0), (0, 1, 0), (1, 1, 1), (−1, −1, 0), (−3, −5, −3)
535952	(3, −5, 6), (1, −2, 4), (1, 0, 0), (0, 1, 0), (−1, 1, −2), (−1, 2, −2)
536013	(0, 1, 1), (0, 0, 1), (0, 1, 0), (2, 1, 1), (−2, −3, −5), (0, −2, −3)
536498	(1, 2, −2), (1, −1, −2), (1, 0, 0), (0, 1, 0), (−2, 1, 5), (1, −1, −3)
537834	(0, 0, 1), (1, 0, 0), (0, 1, 0), (−2, 1, 5), (1, −1, −3), (1, 2, −2)
538356	(−2, −3, −3), (−1, −3, −1), (0, 0, 1), (0, 1, 0), (1, 0, 0), (−1, −1, −3)
539063	(−1, 1, −1), (1, 1, 3), (−3, −2, −2), (1, 0, 0), (0, 1, 0), (1, 1, 2)
539304	(1, 0, 1), (−3, −1, −2), (1, 1, 2), (−2, −1, 0), (1, 0, 0), (1, 2, 0)
539313	(1, −1, −2), (1, 1, −1), (−1, 2, 2), (1, −1, −3), (−2, 1, 5), (1, 0, 0)
540602	(0, 0, 1), (1, 0, 0), (−2, 1, 5), (1, −1, −3), (−1, 2, 2), (1, 1, −1)
540663	(1, 0, 0), (0, 1, 0), (1, 1, 2), (−3, −1, −2), (1, 1, 1), (−3, −2, 0)
474457	(−1, 2, −3), (1, 0, 2), (0, 0, 1), (0, 1, 0), (1, 0, 0), (−1, −1, 0), (−3, −2, −3)
481575	(3, 2, 4), (−1, −1, −2), (−3, −1, −2), (−2, −1, 0), (0, 1, 0), (1, 0, 0), (0, 0, −1)
483109	(3, 0, 2), (1, −2, −2), (0, 0, −1), (−1, −1, 0), (1, 1, 1), (0, 1, 0), (−1, 0, 0)
490478	(1, −1, −2), (1, 1, −1), (−1, 2, 2), (1, −1, −3), (−2, 1, 5), (1, 0, 0), (−1, 0, 2)
490481	(−3, −2, 0), (−5, −3, −2), (1, 0, 0), (0, 1, 0), (1, 1, 2), (−1, −1, −1), (2, 1, 1)
490485	(−1, −1, 0), (1, 2, 0), (1, 0, 0), (−2, −1, 0), (1, 1, 2), (−3, −1, −2), (1, 0, 1)
490511	(1, 0, 0), (0, 1, 0), (−2, −1, 0), (1, 1, 2), (2, 1, 1), (1, 0, 1), (−5, −2, −4)
495687	(0, 0, −1), (1, 1, −1), (−1, 2, 2), (1, −1, −3), (−2, 1, 5), (1, 0, 0), (0, 0, 1)
499287	(1, 1, 1), (−1, −1, −3), (1, 0, 0), (0, 1, 0), (0, 0, 1), (−1, −3, −1), (−2, −3, −3)
499291	(−1, −1, −1), (−1, −1, −3), (1, 0, 0), (0, 1, 0), (0, 0, 1), (−1, −3, −1), (−2, −3, −3)
499470	(1, 0, 0), (0, 1, 0), (−2, −1, 0), (1, 1, 2), (0, 0, 1), (−5, −2, −4), (2, 1, 1)
501298	(3, −6, 8), (−1, 1, −2), (1, −2, 3), (0, 1, 0), (1, 0, 0), (0, 1, −1), (3, −5, 6)
501330	(1, 0, 0), (0, 1, 0), (−2, −1, 0), (1, 1, 2), (1, 1, 1), (0, 0, 1), (−5, −2, −4)
354912	(3, 1, 2), (1, 0, 0), (0, 1, 0), (−2, −1, 0), (1, 1, 2), (2, 1, 1), (1, 0, 1), (−5, −2, −4)
372528	(2, 1, 1), (−1, −1, −1), (1, 1, 2), (0, 1, 0), (1, 0, 0), (−5, −3, −2), (−3, −2, 0), (1, 1, 0)
372973	(−5, −2, −4), (1, 0, 1), (2, 1, 1), (1, 1, 2), (−2, −1, 0), (0, 1, 0), (1, 0, 0), (2, 1, 2)
388701	(1, 1, 1), (−2, −3, −3), (−1, −3, −1), (0, 0, 1), (0, 1, 0), (1, 0, 0), (−1, −1, −3), (−1, −1, −1)

Table 2.8 49 Canonical Fano 3-topes with Fine Interior of Dimension 3. Table contains: vertices vert(Δ^{FI}) of the Fine interior Δ^{FI}

ID	vert(Δ^{FI})
547444	0, (−1/2, −1, −3/2), (0, −1/3, −2/3), (0, 1/3, −1/3)
547465	0, (−1, −1/2, −1/2), (0, 3/4, 1/4), (0, 1/4, 3/4)
547524	0, (0, 1/2, 0), (1/3, 1/3, 0), (−1/3, −1/3, −1)
547525	0, (0, 0, −1/2), (1/3, 0, −1/3), (−1/3, −1, −5/3)
545317	0, (1, −3/2, 2), (2/3, −2/3, 1), (1/2, −1/2, 1), (2/3, −1, 5/3)
545932	0, (−1/2, 1/2, 3/2), (0, 1/3, 2/3), (0, 2/3, 1/3)
546013	0, (1, −3/2, 2), (0, 1/2, 0), (1/2, −1/4, 1/2), (1/2, −3/4, 3/2)
546062	0, (−1/2, −1/2, −1/2), (−2/3, 0, −1/3), (−1/3, 0, 1/3)
546070	0, (0, 1/2, 0), (1/2, 1/4, 0), (0, −1/2, −1), (−1/2, −3/4, −3/2)
546205	0, (−1/2, 1/2, 3/2), (0, 1/3, 2/3), (0, 2/3, 1/3)
546219	0, (−1, −1/2, −1/2), (−1/3, 1/3, 0), (−2/3, −1/3, 0)
546663	0, (0, −1/2, 0), (1/3, −1, −1/3), (−1/3, −1, −2/3)
546862	0, (−1/2, 1/2, 3/2), (0, 1/3, 2/3), (0, 2/3, 1/3)
546863	0, (−1, −1/2, −1/2), (−1/3, 1/3, 0), (−2/3, −1/3, 0)
547240	0, (1, −3/2, 2), (2/3, −2/3, 1), (1/2, −1/2, 1), (2/3, −1, 5/3)
547246	0, (0, 0, −1/2), (1/3, 0, −1/3), (0, −1/2, −1), (−1/3, −2/3, −4/3)
532384	0, (−1/2, 1/2, 3/2), (0, 1/3, 2/3), (0, 2/3, 1/3)
532606	0, (0, 1/2, −1/2), (1/3, 2/3, −1), (−1/3, 1/3, −1)
533513	0, (−1/2, −1/2, −1), (−1/2, 0, 0), (−1/3, 0, −1/3), (−2/3, −2/3, −1)
534667	0, (1/2, 1/2, 1/2), (4/3, 2/3, 1), (2/3, 1/3, 1)
534669	0, (1/2, 1/2, 1/2), (4/3, 1, 2/3), (2/3, 1, 1/3)
534866	0, (0, −1/2, −1/2), (−1/3, −2/3, −1), (−2/3, −4/3, −1)
535952	0, (1, −3/2, 2), (2/3, −2/3, 1), (1/2, −1/2, 1), (2/3, −1, 5/3)
536013	0, (0, 0, −1/2), (1/3, 0, −1/3), (0, −1/2, −1), (−1/3, −2/3, −4/3)
536498	0, (−1/2, 1/2, 3/2), (0, 1/3, 2/3), (0, 2/3, 1/3)
537834	0, (−1/2, 1/2, 3/2), (0, 1/3, 2/3), (0, 2/3, 1/3)
538356	0, (0, −1/2, −1/2), (−1/3, −2/3, −1), (−1/3, −1, −2/3), (−1/2, −1, −1)
539063	0, (−1, −1/2, −1/2), (−2/3, 0, −1/3), (−1/3, 0, 1/3)
539304	0, (0, 1/2, 0), (−1/2, 0, 0), (0, 1/3, 1/3), (−2/3, 0, −1/3)
539313	0, (−1/2, 1/2, 3/2), (0, 1/3, 2/3), (0, 1/2, 1/2), (−1/3, 2/3, 1)
540602	0, (−1/2, 1/2, 3/2), (0, 1/3, 2/3), (0, 1/2, 1/2), (−1/3, 2/3, 1)
540663	0, (−1/2, 0, 0), (−1, −1/2, 0), (−1/3, 0, 1/3), (−1, −1/3, −1/3)
474457	0, (0, 0, −1/2), (−1/3, 1/3, −1), (−2/3, −1/3, −1)
481575	0, (−1/2, 0, 0), (1/2, 1/2, 1), (0, 1/3, 1/3), (−1/3, 0, 1/3)
483109	0, (0, −1/2, 0), (2/3, −1/3, 1/3), (1/3, −2/3, −1/3)
490478	0, (−1/2, 1/2, 3/2), (0, 1/3, 2/3), (0, 1/2, 1/2), (−1/3, 2/3, 1)
490481	0, (−1/2, 0, 0), (−1, −1/2, 0), (−1/3, 0, 1/3), (−4/3, −2/3, −1/3)
490485	0, (0, 1/2, 0), (−1/2, 0, 0), (0, 1/3, 1/3), (−2/3, 0, −1/3)
490511	0, (−3/2, −1/2, −1), (−1/2, 0, 0), (−2/3, 0, −1/3), (−1, −1/3, −1/3)
495687	0, (−1/2, 1/2, 3/2), (0, 1/3, 2/3), (0, 1/2, 1/2), (−1/3, 2/3, 1)
499287	0, (0, −1/2, −1/2), (−1/3, −2/3, −1), (−1/3, −1, −2/3), (−1/2, −1, −1)
499291	0, (0, −1/2, −1/2), (−1/3, −2/3, −1), (−1/3, −1, −2/3), (−1/2, −1, −1)
499470	0, (−3/2, −1/2, −1), (−1/2, 0, 0), (−2/3, 0, −1/3), (−1, −1/3, −1/3)
501298	0, (1/2, −1/2, 1), (2/3, −2/3, 1), (1, −3/2, 2), (1, −5/3, 7/3)
501330	0, (−3/2, −1/2, −1), (−1/2, 0, 0), (−2/3, 0, −1/3), (−1, −1/3, −1/3)
354912	0, (−3/2, −1/2, −1), (−1/2, 0, 0), (−2/3, 0, −1/3), (−1, −1/3, −1/3)
372528	0, (−1/2, 0, 0), (−1, −1/2, 0), (−1/3, 0, 1/3), (−4/3, −2/3, −1/3)
372973	0, (−3/2, −1/2, −1), (−1/2, 0, 0), (−2/3, 0, −1/3), (−1, −1/3, −1/3)
388701	0, (0, −1/2, −1/2), (−1/3, −2/3, −1), (−1/3, −1, −2/3), (−1/2, −1, −1)

Table 2.9 **49 Canonical Fano 3-topes with Fine Interior of Dimension 3**. Table contains: support $\mathrm{supp}(\Delta^{\mathrm{FI}})$ of the Fine interior Δ^{FI}, vertices $\mathrm{vert}(\Delta^{\mathrm{can}})$ of the canonical hull Δ^{can}, and order of fundamental group $|\pi_1(S_\Delta)|$ of the minimal model S_Δ

ID	$\mathrm{supp}(\Delta^{\mathrm{FI}})$	$\mathrm{vert}(\Delta^{\mathrm{can}})$	$\lvert\pi_1(S_\Delta)\rvert$
547444	(−2, −1, 2), (−1, 0, 0), (−1, 2, −1), (1, 1, −1), (3, 0, −1), (5, −1, −1)	vert(Δ)	1
547465	(−1, −1, 3), (−1, 0, 0), (−1, 0, 1), (−1, 0, 2), (−1, 1, 0), (−1, 1, 1), (−1, 2, 0), (−1, 3, −1), (2, −1, −1)	vert(Δ)	2
547524	(−1, −2, 2), (−1, 1, 0), (−1, 2, −1), (0, 0, −1), (0, 1, −1), (0, 2, −1), (1, 0, −1), (1, 1, −1), (2, 0, −1), (3, 0, −1)	vert(Δ), (0, −1, −1)	1
547525	(−1, −2, 2), (−1, 2, −1), (0, −1, 0), (0, 0, −1), (0, 1, −1), (1, −1, −1), (1, −1, 0), (1, 0, −1), (1, 1, −1), (2, −1, −1), (2, −1, 0), (2, 0, −1), (3, −1, −1), (3, −1, 0), (3, 0, −1), (4, −1, −1), (4, 0, −1), (5, −1, −1), (6, −1, −1)	vert(Δ), (1, 1, 1), (−1, −2, −3)	1
545317	(−2, −2, −1), (−1, −1, 0), (−1, 2, 2), (1, −1, −1), (1, 2, 1), (3, 2, 0)	vert(Δ)	1
545932	(−2, −1, −1), (−1, 0, 0), (−1, 2, −1), (0, 1, 0), (1, 1, 0), (2, 0, 1), (3, 0, 1), (5, −1, 2)	vert(Δ), (1, −1, −2), (1, 0, −3)	1
546013	(−2, −2, −1), (−1, 0, 1), (−1, 2, 2), (0, 1, 1), (1, 0, 0), (1, 2, 1), (2, 1, 0), (3, 0, −1), (3, 2, 0)	vert(Δ)	2
546062	(−1, −1, 0), (−1, −1, 1), (−1, −1, 2), (−1, 0, 0), (−1, 0, 1), (−1, 1, 0), (−1, 2, −1), (0, −1, 0), (2, 1, −1)	vert(Δ)	1

(continued)

Table 2.9 (continued)

ID	$\mathrm{supp}(\Delta^{\mathrm{FI}})$	$\mathrm{vert}(\Delta^{\mathrm{can}})$	$\lvert\pi_1(\mathcal{S}_\Delta)\rvert$
546070	(−1, −2, 2), (−1, 2, −1), (0, 0, −1), (0, 1, −1), (0, 2, −1), (1, 0, −1), (1, 1, −1), (2, 0, −1), (3, 0, −1)	vert(Δ)	2
546205	(−2, −1, −1), (−1, 0, 0), (−1, 2, −1), (−1, 3, −1), (0, 1, 0), (1, 1, 0), (1, 2, 0), (2, 0, 1), (3, 0, 1), (3, 1, 1), (5, −1, 2), (5, 0, 2), (7, −1, 3)	vert(Δ)	1
546219	(−1, −1, 3), (−1, 0, 0), (−1, 0, 1), (−1, 0, 2), (−1, 1, −1), (−1, 1, 0), (−1, 1, 1), (−1, 2, 0), (0, 0, −1), (2, −1, −1)	vert(Δ)	1
546663	(−1, −1, −1), (−1, −1, 0), (−1, −1, 1), (−1, −1, 2), (−1, 0, −1), (0, −1, −1), (0, −1, 0), (0, −1, 1), (0, 0, −1), (1, −1, −1), (1, −1, 0), (1, 0, −1), (1, 2, −2), (2, −1, −1), (2, −1, 0), (2, 0, −1), (3, −1, −1)	vert(Δ), (−1, −1, −1)	1
546862	(−2, −1, −1), (−1, 0, 0), (−1, 2, −1), (−1, 3, −1), (0, 1, 0), (0, 4, −1), (1, 1, 0), (1, 2, 0), (2, 0, 1), (2, 3, 0), (3, 0, 1), (3, 1, 1), (4, 2, 1), (5, −1, 2), (5, 0, 2), (6, 1, 2), (7, −1, 3), (8, 0, 3), (10, −1, 4)	vert(Δ), (0, 0, 1)	1
546863	(−1, −1, 3), (−1, 0, 0), (−1, 0, 1), (−1, 0, 2), (−1, 1, −1), (−1, 1, 0), (−1, 1, 1), (−1, 2, 0), (0, 0, −1), (2, −1, −1)	vert(Δ)\{(0, 0, 1)}, (1, 1, 1)	1
547240	(−2, −2, −1), (−1, −1, 0), (−1, 0, 1), (−1, 2, 2), (0, −1, 0), (0, 1, 1), (1, −1, −1), (1, 0, 0), (1, 2, 1), (2, −1, −1), (2, 1, 0), (3, 0, −1), (3, 2, 0)	vert(Δ), (0, 1, −1), (0, 0, 1)	1
547246	(−1, −2, 2), (−1, 2, −1), (0, −1, 0), (0, 0, −1), (0, 1, −1), (0, 2, −1), (1, −1, −1), (1, −1, 0), (1, 0, −1), (1, 1, −1), (2, −1, −1), (2, −1, 0), (2, 0, −1), (3, −1, −1), (3, 0, −1), (4, −1, −1)	vert(Δ), (1, 1, 1), (−1, −1, −2)	1

(continued)

Table 2.9 (continued)

ID	supp(Δ^{FI})	vert(Δ^{can})	$\|\pi_1(\mathcal{S}_\Delta)\|$
532384	(−2, −1, −1), (−1, 0, 0), (−1, 2, −1), (0, 1, 0), (1, 1, 0), (2, 0, 1), (3, 0, 1), (5, −1, 2)	vert(Δ), (1, 0, −3)	1
532606	(−1, −1, −1), (−1, 1, 0), (−1, 2, 1), (0, −1, −1), (0, 1, 0), (1, −2, 0), (1, −1, −1), (2, −1, −1)	vert(Δ), (0, −1, 1)	1
533513	(−1, −1, 1), (−1, 0, 0), (−1, 1, 0), (−1, 2, −1), (0, −1, 0), (0, 1, −1), (0, 3, −2), (2, −2, 1)	vert(Δ), (1, 1, 1), (0, −1, −1)	1
534667	(−1, −1, 2), (−1, −1, 3), (−1, 0, 2), (−1, 1, 1), (−1, 2, 0), (0, −1, 1), (0, −1, 2), (0, 0, 1), (0, 1, 0), (1, −2, −1), (1, −1, 0), (1, −1, 1), (1, 0, 0), (2, −1, −1), (2, −1, 0)	vert(Δ)	1
534669	(−1, 0, 2), (−1, 1, 1), (−1, 2, −1), (−1, 2, 0), (0, 0, 1), (0, 1, −1), (0, 1, 0), (1, −1, −2), (1, 0, −1), (1, 0, 0), (2, −1, −1), (2, −1, 0)	vert(Δ)	1
534866	(−2, 1, 1), (−1, −1, 1), (−1, 0, 0), (−1, 1, −1), (−1, 2, −2), (0, −1, 0), (0, 0, −1), (0, 1, −2), (1, −1, −1), (1, −1, 0), (1, 0, −2), (1, 0, −1), (2, −1, −2), (2, −1, −1), (2, −1, 0)	vert(Δ)	1
535952	(−2, −2, −1), (−1, −1, 0), (−1, 0, 1), (−1, 2, 2), (0, 1, 1), (1, −1, −1), (1, 0, 0), (1, 2, 1), (2, 1, 0), (3, 0, −1), (3, 2, 0)	vert(Δ)	1
536013	(−1, −2, 2), (−1, 2, −1), (0, −1, 0), (0, 0, −1), (0, 1, −1), (0, 2, −1), (1, −1, 0), (1, 0, −1), (1, 1, −1), (2, −1, 0), (2, 0, −1), (3, 0, −1)	vert(Δ)	1
536498	(−2, −1, −1), (−1, 0, 0), (−1, 2, −1), (0, 1, 0), (1, 1, 0), (1, 2, 0), (2, 0, 1), (2, 3, 0), (3, 0, 1), (3, 1, 1), (4, 2, 1), (5, −1, 2), (5, 0, 2), (6, 1, 2), (7, −1, 3), (8, 0, 3), (10, −1, 4)	vert(Δ)	1
537834	(−2, −1, −1), (−1, 0, 0), (−1, 2, −1), (−1, 3, −1), (0, 1, 0), (0, 4, −1), (1, 1, 0), (1, 2, 0), (2, 0, 1), (2, 3, 0), (3, 0, 1), (3, 1, 1), (4, 2, 1), (5, −1, 2), (5, 0, 2), (6, 1, 2), (7, −1, 3), (8, 0, 3), (10, −1, 4)	vert(Δ)	1

(continued)

Table 2.9 (continued)

ID	$\mathrm{supp}(\Delta^{\mathrm{FI}})$	$\mathrm{vert}(\Delta^{\mathrm{can}})$	$\lvert\pi_1(\mathcal{S}_\Delta)\rvert$
538356	(−2, 1, 1), (−1, −1, −1), (−1, −1, 0), (−1, −1, 1), (−1, 0, −1), (−1, 0, 0), (−1, 1, −1), (0, −1, −1), (0, −1, 0), (0, 0, −1), (1, −1, −1), (1, −1, 0), (1, 0, −1), (2, −1, −1), (2, −1, 0), (2, 0, −1), (3, −1, −1)	vert(Δ), (−1, −1, −1)	1
539063	(−1, −1, 1), (−1, 0, 0), (−1, 0, 1), (−1, 0, 2), (−1, 1, 0), (−1, 1, 1), (−1, 2, 0), (−1, 3, −1), (0, −1, 0), (2, −1, −1)	vert(Δ) \ {(0, 1, 0), (1, 1, 2)}, (1, 1, 1)	1
539304	(−1, 0, 0), (−1, 0, 1), (−1, 0, 2), (−1, 1, 0), (−1, 1, 1), (−1, 2, 0), (−1, 2, 1), (−1, 3, 0), (0, 1, −1), (0, 1, 0), (2, −2, −1)	vert(Δ), (−2, −1, −1)	1
539313	(−2, −1, −1), (−1, 0, 0), (−1, 2, −1), (0, 1, 0), (1, −1, 1), (1, 1, 0), (1, 2, 0), (2, 0, 1), (2, 3, 0), (3, 0, 1), (3, 1, 1), (4, 2, 1), (5, 0, 2), (6, 1, 2)	vert(Δ), (−1, 1, 2)	1
540602	(−2, −1, −1), (−1, 0, 0), (−1, 2, −1), (−1, 3, −1), (0, 1, 0), (0, 4, −1), (1, −1, 1), (1, 1, 0), (1, 2, 0), (2, 0, 1), (2, 3, 0), (3, 0, 1), (3, 1, 1), (4, 2, 1), (5, 0, 2), (6, 1, 2)	vert(Δ), (−1, 1, 2)	1
540663	(−1, −1, 1), (−1, −1, 2), (−1, 0, 0), (−1, 0, 1), (−1, 0, 2), (−1, 1, 0), (−1, 1, 1), (−1, 2, −1), (−1, 2, 0), (−1, 2, 1), (0, −1, 0), (0, −1, 1), (2, −2, −1)	vert(Δ), (−1, 0, −1)	1
474457	(−2, 1, 2), (−1, −1, 0), (−1, 0, 0), (−1, 1, 0), (−1, 2, 0), (1, −1, −1), (1, 0, −1), (2, −1, −1)	vert(Δ)	1
481575	(−1, −1, 1), (−1, 0, 1), (−1, 1, 0), (−1, 2, 0), (−1, 3, −1), (0, −1, 1), (0, 1, 0), (2, −2, −1)	vert(Δ), (−1, −1, −1)	1
483109	(−1, −1, 1), (0, −1, 0), (0, −1, 1), (0, 2, −1), (1, −1, −1), (1, −1, 0), (1, −1, 1), (1, 0, −2), (1, 0, −1), (1, 0, 0), (1, 0, 1)	vert(Δ)	1

(continued)

Table 2.9 (continued)

ID	supp(Δ^{FI})	vert(Δ^{can})	$\lvert\pi_1(\mathcal{S}_\Delta)\rvert$
490478	(−2, −1, −1), (−1, 0, 0), (−1, 2, −1), (0, 1, 0), (1, −1, 1), (1, 1, 0), (1, 2, 0), (2, 0, 1), (3, 0, 1), (3, 1, 1), (5, 0, 2)	vert(Δ)	1
490481	(−1, −1, 2), (−1, −1, 3), (−1, 0, 1), (−1, 0, 2), (−1, 1, 0), (−1, 1, 1), (−1, 2, −1), (−1, 2, 0), (0, −1, 0), (0, −1, 1), (0, −1, 2), (2, −2, −1)	vert(Δ)	1
490485	(−1, 0, 0), (−1, 0, 1), (−1, 0, 2), (−1, 1, 0), (−1, 1, 1), (−1, 2, 0), (−1, 2, 1), (0, 1, −1), (0, 1, 0), (2, −2, −1)	vert(Δ)	1
490511	(−1, −1, 2), (−1, 0, 1), (−1, 1, 0), (−1, 1, 1), (−1, 2, 0), (−1, 3, 0), (0, −1, 0), (0, 1, −1), (2, −2, −1)	vert(Δ), (2, 1, 2)	1
495687	(−2, −1, −1), (−1, 0, 0), (−1, 2, −1), (−1, 3, −1), (0, 1, 0), (0, 4, −1), (1, −1, 1), (1, 1, 0), (1, 2, 0), (2, 0, 1), (2, 3, 0), (3, 0, 1), (3, 1, 1), (4, 2, 1)	vert(Δ)	1
499287	(−2, 1, 1), (−1, −1, 1), (−1, 0, 0), (−1, 1, −1), (0, −1, 0), (0, 0, −1), (1, −1, −1), (1, −1, 0), (1, 0, −1), (2, −1, −1), (2, −1, 0), (2, 0, −1), (3, −1, −1)	vert(Δ), (−1, −1, −1)	1
499291	(−2, 1, 1), (−1, −1, −1), (−1, −1, 0), (−1, −1, 1), (−1, 0, −1), (−1, 0, 0), (−1, 1, −1), (0, −1, −1), (0, −1, 0), (0, 0, −1), (1, −1, −1), (1, −1, 0), (1, 0, −1), (2, −1, −1), (2, −1, 0), (2, 0, −1), (3, −1, −1)	vert(Δ)	1
499470	(−1, −1, 2), (−1, 0, 1), (−1, 1, 0), (−1, 1, 1), (−1, 2, −1), (−1, 2, 0), (−1, 3, −1), (−1, 3, 0), (0, −1, 0), (0, 1, −1), (2, −2, −1)	vert(Δ)	1

(continued)

Table 2.9 (continued)

ID	supp(Δ^{FI})	vert(Δ^{can})	$\lvert\pi_1(\mathcal{S}_\Delta)\rvert$
501298	(−2, −2, −1), (−1, −1, 0), (−1, 0, 1), (−1, 2, 2), (0, −1, 0), (0, 1, 1), (1, −1, −1), (1, 0, 0), (1, 2, 1), (2, −1, −1), (2, 1, 0), (3, −1, −2), (3, 0, −1), (4, −1, −2), (5, 0, −2), (6, −1, −3)	vert(Δ)	1
501330	(−1, −1, 1), (−1, −1, 2), (−1, 0, 0), (−1, 0, 1), (−1, 1, 0), (−1, 1, 1), (−1, 2, −1), (−1, 2, 0), (−1, 3, −1), (−1, 3, 0), (0, −1, 0), (0, 1, −1), (2, −2, −1)	vert(Δ)	1
354912	(−1, −1, 2), (−1, 0, 1), (−1, 1, 1), (−1, 2, 0), (−1, 3, 0), (0, −1, 0), (0, 1, −1), (2, −2, −1)	vert(Δ)	1
372528	(−1, 0, 1), (−1, 0, 2), (−1, 1, 0), (−1, 1, 1), (−1, 2, −1), (−1, 2, 0), (0, −1, 0), (0, −1, 1), (0, −1, 2), (2, −2, −1)	vert(Δ)	1
372973	(−1, −1, 2), (−1, 0, 1), (−1, 1, 0), (−1, 1, 1), (−1, 2, 0), (−1, 3, 0), (0, −1, 0), (0, 1, −1), (2, −2, −1)	vert(Δ)	1
388701	(−2, 1, 1), (−1, −1, 1), (−1, 0, 0), (−1, 1, −1), (0, −1, 0), (0, 0, −1), (1, −1, −1), (1, −1, 0), (1, 0, −1), (2, −1, −1), (2, −1, 0), (2, 0, −1), (3, −1, −1)	vert(Δ)	1

Then $(\Delta')^{\mathrm{FI}}$ is the 3-dimensional rational simplex

$$\mathrm{conv}\{(3,1,1,1),(1,3,1,1),(1,1,2,1),(1,1,1,2)\}$$

and $(\Delta')^{\mathrm{FI}} \cap M = \{(1,1,2,1)\}$.

The canonical Fano 3-tope Δ' is the Newton polytope of the μ_4-cyclic quotient $\overline{Z}_{\Delta'}$ of the projective surface of degree 8 defined by the polynomial $z_1^8 + z_2^8 + z_3^4 + z_4^4 = 0$ in the weighted projective space $\mathbb{P}(1,1,2,2)$, where the cyclic group μ_4 acts via $(z_1 : z_2 : z_3 : z_4) \mapsto (z_1 : i^3 z_2 : i z_3 : i^3 z_4)$. The single interior lattice point in this lattice simplex Δ' corresponds to the monomial $z_1 z_2 z_3^2 z_4$. The projective surface $\overline{Z}_{\Delta'}$ has two Gorenstein cyclic quotient singularities of type A_3. The minimal desingularization $S_{\Delta'}$ of $\overline{Z}_{\Delta'}$ is a surface of general type with $K^2 = 2$ and fundamental group $\pi_1(S_\Delta)$ of order 2.

One can identify Δ' with the canonical Fano 3-simplex Δ given as the convex hull of

$$v_1 := (-3,-2,-2),\ v_2 := (1,0,0),\ v_3 := (1,3,1),\ \text{and}\ v_4 := (1,1,3)$$

(ID 547465, Fig. 2.6b, Tables 2.7, 2.8, and 2.9). The primitive inward-pointing facet normals of the facets $\theta_{123}, \theta_{124}, \theta_{234}, \theta_{134} \preceq \Delta$ of this simplex Δ are

$$n_1 := (-1,-1,3),\ n_2 := (-1,3,-1),\ n_3 := (-1,0,0),\ \text{and}\ n_4 := (2,-1,-1),$$

respectively. They satisfy the relation

$$n_1 + n_2 + 2n_3 + 2n_4 = 0.$$

To compute the canonical hull Δ^{can} of Δ, we obtain $\mathrm{supp}(\Delta^{\mathrm{FI}}) = \{s_i \mid 1 \le i \le 9\}$ with $s_1 := (-1,-1,3)$, $s_2 := (-1,0,0)$, $s_3 := (-1,0,1)$, $s_4 := (-1,0,2)$, $s_5 := (-1,1,0)$, $s_6 := (-1,1,1)$, $s_7 := (-1,2,0)$, $s_8 := (-1,3,-1)$, and $s_9 := (2,-1,-1)$, which leads to $\Delta^{\mathrm{can}} = \Delta$.

Remark 37 Information about all 49 canonical Fano 3-topes with $\dim(\Delta^{\mathrm{FI}}) = 3$ can be found in Tables 2.7, 2.8, and 2.9.

2.7 Hollow 3-Topes with Non-empty Fine Interior

A lattice polytope $\Delta \subseteq M_{\mathbb{Q}}$ is called *hollow* if it has no interior lattice points in its relative interior, i.e., $\Delta^{\circ} \cap M = \emptyset$. By [25, Theorem 1.3], any 3-dimensional hollow lattice polytope can be projected to the unimodular 1-simplex, to the double unimodular 2-simplex, or is an exceptional hollow 3-tope, whereas up to unimodular transformation there exist only a finite number of these. This theorem implies that a hollow 3-tope with non-empty Fine interior has to be exceptional because the unimodular 1-simplex and the double unimodular 2-simplex have empty Fine interior. Treutlein has found 9 maximal exceptional hollow polytopes, which was not an complete list. Averkov et al. [1, 2] have found the complete list consisting of 12 maximal exceptional hollow 3-topes Δ_i $(1 \leq i \leq 12)$ (Table 2.10, Fig. 2.7). Computations show that exactly 9 of 12 maximal exceptional hollow 3-topes Δ_i have non-empty Fine interior Δ_i^{FI} (Table 2.10). Moreover, no one of these 9 polytopes contains a proper lattice 3-subpolytope with non-empty Fine interior. Thus, there exist exactly 9 hollow 3-topes Δ_i with non-empty Fine interior Δ_i^{FI}.

It is remarkable that all minimal surfaces $\mathcal{S}_{\Delta_i}$ corresponding to these 9 hollow 3-topes Δ_i have non-trivial fundamental group $\pi_1(\mathcal{S}_\Delta)$ of order 2, 3, or 5 (Table 2.10). There exist exactly 5 hollow 3-topes Δ_i with 0-dimensional Fine interior $\Delta_i^{\mathrm{FI}} = \{R\}$, where $R \in \frac{1}{2}M \setminus M$ is a rational point (Table 2.10). The normal fans Σ^{Δ_i} of these 5 hollow polytopes Δ_i define 5 toric Fano threefolds $X_{\Sigma^{\Delta_i}}$ with at worst canonical singularities (Table 2.11). These Fano threefolds can be obtained as quotients of Gorenstein toric Fano threefolds $X_{\Sigma_{\Delta''}}$ in the following 5 ways:

1. $\mathbb{P}(1, 1, 2, 4)$ with a μ_2-action given by
$$(x_0, x_1, x_2, x_3) \mapsto (x_0, -x_1, -x_2, -x_3);$$
2. $\mathbb{P}^3$ with a μ_4-action given by
$$(x_0, x_1, x_2, x_3) \mapsto (x_0, ix_1, -x_2, -ix_3);$$
3. $\{x_1x_2 - x_3x_4 = 0\} \subseteq \mathbb{P}(2, 1, 1, 1, 1)$ with a μ_2-action given by
$$(x_0, x_1, x_2, x_3, x_4) \mapsto (-x_0, -x_1, -x_2, x_3, x_4);$$
4. $\mathbb{P}^1 \times \mathbb{P}(1, 1, 2)$ with a μ_2-action given by
$$(x_0, x_1, y_0, y_1, y_2) \mapsto (x_0, -x_1, y_0, -y_1, -y_2);$$
5. $\mathbb{P}^1 \times \mathbb{P}^1 \times \mathbb{P}^1$ with a μ_2-action given by
$$(x_0, x_1, y_0, y_1, z_0, z_1) \mapsto (x_0, -x_1, y_0, -y_1, z_0, -z_1).$$

In addition, Table 2.12 contains the support $\mathrm{supp}(\Delta_i^{\mathrm{FI}})$ of the Fine interior Δ_i^{FI} and the vertices of the canonical hull Δ_i^{can} for all 9 hollow polytopes Δ_i with non-empty Fine interior Δ_i^{FI}.

Acknowledgements AK is supported by EPSRC Fellowship EP/N022513/1, *The Combinatorics of Mirror Symmetry*. The computations that underpin this work were performed on the Imperial College London HPC cluster. We thank Andy Thomas, Matt Harvey, and the Imperial College Research Computing Service team for invaluable technical assistance.

Table 2.10 12 Maximal Hollow 3-topes. Table contains: index i of the maximal hollow 3-tope Δ_i, vertices $\mathrm{vert}(\Delta_i)$ of Δ_i, lattice width $\mathrm{w}(\Delta_i)$ of Δ_i, dimension $\dim(\Delta_i^{\mathrm{FI}})$ of Fine interior Δ_i^{FI}, vertices $\mathrm{vert}(\Delta_i^{\mathrm{FI}})$ of Δ_i^{FI}, and order of fundamental group $|\pi_1(S_\Delta)|$ of the minimal model S_{Δ_i}

i	$\mathrm{vert}(\Delta_i)$	$\mathrm{w}(\Delta_i)$	$\dim(\Delta_i^{\mathrm{FI}})$	$\mathrm{vert}(\Delta_i^{\mathrm{FI}})$	$\|\pi_1(S_\Delta)\|$
1	(0, 0, 0), (6, 0, 0), (3, 3, 0), (4, 0, 2)	2	−1	∅	1
2	(0, 0, 0), (4, 0, 0), (0, 4, 0), (2, 0, 2)	2	−1	∅	1
3	(0, 0, 0), (3, 0, 0), (0, 3, 0), (3, 0, 3)	3	−1	∅	1
4	(0, 0, 0), (4, 0, 0), (2, 4, 0), (3, 0, 2)	2	0	1/2 · (5, 1, 2)	2
5	(0, 0, 0), (2, 2, 0), (1, 1, 2), (3, −1, 2)	2	0	1/2 · (3, 1, 2)	2
6	(0, 0, 0), (2, 2, 0), (4, 0, 0), (2, −2, 0), (3, 1, 2)	2	0	1/2 · (5, 1, 2)	2
7	(0, 0, 0), (1, 1, 0), (2, −2, 0), (3, −1, 0), (1, −1, 2), (2, 0, 2)	2	0	1/2 · (3, −1, 2)	2
8	(0, 0, 0), (1, 1, 0), (1, −1, 0), (2, 0, 0), (1, −1, 2), (2, 0, 2), (2, −2, 2), (3, −1, 2)	2	0	1/2 · (3, −1, 2)	2
9	(0, 0, 0), (3, 0, 0), (1, 3, 0), (2, 0, 3)	3	1	(4/3, 1, 1), (5/3, 1, 1)	3
10	(0, 0, 0), (1, 2, 0), (1, −1, 0), (3, 0, 0), (2, 1, 3)	3	1	(4/3, 2/3, 1), (5/3, 1/3, 1)	3
11	(0, 0, 0), (1, 1, 0), (3, 0, 0), (2, −1, 0), (4, 1, 3), (2, 2, 3)	3	1	(5/3, 2/3, 1), (7/3, 1/3, 1)	3
12	(−1, 0, 0), (0, 1, −2), (1, 2, 1), (2, −2, −1)	3	3	(1/5, 1/5, −2/5), (2/5, 2/5, −4/5), (3/5, 3/5, −1/5), (4/5, −1/5, −3/5)	5

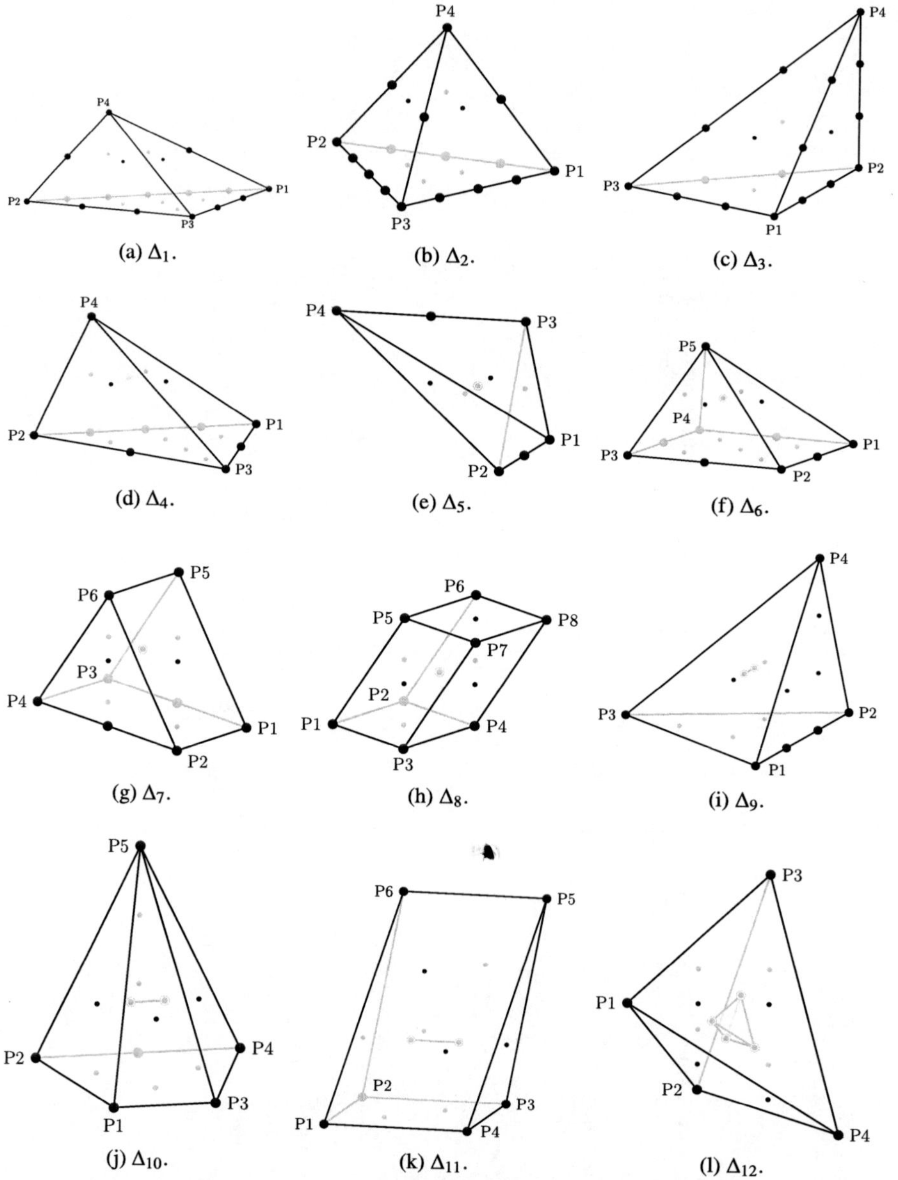

Fig. 2.7 12 Maximal Hollow 3-topes. Shaded faces are occluded. The Fine interior is shown in grey with double borders around its vertices

Table 2.11 5 Hollow 3-topes with 0-dimensional Fine Interior. Table contains: index i of the maximal hollow 3-tope Δ_i, rays of the normal fan Σ^{Δ_i} corresponding to Δ_i, ID of the canonical Fano 3-tope Δ' such that $\Sigma^{\Delta_i} \cong \Sigma_{\Delta'}$, rays of the spanning fan $\Sigma_{\Delta'}$, ID of the reflexive canonical Fano 3-tope Δ'' used to construct the Gorenstein toric Fano threefold $X_{\Sigma_{\Delta''}}$ to obtain the toric Fano threefold $X_{\Sigma_{\Delta'}}$ with at worst canonical singularities as a μ_2 quotient, and reference to the corresponding Gorenstein toric Fano threefold $X_{\Sigma_{\Delta''}}$ including the precise μ_2-action on Sect. 2.7

i	$\Sigma^{\Delta_i}(1)$	ID(Δ')	$\Sigma_{\Delta'}(1)$	ID(Δ'')	$X_{\Sigma_{\Delta''}}$
4	$(2,-1,-3),(0,0,1),(0,1,0),(-2,-1,-1)$	547354	$(-2,-3,-5),(2,1,1),(0,1,0),(0,0,1)$	547363	(i)
5	$(1,-1,0),(1,1,-1),(-1,1,2),(-1,-1,-1)$	547364	$(0,0,1),(0,2,1),(2,1,0),(-2,-3,-2)$	547367	(ii)
6	$(0,0,1),(1,1,-2),(1,-1,-1),(-1,-1,0),(-1,1,-1)$	544353	$(1,0,0),(0,1,0),(-2,-1,0),(1,1,2),(-3,-1,-2)$	544357	(iii)
7	$(0,0,1),(-1,1,-1),(1,-1,-1),(1,1,0),(-1,-1,0)$	544310	$(-1,-1,-2),(1,1,2),(-2,-1,0),(0,1,0),(1,0,0)$	544342	(iv)
8	$(-1,1,1),(0,0,-1),(-1,-1,0),(1,1,0),(1,-1,-1),(0,0,1)$	520134	$(1,0,0),(0,1,0),(1,1,2),(-1,0,0),(0,-1,0),$ $(-1,-1,-2)$	520140	(v)

Table 2.12 9 Hollow 3-topes with Non-empty Fine Interior. Table contains: index i of the maximal hollow 3-tope Δ_i, support $\mathrm{supp}(\Delta_i^{\mathrm{FI}})$ of Δ_i^{FI}, and vertices $\mathrm{vert}(\Delta_i^{\mathrm{can}})$ of the canonical hull Δ_i^{can}

i	$\mathrm{supp}(\Delta_i^{\mathrm{FI}})$	$\mathrm{vert}(\Delta_i^{\mathrm{can}})$
4	$(-2,-1,-1), (0,-1,-2), (2,-1,-3), (0,0,1), (0,0,-1), (0,1,0)$	$\mathrm{vert}(\Delta_i)$
5	$(1,-1,0), (1,1,-1), (0,0,1), (0,0,-1), (-1,-1,-1), (-1,1,2)$	$\mathrm{vert}(\Delta_i)$
6	$(1,1,-2), (1,-1,-1), (-1,-1,0), (-1,1,-1), (0,0,1), (0,0,-1)$	$\mathrm{vert}(\Delta_i)$
7	$(1,1,0), (1,-1,-1), (-1,-1,0), (-1,1,-1), (0,0,1), (0,0,-1)$	$\mathrm{vert}(\Delta_i)$
8	$(1,1,0), (1,-1,-1), (-1,-1,0), (0,0,1), (0,0,-1), (-1,1,1)$	$\mathrm{vert}(\Delta_i)$
9	$(0,-1,-1), (0,0,1), (3,-1,-2), (0,1,0), (-3,-2,-1)$	$\mathrm{vert}(\Delta_i)$
10	$(-1,2,-1), (1,1,-1), (-1,-1,0), (2,-1,-1), (0,0,1)$	$\mathrm{vert}(\Delta_i)$
11	$(1,-1,0), (0,0,1), (-1,-2,1), (-1,1,0), (1,2,-2)$	$\mathrm{vert}(\Delta_i)$
12	$(1,1,1), (1,-1,0), (-2,-1,1), (0,1,-2)$	$\mathrm{vert}(\Delta_i)$

References

1. Averkov, G., Krümpelmann, J., Weltge, S.: Notions of maximality for integral lattice-free polyhedra: the case of dimension three. Math. Oper. Res. **42**(4), 1035–1062 (2017)
2. Averkov, G., Wagner, C., Weismantel, R.: Maximal lattice-free polyhedra: finiteness and an explicit description in dimension three. Math. Oper. Res. **36**(4), 721–742 (2011)
3. Batyrev, V.: The stringy Euler number of Calabi-Yau hypersurfaces in toric varieties and the Mavlyutov duality. Pure Appl. Math. Q. **13**(1), 1–47 (2017)
4. Batyrev, V., Kreuzer, M.: Integral cohomology and mirror symmetry for Calabi-Yau 3-folds. In: Mirror symmetry. V, AMS/IP Studies in Advanced Mathematics, vol. 38, pp. 255–270. American Mathematical Society, Providence, RI (2006)
5. Batyrev, V.V.: Dual polyhedra and mirror symmetry for Calabi-Yau hypersurfaces in toric varieties. J. Algebraic Geom. **3**(3), 493–535 (1994)
6. Beck, M., Robins, S.: Computing the continuous discretely. Undergraduate Texts in Mathematics, 2nd edn. Springer, New York (2015). Integer-point enumeration in polyhedra, With illustrations by David Austin
7. Bosma, W., Cannon, J., Playoust, C.: The Magma algebra system. I. The User Language, pp. 235–265 (1997). Computational algebra and number theory (London, 1993)
8. Brown, G., Kasprzyk, A.M.: Graded Ring Database. http://www.grdb.co.uk
9. Catanese, F.: Surfaces with $K^2 = p_g = 1$ and their period mapping. In: Algebraic Geometry (Proc. Summer Meeting, Univ. Copenhagen, Copenhagen, 1978). Lecture Notes in Mathematics, vol. 732, pp. 1–29. Springer, Berlin (1979)
10. Catanese, F., Debarre, O.: Surfaces with $K^2 = 2,\ p_g = 1,\ q = 0$. J. Reine Angew. Math. **395**, 1–55 (1989)
11. Corti, A., Golyshev, V.: Hypergeometric equations and weighted projective spaces. Sci. China Math. **54**(8), 1577–1590 (2011)
12. Danilov, V.I., Khovanskiĭ, A.G.: Newton polyhedra and an algorithm for calculating Hodge-Deligne numbers. Izv. Akad. Nauk SSSR Ser. Mat. **50**(5), 925–945 (1986)
13. Fine, J.: Resolution and completion of algebraic varieties. Ph.D. thesis, University of Warwick (1983)
14. Hovanskiĭ, A.G.: Newton polyhedra, and the genus of complete intersections. Funktsional. Anal. i Prilozhen. **12**(1), 51–61 (1978)
15. Ishii, S.: The minimal model theorem for divisors of toric varieties. Tohoku Math. J. **51**(2), 213–226 (1999)
16. Kasprzyk, A.M.: Canonical toric Fano threefolds. Canad. J. Math. **62**(6), 1293–1309 (2010)

17. Kreuzer, M., Skarke, H.: Classification of reflexive polyhedra in three dimensions. Adv. Theor. Math. Phys. **2**(4), 853–871 (1998)
18. Kreuzer, M., Skarke, H.: Complete classification of reflexive polyhedra in four dimensions. Adv. Theor. Math. Phys. **4**(6), 1209–1230 (2000)
19. Kynev, V.I.: An example of a simply connected surface of general type for which the local Torelli theorem does not hold. C. R. Acad. Bulgare Sci. **30**(3), 323–325 (1977)
20. Reid, M.: Young person's guide to canonical singularities. In: Algebraic Geometry, Bowdoin, 1985 (Brunswick, Maine, 1985), Proceedings of Symposia in Pure Mathematics, vol. 46, pp. 345–414. American Mathematical Societ, Providence, RI (1987)
21. Skarke, H.: Weight systems for toric Calabi-Yau varieties and reflexivity of Newton polyhedra. Modern Phys. Lett. A **11**(20), 1637–1652 (1996)
22. Stanley, R.P.: Decompositions of rational convex polytopes. Ann. Discrete Math. **6**, 333–342 (1980)
23. Todorov, A.N.: Surfaces of general type with $p_g = 1$ and $(K, K) = 1$. I. Ann. Sci. École Norm. Sup. **13**(1, 4), 1–21 (1980)
24. Todorov, A.N.: A construction of surfaces with $p_g = 1$, $q = 0$ and $2 \leq (K^2) \leq 8$. Counterexamples of the global Torelli theorem. Invent. Math. **63**(2), 287–304 (1981)
25. Treutlein, J.: 3-dimensional lattice polytopes without interior lattice points. Ph.D. thesis, Eberhart Karls Universität Tübingen (2010)

Chapter 3
Lattice Distances in 3-Dimensional Quantum Jumps

Mónica Blanco

Abstract For Q a lattice polytope and $x \notin Q$ a lattice point, we say that (Q, x) is a quantum jump if $\operatorname{conv}(Q \cup \{x\})$ contains exactly one more lattice point than Q. Usually this can only happen when the lattice distance between x and Q is somehow bounded. In this paper I collect several results and information on the bound for that distance in 3-dimensional quantum jumps, and the consequences on the distance between the boundary of a polytope and its interior lattice points.

Keywords Lattice polytope · Lattice distance · Quantum jump · Interior points

3.1 Introduction

Throughout my research on classifying lattice 3-polytopes by their number of lattice points [2–4] there has been a recurrent situation: *suppose there is a lattice polytope Q, and a lattice point $x \notin Q$, what can I say about x with respect to Q so that* $\operatorname{conv}(Q \cup \{x\})$ *does not contain more lattice points other than those of Q and x*? Usually the answer had to do with the distance from x to Q being bounded.

In order to explain things more formally we need to introduce notation and some basic definitions. A *lattice* point is an element of $\mathbb{Z}^d$, and a *lattice polytope* is the convex hull of finitely many lattice points. We write lattice d-polytope if the polytope is d-dimensional. Two polytopes P and Q are *equivalent*, or *unimodularly equivalent*, if there exists a *unimodular transformation* that maps one to the other. That is, if there exists an affine map $t : \mathbb{R}^d \to \mathbb{R}^d$ such that $t(\mathbb{Z}^d) = \mathbb{Z}^d$ and $t(P) = Q$. The *size* of a lattice polytope $P \subset \mathbb{R}^d$ is the number of lattice points in it. An affine functional $f : \mathbb{R}^d \to \mathbb{R}$ is *integer* if $f(\mathbb{Z}^d) \subseteq \mathbb{Z}$ and it is *primitive* if $f(\mathbb{Z}^d) = \mathbb{Z}$. The *(lattice) width* of a lattice d-polytope $P \subset \mathbb{R}^d$ *with respect to* an integer functional f is the length of $f(P) \subset \mathbb{R}$, and the *width* of P is the minimum among those, for f non-constant.

M. Blanco (✉)
Department of Mathematics, Statistics and Computation, University of Cantabria, Avda. Los Castros s/n, 39005 Santander, Cantabria, Spain
e-mail: m.blanco.math@gmail.com

A. M. Kasprzyk and B. Nill (eds.), *Interactions with Lattice Polytopes*, Springer Proceedings in Mathematics & Statistics 386,
https://doi.org/10.1007/978-3-030-98327-7_3

Finally, for $S \subset \mathbb{R}^d$, we denote by $\operatorname{conv}(S)$ and $\operatorname{aff}(S)$ the *convex* and *affine hulls* of S. In particular, an affine subspace $S \subset \mathbb{R}^d$ is *lattice* if $\operatorname{aff}(S \cap \mathbb{Z}^d) = S$.

Let us now introduce the main two definitions in this paper:

Definition 1 The *lattice distance* between a lattice hyperplane $H \subset \mathbb{R}^d$ and a lattice point $x \in \mathbb{Z}^d$ is $\operatorname{dist}(x, H) := |f(x)|$, where f is a primitive functional with $f(H) = 0$.

Definition 2 Let $Q \subset \mathbb{R}^d$ be a lattice polytope, not necessarily full-dimensional, and let $x \in \mathbb{Z}^d \setminus Q$. We say that the pair (Q, x) is a *quantum jump* if

$$\operatorname{conv}\left(Q \cup \{x\}\right) \cap \mathbb{Z}^d = \left(Q \cap \mathbb{Z}^d\right) \cup \{x\}.$$

More generally, let $Q, R \subset \mathbb{R}^d$ be lattice polytopes, not necessarily full-dimensional, with $Q \cap R = \emptyset$. We say that the pair (Q, R) is a *quantum union* if

$$\operatorname{conv}\left(Q \cup R\right) \cap \mathbb{Z}^d = \left(Q \cap \mathbb{Z}^d\right) \cup \left(R \cap \mathbb{Z}^d\right).$$

That is, if the lattice points of $\operatorname{conv}(Q \cup R)$ are either in Q or in R.

The name of *quantum jump* was first used by Bruns, Gubeladze, and Michałek [5]. Notice that they restrict the concept of quantum jump (Q, x) for when both Q and $\operatorname{conv}(Q \cup \{x\})$ are full-dimensional and *normal*. Remember that a lattice d-polytope Q is *normal* if, for all $k \in \mathbb{N}$, every lattice point in kQ can be written as the sum of k lattice points in Q.

Now, if we want to take a look at the distance of quantum jumps, we first need to define the distance between a point and a polytope. Following Definition 1, they are well and naturally defined the following distances:

Definition 3 1. Let $Q \subset \mathbb{R}^d$ be a lattice $(d-1)$-polytope, and let $x \in \mathbb{Z}^d \setminus \operatorname{aff}(Q)$, then

$$\operatorname{dist}(x, Q) := \operatorname{dist}(x, \operatorname{aff}(Q)).$$

2. Let $H_1, H_2 \subset \mathbb{R}^d$ be parallel lattice hyperplanes ($H_1 \cap H_2 = \emptyset$), then

$$\operatorname{dist}(H_1, H_2) := \operatorname{dist}(x, H_2), \qquad \text{for any } x \in H_1.$$

3. Let $\ell_1, \ell_2 \subset \mathbb{R}^3$ be lattice lines such that $\operatorname{aff}(\ell_1 \cup \ell_2) = \mathbb{R}^3$, then

$$\operatorname{dist}(\ell_1, \ell_2) := \operatorname{dist}(H_1, H_2),$$

where H_1, H_2 are the unique pair of parallel lattice hyperplanes such that $\ell_i \subset H_i$.
4. Let $s_1, s_2 \subset \mathbb{R}^3$ be lattice segments such that $\operatorname{aff}(s_1 \cup s_2) = \mathbb{R}^3$, then

$$\operatorname{dist}(s_1, s_2) := \operatorname{dist}\left(\operatorname{aff}(s_1), \operatorname{aff}(s_2)\right).$$

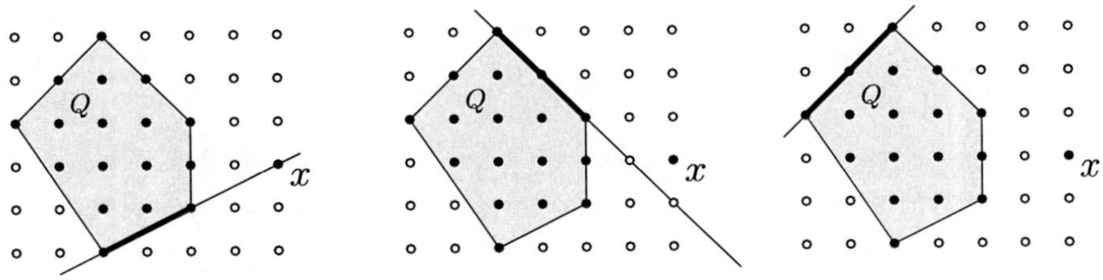

Fig. 3.1 The three figures show different facets of a lattice polygon Q, the hyperplane they are contained in, and the lattice point x. Only the facet in the middle figure is visible from x

Notice that the width and the distance are heavily related. In broad terms, the distance between two lower-dimensional objects $R_1, R_2 \subset \mathbb{R}^d$ with $\text{aff}(R_1 \cup R_2) = \mathbb{R}^d$ is the width of $\text{conv}(R_1 \cup R_2)$ with respect to a specific functional that is determined by the relative position between R_1 and R_2. In general, if $R := \text{conv}(R_1 \cup R_2) \subset \mathbb{R}^d$ is not full-dimensional, the distance between R_1 and R_2 is measured in the lattice $\text{aff}(R) \cap \mathbb{Z}^d \cong \mathbb{Z}^{\dim(R)}$. In the case of lattice segments, we call (lattice) *length* of a segment the distance between its two endpoints (vertices). Notice that a lattice segment of length k has exactly $k + 1$ lattice points. We say that a segment is *primitive* if it has length one.

Now, the distance that is not necessarily well-defined is the distance from a point to a full-dimensional polytope. This notion will be written in terms of the distance to the *visible* facets (see Fig. 3.1):

Definition 4 Let $Q \subset \mathbb{R}^d$ be a lattice d-polytope, $F \subset Q$ a facet of Q and $x \in \mathbb{Z}^d \setminus Q$. F is *visible* from x if $\text{aff}(F)$ strictly separates x from Q.

Definition 5 Let $Q \subset \mathbb{R}^d$ be a lattice d-polytope and let $x \in \mathbb{Z}^d \setminus Q$.

1. The *minimum distance* between x and Q is

$$d_x(Q) := \min\{\text{dist}(x, \text{aff}(F)) \mid F \text{ facet visible from } x\}.$$

2. The *maximum distance* between x and Q is

$$D_x(Q) := \max\{\text{dist}(x, \text{aff}(F)) \mid F \text{ facet visible from } x\}.$$

See Fig. 3.2 for a 2-dimensional example of the maximum and minimum distances between a point and a polygon. Notice that $D_x(Q)$ is the *height of x over Q* as in [5, Definition 4.1].

Let us see what we know about the distance of quantum jumps in each dimension. For this, we can also think of a quantum jump as follows: any d-dimensional quantum jump is of the form (P^v, v), for $P \subset \mathbb{R}^d$ a lattice d-polytope, $v \in \text{vert}(P)$ a vertex of P and $P^v := \text{conv}(P \cap \mathbb{Z}^d \setminus \{v\})$. Notice that the dimension of P^v can be d or $d - 1$.

For example, the lattice distance in quantum jumps of dimension ≤ 2 is always one:

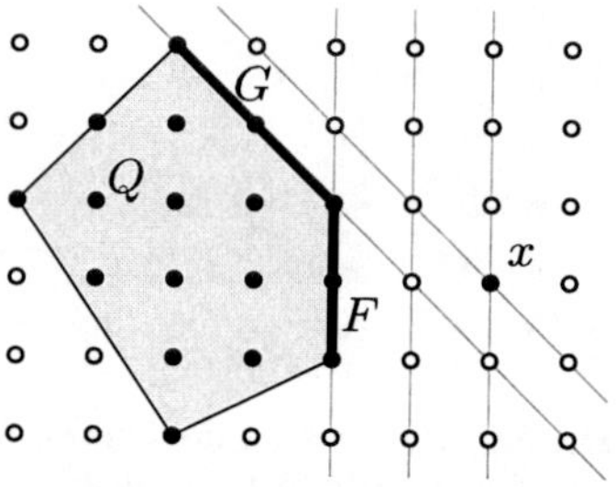

Fig. 3.2 A lattice polygon Q, a lattice point x, and the two facets of Q visible from x, F and G. In this case, $\text{dist}(x, F) = 2$ and $\text{dist}(x, G) = 1$. Hence $d_x(Q) = 1$ and $D_x(Q) = 2$

Lemma 6 *Let P be a lattice polytope of dimension $d \in \{1, 2\}$, and let v be a vertex of P.*

If P^v is of dimension $d - 1$, then $\text{dist}(v, P^v) = 1$ *and, if P^v is of dimension d, we have $d_v(P^v) = D_v(P^v) = 1$.*

To understand the idea in general: let P^v be d-dimensional. For any $(d-1)$-dimensional face of P^v that is visible from v, chose S an empty $(d-1)$-dimensional simplex in it. Since (P^v, v) is a quantum jump, so is (S, v), which implies that the convex hull of S and v is an empty d-simplex. Remember that an *empty simplex of dimension d* is a lattice d-polytope with $d + 1$ vertices and such that those vertices are its only lattice points.

In the cases of $d = 1, 2$, any empty simplex has to be unimodular, hence the *vertex-facet distance* (lattice distance between a vertex and the only facet that does not contain it) is always 1. In dimension 3 things get more complicated since we have empty tetrahedra of arbitrarily high volume, and hence arbitrarily high vertex-facet distance (e.g. Reeve tetrahedra [8]). That is, quantum jumps between a unimodular triangle and a lattice point that is at arbitrarily high lattice distance from it.

In Sect. 3.2 of this paper I put together some information on the lattice distance of 3-dimensional quantum jumps (Q, x) that derives partially from previous research [2–4]. We distinguish when Q is 2 or 3-dimensional:

1. If Q is 2-dimensional (Sect. 3.2.1) it so happens that the classifications of lattice 3-polytopes of size 5 and 6 [2, 3], together with a suitable classification of lattice polygons, give all the information there is to know about the distance from Q to x. It can be summarized as follows:

Theorem 7 (see Corollary 14) *Let $Q \subset \mathbb{R}^3$ be a lattice polygon, and let $x \in \mathbb{Z}^3 \setminus$* $\text{aff}(Q)$ *such that (Q, x) is a quantum jump. Then, the lattice distance from x to Q is at most* 3 *unless Q is a lattice triangle of width one, in which case the distance is unbounded.*

2. As a direct consequence of the results of the previous section, in Sect. 3.2.2 we have the following result on the distance of a quantum union of lattice segments:

Theorem 8 (see Corollary 16) *Let $s, t \subset \mathbb{R}^3$ be lattice segments with* $\text{aff}(s \cup t) =$ $\mathbb{R}^3$ *such that (s, t) is a quantum union. Then, the lattice distance from s to t is one, unless both s and t are primitive, in which case it is unbounded.*

3. For the case of Q being 3-dimensional (Sect. 3.2.3), Example 18 shows that there exist quantum jumps of this type at arbitrarily high distance. We also look at all the lattice 3-polytopes P of size 11 and width > 1 (database of [4]), and for each vertex v of P such that P^v is 3-dimensional we compute the minimum and maximum distances from v to P^v. Looking at the numbers one can easily see that there is no hope in trying to bound these distances, having very high numbers for both the minimum and the maximum distances.

Finally, in Sect. 3.3 I use all the information gathered in the previous section to study the distance between the boundary of a polytope and its interior. More precisely, for $P \subset \mathbb{R}^3$ a lattice 3-polytope we look at the distance between a lattice point or segment in ∂P (the boundary of P) and the *inner lattice polytope of* P, which is $I_P := \text{conv}\{\text{int}(P) \cap \mathbb{Z}^3\}$. Notice that I_P together with a point (or segment) of the boundary is always a quantum jump (or union). The definition of inner polytope also applies to rational polytopes.

We only look at inner polytopes I_P of size ≥ 3 (see Remark 20), and we separate cases according to its dimension:

1. I_P has dimension 1, that is, it is a lattice segment of length ≥ 2. In this case we look at the distance between I_P and a segment in the boundary. By Corollary 16, this distance must be one, leading to:

Theorem 9 (see Theorem 21) *The projection of P in the direction of the segment I_P is a* reflexive *polygon (polygon with a unique interior lattice point).*

2. I_P has dimension 2. In Sect. 3.3.2 we prove a specific property that a polygon has to satisfy in order to appear as the inner polygon of a 3-dimensional lattice polytope (see Theorem 23). Together with the results of Corollary 14 we obtain:

Corollary 10 (see Corollary 24) *For I_P of dimension* 2 *and size* ≥ 12, *the distance from any boundary point of P to I_P is at most* 1.

3. I_P has dimension 3. Again we look at the classification of lattice 3-polytopes of size ≤ 11 and width > 1 [4], take the polytopes with 3-dimensional inner polytope, and look at the minimum and maximum distances from any vertex to the inner polytope. In Sect. 3.3.3 we simply collect some information on the numbers obtained, without exploring it further. This time the values look more promising, since the largest value that appears is a maximum distance of 6, and in very high proportion the maximum and minimum distances are 1.

For future work one could try and complete the results on distances between a lattice point of the boundary of a polytope, and its 2 or 3-dimensional inner polytope. For the inner polytope of dimension 2, it is left to explore the cases when I_P has up to 11 lattice points. This seems perfectly doable with the help of the classification of polygons of Proposition 11, together with the results of Sect. 3.2.1. On the other hand, for I_P of dimension 3, one would have to identify in the used database all the polytopes that yield maximum and minimum distances equal to 1 and try to derive the properties they have as opposed to those that yield larger distances. One would have then to try and extend this to lattice 3-polytopes of size larger than 11.

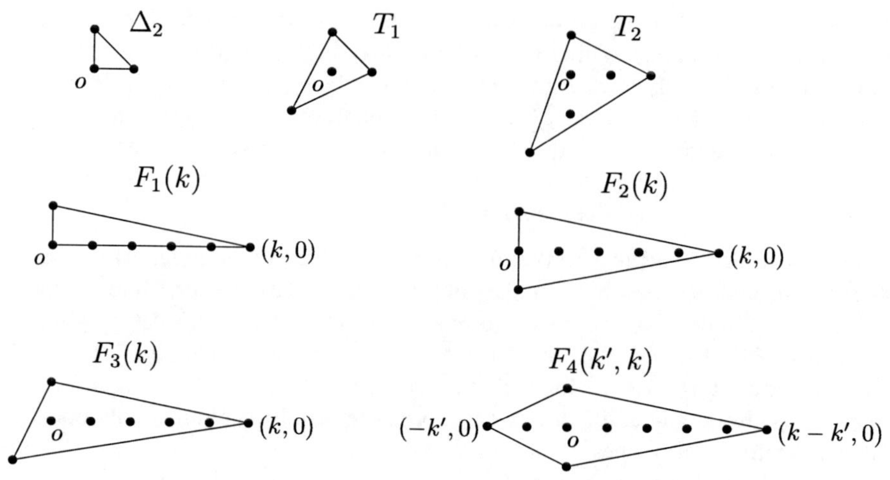

Fig. 3.3 The complete classification of lattice polygons that do not contain a unit square

3.2 Distances in 3-Dimensional Quantum Jumps or Unions

Let $Q \subset \mathbb{Z}^3$ be a lattice polytope and let $x \in \mathbb{Z}^3 \setminus Q$. We study the distance between x and Q, provided that (Q, x) is a quantum jump.

3.2.1 Quantum Jumps (Q, x) with Q of Dimension 2

We first see at what distance can a lattice point be from a lattice polygon, so that they form a quantum jump. For this, we first classify lattice polygons in a way that is suitable distance-wise. In the following lemma, we call *unit square* any lattice polygon unimodularly equivalent to $[0, 1]^2$.

Proposition 11 *Let $Q \subset \mathbb{R}^2$ be a lattice polygon. Then Q either contains a unit square or is equivalent to one of the following configurations:*

1. *Δ_2, the unimodular triangle;*
2. *$T_1 := \mathrm{conv}\{(1, 0), (0, 1), (-1, -1)\}$, the unique terminal triangle;*
3. *$T_2 := \mathrm{conv}\{(2, 0), (0, 1), (-1, -2)\}$, a clean triangle with three non-collinear interior lattice points;*
4. *$F_1(k) := \mathrm{conv}\{(0, 0), (0, 1), (k, 0)\}$, for $k \geq 2$;*
5. *$F_2(k) := \mathrm{conv}\{(0, 1), (0, -1), (k, 0)\}$, for $k \geq 2$;*
6. *$F_3(k) := \mathrm{conv}\{(-1, -1), (0, 1), (k, 0)\}$, for $k \geq 2$; or*
7. *$F_4(k', k) := \mathrm{conv}\{(0, 1), (0, -1), (-k', 0), (k - k', 0)\}$, for $0 < k' < k$.*

See Fig. 3.3 for a depiction of the polygons of Proposition 11. For its proof, let us first establish the following notation.

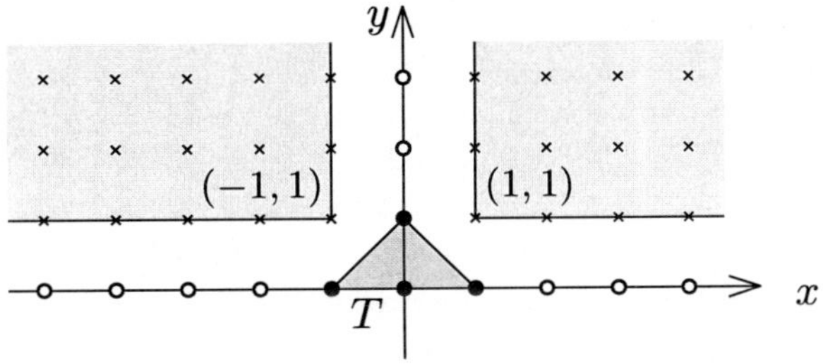

Fig. 3.4 The situation in the proof of Proposition 11 when Q contains three collinear lattice points in a facet

Remark 12 Let $P \subset \mathbb{R}^d$ be a polytope, and let $R \subsetneq P$. If a point $x \in \mathbb{R}^d$ is not a point of P, then $P \cap C_x(R) = \emptyset$, where $C_x(R) := x - \mathbb{R}_{\geq 0}\big(R - x\big)$. This fact follows trivially from convexity of polytopes. We use it whenever we want to determine a polytope P, and we know R a subset of P and x a point not in P.

Proof (Proof of Proposition 11) For Q with 3 lattice points, we have $Q \cong \Delta_2$. If Q has 4 lattice points, then $Q \cong T_1$, $Q \cong F_1(2)$ or Q is equivalent to the unit square.

So assume for the rest of the proof that Q has size at least 5. In particular we know that Q has 3 collinear lattice points and we can assume, without loss of generality, that these are $(-1, 0)$, $(0, 0)$ and $(1, 0)$.

If Q contains a unit square, we have finished. Assume for the rest of the proof that Q does not contain a unit square. That is, we have Q a lattice polygon of size ≥ 5, containing the lattice points $(-1, 0)$, $(0, 0)$ and $(1, 0)$, and not containing a unit square. Since Q is 2-dimensional, it has some lattice point outside of the line $\ell := \{y = 0\}$ and, by Lemma 6, we can choose one in either $\ell_+ := \{y = 1\}$ or $\ell_- := \{y = -1\}$. Without loss of generality, we assume that the point $(0, 1) \in \ell_+$ is in Q. That is, the triangle $T := \operatorname{conv}\{(-1, 0), (1, 0), (0, 1)\} \subset Q$.

Let us now distinguish the cases according to whether Q has three collinear lattice points in a facet or not.

1. Suppose the three collinear points $(-1, 0)$, $(0, 0)$ and $(1, 0)$ are in a facet of Q. Then $Q \subset \{y \geq 0\}$. Since Q does not contain a unit square, the points $(-1, 1)$ and $(1, 1)$ cannot be in Q. Moreover, by Remark 12 this implies that no point in the affine cones $C_{(-1,1)}(T)$ and $C_{(1,1)}(T)$ is in Q. See Fig. 3.4. That means that the only lattice points that can lie in $Q \setminus T$ are in the following sets:
$$A := \big\{(i, 0), i \in \mathbb{Z} \setminus [-1, 1]\big\}, \quad B := \big\{(0, j), j \in \mathbb{Z}, j \geq 2\big\}$$
Q can contain points of A or points of B, but in order for $(1, 1)$ and $(-1, 1)$ not to be in Q, it cannot contain points of A and B at the same time. Adding points of A to T gives rise to polygons of the type $F_1(k)$, and adding points of B gives rise to $F_2(k)$.
2. If no facet of Q contains three collinear points, then the origin, which is in the relative interior of the segment $\operatorname{conv}\{(-1, 0), (1, 0)\} \subset Q$, must be an interior point of Q. So Q must contain some lattice point in $\{y < 0\}$ and, by Lemma 6,

Fig. 3.5 The situation in the proof of Proposition 11 when Q does not contains three collinear lattice points in a facet

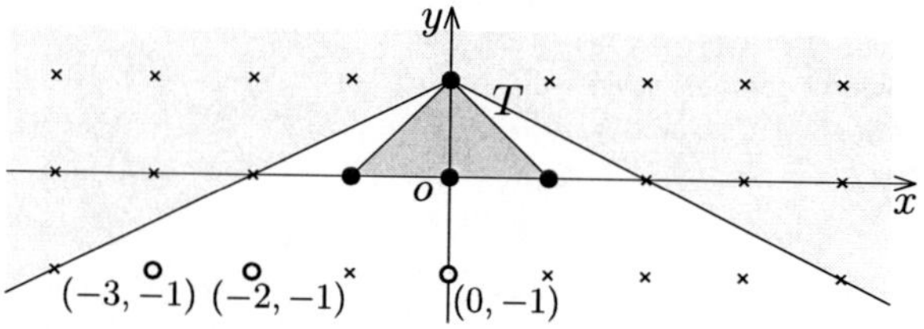

it contains some point of ℓ_-. Let us denote this point by p_-. We can assume without loss of generality that $(-2, 0), (2, 0) \notin \text{conv}(T \cup \{p_-\})$, or else we can simply choose a different triple of collinear lattice points in $\{y = 0\}$. That is, $p_- \in \{\pm 3, \pm 2, \pm 1, 0\} \times \{-1\}$. By symmetry of the already established points with respect to the line $\{x = 0\}$, we can assume that $p_- \in \{x \leq 0\}$. Also we know that $p_- \neq (-1, -1)$ since Q does not contain a unit square. The three remaining possibilities $p_- \in \{(-3, -1), (-2, -1), (0, -1)\}$ are depicted in Fig. 3.5. Let $T' := \text{conv}(T \cup \{p_-\}) \subseteq Q$, and let us study the three options for p_-.

a. $p_- = (-3, -1)$. We can apply the unimodular transformation $(x, y) \mapsto (x - y + 1, y)$ so that T' is mapped to $F_3(2)$ and assume now that $F_3(2) \subseteq Q$. See the left-most picture in Fig. 3.6. The lattice points $(-1, 0)$, $(0, -1)$ and $(1, 1)$ cannot lie in Q, or else it would contain a unit square. This already implies that the cones $C_{(-1,0)}(F_3(2))$, $C_{(0,-1)}(F_3(2))$ and $C_{(1,1)}(F_3(2))$ do not intersect Q. With that, the only lattice points that can be in Q are the points $(r, 0)$, with $r > 2$. We can take as many as wanted and this gives rise to configurations $F_3(k)$.
b. $p_- = (-2, -1)$. Again we apply the same unimodular transformation so that T' is, in this case, mapped to $F_2(2) \subseteq Q$. In order for Q not to have unit squares, no point in $\{-1, 1\}^2$ can lie in Q which, after removing the corresponding lattice cones, leaves the following possibilities for further points of Q:

$$A := \{(-1, 2)\}, \quad B := \{(-1, -2)\},$$

$$C := \{(r, 0), r \in \mathbb{Z}, r \geq 3\}, \quad D := \{(s, 0), s \in \mathbb{Z}, s \leq -1\}$$

The points in A and B cannot be in Q at the same time, and each gives rise to a configuration equivalent to T_2. The points in C or D cannot be in Q at the same time as the points in A or B. If Q has points of D, we have configurations $F_4(k', k)$ and, if Q only has points of C we get configurations $F_2(k)$.
c. $p_- = (0, -1)$. In this case, $T' = F_4(1, 2)$. After excluding the points in the cones with apex in $\{-1, 1\}^2$, Q can have other lattice points in:

$$A := \{(r, 0), r \in \mathbb{Z} \setminus [-1, 1]\}, \quad B := \{(0, s), s \in \mathbb{Z} \setminus [-1, 1]\}$$

Q cannot have points of A and B at the same time, and adding to T' points of either A or B gives rise to configurations equivalent to $F_4(k', k)$. □

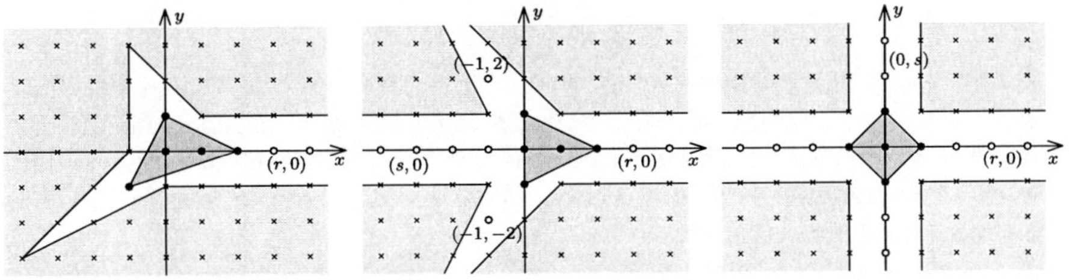

Fig. 3.6 The three possible polygons T' in the proof of Proposition 11. In each figure, the dark gray area is the polygon T' (or equivalent). The light gray area is the union of the cones that do not intersect Q. Black dots are lattice points of T', crosses are lattice points that cannot be in Q, and white dots are the possible lattice points of $Q \setminus T'$

Let us now take that classification and see what are the conditions on the coordinates of a lattice point $x \in \mathbb{Z}^3$ so that a polygon Q and $x \notin \text{aff}(Q)$ form a quantum jump.

Lemma 13 *Let $Q \subset \mathbb{R}^2 \times \{0\}$ be a lattice polygon and let $x = (a, b, c) \in \mathbb{Z}^3$ be a lattice point with $c \neq 0$ and such that (Q, x) is a quantum jump. Then:*

1. *if $Q = \Delta_2$, then at least one of the following happens:*
   - *i.* $a \equiv 1 \pmod{c}$ *and* $\gcd(b, c) = 1$*;*
   - *ii.* $b \equiv 1 \pmod{c}$ *and* $\gcd(a, c) = 1$*;*
   - *iii.* $a + b \equiv 0 \pmod{c}$ *and* $\gcd(a, c) = 1$*;*
2. *if Q contains a unit square, then $c = \pm 1$;*
3. *if $Q = T_1$, then $c = \pm 1$, or $c = \pm 3$ and $a \equiv -b \equiv \pm 1 \mod 3$;*
4. *if $Q = F_1(k)$, for $k \geq 2$, then $b \equiv 1 \mod c$ and $\gcd(a, c) = 1$;*
5. *if $Q = F_3(k)$, for $k \geq 2$, then $c = \pm 1$; or*
6. *if $Q = T_2$, $F_2(k)$, $F_4(k', k)$, for $k > k' > 0$, then $c = \pm 1$, or $c = \pm 2$ and $a \equiv b \equiv 1 \mod 2$.*

For the purpose of simplifying notation in Lemma 13 and its proof, let us denote by Q (resp. R) both the lattice polygon in $\mathbb{R}^2$ and its embedding $Q \times \{0\}$ (resp. $R \times \{0\}$) in $\mathbb{R}^3$.

Proof Part 1 of the statement follows from the classification of empty tetrahedra [9], which states that a lattice tetrahedron is empty if one of the three pairs of opposite edges are at lattice distance one. It is also required that these opposite edges are primitive segments (gcd condition in the statement).

In each of the cases 2–6, we choose a subpolygon R of Q of size 4 or 5:

2. R is the unit square in Q;
3. $R := Q$ of size 4;
4. $R := F_1(2) \subseteq Q$ of size 4;
5. $R := F_3(2) \subseteq Q$, of size 5;

6. $R := F_2(2) \subseteq Q$ or $R := F_4(1, 2) \subseteq Q$, of size 5.

Since (Q, x) is a quantum jump, and $R \subseteq Q$, so is (R, x). That is, the polytope $P := \operatorname{conv}(R \cup \{x\})$ is of size 5 or 6. Let us find the possible equivalences of P in the classification of lattice 3-polytopes of size 5 or 6 [2, 3]. Notice that in all the cases, P is a pyramid over a known polygon with apex x, so it suffices to find these in the mentioned classification:

2. $P \cong \operatorname{conv}\{(0, 0, 0), (1, 0, 0), (0, 1, 0), (1, 1, 0), (0, 0, 1)\}$ (the unique configuration of signature $(2, 2)$ in [2]);
3. $P \cong \operatorname{conv}\{(0, 0, 0), (1, 0, 0), (0, 1, 0), (-1, -1, 0), (0, 0, 1)\}$ or $P \cong \operatorname{conv}\{(0, 0, 0), (1, 0, 0), (0, 1, 0), (-1, -1, 0), (1, 2, 3)\}$ (the two configurations of signature $(3, 1)$ in [2]);
4. $P \cong \operatorname{conv}\{(0, 0, 0), (1, 0, 0), (-1, 0, 0), (0, 1, 0), (p, q, 1)\}$, with $\gcd(p, q) = 1$ (the configurations of signature $(2, 1)$ in [2]);
5. $P \cong \operatorname{conv}\{(-1, -1, 0), (1, 0, 0), (0, 0, 0), (0, 1, 0), (0, 2, 0), (0, 0, 1)\}$ (a pyramid of width one in [3]);
6. $P \cong \operatorname{conv}\{(-1, 0, 0), (1, 0, 0), (0, 0, 0), (0, 1, 0), (0, 2, 0), (0, 0, 1)\}$,
$P \cong \operatorname{conv}\{(-1, 0, 0), (1, 0, 0), (0, 0, 0), (0, 1, 0), (0, 2, 0), (1, 1, 2)\}$,
$P \cong \operatorname{conv}\{(-1, 1, 0), (1, 1, 0), (0, 0, 0), (0, 1, 0), (0, 2, 0), (0, 0, 1)\}$, or
$P \cong \operatorname{conv}\{(-1, 1, 0), (1, 1, 0), (0, 0, 0), (0, 1, 0), (0, 2, 0), (1, 0, 2)\}$ (two pyramids of width one, and configurations $A.1$ and $A.2$ in [3]).

It is left to the reader to see that adding all the lattice points in $Q \setminus R$ does not put any further restrictions on the coordinates of x. That is, for each value of x so that (R, x) is a quantum jump, we also have that (Q, x) is a quantum jump. Notice that the different possibilities for the values of a, b and c that appear in the statement in each of the cases, appear by applying to P all the unimodular transformations in $\mathbb{R}^3$ that are automorphisms of Q. □

In terms of the distance from x to Q, which in Lemma 13 is the value $|c|$, we have the following result.

Corollary 14 *Let $Q \subset \mathbb{R}^3$ be a lattice polygon and let $x \in \mathbb{Z}^3 \setminus \operatorname{aff}(Q)$ be a lattice point such that (Q, x) is a quantum jump. Then exactly one of the following happens:*

1. Q contains a unit square or $Q \cong F_3(k)$, and $\operatorname{dist}(x, Q) = 1$;
2. $Q \cong T_2$, $F_2(k)$ or $F_4(k', k)$, and $\operatorname{dist}(x, Q) = 1$ or 2;
3. $Q \cong T_1$ and $\operatorname{dist}(x, Q) = 1$ or 3; or
4. $Q \cong \Delta_2$ or $F_1(k)$, and the distance from x to Q is unbounded.

The polygons Δ_2, T_i and $F_i(k)$ are as in Proposition 11, for $0 < k' < k$.

Notice that the only cases when the distance is unbounded are $Q \cong \Delta_2$ or $Q \cong F_1(k)$, that is, when *Q is a triangle of width one*.

3.2.2 *Quantum Unions of Lattice Segments*

The two cases where the distance is unbounded in the previous section also have in common that *all the lattice points are along two lattice segments*. Let us think about them as *quantum unions of lattice segments*.

Remark 15 1. In the case of $Q = \Delta_2$, we have that (Q, x) is a quantum jump if $P := \mathrm{conv}(Q \cup \{x\})$ is an empty tetrahedron. We can also write it as $P = \mathrm{conv}(s_1 \cup s_2)$, with (s_1, s_2) a quantum union of primitive segments, where s_1 is an edge from x and one of the three vertices of Δ_2, and s_2 is the opposite edge. Notice that in this case there are three possible choices for the pair of primitive segments.
2. In the case $Q = F_1(k)$, we have that (Q, x) is a quantum jump if $(s_1(k), s_2)$ is a quantum union between the lattice segment $s_1(k) := \mathrm{conv}\{(0, 0, 0), (k, 0, 0)\}$ and the primitive segment $s_2 := \mathrm{conv}\{(0, 1, 0), x\}$.

Cases 1 and 4 of Lemma 13, reformulated in terms of the distance between segments that form a quantum union, are as follows:

Corollary 16 *Let $s, t \subset \mathbb{R}^3$ be lattice segments such that* $\mathrm{aff}(s \cup t) = \mathbb{R}^3$ *and such that (s, t) is a quantum union. Then:*

1. *if one of s or t is not primitive, then* $\mathrm{dist}(s, t) = 1$*;*
2. *if both s and t are primitive, the distance* $\mathrm{dist}(s, t)$ *can be arbitrarily high, but one of the following distances must be one:*

$$\mathrm{dist}(s, t), \quad \mathrm{dist}\big(\mathrm{conv}\{s_1, t_1\}, \mathrm{conv}\{s_2, t_2\}\big), \quad \mathrm{dist}\big(\mathrm{conv}\{s_1, t_2\}, \mathrm{conv}\{s_2, t_1\}\big)$$

where $s_i, t_i \in \mathbb{Z}^3$ are the end-points of s and t, respectively.

3.2.3 *Quantum Jumps (Q, x) with Q of Dimension 3*

For the case when $Q \subset \mathbb{R}^3$ is a lattice 3-polytope, remember that we defined the distance from a point $x \in \mathbb{Z}^3 \setminus Q$ to Q in terms of the distance to the facets of Q that are visible from x (Definition 5). In particular, one can study the distance from x to Q, for (Q, x) a quantum jump, by combining the results of the previous section on the facets of Q that are visible from x.

Remark 17 Let $Q \subset \mathbb{R}^3$ be a lattice 3-polytope. For each facet F of Q, let H_F^- be the open halfspace from which the facet F is visible, and denote by $H_F^+ = \mathbb{R}^3 \setminus H_F^-$ the closed halfspace with $Q \subset H_F^+$. Then subdivide $\mathbb{R}^3 \setminus Q$ into the regions

$$\mathcal{R}_I := \bigcap_{F \in I} H_F^- \cap \bigcap_{F \notin I} H_F^+, \qquad I \neq \emptyset,$$

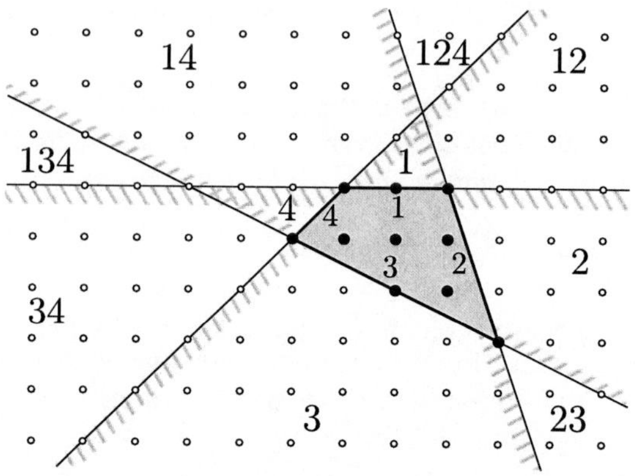

Fig. 3.7 The subdivision of the space of Remark 17 for a lattice polygon. The facets of the polygon are labeled 1 to 4 and each semi-open region of the space is labeled by the facets that are visible from it. Each region can *see* one, two or three facets

so that, for all $x \in \mathcal{R}_{\mathcal{I}}$, the facets of Q that are visible from x are exactly those of $\mathcal{I}$. Notice that the closures of the regions $\mathcal{R}_{\mathcal{I}}$ are rational polytopes or polyhedra. See Fig. 3.7 for a 2-dimensional example of this subdivision of the space.

Let $\mathcal{R}_{\mathcal{I}}$ be one of those regions and suppose that $x \in \mathcal{R}_{\mathcal{I}} \cap \mathbb{Z}^3$ is such that (Q, x) is a quantum jump. We can have three different types of situations.

1. If $\mathcal{R}_{\mathcal{I}}$ is bounded (or if $\mathcal{R}_{\mathcal{I}} \cap \mathbb{Z}^3$ is finite), the distance of x to Q is automatically bounded.
2. If $\mathcal{R}_{\mathcal{I}} \cap \mathbb{Z}^3$ has infinitely many points (in particular $\mathcal{R}_{\mathcal{I}}$ is unbounded), and some facet of $\mathcal{I}$ is not a triangle of width one, then the distance of x to Q is bounded by the results of the previous section (Corollary 14).
3. Finally, if $\mathcal{R}_{\mathcal{I}} \cap \mathbb{Z}^3$ has infinitely many points and all the facets in $\mathcal{I}$ are triangles of width one, the distance from x to Q *may not be bounded.* Notice that, even in this last case the distance from x to Q could still be bounded by combining the restrictions given for the coordinates of x as in parts 1 and 4 of Lemma 13 for all the different facets of $\mathcal{I}$.

For instance, we can find arbitrarily high distance in these types of quantum jumps.

Example 18 Let $h \in \mathbb{Z}$, $h > 0$. Let:

1. $x = (0, 0, 0)$;
2. $F := \text{conv}\{(1, 0, h), (0, 1, h), (1, 1, h)\}$, a unimodular triangle in $\{z = h\}$;
3. $Q \subset \mathbb{R}^3$ be a lattice 3-polytope such that $Q \subset C$, for $C := \mathbb{R}_{\geq 0}(F) = \{xh \leq z\} \cap \{yh \leq z\} \cap \{(x + y)h \geq z\}$ (the triangular cone of F with apex at the origin) and such that $x \notin Q$, $F \subset Q$.

Then F is a facet of Q, it is the only facet that is visible from x, and Q and x are such that (Q, x) is a quantum jump with $D_x(Q) = d_x(Q) = \text{dist}(x, F) = h$. If moreover Q is contained in $C_x := C \cap \{x \leq 1\}$ or $C_y := C \cap \{y \leq 1\}$, then Q and $\text{conv}(Q \cup \{x\})$ are polytopes of width one.

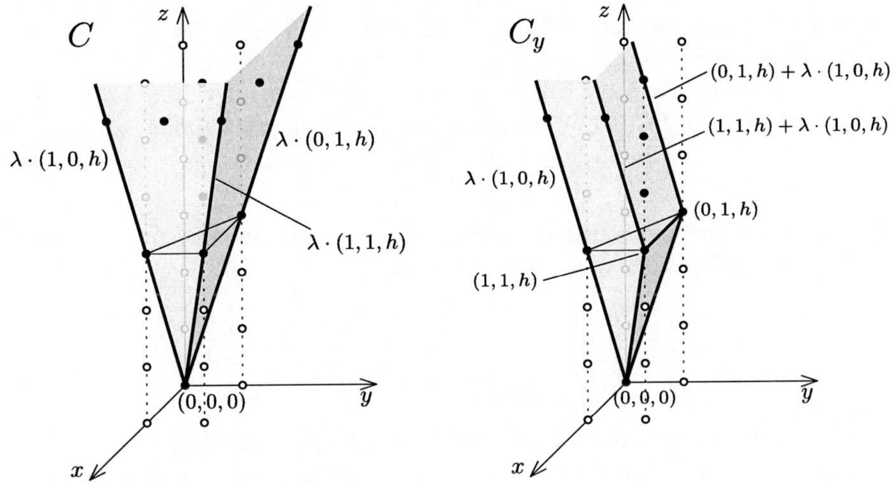

Fig. 3.8 The regions C and C_y of Example 18

We can also choose Q of size n, for any $n \geq 4$, since the regions C, C_x or C_y have infinitely many lattice points. See Fig. 3.8 for a depiction of the polyhedral regions C and C_y.

Even though it is clear that arbitrarily bad examples can occur, how often does this happen? Are they rare or does the general picture look bad? For this, we look at our database of lattice 3-polytopes of size ≤ 11 and width > 1 [4]. For each of those polytopes $P \subset \mathbb{R}^3$ and for each $v \in \text{vert}(P)$ such that P^v is full dimensional, we look at the distance in the quantum jump (P^v, v). Notice that our database contains all the information on the types of quantum jumps when P^v is of size ≤ 10 and extends to a polytope P of width > 1. That is, of size ≤ 10, we do not have the information on polytopes of width one that extend to polytopes of width one, which are infinitely many for each size (and no enumeration exists).

For each quantum jump (P^v, v) we compute the values $d_v(P^v)$ and $D_v(P^v)$ and store the following vectors:

1. $\overline{d}_P := \left(d_v(P^v)\right)_{v \in \text{vert}(P),\, P^v \text{ full-dimensional}}$
2. $\overline{D}_P := \left(D_v(P^v)\right)_{v \in \text{vert}(P),\, P^v \text{ full-dimensional}}$

We separate the 216, 453 polytopes of our database in three different groups. Notice that the entries of each vector $\overline{d}_P$ and $\overline{D}_P$ are positive integers.

1. $\overline{d}_P = (1, 1, \ldots, 1) = \overline{D}_P$. This is the best case scenario we can find, since *every* vertex v of P, with P^v full-dimensional, is at distance one from *all* the facets of P^v that are visible from v. However, only 5,796 polytopes (about 2.7%) fall into this category.
2. $\overline{d}_P = (1, 1, \ldots, 1)$, $\overline{D}_P \neq (1, 1, \ldots, 1)$. In this case, things are not as nice, but we still have that *every* vertex v of P, with P^v full-dimensional, is at distance one

from *at least one* facet of P^v that is visible from v. In this category we have 77,443 polytopes (~35.8%).

3. $\overline{d}_P, \overline{D}_P \neq (1, 1, \ldots, 1)$. This is the worst case we can have, in which some vertex v of P, with P^v full-dimensional, is at distance *larger than one* from *all* the facets of P^v that are visible form v. This is the case for most of the polytopes in our database: 133,214 polytopes, or ~61.5% of the total.

In terms of the magnitudes of the entries, we have that the largest entries in the vectors $\overline{d}_P$ and $\overline{D}_P$, for each n the size of P, are:

n	5	6	7	8	9	10	11
$\max d_v(P^v)$	5	7	13	19	25	31	37
$\max D_v(P^v)$	7	13	19	25	31	37	43

Notice that the maximum values for $d_v(P^v)$ and $D_v(P^v)$, for P of size n, are $6(n-5)+1$ and $6(n-4)+1$, respectively (for $n \neq 5$ in the first case). This has to do with the fact that, as h grows (see Example 18) we need more lattice points to construct a polytope of width > 1 that yields a vertex at distance h.

The average values of the $d_v(I_P)$ and $D_v(I_P)$ are, respectively, 1.42 and 3.35.

Remark 19 If we were to follow the lines of Sect. 3.2.1, we would want to have, in this section, an irredundant list of lattice 3-polytopes Q, and the maximum and minimum distances a point x can be from Q, for (Q, x) a quantum jump.

However, we need to consider that we have 216,453 polytopes and that, for each of those polytopes P and each vertex v of P we have a different polytope P^v. Organizing the information on the distances with no redundancies among the P^v does not seem to be worth undertaking, in light of the distances that appear and the arguments made.

3.3 Distance from the Boundary to the Inner Polytope

Let $P \subset \mathbb{R}^3$ be a lattice 3-polytope with $I_P \neq \emptyset$.

Remark 20 For I_P of size 1 or 2, the classification of lattice 3-polytopes with 1 and 2 interior lattice points was completed, respectively, by Kasprzyk [7] and by Balletti and Kasprzyk [1].

1. The 3-*dimensional* distances that can be measured in the case of I_P consisting of one lattice point are the distances between this point and the facets of P.
2. In the case of I_P having two lattice points, we would have to look at the distance between I_P and a non-coplanar lattice segment in the boundary.

In these two situations the distance is a priori unbounded if we look at it locally: we can have a quantum jump between a unimodular triangle and a point in the first case, and a quantum union of primitive segments in the second, at arbitrarily high distance (see Corollaries 14 and 16). There will be a bound following from the fact

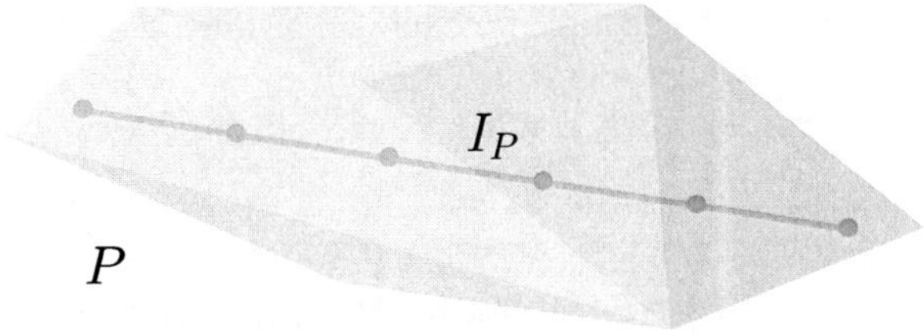

Fig. 3.9 A polytope P having inner polytope I_P of dimension 1

that there are only finitely many lattice 3-polytopes with 1 and 2 interior lattice points, but the author does not believe it is worth exploring the more than 23 million such polytopes.

So let P be such that I_P has size at least three. Let $S \subset \partial P$ be a lattice point or primitive segment in the boundary of P, we look at the distance between S and I_P, relying on the fact that (I_P, S) is a quantum jump (or union). This happens because $\text{conv}(I_P \cup S) \setminus S \subset \text{int}(P)$, and the only interior lattice points of P are those of I_P.

3.3.1 *Inner Polytope of Dimension* 1

If I_P is a lattice segment (see Fig. 3.9), and since I_P has size at least 3, by Corollary 16, the distance from I_P to any lattice segment in the boundary must be one. A consequence of this is the following result[1]:

Theorem 21 (Averkov–Balletti–Blanco–Nill–Soprunov) *Let $P \subset \mathbb{R}^3$ be a lattice 3-polytope with I_P a lattice segment of lattice length k ($k + 1$ collinear lattice points), for $k \geq 2$. If $\pi : \mathbb{R}^3 \to \mathbb{R}^2$ is the lattice projection that maps I_P to the origin then $\pi(P)$ is a reflexive polygon.*

Proof Since $k > 0$, the projection π is well defined and unique, up to unimodular transformation. Because the $k + 1$ collinear lattice points are in the interior of P, their projection, i. e. the origin, is an interior point of $\pi(P)$. Let e be an edge of $\pi(P)$, then there exists a lattice segment e' in the boundary of P such that $\pi(e') = e$. Take the following polytope $R_e := \text{conv}(I_P \cup e') \subset P$. Since $e' \subset \partial P$ and $I_P \subset \text{int}(P)$, then R_e cannot contain any extra lattice points. That is, it is the quantum union of two lattice segments. By Corollary 16, and since I_P is not primitive, the distance between I_P and e' must be one. In the projection, this directly implies that the distance from the edge e and the origin (the respective projections of the segments) is one. Hence $\pi(P)$ is reflexive. □

[1] Discussed in the Oberwolfach mini-workshop *Lattice polytopes: Methods, Advances and Applications*, September 2017.

This result can help, for example, in the full classification of lattice 3-polytopes P with I_P a lattice segment. The projection has 16 possibilities: the 16 reflexive polygons. For one such Q fixed, all the lattice points in P must be in $\pi^{-1}(Q)$.

3.3.2 Inner Polytope of Dimension 2

For P having inner polytope I_P of dimension 2 (see Fig. 3.10), our main result resides in proving a specific property that a polygon must have so that it can actually appear as the inner polytope of a lattice 3-polytope. For this, let us introduce the concept of *front*:

Definition 22 Let $Q \subset \mathbb{Z}^2$ be a lattice polygon and let v be a vertex of Q. A *front* of Q from v is a facet of the polygon $Q^v := \text{conv}(Q \setminus \{v\} \cap \mathbb{Z}^2)$ that is visible from v. See Fig. 3.11 for an example of the fronts of a polygon.

Theorem 23 *If P is a lattice 3-polytope with I_P of dimension 2, then the fronts of I_P have length ≤ 8.*

Proof Let $F \subset I_P$ be the longest front of I_P, of length $\ell > 0$. We can assume without loss of generality that $I_P \subset \{z = 0\}$, $\ell \geq 3$, that $v := (0, 1, 0)$ is a vertex of I_P

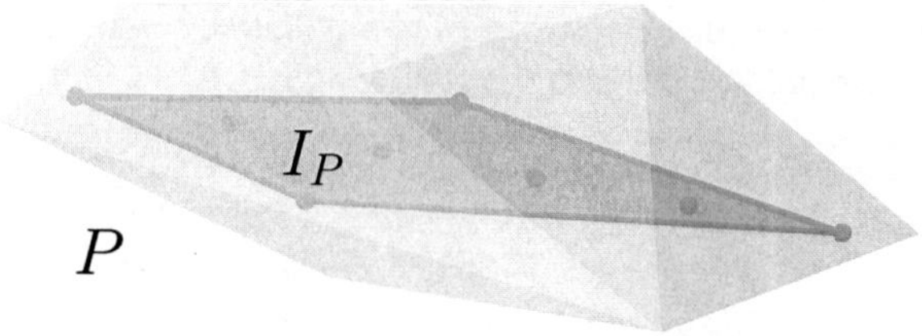

Fig. 3.10 A polytope P having inner polytope I_P of dimension 2

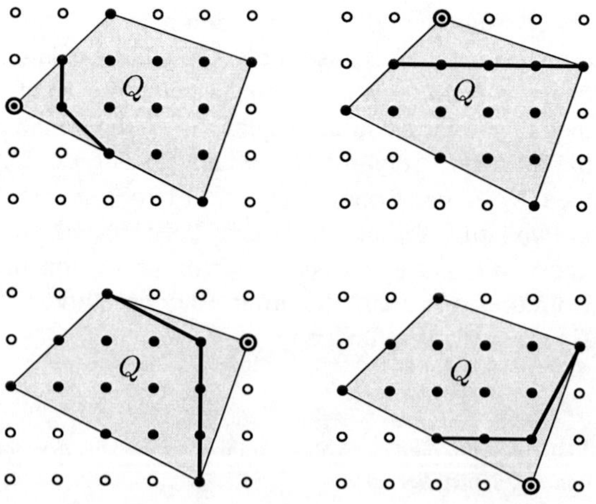

Fig. 3.11 Four copies of a polygon Q, each showing the fronts of Q from a specific vertex

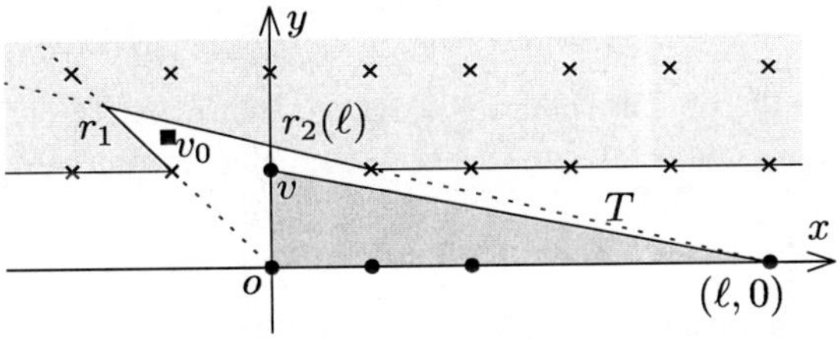

Fig. 3.12 The setting of the proof of Theorem 23

and $F := \operatorname{conv}\{(0,0,0), (\ell,0,0)\}$ (that is, F is a front of I_P from v). In particular, we have that $T := F_1(\ell) \times \{0\} = \operatorname{conv}(F \cup \{v\}) \subseteq I_P$ (for $F_1(\ell)$ as in Proposition 11) and that the only lattice points of I_P in $\{y \geq 0\}$ are those of F and v. We need to prove that $\ell \leq 8$.

The intersection of P with the plane $\{z = 0\}$ is the rational polygon $P_0 := P \cap \{z = 0\}$. We have that $I_P = \operatorname{conv}(\operatorname{relint}(P_0) \cap \mathbb{Z}^3)$. That is, the inner polytope of P coincides with the *relative* inner polygon of P_0. For now let us identify $\mathbb{R}^2 \times \{0\}$ and $\mathbb{R}^2$ in the trivial way, so from now on we simply say *interior of* P_0 for the relative interior of it embedded in the space $\mathbb{R}^3$.

The vertex $v = (0, 1)$ of I_P is an interior point of P_0, so for any line passing through v there must be a vertex of P_0 in each of the open halfspaces determined by this line. In particular, there must be a vertex of P_0 in the open halfspace $\{y > 1\}$. Let us denote this vertex by v_0 and consider the rational polygon $T' := \operatorname{conv}(T \cup \{v_0\})$. Since $T \subseteq I_P \subset \operatorname{int}(P_0)$ and $v_0 \in \partial P_0$, then $T' \setminus \{v_0\} \subset \operatorname{int}(P_0)$. That is, the only lattice points of $T' \setminus \{v_0\}$ are those of T. In particular, $(-1, 1), (1, 1) \notin T' \setminus \{v_0\}$, which implies that $v_0 \notin C_{(-1,1)}(T) \cup C_{(1,1)}(T)$ (see Remark 12).

This in turn implies that v_0 must lie in the open rational triangle R_ℓ determined by the hyperplanes $\{y = 1\}$, $r_1 := \operatorname{aff}\{(0,0), (-1,1)\} = \{x + y = 0\}$ and $r_2(\ell) := \operatorname{aff}\{(\ell, 0), (1, 1)\} = \{x + (\ell - 1)y = \ell\}$ (see Fig. 3.12).

That is,

$$v_0 \in R_\ell = \operatorname{int}\left(\operatorname{conv}\left\{(-1,1), (1,1), \left(\frac{-\ell}{\ell-2}, \frac{\ell}{\ell-2}\right)\right\}\right),$$

which is well defined for $\ell \geq 3$.

Observe that $R_\ell \subset R_3$ for all $\ell > 3$, and that $R_3 \cap \mathbb{Z}^2 = \emptyset$. That is, there is no lattice point in R_ℓ. In particular, $v_0 \notin \mathbb{Z}^3$, and the only possibility is that v_0 is the intersection of a primitive segment $uw := \operatorname{conv}\{u, w\} \subset P$ with the plane $\{z = 0\}$, with neither u nor w in this plane (and one in each of the halfspaces). This segment uw is contained in an edge of P, although it is not necessarily equal to it.

In order to find out more about the coordinates of v_0, we need to know the distances $d_u := \operatorname{dist}(u, H)$ and $d_w := \operatorname{dist}(w, H)$ in the full-dimensional polytope P, for $H := \operatorname{aff}\{P_0\} = \{z = 0\}$.

For this, let us look at a 3-dimensional proper subpolytope of P:

$$K := \text{conv}\{(0,0,0), (1,0,0), (2,0,0), (0,1,0), u, w\} \subset P$$

So far, the information we have is that $K = \text{conv}(K_0 \cup \{u, w\})$, for $K_0 := K \cap \{z = 0\} = \text{conv}\{o, (2,0,0), (0,1,0), v_0\}$ and u and w lying one in $\{z > 0\}$ and the other in $\{z < 0\}$, the edge uw being primitive and cutting the plane $\{z = 0\}$ at the rational point $v_0 \in R_\ell$. Let us prove the following properties of K:

1. *K has size* 6. On one hand, $K \cap \{z = 0\} = K_0$ does not contain more lattice points other than $(0,0,0)$, $(1,0,0)$, $(2,0,0)$ and $(0,1,0)$, since $K_0 \subset T'$ and $T' \cap \mathbb{Z}^3 = T \cap \mathbb{Z}^3$. On the other hand, if K contains an extra lattice point other than those four and u or w, this lattice point would have to lie outside of uw (which is a primitive edge) and outside of $\{z = 0\}$. Since $K \setminus uw \subset \text{int}(P)$, this would be an interior lattice point of P outside of the plane $\{z = 0\}$, which is impossible by hypothesis.
2. *K has width* > 1. Let $f : \mathbb{R}^3 \to \mathbb{R}$ be a linear primitive functional. If f is not constant in the line $\{y = 0 = z\}$, then the width of K with respect to f is > 1 since f will take three different values in the points $(0,0,0), (1,0,0), (2,0,0) \in K$. Take now f to be constant in that line. Then the width of K with respect to f is the width of $\pi(K)$ with respect to f', for π the lattice projection $\pi : \mathbb{R}^3 \to \mathbb{R}^2$, $\pi(x, y, z) = (y, z)$, and f' the primitive functional $f' : \mathbb{R}^2 \to \mathbb{R}$, $f'(y, z) = f(\pi^{-1}(y, z))$. Notice that f' is well defined because f is constant in the fibers of the projection. But under this projection, the point $(0, 1)$ is an interior point of $\pi(K)$, hence the width of $\pi(K)$ with respect to any functional is > 1. See the picture on the right in Fig. 3.13.

That is, K is a lattice 3-polytope of size 6 and width > 1, and the classification of such polytopes appears in [3]. However, not all of these polytopes are a possible candidate for K. To narrow the possibilities, we can figure out the oriented matroid of our configuration K, since the classification in [3] is also organized according to this combinatorial information. Remember that the *oriented matroid* of a finite set of points is the information recording the affine dependencies, in particular coplanarities and collinearities between the points (see [6] for information on oriented matroids).

For this, one extra thing that we can notice is that v cannot be a vertex of K. Suppose otherwise, then the polytope $K' := K^{(0,1,0)} = \text{conv}\{o, (1,0,0), (2,0,0), u, w\}$ is a polytope of size 5 with three collinear lattice points. That is, K' is the convex hull

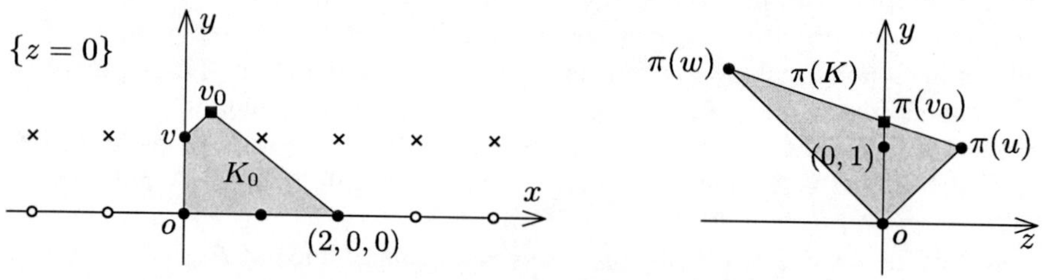

Fig. 3.13 The intersection K_0 of K with the plane $\{z = 0\}$, and the projection of K under π

of a lattice segment of length 2 and a primitive lattice segment uw. By Corollary 16, the lattice distance between these two lattice segments must be one. Taking again projection π, this is equivalent to the segment $\pi(uw)$ being at distance one from the origin. Which is impossible since $(0, 1)$ is a lattice point strictly in between $\pi(uw)$ and the origin (see the picture in the right of Fig. 3.13). In particular, v is not a vertex of K_0 and the picture on the left of Fig. 3.13 is not accurate.

That is, the oriented matroid of the six lattice points of K can be described as follows:

1. four of them are vertices (o, $(2, 0, 0)$, u and w);
2. one non-vertex point is in an edge ($(1, 0, 0) = \frac{1}{2}((0, 0, 0) + (2, 0, 0))$);
3. the hyperplane containing the three collinear points and the other non-vertex point ($\{z = 0\}$) leaves the remaining two vertices (u and w) strictly in opposite sides of it.

Notice that there are four different types (or orbits) of points: the endpoints of the collinearity (o and $(2, 0, 0)$), the middle point of the collinearity ($(1, 0, 0)$), the other two vertices (u and w), and the remaining non-vertex point (v). This sixth point v has three different possibilities, in terms of the oriented matroid. The three possibilities for v are: (I) it is in the relative interior of one of the facets $\text{conv}\{o, u, w\}$ or $\text{conv}\{(2, 0, 0), u, w\}$; (II) it is in the relative interior of the triangle $\text{conv}\{(1, 0, 0), u, w\}$; or (III) it lies in the interior of one of the tetrahedra $\text{conv}\{o, (1, 0, 0), u, w\}$ or $\text{conv}\{(1, 0, 0), (2, 0, 0), u, w\}$. These three options are shown in Fig. 3.14. For each of these three cases, the oriented matroid is fully described. Without going into details of how the oriented matroids are represented and classified in [3], one can derive that the oriented matroid of the three options (I), (II) and (III) are, respectively, oriented matroids 3.6, 3.8 and 4.11 as encoded in [3, Fig. 1].

In [3, Tables 8 and 9] we can see that the only lattice 3-polytopes of size 6, width > 1 and with one of the three specified oriented matroids are B.7 (oriented matroid 3.8), C.1 (oriented matroid 3.6), and F.13 to F.17 (oriented matroid 4.11). The following 3×6 matrices have, as columns, the six lattice points of each of those seven polytopes:

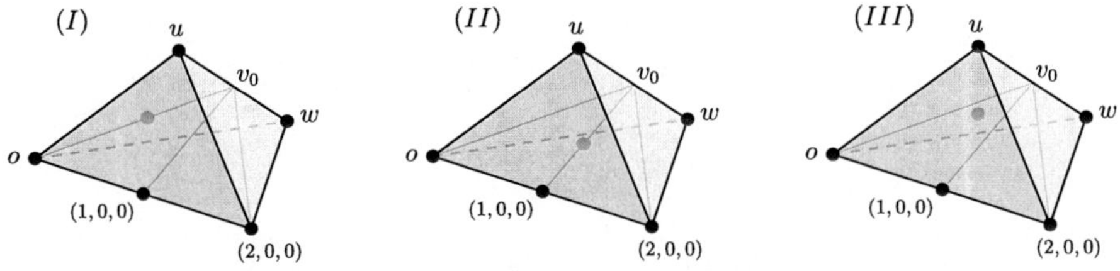

Fig. 3.14 The three possibilities (I), (II), and (III) in the proof of Theorem 23

$$B.7\begin{pmatrix}0\,1\,0\,-1\,0\ \ 0\\0\,0\,1\,-1\,0\ \ 2\\0\,0\,0\ \ 0\ \ 1\,-1\end{pmatrix}\quad F.13\begin{pmatrix}0\,1\,0\,-2\,1\,4\\0\,1\,0\,-1\,0\,1\\0\,1\,1\,-2\,0\,2\end{pmatrix}\quad F.16\begin{pmatrix}0\,1\,-1\,0\,1\,2\\0\,0\,-2\,0\,3\,6\\0\,0\,-1\,1\,1\,1\end{pmatrix}$$

$$C.1\begin{pmatrix}0\,1\,0\,-1\,0\,-1\\0\,0\,1\,-1\,0\ \ 0\\0\,0\,0\ \ 0\ \ 1\ \ 2\end{pmatrix}\quad F.14\begin{pmatrix}0\,0\,-1\,1\,1\ \ 1\\0\,0\,-1\,2\,0\,-2\\0\,1\,-1\,1\,0\,-1\end{pmatrix}\quad F.17\begin{pmatrix}0\,1\,1\,0\,-1\,-2\\0\,0\,3\,0\,-2\,-4\\0\,0\,1\,1\,-1\,-3\end{pmatrix}$$

$$F.15\begin{pmatrix}0\,0\,1\,1\,-1\,-3\\0\,0\,0\,2\,-1\,-4\\0\,1\,0\,1\,-1\,-3\end{pmatrix}$$

That is, our polytope K must be equivalent to one of them, say $\tilde{K}$, and let $t : \mathbb{R}^3 \to \mathbb{R}^3$ be any unimodular transformation that maps K to $\tilde{K}$. Then t will send the edge conv$\{(0, 0, 0), (2, 0, 0)\}$ to the unique collinearity of three lattice points in $\tilde{K}$, and $v = (0, 1, 0)$ to the only non-vertex of the remaining lattice points.

Since unimodular transformations preserve distances, we have that $\{d_u, d_w\} = \{d'_u, d'_w\}$, for $d'_u := \text{dist}(t(u), t(H))$ and $d'_w := \text{dist}(t(w), t(H))$. Moreover, we can assume without loss of generality that $d_u \leq d_w$. Then:

$$(d_u, d_w) = \begin{cases} (1,1) & \text{if } K \cong \text{B.7, C.1, F.13, F.15} \\ (1,2) & \text{if } K \cong \text{F.14, F.17} \\ (1,3) & \text{if } K \cong \text{F.16} \end{cases}$$

The distance in our original coordinates of K, since $H = \{z = 0\}$, is measured on the z-coordinate of the points. That is, let z_u and z_w be the respective z-coordinates of u and w, we have that $d_u = |z_u|$ and $d_w = |z_w|$. Without loss of generality $z_u > 0 > z_w$:

$$(z_u, z_w) = \begin{cases} (1,-1) & \text{if } K \cong \text{B.7, C.1, F.13, F.15} \\ (1,-2) & \text{if } K \cong \text{F.14, F.17} \\ (1,-3) & \text{if } K \cong \text{F.16} \end{cases}$$

Let us now see that the denominator of the rational coordinates of v_0 can only be 2, 3 or 4:

$$v_0 = (1 - \lambda)u + \lambda w, \text{ for some } \lambda \in [0, 1]$$

where λ is such that the z-coordinate of v_0 is 0:

$$0 = (1 - \lambda)z_u + \lambda z_w \Longrightarrow \lambda = \frac{z_u}{z_u - z_w} = \frac{d_u}{d_u + d_w}$$

That is, $\lambda \in \frac{1}{2}\mathbb{Z}$, $\frac{1}{3}\mathbb{Z}$ or $\frac{1}{4}\mathbb{Z}$, hence

$$v_0 = (a, b, 0), \text{ for } (a, b) \in \left(\frac{1}{2}\mathbb{Z}\right)^2 \cup \left(\frac{1}{3}\mathbb{Z}\right)^2 \cup \left(\frac{1}{4}\mathbb{Z}\right)^2.$$

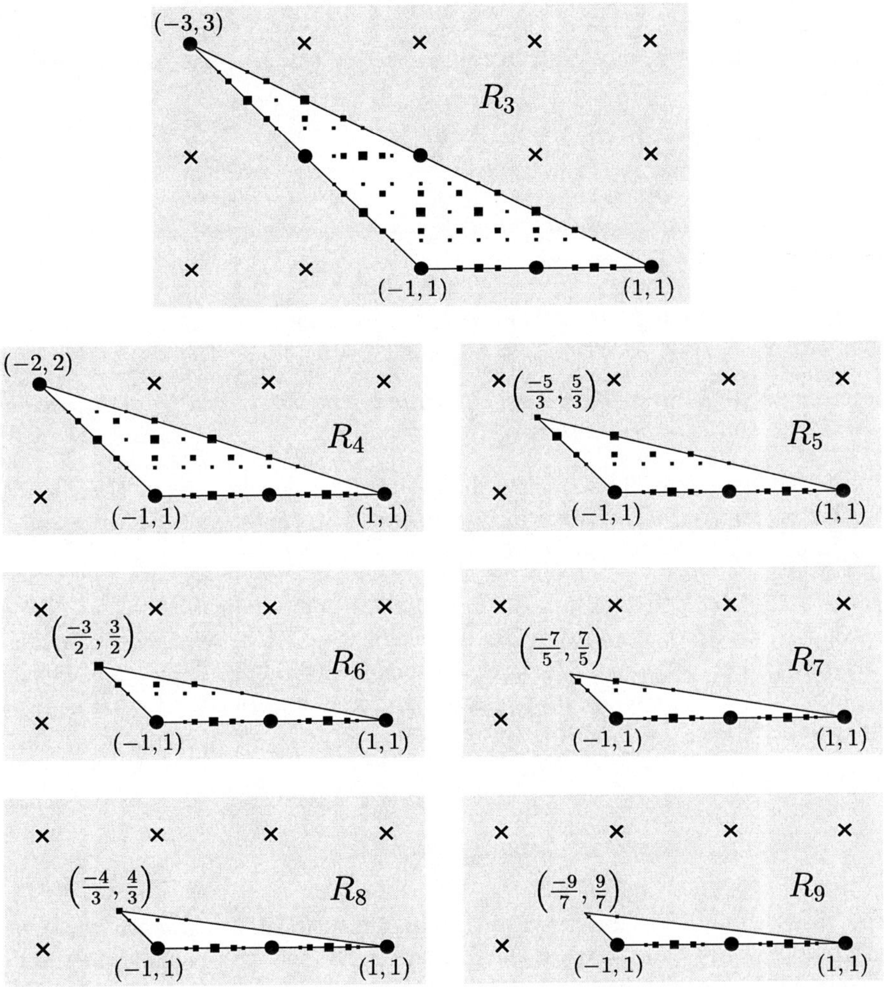

Fig. 3.15 The regions R_3 to R_9, with the points of the lattices L_2, L_3 and L_4 contained in them. Large squares are points of $L_2 \setminus \mathbb{Z}^2$, medium squares are the points of $L_3 \setminus \mathbb{Z}^2$, and small squares the points of $L_4 \setminus L_2$. Black dots and crosses represent points of $\mathbb{Z}^2$ in ∂R_ℓ and $\mathbb{R}^2 \setminus \overline{R_\ell}$, respectively

Remember also that v_0 must lie in the open triangle R_ℓ. To prove the statement of the theorem it remains to see that the intersection of R_ℓ with any of the lattices L_2, L_3 or L_4, for $L_i := \left(\frac{1}{i}\mathbb{Z}\right)^2 \times \{0\}$ is empty for $\ell \geq 9$. This is true since R_9 does not contain any point of those lattices, and since $R_\ell \subseteq R_9$ for $\ell \geq 9$. To help the reader visualize this we have drawn in Fig. 3.15 all the regions R_3 to R_9 with the possible positions for the point v_0. □

As a consequence of the theorem, we find that polygons that do not contain a unit square can only be inner polytopes of lattice 3-polytopes if they have few lattice points:

Corollary 24 *Let* $P \subset \mathbb{R}^3$ *be a lattice* 3*-polytope with* I_P *of dimension* 2*. Then exactly one of the following happens:*

1. I_P *contains a unit square and all its fronts are of length* ≤ 8*;*
2. $I_P \cong T_1$*;*
3. $I_P \cong F_3(k)$*, for* $2 \leq k \leq 8$*;*
4. $I_P \cong T_2$*,* $F_2(k)$ *or* $F_4(k', k)$*, for* $0 < k' < k \leq 8$*; or*
5. $I_P \cong \Delta_2$ *or* $I_P \cong F_1(k)$*, for* $2 \leq k \leq 8$.

In particular, in cases 1–4 any lattice point in the boundary of P *is at distance at most* 1, 3, 1 *and* 2*, respectively, from* I_P.

Proof The first part of the statement follows from Proposition 11 and Theorem 23, considering that the longest fronts in $F_1(k)$, $F_2(k)$, $F_3(k)$ and $F_4(k', k)$ have length k. The second part follows from Corollary 14. □

Remark 25 In case 5, the distance of any boundary lattice point of P to I_P will also be bounded since there are only finitely many lattice 3-polytopes with those particular polygons as inner polytopes. However, this bound can only be found globally, and not locally, since the distance from a single lattice point to I_P is a priori unbounded (see Corollary 14).

3.3.3 Inner Polytope of Dimension 3

Now there is only left the case where I_P is 3-dimensional. In particular, these are quantum jumps of the type considered in Sect. 3.2.3, and we can apply the results of Sect. 3.2.1 as explained in Remark 17. However, notice that we cannot use the results of Sect. 3.3.2, since they heavily rely on I_P being 2-dimensional.

We do the same as we did in Sect. 3.2.3: we check our database of lattice 3-polytopes of size ≤ 11, width > 1 and 3-dimensional inner polytope, of which there are 15,763 polytopes [4]. In this case, since a polytope with interior lattice points cannot have width one, we are not losing cases by having only polytopes of width > 1, but we only have polytopes with at most 11 lattice points in total. Since I_P is 3-dimensional, it has at least size 4, and since $\text{vert}(P) \subset P \setminus I_P$, then I_P has at most 7 lattice points. That is, our database contains the information on I_P of size $k \in \{4, 5, 6, 7\}$, with P of size $n \in \{k + 4, \ldots, 11\}$ and with $n - k \leq 7$ lattice points in the boundary. In particular, both P and I_P are very *clean* (few points in the boundary) among the polytopes being checked.

We consider the following vectors:

1. $d_P := \big(d_v(I_P)\big)_{v \in \mathrm{vert}(P)}$
2. $D_P := \big(D_v(I_P)\big)_{v \in \mathrm{vert}(P)}$

In this case, the results we get are much more hopeful, since most of the polytopes P have the distance from vertices to I_P all ones:

1. 8,786 polytopes (~55.74%) have $d_P = (1, 1, \ldots, 1) = D_P$. That is, *every* vertex v of P is at distance one from *all* the facets of I_P that are visible from v.
2. 5,804 polytopes (~36.82%) have $d_P = (1, 1, \ldots, 1), D_P \neq (1, 1, \ldots, 1)$. In this case, *every* vertex v of P is at distance one from *at least one* facet of I_P that is visible from v.
3. 1,173 polytopes (~7.44%) have $d_P, D_P \neq (1, 1, \ldots, 1)$. That is, there exists a vertex v of P that is at distance *larger than one* from *all* the facets of I_P that are visible form v.

Moreover, the values of the distances are much smaller. The maximum and minimum values for each size n are as follows:

n	8	9	10	11
$\max D_v(I_P)$	3	4	5	6
$\max d_v(I_P)$	3	3	3	4

and the average values of the $D_v(I_P)$ and $d_v(I_P)$ are, respectively, 1.12 and 1.02.

Remark 26 Following the reasonings of Remark 19, in this case we would want to have a list of polytopes Q and the maximum distance we can have a point x so that (Q, x) is a quantum jump and there exists a polytope P such that $Q = I_P$ and $x \in \partial P$. From our database we are only considering 15,763 polytopes and, for each of those polytopes P, we have exactly one polytope I_P.

Putting together all the equivalent inner polytopes, we find out that there are only 39 equivalence classes of inner polytopes. Moreover, around 9,000 polytopes in the database (more than half) have the unimodular tetrahedron as its inner polytope.

The maximum and minimum distances for I_P of size k are as follows:

k	4	5	6	7
$\max D_v(I_P)$	4	4	5	6
$\max d_v(I_P)$	4	3	2	2

Altogether, it seems that we could find manageable bounds for the distance in quantum jumps (I_P, v), although further work is required.

Acknowledgements I would like to thank the referee for the useful comments and suggestions.

References

1. Balletti, G., Kasprzyk, A.M.: Three-dimensional lattice polytopes with two interior lattice points (2016). arXiv:1612.08918 [math.CO]
2. Blanco, M., Santos, F.: Lattice 3-polytopes with few lattice points. SIAM J. Discrete Math. **30**(2), 669–686 (2016)
3. Blanco, M., Santos, F.: Lattice 3-polytopes with six lattice points. SIAM J. Discrete Math. **30**(2), 687–717 (2016)
4. Blanco, M., Santos, F.: Enumeration of lattice 3-polytopes by their number of lattice points. Discrete Comput. Geom. **60**(3), 756–800 (2018)
5. Bruns, W., Gubeladze, J., Michałek, M.: Quantum jumps of normal polytopes. Discrete Comput. Geom. **56**(1), 181–215 (2016)
6. De Loera, J.A., Rambau, J., Santos, F.: Triangulations. Algorithms and Computation in Mathematics, vol. 25. Springer, Berlin (2010). Structures for algorithms and applications
7. Kasprzyk, A.M.: Canonical toric Fano threefolds. Canad. J. Math. **62**(6), 1293–1309 (2010)
8. Reeve, J.E.: On the volume of lattice polyhedra. Proc. Lond. Math. Soc. **3**(7), 378–395 (1957)
9. White, G.K.: Lattice tetrahedra. Can. J. Math. **16**, 389–396 (1964)

Chapter 4
Flag Matroids: Algebra and Geometry

Amanda Cameron, Rodica Dinu, Mateusz Michałek, and Tim Seynnaeve

Abstract Matroids are ubiquitous in modern combinatorics. As discovered by Gel'fand, Goresky, MacPherson and Serganova there is a beautiful connection between matroid theory and the geometry of Grassmannians: representable matroids correspond to torus orbits in Grassmannians. Further, as observed by Fink and Speyer, general matroids correspond to classes in the K-theory of Grassmannians. This yields in particular a geometric description of the Tutte polynomial. In this review we describe all these constructions in detail, and moreover we generalise some of them to polymatroids. More precisely, we study the class of flag matroids and their relations to flag varieties. In this way, we obtain an analogue of the Tutte polynomial for flag matroids.

Keywords Flag matroids · Tutte polynomial · Torus orbits · Grassmannians · K-theory of flag varieties

---

A. Cameron
Max Planck Institute for Mathematics in the Sciences, Inselstr. 22, 04103 Leipzig, Germany
e-mail: a.r.cameron@tue.nl

Department of Mathematics and Computer Science, Eindhoven University of Technology, PO Box 513, 5600 Eindhoven, The Netherlands

R. Dinu · M. Michałek (✉)
University of Konstanz, Fachbereich Mathematik und Statistik, Fach D 197, 78457 Konstanz, Germany
e-mail: mateusz.michalek@uni-konstanz.de

R. Dinu
e-mail: rodica.dinu@uni-konstanz.de

T. Seynnaeve
Mathematisches Institut, Universität Bern, Alpeneggstrasse 22, 3012 Bern, Switzerland
e-mail: tim.seynnaeve@math.unibe.ch

A. M. Kasprzyk and B. Nill (eds.), *Interactions with Lattice Polytopes*,
Springer Proceedings in Mathematics & Statistics 386,
https://doi.org/10.1007/978-3-030-98327-7_4

4.1 Introduction

The aim of this article is to present beautiful interactions among matroids and algebraic varieties. Apart from discussing classical results, we focus on a special class of polymatroids known as *flag matroids*. The ultimate result is a definition of a *Tutte polynomial* for flag matroids. Our construction is geometric in nature and follows the ideas of Fink and Speyer for ordinary matroids [16]. The audience we are aiming at is the union of combinatorists, algebraists and algebraic geometers, not the intersection.

Matroids are nowadays central objects in combinatorics. Just as groups abstract the notion of symmetry, matroids abstract the notion of *independence*. The interplay of matroids and geometry is in fact already a classical subject [21]. Just one of such interactions (central for our article) is the following set of associations:

$$\text{matroids/flag matroids/polymatroids} \rightarrow \text{lattice polytopes} \rightarrow \text{toric varieties.}$$

We describe these constructions in detail. They allow to translate results in combinatorics to and from algebraic geometry. As an example we discuss two ideas due to White:

1. combinatorics of basis covers translates to projective normality of (all maximal) torus orbit closures in arbitrary Grassmannians—Theorem 40;
2. White's conjecture about basis exchanges (Conjecture 38) is a statement about quadratic generation of ideals of toric subvarieties of Grassmannians.

Although the idea to study a matroid through the associated lattice polytope is certainly present already in the works of White and Edmonds, the importance of this approach was only fully discovered by Gel'fand, Goresky, MacPherson and Serganova [21]. The construction of associating a toric variety to a lattice polytope can be found in many sources, we refer the reader e.g. to [9, 17, 49].

The object we focus on is one of the main invariants of a matroid: the Tutte polynomial. It is an inhomogeneous polynomial in two variables. On the geometric side it may be interpreted as a cohomology class (or a class in K-theory or in Chow ring) in a product of two projective spaces.

The applications of algebro-geometric methods are currently flourishing. A beautiful result of Huh confirming a conjecture by Read on unimodality of chromatic polynomials of graphs is based on Lefschetz theorems [28]. This led further to a proof of the general Rota-Heron-Welsh conjecture [1]—which we state in Theorem 93. Although the latter proof is combinatorial in nature, the authors were inspired by geometry, in particular Lefschetz properties. We would like to stress that the varieties and Chow rings studied by Adiprasito, Huh and Katz are not the same as those we introduce in this article. Still, as the focus of both is related to the Tutte polynomial it would be very interesting to know if their results can be viewed in the setting discussed here.

We finish the article with a few open questions. As the construction of the Tutte polynomial we propose is quite involved it would be very nice to know more direct, combinatorial properties and definitions.

Note 1 E will always denote a *finite* set of cardinality n. $\mathcal{P}(E)$ is the set of all subsets of E, and $\binom{E}{k}$ is the set of all subsets of E of cardinality k. We use $[n]$ as a shorthand notation for the set $\{1, 2, \ldots, n\}$. We will denote the difference of two sets X and Y by $X - Y$. This does not imply that $Y \subseteq X$. If Y is a singleton $\{e\}$, we write $X - e$ instead of $X - \{e\}$.

4.2 Matroids: Combinatorics

For a comprehensive monograph on matroids we refer the reader to [43].

4.2.1 Introduction to Matroids

There exist many cryptomorphic definitions of a matroid—it can be defined in terms of its independent sets, or its rank function, or its dependent sets, amongst others. One of the most relevant definitions for us is that of the rank function:

Definition 2 A matroid $M = (E, r)$ consists of a ground set E and a *rank function* $r : \mathcal{P}(E) \to \mathbb{Z}_{\geq 0}$ such that, for $X, Y \in \mathcal{P}(E)$, the following conditions hold:

R1. $r(X) \leq |X|$;
R2. *(monotonicity)* if $Y \subseteq X$, then $r(Y) \leq r(X)$; and
R3. *(submodularity)* $r(X \cup Y) + r(X \cap Y) \leq r(X) + r(Y)$.

We write $r(M)$ for $r(E)$. When $r(X) = |X|$, we say that X is *independent*, and *dependent* otherwise. A minimal dependent set is called a *circuit*. A matroid is *connected* if and only if any two elements are contained in a common circuit. It can be shown that "being contained in a common circuit" is an equivalence relation on E; the equivalence classes are called *connected components*.

If $|X| = r(X) = r(M)$ we call X a *basis* of M. We can use bases to provide an alternative set of axioms with which to define a matroid. We present this as a lemma, but it can just as well be given as the definition. The diligent reader can check that each set of axioms implies the other.

Lemma 3 *A matroid* $M = (E, \mathcal{B})$ *can be described by a set E and a collection of subsets* $\mathcal{B} \subseteq \mathcal{P}(E)$ *such that:*

B1. $\mathcal{B} \neq \emptyset$; *and*

B2. (basis exchange) *if* $B_1, B_2 \in \mathcal{B}$ *and* $e \in B_1 - B_2$, *there exists* $f \in B_2 - B_1$ *such that* $(B_1 - e) \cup f \in \mathcal{B}$.

A reader new to matroid theory should not be surprised by the borrowed terminology from linear algebra: matroids were presented as a generalisation of linear independence in vector spaces in the paper by Whitney [57] initiating matroid theory. Matroids also have a lot in common with graphs, thus explaining even more of the terminology used. For instance, very important matroid operations are that of *minors*. These are analogous to the graph operations of the same names. As there, deletion is very simple, while contraction requires a bit more work.

Definition 4 (*Deletion and Contraction*)

1. We can remove an element e of a matroid $M = (E, r_M)$ by *deleting* it. This yields a matroid $M\backslash e = (E - e, r_{M\backslash e})$, where $r_{M\backslash e}(X) = r_M(X)$ for all $X \subseteq E - e$.
2. We can also remove an element e of a matroid $M = (E, r_M)$ by *contracting* it. This gives a matroid $M/e = (E - e, r_{M/e})$ where $r_{M/e}(X) = r_M(X \cup e) - r_M(\{e\})$ for all $X \subseteq E - e$.

Remark 5 More generally, if $M = (E, r_M)$ is a matroid and S is a subset of E, we can define the deletion $M\backslash S$ (resp. contraction M/S) by deleting (resp. contracting) the elements of S one by one. We have that $r_{M\backslash S}(X) = r_M(X)$ for all $X \subseteq E - S$ and $r_{M/S}(X) = r_M(X \cup S) - r_M(S)$ for all $X \subseteq E - S$.

We will now give two examples of classes of matroids which show exactly the relationship matroids have with linear algebra and graph theory. The first one plays a central role in our article.

Definition 6 Let V be a vector space, and $\phi : E \to V$ a map that assigns to every element in E a vector of V. For every subset X of E, define $r(X)$ to be the dimension of the linear span of $\phi(X)$. We have that (E, r) is a matroid, which we say is *representable*.

Remark 7 Our definition differs slightly from the one found in literature: typically one identifies E with $\phi(E)$. Our definition does not require ϕ to be injective; we can take the same vector several times. We also note that the matroid represented by $\phi : E \to V$ only depends on the underlying map $\phi : E \to \mathbb{P}(V)$, assuming $\phi(E) \subset V\backslash\{0\}$.

If V is defined over a field $\mathbb{F}$, we say that M is $\mathbb{F}$-*representable*. We can describe the bases of a representable matroid: $X \subseteq E$ is a matroid basis if and only if $\phi(X)$ is a vector space basis of the linear span of $\phi(E)$.

Example 8 (*The non-Pappus matroid*) Here is an example of a non-representable matroid: consider the rank-3 matroid R on [9], whose bases are all 3-element subsets of [9] except for the following (see also Fig. 4.1):

Fig. 4.1 The non-Pappus matroid

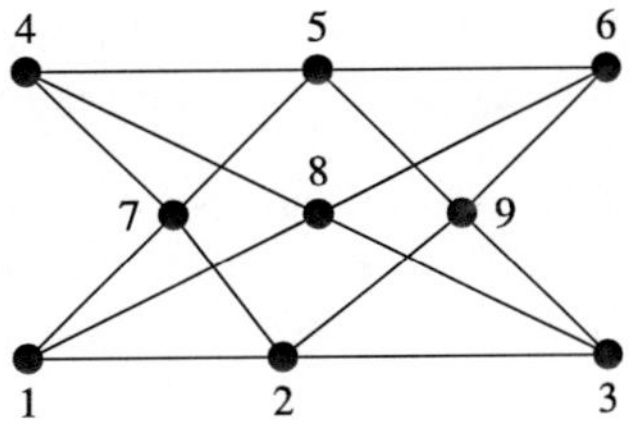

$$\{1, 2, 3\}, \{4, 5, 6\}, \{1, 5, 7\}, \{1, 6, 8\}, \{2, 4, 7\}, \{2, 6, 9\}, \{3, 4, 8\}, \{3, 5, 9\}.$$

If R were representable over a field $\mathbb{F}$, there would be a map $[9] \to \mathbb{P}^2_{\mathbb{F}} : i \to p_i$ such that p_i, p_j, p_k are collinear if and only if $\{i, j, k\}$ is not a basis of R. Now, the classical *Pappus' Theorem* precisely says that this is impossible: if the non-bases are all collinear, then so are p_7, p_8, p_9.

Definition 9 Let $G = (V, E)$ be a graph. The *graphic* (or *cycle*) matroid M of G is formed by taking $E(M) = E(G)$, and setting the rank of a set of edges equal to the cardinality of the largest spanning forest contained within it.

It is easy to see that connectedness of the graph G is not equivalent to connectedness of the matroid $M(G)$. However, there is a correspondence with higher graph connectivity. We say that a graph G is k-connected if we cannot disconnect G by removing less than k vertices. Now, $M(G)$ is connected if and only if G is 2-connected.

4.2.2 *The Tutte Polynomial*

Further matroid definitions will be given later, but we have covered enough to give the major object of our interest in this paper, namely the Tutte polynomial. This is the most famous matroid (and graph) invariant, and, like matroids themselves, has multiple definitions. These will be mentioned where relevant. Here, we give the *corank-nullity* formula, two terms which will be defined below.

Definition 10 Let $M = (E, r)$ be a matroid with ground set E and rank function $r : \mathcal{P}(E) \to \mathbb{Z}_{\geq 0}$. The *Tutte polynomial* of M is

$$T_M(x, y) = \sum_{S \subseteq E} (x-1)^{r(M)-r(S)} (y-1)^{|S|-r(S)}.$$

The term $r(M) - r(S)$ is called the *corank*, while the term $|S| - r(S)$ is called the *nullity*. Readers familiar with matroid theory should be careful not to confuse a mention of corank with dual rank, given the usual naming convention of dual objects. By identifying the rank function of a matroid with the connectivity function of a graph in an appropriate way, one can pass between this formula and the original formulation of the Tutte polynomial which was given for graphs.

Example 11 For the (matroid of the) complete graph K_4, there are four subsets with three elements of rank 2 and all the other subsets with three elements have rank 3. In this case, the Tutte polynomial is

$$T_{M(K_4)}(x, y) = x^3 + 3x^2 + 2x + 4xy + 2y + 3y^2 + y^3.$$

Readers interested in seeing what the Tutte polynomial looks like for a range of different classes of matroids should consult [36].

The prevalence of the Tutte polynomial in the literature is due to the wide range of applications it has. The simplest of these occurs when we evaluate the polynomial at certain points, these being called *Tutte invariants*. For instance, $T(1, 1)$ gives the number of bases in the matroid (or the number of spanning trees in a graph). In this way we can also count the number of independent sets in a matroid or graph, and the number of acyclic orientations of a graph, as well as some other such quantities. Beyond numerics, the Tutte invariants also include other well-known polynomials, appearing in graph theory (the chromatic polynomial, concerned with graph colourings; see also Theorem 93) and network theory (the flow and reliability polynomials). Extending to further disciplines, one can find multivariate versions of the Tutte polynomial which specialise to the Potts model [55] from statistical physics and the Jones polynomial [52] from knot theory. In this paper, we will be looking at the classical Tutte polynomial from an algebraic point of view.

We noted that there are multiple definitions of the Tutte polynomial. One is both so useful and attractive that we would be remiss to not include it. It states that, instead of calculating the full sum above, we can instead simply form a recurrence over minors of our matroid, which can lead to faster calculations. Note that a *coloop* is an element of E which is in every basis of M, while a *loop* is an element which is in no basis.

Lemma 12 ([7]) *Let $T_M(x, y)$ be the Tutte polynomial of a matroid $M = (E, r)$. Then the following statements hold:*

1. $T_M(x, y) = xT_{M/e}(x, y)$ *if e is a coloop;*
2. $T_M(x, y) = yT_{M\setminus e}(x, y)$ *if e is a loop;*
3. $T_M(x, y) = T_{M\setminus e}(x, y) + T_{M/e}(x, y)$ *if e is neither a loop nor a coloop.*

The Tutte polynomial is in fact universal for such formulae: any formula for matroids (or graphs) involving just deletions and contractions will be an evaluation of the Tutte polynomial. There are numerous proofs of this in the literature, and also extensions to related classes of objects. One such reference is [13, Sect. 4].

4.2.3 The Base Polytope

We will now give two more axiom systems for matroids. The first one, via base polytopes, will play a fundamental role in this paper.

We first define what the base polytope of a matroid is: let $\mathcal{B}$ be the set of bases of a matroid $M = (E, r)$. We work in the vector space $\mathbb{R}^E = \{(r_i \mid i \in E)\}$, where $r_i \in \mathbb{R}$. For a set $U \subseteq E$, $\mathbf{e}_U \in \mathbb{R}^E$ is the indicator vector of U, that is, $\mathbf{e}_U$ is the sum of the unit vectors $\mathbf{e}_i$, for all $i \in U$. Note that $\mathbf{e}_{\{i\}} = \mathbf{e}_i$.

Definition 13 The *base polytope* of M is

$$P(M) = \operatorname{conv}\{\mathbf{e}_B \mid B \in \mathcal{B}\}.$$

Note that this is always a lattice polytope. Its dimension is equal to $|E|$ minus the number of connected components of the matroid [15, Proposition 2.4]. We also note that the vertices of $P(M)$ correspond to the bases of M. In particular: given $P(M) \subset \mathbb{R}^E$, we can recover M.

The following theorem gives a characterisation of which lattice polytopes appear as the base polytope of a matroid. It can be used as an axiom system to define matroids:

Theorem 14 ([12], see also [21, Theorem 4.1]) *A polytope $P \subset \mathbb{R}^E$ is the base polytope of matroid on E if and only if the following two conditions hold:*

P1. every vertex of P is a $0, 1$-vector; and
P2. every edge of P is parallel to $\mathbf{e}_i - \mathbf{e}_j$ for some $i, j \in E$.

More generally the description of faces of matroid base polytopes is provided in [15, 30]. The base polytope is a face of the *independent set polytope* of M, which is the convex hull of indicator vectors of the independent sets of M.

4.2.4 Definition via Gale Orderings

We move on to another axiom system: via Gale orderings. This definition is orginally due to Gale [20]; our formulation is based on lecture notes by Reiner [45].

Definition 15 Let ω be a linear ordering on E, which we will denote by $\leq$. Then the *dominance ordering* $\leq_\omega$ on $\binom{E}{k}$, also called *Gale ordering*, is defined as follows. Let $A, B \in \binom{E}{k}$, where

$$A = \{i_1, \ldots, i_k\}, i_1 < \cdots < i_k$$

and

$$B = \{j_1, \ldots, j_k\}, j_1 < \cdots < j_k.$$

Then we set

$$A \leq_\omega B \text{ if and only if } i_1 \leq j_1, \ldots, i_k \leq j_k.$$

Theorem 16 ([20]) *Let $\mathcal{B} \subseteq \binom{E}{k}$. We have that $\mathcal{B}$ is the set of bases of a matroid if and only if for every linear ordering ω on E, the collection $\mathcal{B}$ has a maximal element under the Gale ordering $\leq_\omega$ (i.e. there is a unique member $A \in \mathcal{B}$ such that $B \leq_\omega A$, for all $B \in \mathcal{B}$).*

In Sect. 4.6.1, we will introduce a generalisation of matroids, called *flag matroids*, via Gale orderings. This will generalise the above characterisation of matroids. In this paper, they will arise quite naturally when generalising our geometric description of matroids given in Sect. 4.5.

4.2.5 The Matroid Union Theorem

Next we present one of the central theorems in matroid theory.

Theorem 17 (The matroid union theorem) *Let $M_1, \ldots, M_k$ be matroids on the same ground set E with respective families of independent sets $\mathfrak{l}_1, \ldots, \mathfrak{l}_k$ and rank functions $r_1, \ldots, r_k$. Let*

$$\mathfrak{l} := \{I \subset E : I = \bigcup_{i=1}^{k} I_i \text{ for } I_i \in \mathfrak{l}_i\}.$$

Then $\mathfrak{l}$ is also a family of independent sets for a matroid, known as the union of $M_1, \ldots, M_k$. Further, the rank of any set $A \subset E$ for the union matroid is given by:

$$r(A) = \min_{B \subset A}\{|A \backslash B| + \sum_{i=1}^{k} r_i(B)\}.$$

The proof can be found e.g. in [43, 12.3.1]. The following corollary is essentially due to Edmonds.

Corollary 18 *Let $M_1, \ldots, M_k$ be matroids on a ground set E with rank functions respectively $r_1, \ldots, r_k$. E can be partitioned into independent sets, one for each matroid, if and only if for all subsets $A \subset E$ we have $|A| \leq \sum_{i=1}^{k} r_i(A)$.*

Proof The implication $\Rightarrow$ is straightforward.

For the other implication let U be a matroid that is the union of $M_1, \ldots, M_k$. We compute the rank of E in U, applying the matroid union Theorem 17:

$$r_U(E) = \min\left\{|E| - |B| + \sum_{i=1}^{k} r_i(B)\right\}.$$

By assumption, for any $B \subset E$ we have $|E| - |B| + \sum_{i=1}^{k} r_i(B) \geq |E|$. Further, equality holds for $B = \varnothing$. Hence, $r_U(E) = |E|$. Thus, E is an independent set of U. By definition it is a union of k independent sets, one in each of the M_i's. □

4.3 Polymatroids: Combinatorics

Consider what happens if we drop one of the rank axioms, namely that which states $r(X) \leq |X|$. What object do we get, and what relation does it have to matroids? This object was originally studied by Edmonds [12] (although in a different guise, see Definition 23), and dubbed a *polymatroid*. The class of polymatroids includes, naturally, the class of matroids, and is greatly important in the field of combinatorial optimisation.

Definition 19 A *polymatroid* $\mathcal{M} = (E, r)$ consists of a ground set E and a rank function $r : \mathcal{P}(E) \to \mathbb{Z}_{\geq 0}$. The rank function r satisfies conditions R2 (monotonicity) and R3 (submodularity) of Definition 2, while condition R1 is relaxed to $r(\varnothing) = 0$.

A polymatroid is called a *k-polymatroid* if all singletons have rank at most k. In particular, a matroid is a 1-polymatroid.

Remark 20 As we assume that our rank function take only integral values, the object we defined is sometimes referred to in the literature as a *discrete polymatroid* [27].

One vital difference between matroids and polymatroids is that polymatroids do not have well-defined properties of deletion and contraction.

One problem behind this is the following. Take any element e in the ground set of a given matroid. All the bases not containing e are bases in the deletion of e, while all the bases containing e exactly correspond to bases in the contraction of e. There is no such partition of bases among minors in a polymatroid. In consequence, the Tutte polynomial is not directly applicable to polymatroids. In restricted cases, this can be somewhat solved: this is done by Oxley and Whittle [44] for 2-polymatroids, where the corank-nullity polynomial is still universal for a form of deletion-contraction recurrence. In [11], the authors strengthen the notion of "deletion-contraction invariant" to more general combinatorial objects via the use of coalgebras, and compare their results to that of Oxley and Whittle. In their strengthening, the corank-nullity polynomial is indeed still universal, and furthermore, out of the polynomials found in [44], the corank-nullity one is optimal, under the norms used in [11].

Cameron and Fink [8] construct a version of the Tutte polynomial for all polymatroids which specialises to an evaluation of the classical Tutte polynomial when applied to a matroid. This will be discussed below. In order to describe it, we first have to explain bases and base polytopes for polymatroids.

Definition 21 Let $\mathcal{M} = (E, r)$ be a polymatroid. An integer vector $\mathbf{x} \in \mathbb{Z}_{\geq 0}^{E}$ is called an *independent vector* if $\mathbf{x} \cdot \mathbf{e}_U \leq r(U)$ for all $U \subseteq E$. If in addition $\mathbf{x} \cdot \mathbf{e}_E = r(E)$, then $\mathbf{x}$ is called a *basis*.

Analogously to the matroid case, we can give an axiom system for polymatroids in terms of their bases.

Lemma 22 ([27, Theorem 2.3]) *A nonempty finite set $\mathcal{B} \subset \mathbb{Z}^E_{\geq 0}$ is the set of bases of a polymatroid on E if and only if $\mathcal{B}$ satisfies*

1. *all $\mathbf{u} \in \mathcal{B}$ have the same modulus (sum of entries); and*
2. *if $\mathbf{u} = (u_1, \ldots, u_n)$ and $\mathbf{v} = (v_1, \ldots, v_n)$ belong to $\mathcal{B}$ with $u_i > v_i$ then there is $j \in E$ with $u_j < v_j$ such that $\mathbf{u} - \mathbf{e}_i + \mathbf{e}_j \in \mathcal{B}$.*

Definition 23 Let $\mathcal{M} = (E, r)$ be a polymatroid. Let $\mathcal{I} \subseteq \mathbb{Z}^E_{\geq 0}$ be the set of independent vectors, and $\mathcal{B} \subseteq \mathbb{Z}^E_{\geq 0}$ be the set of bases. We have the independent set polytope, which is also referred to as the *extended polymatroid* of r:

$$EP(\mathcal{M}) = \operatorname{conv} \mathcal{I} = \{\mathbf{x} \in \mathbb{R}^E_{\geq 0} \mid x \cdot \mathbf{e}_U \leq r(U) \text{ for all } U \subseteq E\}.$$

This is in fact what was originally defined to be a polymatroid, by Edmonds [12]. We also have the polymatroid *base polytope*:

$$P(\mathcal{M}) = \operatorname{conv} \mathcal{B} = EP(\mathcal{M}) \cap \{x \in \mathbb{R}^E \mid x \cdot \mathbf{e}_E = r(E)\}.$$

As before, the base polytope is a face of the extended polymatroid. When the polymatroid considered is a matroid, these definitions coincide exactly with those from Sect. 4.2.3.

Theorem 14 generalises to the case of polymatroids, giving us another equivalent definition of polymatroids in terms of their base polytopes:

Theorem 24 ([27, Theorem 3.4]) *A polytope $P \subset \mathbb{R}^n$ is the base polytope of a polymatroid on $[n]$ if and only if the following two conditions hold:*

1. *every vertex of P has coordinates in $\mathbb{Z}_{\geq 0}$; and*
2. *every edge of P is parallel to $\mathbf{e}_i - \mathbf{e}_j$, for some $i, j \in [n]$.*

If $\mathcal{M}$ is a polymatroid, then the bases (resp. independent vectors) of $\mathcal{M}$ are precisely the lattice points of $P(\mathcal{M})$ (resp. $EP(\mathcal{M})$). The following proposition describes which bases of $\mathcal{M}$ correspond to vertices of $P(\mathcal{M})$.

Proposition 25 ([12], see also [27, Proposition 1.3]) *Let $\mathcal{M} = ([n], r)$ be a polymatroid and assign an ordering S to the ground set $[n]$. Let S_i be the first i elements according to this ordering. Every possible S corresponds (not necessarily uniquely) to a vertex of $P(M)$; $\mathbf{x} = \mathbf{x}_S$, where $\mathbf{x} = (x_1, \ldots x_n)$, and*

$$x_i = r(S_i) - r(S_{i-1}).$$

In particular, a polymatroid base polytope has at most $n!$ vertices.

We finish by slightly generalising the ideas of White [56]. As we will see later in Theorem 40 the statement below has geometric consequences. It was proven in [46, Corollary 46.2c] using different methods.

Theorem 26 *Let $\mathcal{M}_1, \ldots, \mathcal{M}_k$ be polymatroids on a ground set E with respective polytopes $P(\mathcal{M}_1), \ldots, P(\mathcal{M}_k)$. Then every lattice point $p \in P(\mathcal{M}_1) + \cdots + P(\mathcal{M}_k)$ is a sum $p = s_1 + \cdots + s_k$, where each s_i is a lattice point of $P(\mathcal{M}_i)$.*

Proof Proceeding by induction on k, it is enough to prove the theorem for $k = 2$.

Let us choose r large enough, so that $\mathcal{M}_1$ and $\mathcal{M}_2$ are r-polymatroids. We define a matroid $\tilde{M}_1$ (resp. $\tilde{M}_2$) on $E \times [r]$ as follows. Let $\pi_1 : E \times [r] \to E$ be the projection. A subset $A \subset E \times [r]$ is independent in $\tilde{M}_1$ (resp. $\tilde{M}_2$) if and only if for every subset $B \subset \pi_1(A)$ we have $r_1(B) \geq |A \cap (B \times [r])|$ (resp. $r_2(B) \geq |A \cap (B \times [r])|$), where r_1 (resp. r_2) is the rank function of the polymatroid $\mathcal{M}_1$ (resp. $\mathcal{M}_2$). Intuitively, an independent set in $\tilde{M}_j$ is an independent set I in $\mathcal{M}_j$ where we replace one point in E by as many points as the rank function dictates. We have natural surjections, $P(\tilde{M}_j) \to P(\mathcal{M}_j)$ and $P(\tilde{M}_1) + P(\tilde{M}_2) \to P(\mathcal{M}_1) + P(\mathcal{M}_2)$, coming from the projection π_1. Thus, it is enough to prove the statement for two matroids. From now on we assume that $\mathcal{M}_1$ and $\mathcal{M}_2$ are matroids.

Let $p \in P(M_1) + P(M_2)$ be a lattice point. We know that $p = \sum_i \lambda_i t_i + \sum_j \mu_j q_j$ with $\sum \lambda_i = 1, \sum \mu_j = 1$, for $\lambda_i, \mu_j \in \mathbb{Q}_{\geq 0}$, and that t_i (resp. q_j) are lattice points of $P(M_1)$ (resp. $P(M_2)$). After clearing the denominators we have

$$dp = \sum_i \lambda'_i t_i + \sum_j \mu'_j q_j,$$

where $\sum \lambda'_i = d, \sum \mu'_j = d$ and $\lambda'_i, \mu'_j \in \mathbb{Z}_{\geq 0}$.

By restricting the set E we may assume that all coordinates of $p = (p_1, \ldots, p_n)$ are nonzero (i.e. $p_i \in \{1, 2\}$), where we identify E with $[n]$. Dually, by contracting the elements of E that belong to all bases corresponding to any t_i and q_j, we may assume $p = (1, \ldots, 1)$.

We want to prove that the ground set E is covered by a basis of M_1 and a basis of M_2. Hence, by Corollary 18 it is sufficient to prove the following:

For any $A \subset E$ we have $|A| \leq r_{M_1}(A) + r_{M_2}(A)$.

We define a matroid N_1 (resp. N_2) on the ground set $E_N := \{(i, j) : i \in E, 1 \leq j \leq d\}$. In other words we replace any point of E by d equivalent points. A subset $\{(i_1, j_1), \ldots, (i_s, j_s)\} \subset E_N$ is independent in N_1 (resp. N_2) if only if

1. all i_q's are distinct, and
2. $\{i_1, \ldots, i_s\}$ is an independent set in M_1 (resp. M_2).

We have a natural projection $\pi : E_N \to E$ given by forgetting the second coordinate. We note that $r_{N_j}(\pi^{-1}(A)) = r_{M_j}(A)$ for $j = 1, 2$. As the point dp is decomposable we know that the set E_N can be covered by d bases of N_1 and d bases of N_2. Hence, for any $B \subset E_N$ we have: $|B| \leq dr_{N_1}(B) + dr_{N_2}(B)$. Applying this to $\pi^{-1}(A)$ we obtain:

$$d|A| = |\pi^{-1}(A)| \leq d \cdot r_{N_1}(\pi^{-1}(A)) + d \cdot r_{N_2}(\pi^{-1}(A)) = d \cdot \left(r_{M_1}(A) + r_{M_2}(A)\right).$$

After dividing by d we obtain the statement we wanted to prove. □

4.3.1 The Tutte Polynomial for Polymatroids

The Tutte polynomial for polymatroids is not nearly as well-studied as in the matroid case. We gave two examples of where it was considered in certain classes of polymatroids. We will now go into detail about one suggestion how to construct Tutte polynomial in full generality.

As mentioned, Cameron and Fink [8] form a polynomial having Tutte-like properties for polymatroids, which specialises to an evaluation of the Tutte polynomial when applied only to matroids. This is a construction which takes a polytopal, lattice-point-counting, approach as opposed to a straight combinatorial one. It is motivated by an alternative definition of the Tutte polynomial to those we have discussed so far.

Definition 27 Take a matroid $M = (E, r)$, and give E some ordering. Let B be a basis of M.

1. We say that $e \in E - B$ is *externally active* with respect to B if e is the smallest element in the unique circuit contained in $B \cup e$, with respect to the ordering on E.
2. We say that $e \in B$ is *internally active* with respect to B if e is the smallest element in the unique cocircuit in $(E - B) \cup e$.

A *cocircuit* is a minimal set among sets intersecting every basis. We will not be using this notion again in the article.

We will denote the number of internally active elements with respect to B with $I(B)$ and the number of externally active elements by $E(B)$. Then we have the following result.

Theorem 28 ([53])

$$T_M(x, y) = \sum_{B \in \mathcal{B}} x^{I(B)} y^{E(B)}.$$

Activity was generalised to hypergraphs by Kálmán in [29], where he proved that a formula similar to the one above does not hold for hypergraphs. This is due to the above sum not being independent of the edge ordering chosen, as is the case for matroids. The one-variable specialisations are, however, consistent. That is, $T(x, 0)$, $T(0, y)$ can be written in terms of activity generating functions for hypergraphic polymatroids. In [8], the authors show that this behaviour extends to all polymatroids given their own Tutte-like polynomial for polymatroids. Their construction is as follows.

Let Δ be the standard simplex in $\mathbb{R}^E$ of dimension equal to $|E| - 1$, and ∇ be its reflection through the origin. Construct the polytope given by the Minkowski sum $P(\mathcal{M}) + u\Delta + t\nabla$ where $\mathcal{M} = (E, r)$ is any polymatroid and $u, t \in \mathbb{Z}_{\geq 0}$. By [35, Theorem 7], the number of lattice points inside the polytope is a polynomial in t and u, of degree $\dim(P(\mathcal{M}) + u\Delta + t\nabla) = |E| - 1$. This polynomial is written in the form

$$Q_{\mathcal{M}}(t,u) := \#(P(\mathcal{M}) + u\Delta + t\nabla) = \sum_{i,j} c_{ij} \binom{u}{j}\binom{t}{i}.$$

Changing the basis of the vector space of rational polynomials gives the polynomial

$$Q'_{\mathcal{M}}(x,y) = \sum_{ij} c_{ij}(x-1)^i(y-1)^j$$

As mentioned, this specialises to the Tutte polynomial:

Theorem 29 ([8, Theorem 3.2]) *Let $M = (E, r)$ be a matroid. Then we have*

$$Q'_M(x,y) = \frac{x^{|E|-r(M)}y^{r(M)}}{x+y-1} \cdot T_M\left(\frac{x+y-1}{y}, \frac{x+y-1}{x}\right)$$

as an identification of rational functions.

Formulae can be given for $Q'_{\mathcal{M}}$ under the polymatroid generalisation of many standard matroid operations such as direct sum and duality, and in many cases the results are versions of those true for the Tutte polynomial. Most importantly, this polynomial also satisfies a form of deletion-contraction recurrence.

Theorem 30 ([8, Theorem 5.6]) *Let $\mathcal{M} = (E, r)$ be a polymatroid and take $a \in E(\mathcal{M})$. Let $\mathcal{N}_k$ be the convex hull of $\{p \in P(\mathcal{M}) \mid p_a = k\}$ for some $k \in \{0, \ldots, r(\mathcal{M})\}$, and $Q'_{\mathcal{N}}$ the polynomial formed by replacing the $\mathcal{M}$ in the above definition by $\mathcal{N}$. Then*

$$Q'_{\mathcal{M}}(x,y) = (x-1)Q'_{\mathcal{M}\backslash a}(x,y) + (y-1)Q'_{\mathcal{M}/a}(x,y) + \sum_k Q'_{\mathcal{N}_k}(x,y).$$

Remark 31 We mention another approach of generalising the Tutte polynomial to polymatroids. In the proof of Theorem 26, we explained how to associate to an r-polymatroid $\mathcal{M}$ on E a matroid M on $E \times [r]$. We could define the Tutte polynomial of $\mathcal{M}$ to be simply the usual Tutte polynomial of M. Of course, the result might depend on the chosen r. Still, it is natural to ask if there is any relation between this construction and the one described above.

One possible strategy for getting rid of this dependence on r is as follows: There is an action of S_r permuting the second factor of $E \times [r]$, and the sum in Definition 10 is constant on orbits, so the whole polynomial can be enriched to an S_r-representation. Our hope is that as r increases, this representation stabilises in some way, removing the dependency on r.

4.4 Flag Varieties: Geometry

In this section, we always work over the field of complex numbers.

4.4.1 Representations and Characters

We begin by fixing notation regarding the representation theory of GL_n. We refer the reader to [50, Chap. 4] for a brief introduction to the representation theory of GL_n, or to [19] for a more detailed account.

The (polynomial) irreducible representations of GL_n are in bijection with Young diagrams λ with at most n rows. We write $\lambda = (a_1, \ldots, a_k)$ for the Young diagram with rows of length $a_1 \geq a_2 \geq \ldots \geq a_k > 0$. The associated GL_n-representation is called a *Weyl module of highest weight* λ, and will be denoted by $\mathbb{S}^\lambda V$ (here V refers to the natural representation of GL_n, i.e. an n dimensional vector space with the linear $GL_n = GL(V)$ action). The usual construction of Weyl modules goes via Young symmetrisers, which we will not recall here. For readers not familiar with them, it suffices for now to know that $\mathbb{S}^{(a)} V = S^a V$, the a-th symmetric power of the natural representation, and that $\mathbb{S}^{(1,\ldots,1)} V$ (where 1 appears a times) is the exterior power $\bigwedge^a V$. We will give a full description later: see after Example 33.

In this article we will be interested in the action of a maximal torus $T \subset GL(V)$ on flag varieties. Such a torus $T \simeq (\mathbb{C}^*)^n$ may be identified with the diagonal non-degenerate matrices, after fixing a basis of V. We recall that a torus T acting on any vector space W induces a weight decomposition:

$$W = \bigoplus_{\mathbf{c} \in \mathbb{Z}^n} W_{\mathbf{c}},$$

where $(t_1 \ldots, t_n) \in T$ acts on $v \in W_{\mathbf{c}}$ by scaling as follows:

$$(t_1, \ldots, t_n)v = t_1^{c_1} \cdots t_n^{c_n} v.$$

In particular, an action of the torus T on a one-dimensional vector space $\mathbb{C}$ may be identified with a lattice point in $\mathbb{Z}^n$. We call $\mathbb{Z}^n = M$ the lattice of *characters* of T. An element $(a_1, \ldots, a_n) \in M$ is a character identified with the map:

$$T \ni (t_1, \ldots, t_n) \mapsto t_1^{a_1} \cdots t_n^{a_n} \in \mathbb{C}^*.$$

Any irreducible $GL(V)$-representation $W = \mathbb{S}^{c_1,\ldots,c_n} V$ decomposes as above under the action of T with a one-dimensional component $W_{c_1,\ldots,c_n}$; moreover all other components have a lexicographically smaller weight. This explains the name "Weyl module of highest weight $(c_1, \ldots, c_n)$."

4.4.2 Grassmannians

A basic example of a flag variety is a Grassmannian $G(k, V)$, which as a set parameterises k-dimensional subspaces of an n-dimensional space V. A point in $G(k, n)$ can be represented by a full-rank $k \times n$ matrix A, where our k-dimensional subspace is the row span of A. Two matrices A and B represent the same point in $G(k, n)$ if and only if they are the same up to elementary row operations.

$G(k, n)$ can be realised as an algebraic variety as follows:

$$G(k, n) = G(k, V) = \{[v_1 \wedge \ldots \wedge v_k] \subset \mathbb{P}(\bigwedge^k V)\}.$$

Here, $v_1, \ldots, v_k$ are the rows of the aforementioned matrix A, and thus a point of $G(k, n)$ is identified with the subspace $\langle v_1, \ldots, v_k \rangle$. The embedding presented above is known as the *Plücker embedding* and the Grassmannian is defined by quadratic polynomials known as *Plücker relations* [34]. Explicitly, in coordinates, the map associates to the matrix A the value of all $k \times k$ minors. We refer the readers not familiar with algebraic geometry, and in particular Grassmannians, to a short introduction in [39, Chap. 5].

The Plücker embedding may be identified with a very ample line bundle on $G(k, n)$, which we will denote by $\mathcal{O}(1)$. Other very ample line bundles on $G(k, n)$ are the d-th tensor powers $\mathcal{O}(d)$. They can be realised as a composition of the Plücker embedding with the d-th Veronese map $\mathbb{P}(\bigwedge^k V) \to \mathbb{P}(\mathrm{Sym}^d \bigwedge^k V)$.

Remark 32 A reader not familiar at all with very ample line bundles may think about them as maps into projective spaces. Let us present this with the example of the projective space $\mathbb{P}^n$ (which also equals $G(1, n+1)$). We have an identity map $\mathbb{P}^n \to \mathbb{P}^n$, which corresponds to $\mathcal{O}(1)$. The r-th *Veronese map* embeds $\mathbb{P}^n$ in a larger projective space $\mathbb{P}^{\binom{n+r}{n}-1}$ by evaluating on a point all degree r monomials. The associated map is given by $\mathcal{O}(r)$. For $n = 1$ and $r = 2$ we get:

$$\mathbb{P}^1 \ni [x : y] \to [x^2 : xy : y^2] \in \mathbb{P}^2.$$

It will follow from Proposition 34 that the embedding of $G(k, n)$ by $\mathcal{O}(d)$ spans a projectivisation of an irreducible representation V_{λ_0} of GL_n. The Young diagram $\lambda_0 = (d, \ldots, d)$ consists of k rows of length d.

4.4.3 Flag Varieties

More generally, for any irreducible representation V_λ of GL_n the projective space $\mathbb{P}(V_\lambda)$ contains a unique closed orbit, known as a homogeneous variety or more precisely a flag variety. To describe it, let us fix a sequence of s positive integers $0 < k_1 < \cdots < k_s < n$. The flag variety is defined as follows:

$$Fl(k_1, \dots, k_s; n) = \{V_1 \subset \dots \subset V_s \subset V : \dim V_i = k_i\}$$
$$\subset G(k_1, V) \times \dots \times G(k_s, V).$$

Hence, flag varieties are in bijection with (nonempty) subsets of $\{1, \dots, n-1\}$, while Grassmannians correspond to singletons.

From now on we will abbreviate the tuple $(k_1, \dots, k_s)$ to $\mathbf{k}$, and the flag variety $Fl(k_1, \dots, k_s; n)$ to $Fl(\mathbf{k}, n)$. A point in $Fl(\mathbf{k}, n)$ can be represented by a full-rank $n \times n$-matrix A: the row span of the first d_i rows is V_i. (Although note that only the first k_s rows of the matrix are relevant.) As with Grassmannians, different matrices can represent the same point in $Fl(\mathbf{k}, n)$. More precisely, if we partition the rows of A into blocks of size $k_1, k_2 - k_1, \dots, n - k_s$, then we are allowed to do row operations on A, with the restriction that to a certain row we can only add a multiple of a row in the same block or a block above. Another way to think about this is the following: let $P_{\mathbf{k}} \subset GL_n(\mathbb{C})$ be the parabolic subgroup of all invertible matrices A with $A_{ij} = 0$ if $i \leq k_r < j$, for some r. Then two $n \times n$ matrices represent the same flag if and only if they are the same up to left multiplication with an element of $P_{\mathbf{k}}$. Hence $Fl(\mathbf{k}, n)$ can also be described as the quotient ${}_{P_{\mathbf{k}}}\backslash^{GL_n(\mathbb{C})}$ (a homogeneous variety).

Just as for Grassmannians we may study different embeddings of flag varieties. The natural one is given by the containment

$$G(k_1, V) \times \dots \times G(k_s, V) \subset \mathbb{P}(\bigwedge^{k_1} V) \times \dots \times \mathbb{P}(\bigwedge^{k_s} V) \subset \mathbb{P}(\bigwedge^{k_1} V \otimes \dots \otimes \bigwedge^{k_s} V),$$

where the last map is the Segre embedding. The representation $\bigwedge^{k_1} V \otimes \dots \otimes \bigwedge^{k_s} V$ in general is reducible—a precise decomposition is known by Pieri's rule (or more generally by the Littlewood-Richardson rule), see for example [19, Proposition 15.25]. As we will prove below, the flag variety spans an irreducible representation with the corresponding Young diagram with s columns of lengths $k_s, k_{s-1}, \dots, k_1$ respectively.

Other embeddings can be obtained as follows. We replace the Segre map by the Segre–Veronese, i.e. we first re-embed a Grassmannian with a Veronese map. Thus, a flag variety with an embedding can be specified by a function:

$$f : \{1, \dots, n-1\} \to \mathbb{Z}_{\geq 0}.$$

To abstractly obtain a flag variety from f we first consider a subset $\{a \in [n-1] : f(a) > 0\}$. The Segre–Veronese map is specified by the values of the function f—the case $f(a) = 1$ corresponds to the usual Segre. The irreducible representation we obtain has an associated Young diagram with $f(a)$ columns of length a. Before proving all these statements we present an example.

Example 33 Let us fix $n = 4$ and a function f that on $1, 2, 3$ takes values $2, 1, 0$ respectively. The corresponding flag variety equals $Fl(1, 2; 4)$, i.e. it parameterises one-dimensional subspaces, l, inside two-dimensional subspaces, S, in a fixed four-dimensional space, V. The line l corresponds to a point in a projective space $\mathbb{P}(V) = G(1, V)$ and the space S to a point in $G(2, V)$. If we suppose $l = [v_1]$, then we may always find $v_2 \in V$ such that $S = \langle v_1, v_2 \rangle$. Hence

$$Fl(1, 2) = \{[v_1] \times [v_1 \wedge v_2] \in \mathbb{P}(V) \times G(2, V)\}.$$

We now pass to the embedding. As $f(1) = 2$ we have to consider the second Veronese map $\mathbb{P}(V) \to \mathbb{P}(S^2(V))$ given by $[v] \to [v \cdot v]$. We obtain

$$Fl(1, 2) = \{[v_1 \cdot v_1] \times [v_1 \wedge v_2] \in \mathbb{P}(S^2V) \times G(2, V)\} \subset \mathbb{P}(S^2V \otimes \bigwedge^2 V).$$

By Pieri's rule, we have a decomposition of $GL(V)$ representations:

$$S^2V \otimes \bigwedge^2 V = \mathbb{S}^{3,1}V \oplus \mathbb{S}^{2,1,1}V.$$

Hence, $\mathbb{S}^{3,1}(V)$ corresponds to the Young diagram with the first row of length three and the second row of length one. We note that this diagram indeed has 2 columns of length 1, 1 column of length 2, and 0 columns of length 3.

The flag variety is always contained in the lexicographically-first (highest weight) irreducible component—cf. Proposition 34 below; in our example this is $\mathbb{S}^{3,1}(V)$. In particular, we may realise the representation $\mathbb{S}^{3,1}(V)$ as a linear span of the affine cone over the flag variety:

$$\left\langle \widehat{Fl(1, 2)} = \{(v_1 \cdot v_1) \otimes (v_1 \wedge v_2) : v_1, v_2 \in V\}\right\rangle \subset S^2V \otimes \bigwedge^2 V.$$

For readers not familiar with the construction of $\mathbb{S}^\lambda V$, this can be taken as a definition. For a proof that this definition is equivalent to the usual construction, see Proposition 34 below.

The function f is often represented on a Dynkin diagram as shown in Fig. 4.2. For Example 33 this would be as shown in Fig. 4.3.

We note that if f is positive we obtain a complete flag variety, i.e. the variety parametrizing complete flags:

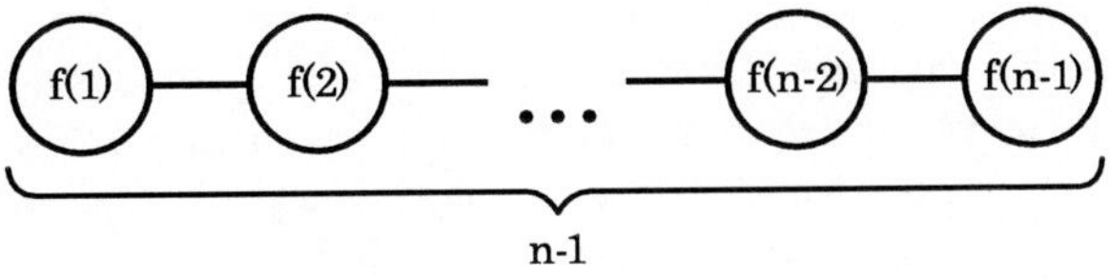

Fig. 4.2 The Dynkin diagram for the function f

Fig. 4.3 The Dynkin diagram for Example 33

$$V_1 \subset V_2 \subset \cdots \subset V_{n-1} \subset \mathbb{C}^n.$$

The complete flag variety maps to any other flag variety, simply by forgetting the appropriate vector spaces. We note that all our constructions are explicit and only use exterior (for Grassmannians), symmetric (for Veronese) and usual (for Segre) tensor products, as in Example 33.

We are now ready to prove a special case of the Borel-Weil-Bott Theorem relating representations and embeddings of flag varieties.

Proposition 34 (Borel-Weil) *Any flag variety $Fl(k_1, \ldots, k_s; n)$ with an embedding given by a function f spans the irreducible $GL(V)$-representation $\mathbb{S}^\lambda V$, where the Young diagram λ has $f(j)$ columns of length j.*

Proof Fix a basis $e_1, \ldots, e_n$ of V. Let us consider the flag of subspaces

$$\langle e_1, \ldots, e_{k_1} \rangle \subset \langle e_1, \ldots, e_{k_2} \rangle \subset \cdots \subset \langle e_1, \ldots, e_{k_s} \rangle$$

and the corresponding point $p \in Fl(k_1, \ldots, k_s)$. Under the embedding specified by f it is mapped to

$$(e_1 \wedge \cdots \wedge e_{k_1})^{\circ f(k_1)} \otimes \cdots \otimes (e_1 \wedge \cdots \wedge e_{k_s})^{\circ f(k_s)}$$
$$\subset S^{f(k_1)}\Big(\bigwedge^{k_1} V\Big) \otimes \cdots \otimes S^{f(k_s)}\Big(\bigwedge^{k_s} V\Big).$$

The $GL(V)$-decomposition of the ambient space is highly non-trivial. However, looking directly at the T decomposition we see that, up to scaling, the image of p is the unique lexicographically-highest vector. Hence, in particular, the image of p belongs to $\mathbb{S}^\lambda V$, as all other $GL(V)$-representations appearing in the decomposition have strictly smaller highest weights. Furthermore, the flag variety is an orbit under the $GL(V)$-action—one can explicitly write a matrix mapping any flag to any other given flag. Thus, if one point is contained in the irreducible representation, the whole variety must be contained in it.

It remains to show that the span of the flag variety is indeed the whole irreducible representation. This is true, as the flag variety is $GL(V)$-invariant, and thus its linear span is a representation of $GL(V)$. As $\mathbb{S}^\lambda V$ is irreducible, the linear span must coincide with it. □

The above theorem may be regarded as a realisation of irreducible $GL(V)$-representations as spaces of sections of a very ample line bundle on a flag variety. A more general Borel-Weil-Bott theorem provides not only a description of global sections—zeroth cohomology—but also higher, arbitrary cohomology.

Later on, we will denote a flag variety together with the embedding given by f as $Fl(\mathbf{k}; n)$, where $\mathbf{k} = (k_1, \ldots, k_s)$ satisfies $0 < k_1 \leq \ldots \leq k_s < n$ and has $f(a)$ entries equal to a, for all $a \in [n-1]$. For example, the embedding of Example 33 will be written as $Fl(1, 1, 2; 4)$.

4.5 Representable Matroids: Combinatorics and Geometry

Let us consider a representable matroid M given by $n = |E|$ vectors spanning a k-dimensional vector space V. By fixing a basis of V we may represent this matroid as a $k \times n$ matrix A. On the other hand the matrix A may be regarded as defining a k-dimensional subspace of an n-dimensional vector space, i.e. a point in $G(k, n)$. Since applying elementary row operations to A does not change which of the maximal minors of A vanish, the matroid M only depends on the k-dimensional subspace, and not on the specific matrix A representing our subspace. In this way we have associated to any point $p \in G(k, n)$ a representable rank-k matroid M_p on $[n]$.

Remark 35 In the literature, the correspondence between points in $G(k, n)$ and vector arrangements in $\mathbb{C}^k$ is known as the *Gel'fand-MacPherson correspondence*. The way we just constructed it is very explicit, but has the disadvantage of not being canonical (it depends on a chosen basis of $\mathbb{C}^n$). There are several ways to fix this.

One way of obtaining a more intrinsic construction is to replace the Grassmannian $G(k, n)$ by the Grassmannian $G(n-k, n)$. If A represents a surjective linear map from $\mathbb{C}^n$ to V, then to A one can associate the $(n-k)$-dimensional kernel of this map, i.e. a point in $G(n-k, n)$. As this construction requires dual matroids we decided to present the one above in coordinates.

A different (but closely related) intrinsic construction would be to define $G(k, \mathbb{C}^n)$ as the space of k-dimensional quotients of $\mathbb{C}^n$ (instead of k-dimensional subspaces). Then A maps the standard basis of $\mathbb{C}^n$ to n vectors in a smaller space $V \in G(k, \mathbb{C}^n)$; these n vectors represent a matroid.

A third solution would be to talk about hyperplane arrangements instead of vector arrangements.

The vector space $\mathbb{C}^n$ comes with the action of a torus $T = (\mathbb{C}^*)^n$. We have associated a point $p \in G(k, \mathbb{C}^n)$ to a representation of a matroid. If we change the representation by rescaling the vectors we do not change the matroid and the associated point belongs to the orbit Tp. Hence, the intrinsic properties of the matroid M_p should be related to the geometry of Tp—a feature we will examine in detail throughout the article. The closure $\overline{Tp}$ is a projective toric variety. For more information about toric geometry we refer to [9, 17, 38, 49].

Remark 36 Of course it can happen that different torus orbits give rise to the same matroid: there are only finitely many matroids on $[n]$, but if $1 < k < n-1$ there are infinitely many torus orbits in $G(k, n)$. In fact, the set of all points in $G(k, n)$ giving rise to the same matroid forms a so-called *thin Schubert cell* or *matroid stratum*, which typically is a union of infinitely many torus orbits. Thin Schubert cells were first introduced in [21]. Thin Schubert cells are badly behaved in general: for fixed $k \geq 3$ the thin Schubert cells of $G(k, n)$ exhibit arbitrary singularities if n is large enough. This is a consequence of Mnëv's theorem [41]. See [31, Sect. 1.8] for a more detailed discussion.

Theorem 37 ([21]) *The lattice polytope representing the projective toric variety $\overline{Tp}$ described above is equal to $P(M_p)$.*

Proof Let A be the matrix whose rows span the space corresponding to p. The parameterisation of $\overline{Tp}$ is given by:

$$\phi : T \to \mathbb{P}(\bigwedge^k \mathbb{C}^n).$$

The coordinates of the ambient space are indexed by k-element subsets of the n columns of the matrix A. The Plücker coordinate indexed by I of $\phi(t_1, \ldots, t_n)$ equals $\prod_{i \in I} t_i$ times the $k \times k$ minor of A determined by I, which we will denote by $\det(A_I)$. In other words, the map ϕ in Plücker coordinates is given as follows:

$$\phi(t_1, \ldots, t_n) = (\det(A_I) \cdot \prod_{i \in I} t_i)_{I \in \binom{[n]}{k}}.$$

The I-th coordinate is nonzero if and only if I is a basis of M_p. Hence, the ambient space of $\overline{Tp}$ has coordinates indexed by basis elements of M_p. After restricting to this ambient space and composing with the isomorphism inverting the nonzero minors $\det(A_I)$, our map can be written as

$$\phi(t_1, \ldots, t_n) = \left(\prod_{i \in I} t_i \right)_{I \in P(M_p)}.$$

This is exactly the construction of the toric variety represented by $P(M_p)$. □

It is a major problem to provide the algebraic equations of $\overline{Tp}$. This is equivalent to finding integral relations among the basis of a matroid. We point out that matroids satisfy a 'stronger' property then one could expect from the basis exchange axiom B2 of Lemma 3. Precisely, for any two bases $B_1, B_2 \in \mathcal{B}$ and a subset $A \subset B_1 - B_2$, there exists $A' \subset B_2 - B_1$ such that $(B_1 - A) \cup A'$ and $(B_2 - A') \cup A$ are in $\mathcal{B}$ [26]. This exactly translates to a binomial quadric (degree 2 polynomial) in the ideal of $\overline{Tp}$: $x_{B_1}x_{B_2} - x_{(B_1-A)\cup A'}x_{(B_2-A')\cup A}$, where, as in the proof of Theorem 37, we label each coordinate by a basis of the matroid. Further, if $|A| = 1$ we obtain special quadrics corresponding to exchanging one element in a pair of bases. The following conjecture due to White provides a full set of generators for any matroid M.

Conjecture 38 The ideal of the toric variety represented by $P(M)$ is generated by the special quadrics corresponding to exchanging one element in a pair of bases.

We note that it is unknown whether the ideal of this toric variety is generated by quadrics, or that all quadrics are spanned by the special quadrics described above. However, it is known that the special quadrics define the variety as a set (or more precisely as a projective scheme) [32, 33].

The combinatorial methods can be used to prove geometric properties of torus orbit closures in Grassmannians. Below we recall a definition of a normal lattice polytope.

Definition 39 A lattice polytope P, containing 0 and spanning (as a lattice) the lattice N, is *normal* if and only if for any $k \in \mathbb{Z}_{\geq 0}$ and any $p \in kP \cap N$ we have $p = p_1 + \cdots + p_k$ for some $p_i \in P \cap N$.

Normality of a polytope is a very important notion as it corresponds to *projective normality* of the associated toric variety [9, Chap. 2], [49] (less formally, the associated toric variety is not very singular and is embedded in a particularly nice way in the projective space).

Theorem 40 (White) *For any matroid M the polytope $P(M)$ is normal. In particular, any torus orbit closure in any Grassmannian is projectively normal.*

Proof This is a special case of Theorem 26, where we take all M_i equal to M. □

4.6 Introduction to Flag Matroids

In Sect. 4.5 we explained a correspondence between torus orbits in a Grassmannian (geometric objects) and representable matroids (combinatorial objects). We will generalise this correspondence in different ways. For instance, on the geometry side, we can replace Grassmannians with flag varieties. On the combinatorics side, this naturally leads to the notion of a (representable) *flag matroid*. Flag matroids first arose as a special case of the so-called *Coxeter matroids*, introduced by Gel'fand and Serganova [22, 23]. In this section we first give a combinatorial introduction to flag matroids. Afterwards, we explain how they are related to flag varieties. The exposition is largely based on [3, Chap. 1].

4.6.1 Flag Matroids: Definition

We start by defining flag matroids in the way they are usually defined in the literature: using Gale orderings.

Definition 41 Let $0 < k_1 \leq \ldots \leq k_s < n$ be natural numbers. Let $\mathbf{k} = (k_1, \ldots, k_s)$. A *flag F of rank* $\mathbf{k}$ *on E* is an increasing sequence

$$F^1 \subseteq F^2 \subset \cdots \subseteq F^s$$

of subsets of E such that $|F^i| = k_i$ for all i. The set of all such flags will be denoted by $\mathcal{F}_E^{\mathbf{k}}$.

Let ω be a linear ordering on E. We can extend the Gale ordering $\leq_\omega$ to flags:

$$(F^1, \ldots, F^s) \leq_\omega (G^1, \ldots, G^s) \text{ if and only if } F^i \leq_\omega G^i \text{ for all } i.$$

Definition 42 A *flag matroid of rank* $\mathbf{k}$ *on* E is a collection $\mathcal{F}$ of flags in $\mathcal{F}_E^{\mathbf{k}}$, which we call *bases*, satisfying the following property: for every linear ordering ω on E, the collection $\mathcal{F}$ contains a unique element which is maximal in $\mathcal{F}$ with respect to the Gale ordering $\leq_\omega$.

If $\mathcal{F}$ is a flag matroid, the collection $\{F^i \mid F \in \mathcal{F}\}$ is called the i*-th constituent* of $\mathcal{F}$. This is clearly a matroid (of rank k_i).

Remark 43 In the literature it is usually required that we have strict inequalities $0 < k_1 < \cdots < k_s < n$. From a combinatorial point of view this does not make a difference, but when we later consider flag matroid polytopes this restriction would appear artificial. This is also the reason why in Sect. 4.4.3 we did not just consider flag varieties, but also their Veronese re-embeddings.

Next, we want to describe which tuples of matroids can arise as the constituents of a flag matroid. In order to give this characterisation, we first need to recall *matroid quotients*.

4.6.2 Matroid Quotients

Definition 44 Let M and N be matroids on the same ground set E. We say that N is a *quotient* of M if one of the following equivalent statements holds:

1. every circuit of M is a union of circuits of N;
2. if $X \subseteq Y \subseteq E$, then $r_M(Y) - r_M(X) \geq r_N(Y) - r_N(X)$;
3. there exists a matroid R and a subset X of $E(R)$ such that $M = R\backslash X$ and $N = R/X$;
4. for all bases B of M and all $x \notin B$, there is a basis B' of N with $B' \subseteq B$ and such that $\{y : (B' - y) \cup x \in \mathcal{B}(N)\} \subseteq \{y : (B - y) \cup x \in \mathcal{B}(M)\}$.

For the equivalence of 1, 2 and 3, we refer to [43, Proposition 7.3.6]. Part 4 is left to the reader.

Here are some basic properties of matroid quotients:

Proposition 45 *Let* N *be a quotient of* M.

1. *Every basis of* N *is contained in a basis of* M*, and every basis of* M *contains a basis of* N.
2. $\operatorname{rk}(N) \leq \operatorname{rk}(M)$ *and in case of equality* $N = M$.

Proof Both statements can be easily deduced by plugging in $Y = E$ or $X = \varnothing$ in Definition 44 2. □

The next result will be essential for defining representable flag matroids. It also explains where the term "matroid quotient" comes from—below we think of W as a vector space quotient of V.

Proposition 46 ([6, Proposition 7.4.8 (2)]) *Let V and W be vector spaces and $\psi : E \to V$ be a map. Furthermore, let $f : V \to W$ be a linear map. Consider the matroid M represented by ψ, and the matroid N represented by $f \circ \psi$. Then N is a matroid quotient of M.*

Example 47 If R is a representable matroid on E and X is a subset of E, then $M := R \backslash X$ and $N := R/X$ are representable matroids, and there is a linear map as in Proposition 46. Indeed, if R is represented by $\psi : E \to V$, then consider the projection $\pi : V \to V/\langle \psi(X) \rangle$. It is not hard to see that M is represented by $\psi|_{E-X}$ and that N is represented by $\pi \circ \psi|_{E-X}$.

Example 48 The converse of Proposition 46 is false: we now give an example (taken from [3, Sect. 1.7.5]) of two representable matroids M and N such that N is a quotient of M, but there is no map as in Proposition 46.

Let M be the rank-3 matroid on [8] represented by the following matrix

$$\begin{pmatrix} 1&0&1&0&1&1&0&1 \\ 0&1&1&0&2&2&2&1 \\ 0&0&0&1&1&2&1&1 \end{pmatrix}$$

and let N be the rank-2 matroid on [8] whose bases are all 2-element subsets except for $\{2, 6\}$ and $\{3, 5\}$. It is easy to see that N is a representable matroid: just pick six pairwise independent vectors in the plane, and map 2 and 6, as well as 3 and 5, to the same vector. Now N is a matroid quotient of M, since the matroid R from Example 8 satisfies $M = R\backslash 9$ and $N = R/9$. However, it is not possible to find representations V (resp. W) of M (resp. N) such that there is a map $f : V \to W$ as in Proposition 46. Roughly speaking, the problem is that the "big" matroid R is not representable. For a more precise argument, see [3, Sect. 1.7.5].

4.6.3 Representable Flag Matroids

We now come to the promised characterisation of constituents of flag matroids. In fact, it will turn out we can use it as an alternative definition of flag matroids.

A collection $(M_1, \dots, M_s)$ of matroids is called *concordant* if, for every pair (M_i, M_j), either M_i is a quotient of M_j or vice versa. Note that this is equivalent to the fact that they can be ordered in such a way that M_i is a quotient of M_{i+1}, because "being a quotient of" is transitive.

Theorem 49 ([3, Theorem 1.7.1]) *A collection $\mathcal{F}$ of flags in $\mathcal{F}_E^{\mathbf{k}}$ is a flag matroid if and only if the following three conditions hold:*

1. *every constituent $M_i := \{F^i \mid F \in \mathcal{F}\}$ is a matroid;*
2. *the matroids $M_1, \dots, M_s$ are concordant;*
3. *every flag $B_1 \subseteq \dots \subseteq B_s$, with B_i a basis of M_i, is in $\mathcal{F}$.*

In other words, flag matroids on E are in one-to-one correspondence with tuples of concordant matroids on E.

We can now define representable flag matroids: let $0 \subsetneq V_1 \subseteq \cdots \subseteq V_s \subsetneq \mathbb{C}^n$ be a flag of subspaces of $\mathbb{C}^n$. Then, viewing V_i as a point in $G(k_i, n)$, it holds that $(M_{V_1}, \ldots, M_{V_s})$ is a concordant collection of matroids by Proposition 46. Indeed: if $V_i \subseteq V_j$, we can pick a $k_j \times n$ matrix A_j representing V_j such that the first k_i rows of A_j span V_i. Then the columns of A_i are obtained from the columns of A_j by deleting the last $k_j - k_i$ entries.

Definition 50 The *representable flag matroid* $\mathcal{F}(V_1 \subseteq \cdots \subseteq V_s)$ is the unique flag matroid whose constituents are $M_{V_1}, \ldots, M_{V_s}$.

Remark 51 Example 48 shows that it can happen that all constituents of a flag matroid are representable matroids, but still the flag matroid is not representable (because the matroid representations are "not compatible").

4.6.4 Flag Matroid Polytopes

Definition 52 Given a flag F on $[n]$, we can identify each constituent with a 0, 1-vector and then add them all up to a vector $\mathbf{e}_F \in \mathbb{Z}^n_{\geq 0}$. In this way we have identified the flags in $[n]$ of rank $\mathbf{k}$ with integer vectors with k_1 entries equal to m, $k_2 - k_1$ entries equal to $m - 1$, ..., $k_{i+1} - k_i$ entries equal to $m - i$, ..., $k_s - k_{s-1}$ entries equal to 1, and $n - k_s$ entries equal to 0. We will refer to such vectors as *rank-*$\mathbf{k}$ *vectors*. In other words, if we think of $\mathbf{k}$ as a partition of length s, we can consider the conjugate partition $\mathbf{k}^*$ of length $\leq n$. Then a rank-$\mathbf{k}$ vector is a vector $v \in \mathbb{Z}^n_{\geq 0}$ obtained from $\mathbf{k}^*$ by adding 0's and permuting the entries.

Definition 53 The *base polytope* of a flag matroid $\mathcal{F}$ on $[n]$ is the convex hull of the set $\{\mathbf{e}_F \mid F \in \mathcal{F}\} \subset \mathbb{R}^n$.

Example 54 Let $\mathcal{F}$ be the rank $(1, 2)$ flag matroid on [3] whose bases are $1 \subseteq 12$, $1 \subseteq 13$, $2 \subseteq 12$ and $3 \subseteq 13$. Then its base polytope is the convex hull of the points $(2, 1, 0)$, $(2, 0, 1)$, $(1, 2, 0)$, $(1, 0, 2)$. Its constituents are the uniform rank 1 matroid on [3], and the rank 2 matroid with bases 12 and 13. The base polytope of this flag matroid is depicted in Fig. 4.4. $\mathcal{F}$ is a representable flag matroid: a representing flag is for example $\langle \mathbf{e}_1 + \mathbf{e}_2 + \mathbf{e}_3 \rangle \subset \langle \mathbf{e}_1, \mathbf{e}_2 + \mathbf{e}_3 \rangle \subset \mathbb{C}^3$.

Theorem 55 ([3, Theorem 1.11.1]) *A lattice polytope* $P \subset \mathbb{R}^n$ *is the base polytope of a rank-*$\mathbf{k}$ *flag matroid on* $[n]$ *if and only if the following two conditions hold:*

1. *every vertex of* P *is a rank-*$\mathbf{k}$ *vector;*
2. *every edge of* P *parallel to* $\mathbf{e}_i - \mathbf{e}_j$ *for some* $i, j \in [n]$.

Theorem 56 ([3, Corollary 1.13.5]) *The polytope of a flag matroid is the Minkowski sum of the matroid polytopes of its constituent matroids.*

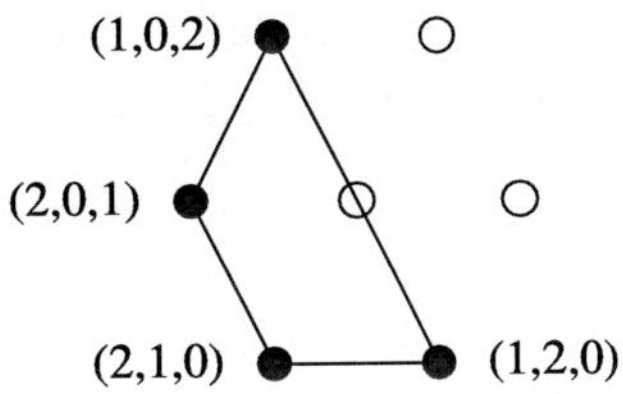

Fig. 4.4 A flag matroid base polytope

Thus, each flag matroid defines a polytope that is a base polytope of a polymatroid.

Remark 57 It follows from Theorem 26 and the previous theorem that the lattice points of a flag matroid polytope correspond to tuples (not necessarily flags) $(B_1, \ldots, B_s)$, where B_i is a basis of M_i. For example point $(1, 1, 1)$ in Fig. 4.4 corresponds to a basis of a polymatroid, but not to a flag, i.e. not to a basis of the flag matroid.

4.6.5 Flag Matroids and Torus Orbits

Consider the flag variety $Fl(\mathbf{k}, n)$, as described in Sect. 4.4. The action of the torus $T = (\mathbb{C}^*)^n$ on $\mathbb{C}^n$ induces an action of T on $Fl(\mathbf{k}, n)$. A point $p \in Fl(\mathbf{k}, n)$ gives rise to a representable flag matroid M on $[n]$, as in Definition 50. All points in the orbit Tp give rise to the same flag matroid. This last statement follows easily from the analogous fact for matroids and the fact that a flag matroid is determined by its constituent matroids. The analogue of Theorem 37 holds:

Theorem 58 *The lattice polytope representing the toric variety $\overline{Tp}$ is equal to the flag matroid polytope of M.*

Proof The proof is a straightforward generalisation of the proof of Theorem 37, with the parameterisation of $\overline{Tp}$ given by:

$$\phi : T \to \mathbb{P}(\bigwedge^{k_1} \mathbb{C}^n) \times \cdots \times \mathbb{P}(\bigwedge^{k_s} \mathbb{C}^n).$$

4.7 Representable Polymatroids

Representable polymatroids generalise representable matroids, replacing vectors with subspaces. While they do not appear as frequently in pure mathematics as matroids, one of the possible references for an interested reader is [4, Sect. 3].

We start by giving a precise definition.

Definition 59 Let V be a vector space and denote by $S(V)$ the set of all subspaces of V. Suppose we have a map $\phi : E \to S(V)$ and assume, without loss of generality, that $\sum_{e \in E} \phi(e) = V$. The *representable polymatroid* $\mathcal{M}(\phi)$ is defined by the rank function: for $A \subseteq E$, define $r(A)$ as the dimension of $\phi(A) := \sum_{a \in A} \phi(a)$.

Example 60 Consider the map $\phi : [3] \to S(\mathbb{C}^3)$ defined by $\phi(1) = \langle \mathbf{e}_1, \mathbf{e}_2 \rangle$ and $\phi(2) = \phi(3) = \langle \mathbf{e}_1, \mathbf{e}_3 \rangle$. Then $\mathcal{M}(\phi)$ is a rank 3 polymatroid on [3] with 5 bases: $(2, 1, 0), (2, 0, 1), (1, 2, 0), (1, 0, 2), (1, 1, 1)$. Its base polytope is given in Fig. 4.4.

Now we want to define the polymatroid analogue of a matroid associated to a subspace. For this, we fix a number $r \in \mathbb{Z}_{>0}$, and we consider the vector space $\mathbb{C}^{rn}$, with a fixed basis indexed by $[r] \times [n]$. If we consider a full rank $k \times rn$ matrix A, it represents a k-dimensional subspace of $\mathbb{C}^{rn}$. Moreover, the r columns indexed by (i, j) for some fixed j span a subspace W_j of $\mathbb{C}^k$. Then the map $[n] \to S(\mathbb{C}^k) : j \to W_j$ defines a representable polymatroid. Note that this is an r-polymatroid, and that every representable r-polymatroid can be obtained in this way.

As before, applying elementary row operations to A does not change the polymatroid. Hence, we have described how to associate an r-polymatroid $\mathcal{M}(W)$ to a subspace W of $\mathbb{C}^{rn}$.

Remark 61 If we defined Grassmannians via quotients instead of subspaces, we could present this story more invariantly, as follows: Let V be a vector space of dimension rn, with a fixed collection of n independent r-dimensional subspaces $V_1, \ldots V_n$. Then any k-dimensional quotient $\phi : V \to W$ gives rise to an r-polymatroid via the map $[n] \to S(W) : i \mapsto \phi(V_i)$.

We consider the action of $T = (\mathbb{C}^*)^n$ on $\mathbb{C}^{rn}$, where $\mathbf{t} = (t_1, \ldots, t_n)$ acts by multiplying the (i, j)-th coordinate by t_j. If we consider the action of $T = (\mathbb{C}^*)^n$ on $G(k, \mathbb{C}^{rn})$ induced by the action on $\mathbb{C}^{rn}$, then every point in the orbit Tp gives the same polymatroid. This is very similar to the situation for representable matroids, so it should be no surprise that we also have an analogue of Theorem 37 in this setting:

Theorem 62 *The lattice polytope representing the projective toric variety* $\overline{Tp}$ *described above is equal to the base polytope of the polymatroid* $\mathcal{M}$.

Proof It is a straightforward generalisation of the proof of Theorem 37. □

4.7.1 Comparison Between Polymatroids and Flag Matroids

A flag matroid polytope is a special case of a polymatroid polytope, by Theorem 55. It is tempting to think about flag matroids as a special case of polymatroids, but we need to be careful when doing this: the notion of a basis of a flag matroid is not compatible with the notion of a basis for a polymatroid. More precisely, the bases of a polymatroid are all lattice points of its base polytope, while the bases of a flag matroid are only the vertices of the associated polytope—cf. Remark 57 and Fig. 4.4.

For a flag matroid $\mathcal{F}$, let $\mathcal{M}(\mathcal{F})$ be the associated polymatroid.

The definitions of representable flag matroid and representable polymatroid look unrelated at first sight, but there is a connection.

Proposition 63 *If $\mathcal{F}$ is a representable flag matroid, then $\mathcal{M}(\mathcal{F})$ is a representable polymatroid.*

Proof For this proof we will use the more intrinsic definition of representable flag matroids and polymatroids using quotients. Let $\mathcal{F}$ be the flag matroid represented by the flag of quotients $\mathbb{C}^n \to V_1 \to \cdots \to V_r$. Then the quotient $\mathbb{C}^{kn} \to V_1 \bigoplus \cdots \bigoplus V_r$ represents a polymatroid $\mathcal{M}$ on $[n]$. We need to argue that $\mathcal{F}$ and $\mathcal{M}$ have the same base polytope.

Bases of $\mathcal{F}$ correspond to flags $[n] \supsetneq F^1 \supseteq \ldots \supseteq F^r$ such that F^i gives a basis of V_i. On the other hand, choosing a basis of $\mathcal{M}$ corresponds to choosing for every i a subset $F^i \subsetneq [n]$ such that F^i gives a basis of V_i. From this it follows that every vertex of $P(\mathcal{F})$ is a lattice point of $P(\mathcal{M})$, and (using Remark 57) that every lattice point of $P(\mathcal{M})$ is a lattice point of $P(\mathcal{F})$. So $P(\mathcal{M}) = P(\mathcal{F})$ as desired. □

Remark 64 Thinking again about flag varieties in terms of subspaces, we have that if $\mathcal{F}$ is represented by a flag $V_1 \subseteq \ldots \subseteq V_r \subsetneq \mathbb{C}^n$, then $\mathcal{M}(\mathcal{F})$ is represented by a subspace $V_1 \oplus \cdots \oplus V_r \subsetneq \mathbb{C}^{rn}$. Geometrically, this construction corresponds to an algebraic map

$$Fl(k_1, \ldots, k_r; n) \to G(\sum_i k_i, rn).$$

Remark 65 The converse of Proposition 63 is not true: given a representable polymatroid that is also a flag matroid, it is not always a representable flag matroid. One way to construct a counterexample is as follows. If M_1, M_2 are representable matroids with corresponding base polytopes P_1, P_2, then $P_1 + P_2$ corresponds to a representable polymatroid, by the same argument as in the proof of Proposition 63. However, Example 48 gives an example of two such (concordant) matroids such that $P_1 + P_2$ corresponds to a flag matroid that is not representable.

4.8 Equivariant K-theory

We have presented a correspondence between representable matroids and torus orbits in Grassmannians, and generalisations of this correspondence to representable flag matroids and representable polymatroids. We would like to drop the word "representable" from all of those. As we will see, one way to do this is by replacing "torus orbits" with "classes in equivariant K-theory". This was done for matroids by Fink and Speyer [16]. In this section, we review their construction, and consider generalisations to flag matroids. We start with an introduction to non-equivariant and equivariant K-theory.

4.8.1 A Very Brief Introduction to K-theory

This section is based on [18, Sect. 15.1].

Let X be an algebraic variety. We define $K^0(X)$ to be the free abelian group generated by vector bundles on X, subject to relations $[E] = [E'] + [E'']$ whenever E' is a subbundle of E, with quotient bundle $E'' = E/E'$. The group $K^0(X)$ inherits a ring structure from the tensor product: $[E] \cdot [F] = [E \otimes F]$.

Similarily, we can define $K_0(X)$ to be the free abelian group generated by isomorphism classes of coherent sheaves on X, subject to relations $[A] + [C] = [B]$ whenever there is a short exact sequence $0 \to A \to B \to C \to 0$. There is an inclusion $K^0(X) \hookrightarrow K_0(X)$. From now on, we will always assume that X is a smooth variety. In this case, the inclusion is an isomorphism, allowing us to identify $K^0(X)$ and $K_0(X)$.

Let $f : X \to Y$ be a map of (smooth) varieties. Then there is a pullback map $f^* : K^0(Y) \to K^0(X)$ defined by $f^*[E] = [f^*E]$ (where E is a vector bundle on Y). If f is a proper map, there is also a pushforward map $f_* : K_0(X) \to K_0(Y)$ given by $f_*[A] = \sum (-1)^i [R^i f_* A]$. Here $R^i f_*$ are right derived functors of the pushforward. An interested reader is advised to find the details in [18, Sect. 15]. In this paper, we will not be using the formal definitions of $K^0(X)$, f^* or f_*. Instead, we will refer to explicit descriptions of those in the cases that we need, each time providing a theorem we build upon.

Remark 66 In all the cases we study the ring $K^0(X)$ is isomorphic to the cohomology ring and to the Chow ring (after tensoring with $\mathbb{Q}$). Note however that the map from $K^0(X)$ to the Chow ring is nontrivial and given by the Chern character.

Example 67 Consider the projective space $\mathbb{P}^n$. The (rational) Chow ring is $A(\mathbb{P}^n) = \mathbb{Q}[H]/(H^{n+1})$. Here one should think about H as a hyperplane in $\mathbb{P}^n$ and H^k as a codimension k projective subspace. The most important line bundle is $\mathcal{O}(1)$. The Chern character $ch : K^0(\mathbb{P}^n) \to A(\mathbb{P}^n)$ sends $[\mathcal{O}(1)]$ to $\sum_{i=0}^n H^i/i!$. Note that $K^0(\mathbb{P}^n)$ can be written as $\mathbb{Z}[\alpha]/(\alpha^{n+1})$, where $\alpha = 1 - [\mathcal{O}(-1)]$ is the class of the structure sheaf of a hyperplane. As a special case, the K-theory of a point is $\mathbb{Z}$.

If X is a smooth variety equipped with an action of a torus T, we can define its *equivariant K-theory* $K_T^0(X) \cong K_0^T(X)$. The construction and properties are exactly the same as in the previous paragraphs, if we replace "vector bundles" and "coherent sheaves" by "T-equivariant vector bundles" and "T-equivariant coherent sheaves".

For later reference, we describe the equivariant K-theory of a point: $K_T^0(pt) = \mathbb{Z}[\mathrm{Char}(T)]$, where $\mathrm{Char}(T) = \mathrm{Hom}(T, \mathbb{C}^*)$ is the lattice of characters of T. Here $\mathbb{Z}[\mathrm{Char}(T)]$ is the *group ring* of $\mathrm{Char}(T)$, i.e. as a module over $\mathbb{Z}$ it has a basis given by $\mathrm{Char}(T)$, and multiplication is induced from addition in $\mathrm{Char}(T)$. It is the ring of Laurent polynomials in $\dim T$ variables. Explicitly, a T-equivariant sheaf on pt is just a vector space W with a T-action. We may decompose $W = \bigoplus_{\mathbf{c} \in \mathrm{Char}(T)} W_{\mathbf{c}}$ as in Sect. 4.4.1. The corresponding element of $\mathbb{Z}[\mathrm{Char}(T)]$ is the character (also called *Hilbert series*) $\mathrm{Hilb}(W) := \sum_{\mathbf{c}\in\mathrm{Char}(T)}(\dim W_{\mathbf{c}})\mathbf{c}$. We point out that even for infinite-dimensional T-modules, $\mathrm{Hilb}(W)$ makes sense as a formal power series, as long as $W_{\mathbf{c}}$ is finite-dimensional for all $\mathbf{c}$.

We finish this section by describing the relation between ordinary and T-equivariant K-theory:

Theorem 68 ([37, Theorem 4.3]) *Let X be a smooth projective variety with an action of a torus T. Let $S \subseteq T$ be a subtorus. Then the natural map*

$$K_T^0(X) \otimes_{\mathbb{Z}[\mathrm{Char}(T)]} \mathbb{Z}[\mathrm{Char}(S)] \to K_S^0(X)$$

is an isomorphism. In particular, taking S to be the trivial group, the natural map

$$K_T^0(X) \otimes_{\mathbb{Z}[\mathrm{Char}(T)]} \mathbb{Z} \to K^0(X)$$

is an isomorphism.

We note that the map $\mathbb{Z}[\mathrm{Char}(T)] \to \mathbb{Z}$ above is given in coordinates by sending each generator t_i of T to 1.

4.8.2 Explicit Construction via Equivariant Localisation

Let X be a smooth projective variety over $\mathbb{C}$, and T a torus acting on it. If X has only finitely many torus-fixed points, we can use the method of *equivariant localisation* to give an explicit combinatorial description of classes in $K_T^0(X)$. Our exposition here is largely based on the one in [16]. The following theorem is central to our discussion.

Theorem 69 ([42, Theorem 3.2], [16, Theorem 2.5] and references therein) *If X is a smooth projective variety with a torus action, then the restriction map $K_T^0(X) \to K_T^0(X^T)$ is an injection.*

From now on we will always assume that X has only finitely many torus-fixed points. In this case $K_T^0(X^T)$ is simply the ring of functions from X^T to $\mathbb{Z}[\mathrm{Char}(T)]$. In other words, we can describe a class in $K_T^0(X)$ just by giving a finite collection of Laurent polynomials in $\mathbb{Z}[\mathrm{Char}(T)]$.

Remark 70 In the literature, a variety X for which $K_T^0(X)$ is a free $\mathbb{Z}[\mathrm{Char}(T)]$-module, and has a $\mathbb{Z}[\mathrm{Char}(T)]$-basis that restricts to a $\mathbb{Z}$-basis of $K^0(X)$, is called *equivariantly formal*. This notion was first introduced in [24]. In [2, Sect. 2.4], it is noted that smooth projective varieties with finitely many T-fixed points are equivariantly formal.

We now explicitly describe the class of a T-equivariant coherent sheaf on X. We will do this under the following additional assumption (which is not essential but makes notation easier and will hold for all varieties of interest):

Definition 71 A finite-dimensional representation of T is called *contracting* if all characters lie in an open halfspace, or equivalently if the characters generate a pointed cone (see Sect. 4.8.3). The action of T on a variety X is *contracting*, if for every torus-fixed point $x \in X$, there exists an open neighbourhood U_x isomorphic to $\mathbb{A}^N$ such that the action of T on U_x is a contracting representation.

Let E be a T-equivariant coherent sheaf on X. We will construct a map $[E]^T : X^T \to \mathbb{Z}[\mathrm{Char}(T)]$. For every $x \in X^T$, we have an open neighbourhood U_x as in Definition 71. Let $\chi_1, \ldots, \chi_N$ be the characters by which T acts on U_x (so $\mathcal{O}(U_x)$ is a polynomial ring multigraded by T in the sense of [40, Definition 8.1], with characters $\chi_1^{-1}, \ldots, \chi_N^{-1}$). Our sheaf E, restricted to U_x, corresponds to a graded, finitely generated $\mathcal{O}(U_x)$-module $E(U_x)$.

Since $E(U_x)$ is a graded module over the polynomial ring $\mathcal{O}(U_x)$, which is multigraded by T, it follows from [40, Theorem 8.20] that $E(U_x)$ is a T-module, and its Hilbert series is of the form

$$\frac{K(E(U_x), \mathbf{t})}{\prod_{i=1}^{N} (1 - \chi_i^{-1})}, \tag{4.1}$$

for some $K(E(U_x), \mathbf{t}) \in \mathbb{Z}[\mathrm{Char}(T)]$.

Definition 72 For E a T-equivariant coherent sheaf on X, we define $[E]^T$ to be the map that sends $x \in X^T$ to $K(E(U_x), \mathbf{t}) \in \mathbb{Z}[\mathrm{Char}(T)]$, the numerator in (4.1).

Theorem 73 ([16, Theorem 2.6]) *The map $[E]^T$ defined above is the image of the class of E under the injection $K_T^0(X) \hookrightarrow K_T^0(X^T)$ of Theorem 69.*

Example 74 Let $X = \mathbb{P}^n$, equipped with the natural torus action $\mathbf{t} \cdot [a_0 : \ldots : a_n] = [t_0^{-1} a_0 : \ldots : t_n^{-1} a_n]$. Then $\mathcal{O}(d)$ is a T-equivariant sheaf. The torus action on $\mathbb{P}^n$ has $n + 1$ fixed points, namely $p_i = [0 : \ldots : 1 : \ldots : 0]$, where the 1 is at position i. We use equivariant localisation to describe the class $[\mathcal{O}(d)]^T$.

Every p_i has an open neighbourhood $U_i = \mathrm{Spec}\, A_i$, where

$$A_i = \mathbb{C}[x_0, \ldots, \hat{x}_i, \ldots, x_n]$$

is multigraded by T via $\deg(x_j) = t_i^{-1} t_j$. The A_i-module $\mathcal{O}(d)(U_i)$ is a copy of A_i generated in degree t_i^d. So its Hilbert series is $t_i^d / \prod_j (1 - t_i^{-1} t_j)$. Hence $[\mathcal{O}(d)]^T$ can be represented by the map $(\mathbb{P}^n)^T \to \mathbb{Z}[\mathrm{Char}(T)] : p_i \mapsto t_i^d$.

We can describe the image of the map from Theorem 69 explicitly, if we impose an additional condition on X.

Theorem 75 ([54, Corollary 5.12], [16, Theorem 2.9] and references therein) *Suppose X is a projective variety with an action of a torus T, such that X has finitely many T-fixed points and finitely many 1-dimensional T-orbits, each of which has closure isomorphic to $\mathbb{P}^1$. Then a map $f : X^T \to \mathbb{Z}[\mathrm{Char}(T)]$ is in the image of the map $K_T^0(X) \to K_T^0(X^T)$ of Theorem 69 if and only if the following condition holds:*

For every one-dimensional orbit, on which T acts by character χ and for which x and y are the T-fixed points in the orbit closure, we have

$$f(x) \equiv f(y) \mod 1 - \chi.$$

Example 76 (*Example* 74 *continued*) Note that $\mathbb{P}^n$ has only finitely many one-dimensional torus orbits: for every pair p_i, p_j of T-fixed points, there is a unique T-orbit whose closure contains p_i and p_j. Furthermore, T acts on this orbit with character $t_j^{-1}t_i$. We see that $t_i^d \equiv t_j^d \mod 1 - t_j^{-1}t_i$, so that the class $[\mathcal{O}(d)]^T$ indeed fulfills the condition of Theorem 75.

We also can describe pullback and pushforward in the language of equivariant localisation. Let $\pi : X \to Y$ be a T-equivariant map of smooth projective varieties with finitely many T-fixed points, then for $[E]^T \in K_T^0(Y)$, its pullback can be computed by

$$(\pi^*[E]^T)(x) = [E]^T(\pi(x)) \tag{4.2}$$

for $x \in X^T$.

Describing the pushforward of $[F]^T \in K_T^0(X)$ is a bit more complicated. Suppose that X and Y are contracting. For every point $x \in X^T$ (resp. $y \in Y^T$), we pick as before an open neighbourhood U_x (resp. V_y) on which T acts by characters $\chi_1(x), \dots, \chi_r(x)$ (resp. $\eta_1(y), \dots, \eta_s(y)$). Then the pushforward of $[F]^T$ is determined by the formula

$$\frac{(\pi_*[F]^T)(y)}{\prod(1-\eta_j(y)^{-1})} = \sum_{x \in \pi^{-1}(y) \cap X^T} \frac{[F]^T(x)}{\prod(1-\chi_i(x)^{-1})}. \tag{4.3}$$

Remark 77 We can use Theorem 68 to obtain a description of the ordinary K-theory using equivariant localisation. However, one should be careful when using this for computations in practice. Here is a toy example: let $X = \mathbb{P}^2$ with the usual action of $(\mathbb{C}^*)^2$. Then $X^T = \{[1:0], [0:1]\}$, and we can write the elements of $K_T^0(X^T) \cong \mathrm{Maps}(X^T, \mathbb{Z}[t_0^{\pm}, t_1^{\pm}]) \ni f$ as pairs $(f([1:0]), f([0:1]))$. Then $(t_0 - t_1, 0)$ satisfies the condition from Theorem 75, hence it gives a class in $K_T^0(X)$. It is tempting to do the following computation in $K^0(X) \cong K_T^0(X) \otimes_{\mathbb{Z}[\mathrm{Char}(T)]} \mathbb{Z}$:

$$(t_0 - t_1, 0) \otimes 1 = (1, 0) \otimes (1 - 1) = 0$$

but this is wrong! Indeed, $(1, 0)$ does not satisfy the condition from Theorem 75, hence is not in $K_T^0(X)$. In fact, one can check that $(t_0 - t_1, 0)$ is the equivariant class of the torus-fixed point $[1:0] \in \mathbb{P}^2$.

4.8.3 A Short Review on Cones and Their Hilbert Series

For more details about the topic of this subsection we refer to [9, Sect. 1.2] and [48, Sect. 4.5]. Recall that a *convex polyhedral rational cone* is a subset of $\mathbb{R}^n$ of the form $C = \mathrm{cone}(S) := \{\sum_{\mathbf{u} \in S} \lambda_{\mathbf{u}} \mathbf{u} \mid \lambda_{\mathbf{u}} \in \mathbb{R}_{\geq 0}\}$, where $S \subset \mathbb{Z}^n \subset \mathbb{R}^n$ is finite. A cone is called *pointed* if it does not contain a line. If C is a pointed rational cone, then every one-dimensional face ρ contains a unique lattice point $\mathbf{u}_\rho$ that is closest to the origin. It is not hard to see that $MG(C) := \{\mathbf{u}_\rho \mid \rho \text{a one-dimensional face of} C\}$ is a minimal generating set of C. If the minimal generators are linearly independent over $\mathbb{R}$, we call C *simplicial*. If they are part of a $\mathbb{Z}$-basis of $\mathbb{Z}^n$, we call C *regular*.

For a pointed cone C in $\mathbb{R}^n$, we define its *Hilbert series* Hilb(C) by:

$$\text{Hilb}(C) := \sum_{\mathbf{a}\in C\cap\mathbb{Z}^n} \mathbf{t}^{\mathbf{a}}.$$

This is always a rational function [48, Theorem 4.5.11] whose denominator is equal to $\prod_{\mathbf{u}\in MG(C)} (1 - \mathbf{t}^{\mathbf{u}})$. If C is a regular cone, then its Hilbert series is easy to compute: $\text{Hilb}(C) = \prod_{\mathbf{u}\in MG(C)} 1/(1 - \mathbf{t}^{\mathbf{u}})$. If C is a simplicial cone, we can compute its Hilbert series as follows. First compute the finite set $D_C := \{\mathbf{b} \in C \cap \mathbb{Z}^n : \mathbf{b} = \sum_{\mathbf{b}\in MG(C)} \lambda_{\mathbf{u}}\mathbf{u} \mid 0 \le \lambda_u < 1\}$. Then

$$\text{Hilb}(C) = \left(\sum_{\mathbf{b}\in D_C} \mathbf{t}^{\mathbf{b}}\right) \prod_{\mathbf{u}\in MG(C)} \frac{1}{1 - \mathbf{t}^{\mathbf{u}}}.$$

For a general rational polyhedral cone, we can compute its Hilbert series by triangulating it.

4.8.4 Matroids and the K-theory of Grassmannians

In this subsection we compute the class in equivariant K-theory of a torus orbit closure in a Grassmannian. We then note that this class only depends on the underlying matroid, and give a combinatorial algorithm to get the class in K-theory directly from the matroid. This algorithm can then be used as a definition to associate a class in K-theory to an arbitrary (not necessarily representable) matroid. This was first done by Fink and Speyer [16].

Let us first fix the following sign conventions. The torus $T = (\mathbb{C}^*)^n$ acts on $\mathbb{C}^n$ as follows: $\mathbf{t} \cdot (x_1, \ldots, x_n) = (t_1^{-1}x_1, \ldots, t_n^{-1}x_n)$. The action of T on $G(k, n)$ is induced from this action. Explicitly, if $p \in G(k, n)$ has Plücker coordinates $[P_I]_{I\in\binom{[n]}{k}}$, then $\mathbf{t} \cdot p$ has Plücker coordinates $[(\prod_{i\in I} t_i^{-1})P_I]_{I\in\binom{[n]}{k}}$.

We begin by describing the T-equivariant K-theory of the Grassmannian $G(k, n)$ using equivariant localisation.

The torus-fixed points of $G(k, n)$ are easy to describe: for every size k subset $I \subset [n]$, we define the k-plane $V_I = \text{span}(\{e_i \mid i \in I\}) \subset \mathbb{C}^n$, and denote the corresponding point in $G(k, n)$ by p_I. In Plücker coordinates, p_I is given by $P_J = 0$ if $J \neq I$. It is easy to see that the $\binom{n}{k}$ points p_I are precisely the torus-fixed points of $G(k, n)$.

We can also describe the one-dimensional torus orbits: there is a (unique) one-dimensional torus orbit between p_I and p_J if and only if $|I \cap J| = k - 1$. In this case, we write $I - J = \{i\}$, $J - I = \{j\}$. If we identify the one-dimensional orbit from p_I to p_J with $\mathbb{A}^1 \backslash 0$ in such a way that the origin corresponds to the torus-fixed point p_I (and so p_J corresponds to the point at infinity), then T acts on the orbit with character $t_j^{-1} t_i$.

Let us now check that the action of T is contracting. We fix a torus-fixed point p_I, and consider the open neighbourhood U_I given by $P_I = 1$. Then $U_I \cong \mathbb{A}^{k(n-k)}$. For $p \in U_I$, we will denote its coordinates with $(u_{i,j})_{i \in I, j \notin I}$, where $u_{i,j} = \frac{P_{I-i \cup j}}{P_I}$. Then $\mathbf{t} \cdot p$ has coordinates $(t_j^{-1} t_i u_{i,j})_{i \in I, j \notin I}$. Thus, T acts on this space with characters $t_j^{-1} t_i$, where $i \in I$, $j \notin I$. Identifying $t_1^{a_1} \cdots t_n^{a_n}$ with $(a_1, \ldots, a_n)$, all these points lie in the open halfspace defined by $\sum_{i \in I} a_i > 0$.

Example 78 We compute the class of $\mathcal{O}(1)$. The sheaf $\mathcal{O}(1)$ on $G(k, n)$ was already mentioned in Sect. 4.4.2: it is the pullback of $\mathcal{O}(1)$ on $\mathbb{P}^{\binom{n}{k}-1}$ via the Plücker embedding. We can also describe $\mathcal{O}(1)$ as $\bigwedge^k S^\vee$, where S is the tautological vector bundle on $G(k, n)$ whose fiber over a point is the corresponding k-plane.

We can apply Theorem 68 to the result from Example 74 to replace the torus action there with a different torus action, induced from the action on the Plücker coordinates. By applying pullback formula (4.2), we find that the class $[\mathcal{O}(1)]^T$ in equivariant K-theory is the map

$$[\mathcal{O}(1)]^T : Gr(k, n)^T \to \mathbb{Z}[\mathrm{Char}(T)] : p_I \mapsto t_{i_1} \cdots t_{i_k},$$

where we wrote $I = \{i_1, \ldots, i_k\}$.

Let p be a point in $Gr(k, n)$ and $M = M_p$ be the corresponding matroid. Then $\overline{Tp}$ is a closed subvariety of $Gr(k, n)$; in particular, it is given by a coherent sheaf. We want to compute its class in T-equivariant K-theory, which is a map $[\overline{Tp}]^T : Gr(k, n)^T \to \mathbb{Z}[\mathrm{Char}(T)]$. As above, let $p_I \in Gr(k, n)^T$ be the torus-invariant point given by $P_J = 0$ for $J \neq I$, and let U_I be the affine open neighbourhood U_I of p_I defined by $P_I = 1$.

If I is not a basis of M, then $\overline{Tp}$ does not intersect U_I, hence $[\overline{Tp}]^T(p_I) = 0$. Hence, we will assume that I is a basis of M, i.e. that $p \in U_I$.

The coordinate ring of $\overline{Tp} \cap U_I$ is isomorphic to $\mathbb{C}[s_i^{-1} s_j]$, where $s_i^{-1} s_j$ is a generator if and only if $(I - i) \cup j$ is a basis of M. We will denote this ring by $R_{M,I}$. This ring should be viewed as a T-module, with $\mathbf{t} \cdot s_i^{-1} s_j = t_i^{-1} t_j s_i^{-1} s_j$. The Hilbert series of $R_{M,I}$ is a rational function with denominator dividing $\prod_{i \in I} \prod_{j \notin I} (1 - t_i^{-1} t_j)$. Thus, by (4.1),

$$[\overline{Tp}]^T(p_I) = \mathrm{Hilb}(R_{M,I}) \prod_{i \in I} \prod_{j \notin I} (1 - t_i^{-1} t_j). \tag{4.4}$$

Definition 79 For any lattice polytope P and v a vertex of P, we define $\mathrm{cone}_v(P)$ to be the cone spanned by all vectors of the form $u - v$ with $u \in P$. For $I \in \binom{[n]}{k}$, we write $\mathrm{cone}_I(M) := \mathrm{cone}_{\mathbf{e}_I}(P(M))$ if I is a basis of M, and $\mathrm{cone}_I(M) := \varnothing$ otherwise.

Since $\mathrm{cone}_I(M)$ is the positive real span of all vectors $e_J - e_I$, where $J \in \mathcal{B}(M)$, we find that $\mathrm{Hilb}(R_{M,I}) = \mathrm{Hilb}(\mathrm{cone}_I(M))$. So (4.4) can also be written as

$$[\overline{Tp}]^T(p_I) = \mathrm{Hilb}(\mathrm{cone}_I(M)) \prod_{i \in I} \prod_{j \notin I} (1 - t_i^{-1} t_j). \tag{4.5}$$

We note that (4.5) only depends on the matroid M and not on the chosen point p or even the torus orbit $\overline{Tp}$. Moreover, the formulas make sense even for non-representable matroids. Thus we can use them as a definition for the class in K-theory for a matroid:

Definition 80 ([16]) For any rank k matroid M on $[n]$, we define $y(M) : Gr(k,n)^T \to \mathbb{Z}[\mathrm{Char}(T)]$ by

$$y(M)(p_I) = \mathrm{Hilb}(\mathrm{cone}_I(M)) \prod_{i \in I} \prod_{j \notin I} (1 - t_i^{-1} t_j).$$

Theorem 81 ([16, Proposition 3.3]) *The class $y(M) \in K_T^0(Gr(k,n)^T)$ satisfies the condition of Theorem 75, and hence defines a class in $K_T^0(Gr(k,n))$.*

4.8.5 Flag Matroids and the K-theory of Flag Varieties

In this section, we generalize the results from the previous section replacing "matroids" by "flag matroids" and "Grassmannians" by "flag varieties".

We first describe the equivariant K-theory of a flag variety $Fl(\mathbf{k}, n)$. The torus-fixed points are given as follows: for every (set-theoretic) flag $F = (F_1 \subseteq \ldots \subseteq F_s)$ of rank $\mathbf{k}$ on $[n]$, we define a (vector space) flag $V_1 \subseteq \ldots \subseteq V_s$ by $V_i = \mathrm{span}(\{e_j \mid j \in F_i\})$. We will denote the corresponding point in $Fl(\mathbf{k}, n)$ by p_F. The Plücker coordinates of p_F are given by $P_S = 1$ if S is a constituent of F and $P_S = 0$ otherwise. Here, the Plücker coordinates of a point in $Fl(\mathbf{k}, n)$ are the ones induced from the embedding $Fl(\mathbf{k}, n) \hookrightarrow \prod G(k_i, n)$.

We can also describe the one-dimensional torus orbits: let p_F be a torus-fixed point. We define $S(F)$ to be the set of all pairs $(i, j) \in [n] \times [n]$ for which there exists an ℓ such that $i \in F_\ell$ and $j \notin F_\ell$. For every $(i, j) \in S(F)$, we define a new flag $F' = F_{i \to j}$ by switching the roles of i and j. More precisely: if $i \in F_r$ but $j \notin F_r$, then $F'_r := (F_r - i) \cup j$; in any other case $F'_r := F_r$. Then there is a unique one-dimensional torus orbit between p_F and $p_{F'}$, and all one-dimensional torus orbits arise in this way. The torus T acts on this orbit with character $t_j^{-1} t_i$.

Lemma 82 *The action of T on $Fl(\mathbf{k}, n)$ is contracting.*

Proof For every torus-fixed point p_F, we consider the open neighbourhood U_F given by $P_{F_r} \neq 0$ for all r. Then $U_F \cong \mathbb{A}^N$, where $N = \dim(Fl(\mathbf{k}, n)) = \sum_{i=1}^{s} k_i(k_{i+1} - k_i)$ (here $k_{s+1} := n$). We will denote the coordinates of a point q in U_F by $(u_{i,j})_{(i,j)\in S(F)}$, where $u_{i,j} = \frac{P_{F_r - i \cup j}}{P_{F_r}}$ for any r which satisfies $i \in F_r$ and $j \notin F_r$. Then $\mathbf{t} \cdot q$ has coordinates $(t_j^{-1} t_i u_{i,j})_{(i,j)\in S(F)}$. So T acts on U_F with characters $t_j^{-1} t_i$, $(i, j) \in S(F)$. As before, identifying $t_1^{a_1} \cdots t_n^{a_n}$ with $(a_1, \ldots, a_n)$, all these characters lie on the open halfspace $\sum_{r=1}^{s} \sum_{i\in F_r} a_i > 0$. □

Let p be a point in $Fl(\mathbf{k}, n)$, and let $\mathcal{F}$ be the corresponding flag matroid. We want to compute the T-equivariant class of $\overline{Tp}$, which is a map $[\overline{Tp}]^T : Fl(\mathbf{k}, n)^T \to \mathbb{Z}[\mathrm{Char}(T)]$. We fix a point $p_F \in Fl(\mathbf{k}, n)^T$, and consider the affine neighbourhood U_F described above.

If F is not a basis of $\mathcal{F}$, then $\overline{Tp}$ does not intersect U_F, hence $[\overline{Tp}]^T(p_F) = 0$. Thus, we will assume that F is a basis of $\mathcal{F}$, i.e. that $p \in U_F$.

The coordinate ring of $\overline{Tp} \cap U_F$ is isomorphic to $k[s_i^{-1} s_j]$, where $s_i^{-1} s_j$ is a generator if and only if $F_{i\to j} \in \mathcal{F}$. We will denote this ring by $R_{\mathcal{F},F}$. This ring should be viewed as a T-module, with $\mathbf{t} \cdot s_i^{-1} s_j = t_i^{-1} t_j s_i^{-1} s_j$. The Hilbert series of this T-module is a rational function with denominator dividing $\prod_{(i,j)\in S(F)} (1 - t_i^{-1} t_j)$. Thus, by (4.1),

$$[\overline{Tp}]^T(p_F) = \mathrm{Hilb}(R_{\mathcal{F},F}) \prod_{(i,j)\in S(F)} (1 - t_i^{-1} t_j). \tag{4.6}$$

Definition 83 We define $\mathrm{cone}_F(\mathcal{F})$ to be the cone $\mathrm{cone}_{\mathbf{e}_F}(P(\mathcal{F}))$, as in Definition 79.

As before, we find that $\mathrm{Hilb}(R_{\mathcal{F},F}) = \mathrm{Hilb}(\mathrm{cone}_F(\mathcal{F}))$. Hence, (4.6) can also be written as

$$[\overline{Tp}]^T(p_F) = \mathrm{Hilb}(\mathrm{cone}_F(\mathcal{F})) \prod_{(i,j)\in S(F)} (1 - t_i^{-1} t_j). \tag{4.7}$$

As before, (4.7) only depends on the flag matroid $\mathcal{F}$ and not on the chosen point p or even the torus orbit $\overline{Tp}$. Moreover the formulas make sense even for non-representable flag matroids. Thus we can use them as a definition for the class in K-theory for a flag matroid:

Definition 84 For any rank $\mathbf{k}$ flag matroid $\mathcal{F}$ on $[n]$, we define $y(\mathcal{F}) : Fl(\mathbf{k}, n)^T \to \mathbb{Z}[\mathrm{Char}(T)]$ by

$$y(\mathcal{F})(p_F) = \mathrm{Hilb}(\mathrm{cone}_F(\mathcal{F})) \prod_{(i,j)\in S(F)} (1 - t_i^{-1} t_j).$$

Proposition 85 *The class $y(\mathcal{F}) \in K_T^0(Fl(\mathbf{k}, n)^T)$ satisfies the condition of Theorem 75, and hence defines a class in $K_T^0(Fl(\mathbf{k}, n))$.*

Proof The proof is a straightforward generalisation of the proof of [16, Proposition 3.3]. □

Fig. 4.5 The class in K-theory of a flag matroid

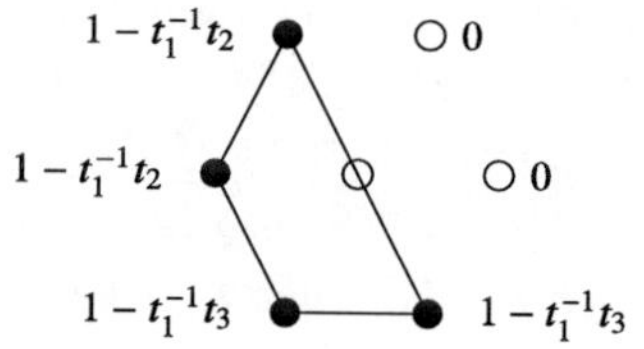

Example 86 Let $\mathcal{F}$ be the flag matroid of Example 54. We first compute $y(\mathcal{F})(p_F)$, where F is the flag $2 \subseteq 12$ (so $e_F = (1, 2, 0)$). From Fig. 4.4, it is clear that $\mathrm{cone}_F(\mathcal{F}) = \mathrm{cone}((1, -1, 0), (0, -1, 1))$, which has Hilbert series $\frac{1}{(1-t_2^{-1}t_1)(1-t_2^{-1}t_3)}$. Furthermore, we have $S(F) = \{(2, 1), (2, 3), (1, 3)\}$. We find that $y(\mathcal{F})(p_F) = 1 - t_1^{-1}t_3$. We can do the same for the other torus-fixed points. The result is summarised in Fig. 4.5.

4.8.6 The Tutte Polynomial via K-theory

In [16], a geometric description of the Tutte polynomial of a matroid is given. Consider the following diagram, where all the maps are natural projections or inclusions:

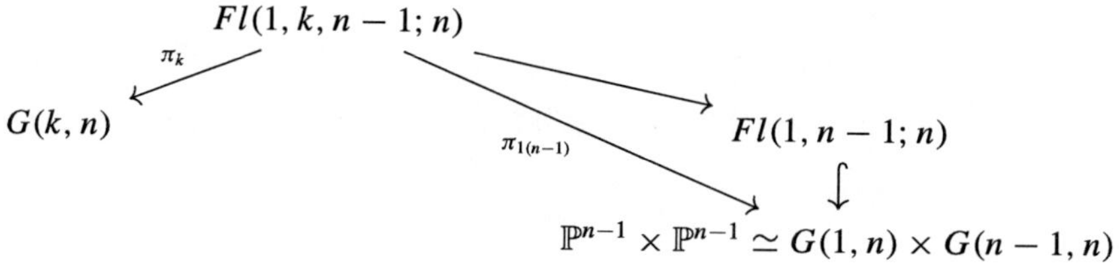

Generalising Example 67, one can show that $K^0(\mathbb{P}^{n-1} \times \mathbb{P}^{n-1}) \cong \mathbb{Z}[\alpha, \beta]/(\alpha^n, \beta^n)$, where α and β are the structure sheaves of hyperplanes.

Theorem 87 ([16, Theorem 7.1]) *The following equality holds:*

$$(\pi_{1(n-1)})_* \pi_d^*(Y(M) \cdot [\mathcal{O}(1)]) = T_M(\alpha, \beta),$$

where $Y(M)$ is the class associated to the matroid M in the non-equivariant K-theory of the Grassmannian.

In other words, the Tutte polynomial of a matroid can be viewed as a Fourier-Mukai transform of its associated class in K-theory.

We can now generalize this construction to get a *definition* of the Tutte polynomial of a flag matroid.

Definition 88 Consider the following diagram.

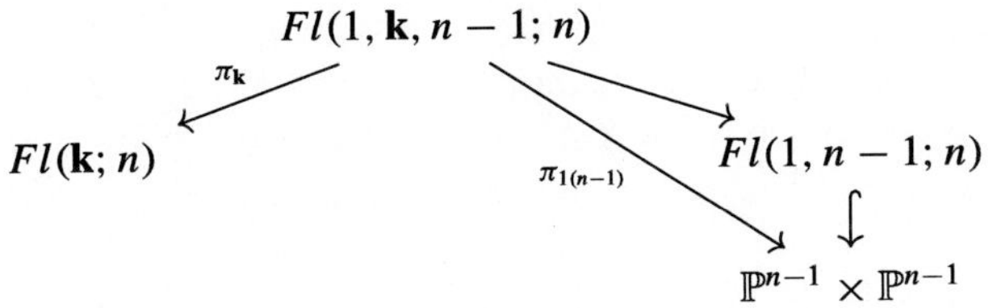

Let $\mathcal{F}$ be a flag matroid on $[n]$ of rank $\mathbf{k}$, and let $Y(\mathcal{F}) \in K^0(Fl(\mathbf{k}; n))$ be its class in non-equivariant K-theory, as in Definition 84. Then the *K-theoretic Tutte polynomial* of $\mathcal{F}$ is defined to be

$$T_{\mathcal{F}}(\alpha, \beta) := (\pi_{1(n-1)})_* \pi_d^*(Y(\mathcal{F}) \cdot [\mathcal{O}(1)]).$$

We computed the Tutte polynomial for some small examples using SAGE [51], Macaulay2 [25], and Normaliz [5]. Our code is available at [47]. The program first computes the equivariant class $(\pi_{1(n-1)})_* \pi_d^*(y(M) \cdot [\mathcal{O}(1)]) \in K_T^0(\mathbb{P}^{n-1} \times \mathbb{P}^{n-1})$ using equivariant localisation, and then computes the underlying non-equivariant class.

Remark 89 After we finished the article Christopher Eur implemented the algorithm in Macaulay2 [14].

Example 90 We consider again the flag matroid of Examples 54 and 86. We first compute $y(\mathcal{F}) \cdot [\mathcal{O}(1)]$, which is displayed in Fig. 4.6. The two projections from $Fl(1, \mathbf{k}, 2; 3) = Fl(1, 1, 2, 2; 3)$ to $Fl(1, 2; 3)$ are isomorphisms, hence pulling back and pushing forward along them does nothing. Next we need to push our class $X = y(\mathcal{F}) \cdot [\mathcal{O}(1)] \in K_T^0(Fl(1, 2; 3))$ to a class $Y \in K_T^0(\mathbb{P}^{n-1} \times \mathbb{P}^{n-1})$, using formula (4.3).

The T-fixed points of $\mathbb{P}^{n-1} \times \mathbb{P}^{n-1}$ are given by pairs $p = (\ell, H)$, where $\ell \in G(1, 3)^T = \{\langle e_1 \rangle, \langle e_2 \rangle, \langle e_3 \rangle\}$ and $H \in G(2, 3)^T = \{\langle e_1, e_2 \rangle, \langle e_1, e_3 \rangle, \langle e_2, e_3 \rangle\}$. If $\ell \not\subset H$, then $Y(p) = 0$. If $\ell \subset H$, then $p \in Fl(1, 2; 3) \subset \mathbb{P}^2 \times \mathbb{P}^2$. Since we are pushing forward along an embedding, the formula (4.3) has a simple form: we can find characters $\chi_1, \chi_2, \chi_3, \eta$ and open neighbourhoods $p \ni U_1 \subset Fl(1, 2; 3)$ and $p \ni U_2 \subset \mathbb{P}^2 \times \mathbb{P}^2$, such that T acts on U_1 with characters χ_1, χ_2, χ_3, and on U_2 with characters $\chi_1, \chi_2, \chi_3, \eta$. Then (4.3) becomes:

$$Y(p) = (1 - \eta^{-1}) X(p).$$

Fig. 4.6 $y(\mathcal{F}) \cdot [\mathcal{O}(1)]$

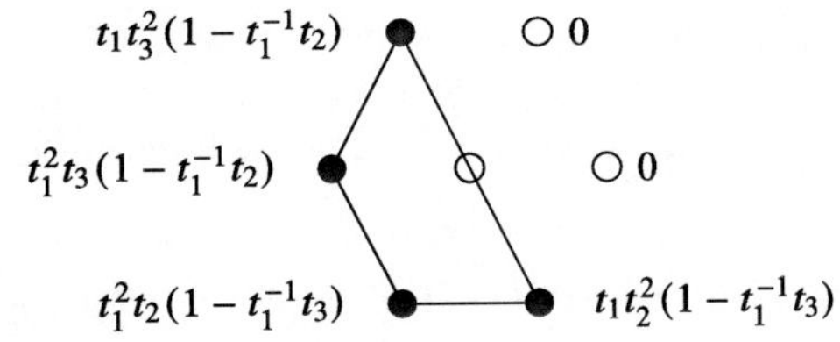

Consider for example $p = (\langle e_1\rangle, \langle e_1, e_2\rangle)$. Then $p \in Fl(1, 2; 3)$ has an open neighbourhood where T acts by characters $t_2t_3^{-1}, t_1t_3^{-1}, t_1t_2^{-1}$, while $p \in \mathbb{P}^2 \times \mathbb{P}^2$ has an open neighbourhood where T acts by characters $t_2t_3^{-1}, t_1t_3^{-1}, t_1t_3^{-1}, t_1t_2^{-1}$. We compute that

$$Y((\langle e_1\rangle, \langle e_1, e_2\rangle)) = t_1^2t_2(1 - t_1^{-1}t_3)(1 - t_1^{-1}t_3) = t_2(t_1 - t_3)^2.$$

Similarily, we find that

$$\begin{aligned} Y((\langle e_1\rangle, \langle e_1, e_3\rangle)) &= t_3(t_1 - t_2)^2 \\ Y((\langle e_3\rangle, \langle e_1, e_3\rangle)) &= t_3(t_1 - t_2)(t_3 - t_2) \\ Y((\langle e_2\rangle, \langle e_1, e_2\rangle)) &= t_2(t_1 - t_3)(t_2 - t_3) \end{aligned}$$

and $Y(p) = 0$ in all other cases.

Finally, we need to find the underlying class in non-equivariant K-theory. This is quite tedious to do by hand, so we just refer to the algorithm provided at [47] for this. In the end, we find that

$$T_{\mathcal{F}}(x, y) = x^2y^2 + x^2y + xy^2 + x^2 + xy.$$

Example 91 As another example, consider the uniform flag matroid $\mathcal{U}_{(2,3);5}$ of rank $(2, 3)$ on $[5]$ (that is, the constituents of $\mathcal{U}_{(2,3);5}$ are the uniform matroids $U_{2,5}$ and $U_{3,5}$). Using our program, we find that its K-theoretic Tutte polynomial $T_{\mathcal{U}_{(2,3):5}}(x, y)$ equals

$$\begin{aligned} x^3y^3 + 2x^3y^2 + 2x^2y^3 + 3x^3y + 8x^2y^2 + 3xy^3 + 4x^3 + \\ 8x^2y + 8xy^2 + 4y^3 + 2x^2 + 4xy + 2y^2. \end{aligned}$$

After the preprint of the article appeared Christopher Eur found an example of a K-theoretic Tutte polynomial $T_{\mathcal{U}_{(1,3):5}}(x, y)$ with a strictly negative coefficient.

4.9 Open Problems

Our definition of the Tutte polynomial of a flag matroid is admittedly quite involved. It is natural to wonder whether there is an easier definition, avoiding geometry:

Problem 92 Is there a purely combinatorial description of the K-theoretic Tutte polynomial of a flag matroid? In particular, can one obtain the K-theoretic Tutte polynomial from the Tutte polynomials of the constituents?

For matroids, one can define the *characteristic polynomial* (also called *chromatic polynomial*, as it generalises the chromatic polynomial of a graph) by

$$\chi_M(\lambda) = (-1)^{r(M)} T_M(1-\lambda, 0).$$

In 2015, Adiprasito, Huh and Katz proved the following conjecture by Rota-Heron-Welsh:

Theorem 93 ([1]) *Let $w_i(M)$ be the absolute value of the coefficient of $\lambda^{r(M)-i}$ in the characteristic polynomial of M. Then the sequence $w_i(M)$ is log-concave:*

$$w_{i-1}(M)w_{i+1}(M) \leq w_i(M)^2 \text{ for all } 1 \leq i < r(M).$$

Since we now have a definition for the Tutte polynomial of a flag matroid, we can define the *characteristic polynomial* of a rank $\mathbf{k}$ flag matroid $\mathcal{F}$ by

$$\chi_{\mathcal{F}}(\lambda) = (-1)^{r(\mathcal{F})} T_{\mathcal{F}}(1-\lambda, 0),$$

where $r(\mathcal{F}) := |\mathbf{k}| := \sum k_i$.

Conjecture 94 Theorem 93 holds for the characteristic polynomial of an arbitrary flag matroid.

In Examples 90 and 91, the characteristic polynomials are $-\lambda^2 + 2\lambda - 1$ and $4\lambda^3 - 14\lambda^2 + 16\lambda - 6$, respectively. Thus, we see that Conjecture 94 holds for these examples.

Flag matroids are a special class of Coxeter matroids. Hence, another possible direction of research would be:

Problem 95 Explore how our constructions and results could be generalised to arbitrary Coxeter matroids.

Flag matroids can also be viewed as a special class of polymatroids. In particular, we can apply the construction of Sect. 4.3.1 to them.

Problem 96 Is there a connection between the Tutte polynomial of a polymatroid, as defined by Cameron and Fink, and our construction of the Tutte polynomial of flag matroid? How about the construction from Remark 31?

Next, we note that by Sect. 4.7 we know how to associate to a rank k representable r-polymatroid on $[n]$ a class in $K_T^0(G(k, rn))$.

Problem 97 Can we associate a class in $K_T^0(G(k, rn))$ to a non-representable r-polymatroid? Can this be used to define a K-theoretic Tutte polynomial for r-polymatroids?

Another possible generalisation, brought to our attention by Alastair Craw after we wrote the article, would be the class of quiver flag varieties [10]. We have not yet pursued this direction.

Finally, we could apply the construction of Sect. 4.8.6 to *any* subvariety of a Grassmannian (or even a flag variety), not just to torus orbits. It could be interesting to study the properties of this invariant.

Acknowledgements We would like to thank Christopher Eur for useful comments and sharing with us interesting examples. We thank Alastair Craw and the anonymous reviewers for their careful reading and suggestions to improve the manuscript. The third author was supported by the Polish National Science Centre grant no. 2015/19/D/ST1/01180.

References

1. Adiprasito, K., Huh, J., Katz, E.: Hodge theory for combinatorial geometries. Ann. Math. **188**(2), 381–452 (2018)
2. Anderson, D.: Introduction to equivariant cohomology in algebraic geometry. In: Contributions to Algebraic Geometry, EMS Series of Congress Reports, pp. 71–92. European Mathematical Society, Zürich (2012)
3. Borovik, A.V., Gel'fand, I.M., White, N.: Coxeter Matroids, Progress in Mathematics, vol. 216. Birkhäuser Boston Inc, Boston, MA (2003)
4. Brändén, P.: Obstructions to determinantal representability. Adv. Math. **226**(2), 1202–1212 (2011)
5. Bruns, W., Ichim, B.: Normaliz: algorithms for affine monoids and rational cones. J. Algebra **324**(5), 1098–1113 (2010)
6. Brylawski, T.: Constructions. In: Theory of Matroids, The Encyclopedia of Mathematics and its Applications, vol. 26, pp. 127–223. Cambridge University Press, Cambridge (1986)
7. Brylawski, T., Oxley, J.: The Tutte polynomial and its applications. In: Matroid Applications, the Encyclopedia of Mathematics and Its Applications, vol. 40, pp. 123–225. Cambridge University Press, Cambridge (1992)
8. Cameron, A., Fink, A.: The Tutte polynomial via lattice point counting (2018). arXiv:1802.09859
9. Cox, D.A., Little, J.B., Schenck, H.K.: Toric Varieties, Graduate Studies in Mathematics, vol. 124. American Mathematical Society, Providence, RI (2011)
10. Craw, A.: Quiver flag varieties and multigraded linear series. Duke Math. J. **156**(3), 469–500 (2011)
11. Dupont, C., Fink, A., Moci, L.: Universal Tutte characters via combinatorial coalgebras. Algebr. Comb. **1**(5), 603–651 (2018)
12. Edmonds, J.: Submodular functions, matroids, and certain polyhedra. In: Combinatorial Structures and their Applications (Proc. Calgary Internat. Conf., Calgary, Alta., 1969), pp. 69–87. Gordon and Breach, New York (1970)
13. Ellis-Monaghan, J.A., Merino, C.: Graph polynomials and their applications I: the Tutte polynomial. In: Structural Analysis of Complex Networks, pp. 219–255. Birkhäuser/Springer, New York (2011)
14. Eur, C.: Webpage. https://github.com/chrisweur
15. Feichtner, E.M., Sturmfels, B.: Matroid polytopes, nested sets and Bergman fans. Port. Math. (N.S.) **62**(4), 437–468 (2005)
16. Fink, A., Speyer, D.E.: K-classes for matroids and equivariant localization. Duke Math. J. **161**(14), 2699–2723 (2012)
17. Fulton, W.: Introduction to Toric Varieties, Annals of Mathematics Studies, vol. 131. Princeton University Press, Princeton, NJ (1993). The William H. Roever Lectures in Geometry
18. Fulton, W.: Intersection theory, Ergebnisse der Mathematik und ihrer Grenzgebiete. 3. Folge. A Series of Modern Surveys in Mathematics [Results in Mathematics and Related Areas. 3rd Series. A Series of Modern Surveys in Mathematics], vol. 2, 2nd edn. Springer, Berlin (1998)
19. Fulton, W., Harris, J.: Representation Theory, Graduate Texts in Mathematics, vol. 129. Springer, New York (1991). A first course, Readings in Mathematics
20. Gale, D.: Optimal assignments in an ordered set: an application of matroid theory. J. Comb. Theory **4**, 176–180 (1968)

21. Gel'fand, I.M., Goresky, R.M., MacPherson, R.D., Serganova, V.V.: Combinatorial geometries, convex polyhedra, and Schubert cells. Adv. Math. **63**(3), 301–316 (1987)
22. Gel'fand, I.M., Serganova, V.V.: Combinatorial geometries and the strata of a torus on homogeneous compact manifolds. Uspekhi Mat. Nauk **42**(2(254)), 107–134, 287 (1987)
23. Gel'fand, I.M., Serganova, V.V.: On the general definition of a matroid and a greedoid. Dokl. Akad. Nauk SSSR **292**(1), 15–20 (1987)
24. Goresky, M., Kottwitz, R., MacPherson, R.: Equivariant cohomology, Koszul duality, and the localization theorem. Invent. Math. **131**(1), 25–83 (1998)
25. Grayson, D.R., Stillman, M.E.: Macaulay2, a software system for research in algebraic geometry. http://www.math.uiuc.edu/Macaulay2/
26. Greene, C.: A multiple exchange property for bases. Proc. Am. Math. Soc. **39**, 45–50 (1973)
27. Herzog, J., Hibi, T.: Discrete polymatroids. J. Algebr. Combin. **16**(3), 239–268 (2002, 2003)
28. Huh, J.: Milnor numbers of projective hypersurfaces and the chromatic polynomial of graphs. J. Am. Math. Soc. **25**(3), 907–927 (2012)
29. Kálmán, T.: A version of Tutte's polynomial for hypergraphs. Adv. Math. **244**, 823–873 (2013)
30. Kim, S.: Flag enumerations of matroid base polytopes. J. Combin. Theory Ser. A **117**(7), 928–942 (2010)
31. Lafforgue, L.: Chirurgie Des Grassmanniennes. CRM Monograph Series, vol. 19. American Mathematical Society, Providence, RI (2003)
32. Lasoń, M.: On the toric ideals of matroids of a fixed rank. Selecta Math. (N.S.) **27**(2), Paper No. 18, 17 (2021)
33. Lasoń, M., Michałek, M.: On the toric ideal of a matroid. Adv. Math. **259**, 1–12 (2014)
34. Manivel, L.: Symmetric functions, Schubert polynomials and degeneracy loci. SMF/AMS Texts and Monographs, vol. 6. American Mathematical Society, Providence, RI; Société Mathématique de France, Paris (2001). Translated from the 1998 French original by John R. Swallow, Cours Spécialisés [Specialized Courses], 3
35. McMullen, P.: Lattice invariant valuations on rational polytopes. Arch. Math. (Basel) **31**(5), 509–516 (1978/79)
36. Merino, C., Ramírez-Ibáñez, M., Rodríguez-Sánchez, G.: The Tutte polynomial of some matroids. Int. J. Comb. Art. ID 430859, 40 (2012)
37. Merkur'ev, A.S.: Comparison of the equivariant and the standard K-theory of algebraic varieties. Algebra i Analiz **9**(4), 175–214 (1997)
38. Michałek, M.: Selected topics on toric varieties. In: The 50th anniversary of Gröbner bases. Adv. Stud. Pure Math. Math. Soc. Japan, Tokyo **77**, 207–252 (2018)
39. Michałek, M., Sturmfels, B.: Invitation to nonlinear algebra, vol. 211. American Mathematical Society (2021)
40. Miller, E., Sturmfels, B.: Combinatorial Commutative Algebra, Graduate Texts in Mathematics, vol. 227. Springer, New York (2005)
41. Mnëv, N.E.: The universality theorems on the classification problem of configuration varieties and convex polytopes varieties. In: Topology and geometry—Rohlin Seminar. Lecture Notes in Mathematics, vol. 1346, pp. 527–543. Springer, Berlin (1988)
42. Nielsen, H.A.: Diagonalizably linearized coherent sheaves. Bull. Soc. Math. France **102**, 85–97 (1974)
43. Oxley, J.: Matroid theory. Oxford Graduate Texts in Mathematics, vol. 21, 2nd edn. Oxford University Press, Oxford (2011)
44. Oxley, J., Whittle, G.: A characterization of Tutte invariants of 2-polymatroids. J. Combin. Theory Ser. B **59**(2), 210–244 (1993)
45. Reiner, V.: Lectures on Matroids and Oriented Matroids (2005). https://www-users.math.umn.edu/~reiner/Talks/Vienna05/Lectures.pdf
46. Schrijver, A.: Combinatorial optimization. Polyhedra and efficiency. Algorithms and Combinatorics, vol. C, 24. Springer, Berlin (2003). Disjoint paths, hypergraphs, Chapters 70–83
47. Seynnaeve, T.: Webpage. https://mathsites.unibe.ch/seynnaeve/index.html
48. Stanley, R.P.: Enumerative combinatorics. Cambridge Studies in Advanced Mathematics, vol. 1, 49. Cambridge University Press, Cambridge (1997). With a foreword by Gian-Carlo Rota, Corrected reprint of the 1986 original

49. Sturmfels, B.: Gröbner Bases and Convex Polytopes. University Lecture Series, vol. 8. American Mathematical Society, Providence, RI (1996)
50. Sturmfels, B.: Algorithms in Invariant Theory. Texts and Monographs in Symbolic Computation, 2nd edn. Springer, Wien, NewYork, Vienna (2008)
51. The Sage Developers: SageMath, the Sage Mathematics Software System (Version 8.4.0) (2018). http://www.sagemath.org
52. Thistlethwaite, M.B.: A spanning tree expansion of the Jones polynomial. Topology **26**(3), 297–309 (1987)
53. Tutte, W.T.: A contribution to the theory of chromatic polynomials. Canad. J. Math. **6**, 80–91 (1954)
54. Vezzosi, G., Vistoli, A.: Higher algebraic K-theory for actions of diagonalizable groups. Invent. Math. **153**(1), 1–44 (2003)
55. Welsh, D.J.A., Merino, C.: The Potts Model and the Tutte Polynomial. pp. 1127–1152 (2000). Probabilistic techniques in equilibrium and nonequilibrium statistical physics
56. White, N.L.: The basis monomial ring of a matroid. Adv. Math. **24**(3), 292–297 (1977)
57. Whitney, H.: On the abstract properties of linear dependence. Am. J. Math. **57**(3), 509–533 (1935)

Chapter 5
Classification of Minimal Polygons with Specified Singularity Content

Daniel Cavey and Edwin Kutas

Abstract It is known that there are only finitely many mutation-equivalence classes with a given singularity content, and each of these equivalence classes contains only finitely many minimal polygons. We describe an efficient algorithm to classify these minimal polygons. To illustrate this algorithm we compute all mutation-equivalence classes of Fano polygons with basket of singularities given by (i) $\mathcal{B} = \{m_1 \times \frac{1}{3}(1, 1), m_2 \times \frac{1}{6}(1, 1)\}$ and (ii) $\mathcal{B} = \{m \times \frac{1}{5}(1, 1)\}$.

Keywords Fano variety · qG-deformation · Mutation · Singularity content

5.1 Introduction

Let N be a lattice. A *polytope* P in $N_{\mathbb{R}} = N \otimes \mathbb{R}$ is a set of the form

$$P = \left\{ \sum_{u \in S} \lambda_u u : \lambda_u \geq 0 \text{ and } \sum_{u \in S} \lambda_u = 1 \right\},$$

where $S \subset N_{\mathbb{R}}$ is a finite set of points. A *Fano polytope* is a full-dimensional convex polytope such that the vertices $\text{vert}(P) \subset N$ are all primitive, and that the origin lies in the strict interior of P. Polytopes are considered up to $\text{GL}(N)$-equivalence. When N is a rank-two lattice, P is known as a *Fano polygon*.

The span of each face E of a Fano polygon P, that is $\mathbb{R}_{\geq 0} E$, defines a cone. Equivalently this is the polyhedral cone whose primitive generating vertices are given by the endpoints of E. This describes a fan in $N_{\mathbb{R}}$ corresponding to P, which in turn determines a toric del Pezzo surface X_P. Many properties of X_P have combinato-

D. Cavey (✉)
School of Mathematical Sciences, University of Nottingham, Nottingham NG7 2RD, UK
e-mail: Daniel.Cavey2@nottingham.ac.uk

E. Kutas
Mathematics Institute, University of Warwick, Coventry CV4 7AL, UK
e-mail: E.Kutas@warwick.ac.uk

A. M. Kasprzyk and B. Nill (eds.), *Interactions with Lattice Polytopes*,
Springer Proceedings in Mathematics & Statistics 386,
https://doi.org/10.1007/978-3-030-98327-7_5

rial analogues in the Fano polygon P; examples include the singularities and the anticanonical degree $(-K_{X_P})^2$. Toric geometry can be further studied in [6, 8].

The dual lattice of N is $M = \mathrm{Hom}(N, \mathbb{Z})$ with the pairing $\langle \cdot, \cdot \rangle : N \times M \to \mathbb{Z}$. The *lattice length* of a line segment $E \subset N_{\mathbb{R}}$ is given by the value $|E \cap N| - 1$. The *lattice height* of a line segment is given by the lattice distance from the origin: that is, given the unique primitive inward pointing normal n_E of E belonging to M, the height is given by $|\langle v, n_E \rangle|$, for any $v \in E$.

The motivation for the paper comes from an attempt to classify Fano varieties using mirror symmetry. An understanding of mirror symmetry can be gained from Coates–Corti–Galkin–Golyshev–Kasprzyk [4]. Mirror symmetry suggests that classifying Fano polytopes could help to classify Fano varieties.

More specifically we study Fano polytopes up to mutation-equivalence. A mutation is a combinatorial operation on a Fano polygon $P \subset N_{\mathbb{R}}$ introduced by Akhtar–Coates–Galkin–Kasprzyk [2], and is described in Sect. 5.2. An important mutation invariant of Fano polygons known as singularity content was introduced in [3] by Akhtar–Kasprzyk. This is a combinatorial property of P that describes the singularities of the corresponding Fano toric variety X_P. In this setting Fano varieties are considered up to qG-deformation, see Coates–Corti–Kasprzyk et al. and Kollár–Shepherd-Barron [1, 12].

Definition 1 ([1]) A del Pezzo surface with cyclic quotient singularities that admits a qG-deformation to a normal toric del Pezzo surface is said to be of *class TG*.

The reason we consider Fano polytopes and Fano varieties up to their respective equivalence classes lies in the following conjecture.

Conjecture 2 ([1, Conjecture A]) There exists a bijective correspondence between the set of mutation-equivalence classes of Fano polygons and the set of qG-deformation equivalence classes of locally qG-rigid class TG del Pezzo surfaces with cyclic quotient singularities.

Recent results from Corti–Heuberger [5] and Kasprzyk–Nill–Prince [11] support this conjecture.

Theorem 3 ([11, Theorem 1.2]) *There are precisely ten mutation-equivalence classes of Fano polygons such that the toric del Pezzo surface X_P has only T-singularities. They are in bijective correspondence with the ten families of smooth del Pezzo surfaces.*

Theorem 4 ([5, 11]) *There are precisely 29 qG-deformation families of del Pezzo surfaces with $m \geq 1$ singular points of type $\frac{1}{3}(1, 1)$. Precisely 26 of these are of class TG, and furthermore are in bijective correspondence with 26 mutation-equivalence classes of Fano polygons with singularity content $(n, \{m \times \frac{1}{3}(1, 1)\})$, where $m \geq 1$.*

Within this context, the aim is to classify Fano polygons with a given singularity content. Assuming Conjecture 2 holds, this is equivalent to a classification of del Pezzo surfaces admitting a toric degeneration which have the singularities described by the specified singularity content. We will use our work to provide more examples for understanding Conjecture 2. The main result of this paper is an efficient algorithm (described in Sect. 5.4) to produce representations (called minimal polygons) for the mutation-equivalence classes in a classification of Fano polygons with a specified singularity content in the case where the maximal height of the edges of each Fano polygon is given by an edge representing a non qG-smoothable singularity. As a corollary to the algorithm the following classifications (derived in Sects. 5.5 and 5.6 respectively) have been completed:

Theorem 5 *There are precisely 14 mutation-equivalence classes of Fano polygons with singularity content* $(n, \{m_1 \times \frac{1}{3}(1,1), m_2 \times \frac{1}{6}(1,1)\})$ *with* $m_1 \geq 0, m_2 > 0$.

Theorem 6 *There are precisely 12 mutation-equivalence classes of Fano polygons with singularity content* $(n, \{m \times \frac{1}{5}(1,1)\})$ *with* $m > 0$.

The reason for computing these classifications is that they are both in some sense the simplest cases after the $\frac{1}{3}(1,1)$ classification of Theorem 4. A $\frac{1}{6}(1,1)$ singularity, like a $\frac{1}{3}(1,1)$, is represented by an edge of a Fano polygon P of height 3. This is the smallest possible height for an edge representing any singularity that is not smoothable by a qG-deformation. Whereas 5 is the smallest value for r after 3 for which a $\frac{1}{r}(1,1)$ singularity is not smoothable by a qG-deformation.

If one of the Fano polygons appearing in the classifications is (via mutation if necessary) a triangle, then the corresponding toric variety will be the quotient of a weighted projective space as described in [9]. In particular let $\rho_0, \rho_1, \rho_2 \in N$ be the primitive generators of the rays for the fan of a Fano polygon P. Suppose that ρ_i satisfy the equation

$$\lambda_0\rho_0 + \lambda_1\rho_1 + \lambda_2\rho_2 = 0,$$

and span the lattice N. Then $X_P = \mathbb{P}(\lambda_0, \lambda_1, \lambda_2)$.

5.2 Mutations of Fano Polygons and Singularity Content

5.2.1 Mutations

Recall the definition of the Minkowski sum of lattice polygons.

Definition 7 Let $P, Q \subset N_{\mathbb{R}}$ be two lattice polytopes. Define the *Minkowski sum* of P and Q by

$$P + Q = \{p + q : p \in P, q \in Q\}.$$

By convention $P + \varnothing = \varnothing$.

Let $P \subset N_{\mathbb{R}}$ be a Fano polygon, and E be an edge of P. Consider the primitive inward pointing normal $n_E \in M$ of this edge. This vector acts as a grading function on P. For $h \in \mathbb{Z}$, define

$$\omega_h(P) = \mathrm{conv}\{v \in N \cap P : n_E(v) = h\}.$$

Note that $\omega_h(P)$ may be empty (indeed it will be for infinitely many values of h) and that $\omega_{-r_E}(P) = E$, where r_E is the height of E. Choose v_E to be a primitive vector of N such that $n_E(v_E) = 0$. Note this is uniquely defined up to sign in the two dimensional case. Set $F = \mathrm{conv}\{\mathbf{0}, v_E\}$; a line of lattice length 1 and height 0, that is parallel to E.

Definition 8 For all $h < 0$, suppose that there exists $G_h \subset N_{\mathbb{R}}$ such that

$$\{v \in \mathrm{vert}(P) : n_E(v) = h\} \subseteq G_h + |h|F \subseteq \omega_h(P).$$

In the case $\omega_h(P) = \varnothing$ the inclusion holds taking $G_h = \varnothing$. Then define the *mutation* of P with respect to n_E, F and G_h to be

$$\mathrm{mut}_{(n_E,F)}(P) = \mathrm{conv}\left(\bigcup_{h<0} G_h \cup \bigcup_{h\geq 0}(\omega_h(P) + hF)\right) \subset N_{\mathbb{R}}.$$

Example 9 Consider the Fano polygon $P = \mathrm{conv}\{(1, 0), (0, 1), (-5, -1)\}$ corresponding to the weighted projective space $\mathbb{P}(1, 1, 5)$. This polygon will be used as a running example throughout the paper. Mutate P with respect to the edge $E = \mathrm{conv}\left\{(1, 0), (0, 1)\right\}$. The primitive inner pointing normal is given by $n_E = (-1, -1) \in M$. This describes a grading on the points of N as shown in Fig. 5.1.

Set $F = \mathrm{conv}\left\{\mathbf{0}, (1, -1)\right\}$, a primitive slice at height 0. Choose $G_{-1} = \{(0, 1)\}$ which satisfies the required inclusion:

$$\left\{(0, 1), (1, 0)\right\} \subseteq G_{-1} + F \subseteq \mathrm{conv}\left\{(0, 1), (1, 0)\right\}.$$

For $h < -1$, $\omega_h(P) = \varnothing$, so trivially choose $G_h = \varnothing$. Calculating the mutation of P with respect to the primitive inner point normal n_E, the factor F and the polygon G_{-1} obtain:

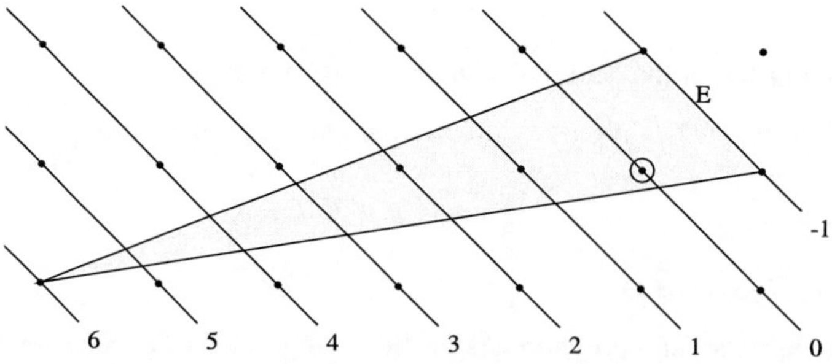

Fig. 5.1 The grading induced by $n_E = (-1, -1) \in M$

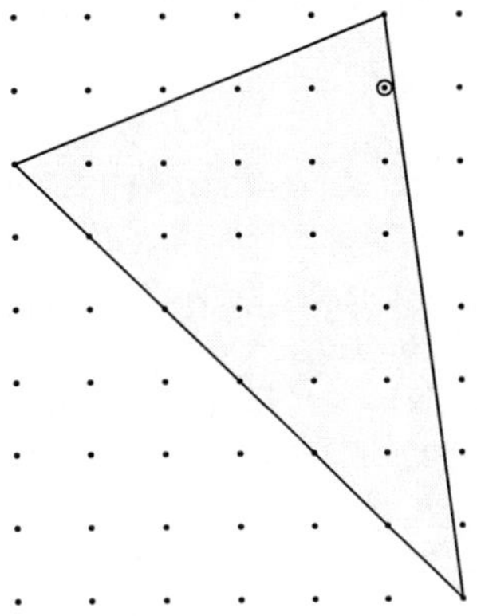

Fig. 5.2 The polygon Q associated with $\mathbb{P}(1,5,36)$

$$\begin{aligned} Q &= \mathrm{mut}_{(n_E,F)}(P) \\ &= \mathrm{conv}\left\{(G_{-1}) \cup (\omega_0(P)) \cup (\omega_1(P)+F) \cup (\omega_2(P)+2F) \cup \cdots \cup (\omega_6(P)+6F)\right\} \\ &= \mathrm{conv}\left\{(0,1),(-5,-1),(1,-7)\right\}. \end{aligned}$$

Q corresponds to the toric variety $\mathbb{P}(1,5,36)$. Informally the mutation subtracts one copy of F from P along the edge E, and adds six copies of F at the opposite vertex of P which is at $(-5,-1)$. This is illustrated in Fig. 5.2.

Note $\mathrm{mut}_{(n_E,F)}(P)$ is independent of the choice for G_h. If there is no possible choice of G_h, then the mutation with respect to n_E does not exist.

Lemma 10 ([11, Lemma 2.3]) *Let E be an edge of a Fano polygon P with primitive inner normal vector $n_E \in M$. Then P admits a mutation with respect to n_E if and only if*

$$|E \cap N| - 1 \geq r_E.$$

Applying Lemma 10 to the polygon P in Example 9 gives that the edge conv$\{(0,1),(-5,-1)\}$ does not admit a mutation: it has lattice length 1 and height 5.

There are a number of additional properties of mutations.

1. The choice of v_E is not important: $\mathrm{mut}_{(n_E,F)}(P)$ is isomorphic to $\mathrm{mut}_{(n_E,-F)}(P)$ via a $\mathrm{GL}(N)$-equivalence.
2. Mutation is invertible: If $Q = \mathrm{mut}_{(n_E,F)}(P)$, then $P = \mathrm{mut}_{(-n_E,F)}(Q)$.
3. P is a Fano polytope if and only if $\mathrm{mut}_{(n_E,F)}(P)$ is a Fano polytope [2, Proposition 2].

Definition 11 Let $P, Q \subset N_{\mathbb{R}}$ be two Fano polygons. Then P and Q are *mutation-equivalent* if there exists a finite sequence of polygons $P_0, P_1, \ldots, P_n$ such that $P_0 \cong P$, $P_n \cong Q$ and, $P_{i+1} = \mathrm{mut}_{(n_i,F_i)}(P_i)$ for some n_i and F_i, for all $i \in \{0,\ldots,n-1\}$.

Mutation-equivalence defines an equivalence relation.

5.2.2 Singularity Content

Recall the definition of singularity content introduced in [3].

Consider the action of μ_R, the cyclic group of order R, on $\mathbb{C}^2$ by $(x, y) \mapsto (\epsilon^a x, \epsilon^b y)$ where ϵ is an Rth root of unity. Recall that a *quotient singularity* $\frac{1}{R}(a, b)$ is given by the germ of the origin of $Z = \mathrm{Spec}(\mathbb{C}[x, y]^{\mu_R})$.

A *cyclic quotient singularity* is a quotient singularity $\frac{1}{R}(a, b)$ such that $\gcd(R, a) = 1$ and $\gcd(R, b) = 1$. In this situation, set $k = \gcd(a + b, R)$ so that $a + b = k\tilde{c}$ and $R = kr$, and let $c = \tilde{c}a^{-1}$. The cyclic quotient singularity can be written as $\frac{1}{kr}(1, kc - 1)$.

Definition 12 A cyclic quotient singularity $\frac{1}{kr}(1, kc - 1)$ is a *T-singularity* if $r \mid k$.

For an arbitrary T-singularity, let $k = nr$, so the singularity can be written in the form $\frac{1}{nr^2}(1, nrc - 1)$. If $n = 1$, the singularity is known as a *primitive T-singularity*. Kollár–Shepherd-Barron [12] show a cyclic quotient singularity is a T-singularity if and only if it admits a qG-smoothing.

Definition 13 A cyclic quotient singularity $\frac{1}{kr}(1, kc - 1)$ is an *R-singularity* if $k < r$.

A singularity is an R-singularity if and only if it is rigid under qG-deformation.

Consider an arbitrary cyclic quotient singularity $\frac{1}{kr}(1, kc - 1)$. By the Euclidean algorithm there exists unique non-negative integers n and k_0 such that $k = nr + k_0$. If $k_0 = 0$, then the singularity is a T-singularity and is qG-smoothable. If $k_0 > 0$, then the singularity is qG-deformation equivalent to a $\frac{1}{k_0 r}(1, k_0 c - 1)$ cyclic quotient singularity.

Definition 14 ([3]) Let $\sigma = \frac{1}{kr}(1, kc - 1)$ be a cyclic quotient singularity, and $k = nr + k_0$. The *residue* of σ is given by

$$\mathrm{res}(\sigma) = \begin{cases} \varnothing, & \text{if } k_0 = 0, \\ \frac{1}{k_0 r}(1, k_0 c - 1), & \text{otherwise.} \end{cases}$$

The *singularity content* of σ is the pair

$$\mathrm{SC}(\sigma) = \big(n, \mathrm{res}(\sigma)\big).$$

This discussion of T-singularities and R-singularities has a natural description in the language of cones. Each cone of a Fano polygon corresponds to a singularity of the corresponding toric variety. Specifically a cone C defines an edge $E = \mathrm{conv}\,\{u_1, u_2\}$, where u_1 and u_2 are primitive generators for the rays of C. Choose a point v in the interior of conv $\{\mathbf{0}, u_1, u_2\}$ such that

$$\text{conv}\left\{\mathbf{0}, u_i, v\right\} \cap N = \{\mathbf{0}, u_i, v\}, \qquad \text{for some } i \in \{1, 2\}.$$

Expressing v in terms of u_1 and u_2, obtain $v = pu_1 + qu_2$ where $p = a/R$ and $q = b/R$. The cone corresponds to the singularity $\sigma_C = \frac{1}{R}(a, b)$ on X_P.

Let $C \subset N_{\mathbb{R}}$ be a cone with generating rays described by the primitive lattice points p_0 and p_1. Consider $E = \text{conv}\{p_0, p_1\}$. Let r be the height of E and k be the lattice length. By the Euclidean algorithm $k = nr + k_0$. Divide C into separate sub-cones $C_0, \ldots, C_n$ where $C_1, \ldots, C_n$ (known as T-cones) have lattice length r, and C_0 has lattice length k_0 and is known as an R-cone. Importantly T-cones correspond to T-singularities on X_P and R-cones to R-singularities. The decomposition of C into sub-cones is analogous to the fact that the cyclic quotient singularity $\sigma_C = \frac{1}{kr}(1, kc - 1)$ can be qG-smoothed to the cyclic quotient singularity $\frac{1}{k_0 r}(1, k_0 c - 1)$ which is described by the sub-cone C_0. This discussion is independent of the choice of decomposition of C.

Definition 15 Let $P \subset N_{\mathbb{R}}$ be a Fano polygon. Label the edges of P in clockwise order $E_1, \ldots E_k$. Each edge E_i describes a cone which corresponds to a cyclic quotient singularity σ_i. Define the singularity content of an edge to be $\text{SC}(E_i) = \text{SC}(\sigma_i) = \left(n_i, \text{res}(\sigma_i)\right)$. Set $n = \sum_{i=1}^{k} n_i$ and $\mathcal{B} = \{\text{res}(\sigma_1), \ldots, \text{res}(\sigma_k)\}$, where $\mathcal{B}$ is an ordered set known as the *basket of R-singularities*. Then the *singularity content* of P is defined to be

$$\text{SC}(P) = \left(n, \mathcal{B}\right).$$

Example 16 Consider $P = \text{conv}\left\{(0, 1), (1, 0), (-5, -1)\right\}$ from Example 9. The edges conv $\left\{(1, 0), (0, 1)\right\}$ and conv $\left\{(1, 0), (-5, -1)\right\}$ are both of length 1 and height 1 and so contribute a single T-cone each. The final edge conv $\left\{(1, 0), (0, 1)\right\}$ is of height 5 and length 1 and so is a single R-cone. It remains to calculate the singularity corresponding to this edge. Using the earlier notation, the only choice for v is $(-1, 0)$, and finding the unique solution to the equation $(-1, 0) = p(0, 1) + q(-5, -1)$ as $p = q = \frac{1}{5}$ shows that the edge represents a $\frac{1}{5}(1, 1)$ R-singularity. Therefore $\text{SC}(P) = (2, \{\frac{1}{5}(1, 1)\})$.

Proposition 17 ([3, Proposition 3.6]) *The singularity content is an invariant of Fano polygons under mutation.*

Indeed in Example 9, it is routine to check that $\text{SC}(Q) = \left(2, \{\frac{1}{5}(1, 1)\}\right)$ which is the same as the singularity content for P.

5.2.3 Hirzebruch–Jung Continued Fractions and Applications to Algebraic Geometry

There is information about the del Pezzo surface X_P corresponding to a polygon P written into the singularity content; X_P is qG-deformation equivalent to a del Pezzo surface X such that the topological Euler number $\chi\left(X\backslash\text{Sing}(X)\right) = n$ and the singular points are given by $\text{Sing}(X) = \mathcal{B}$. The anticanonical degree and Hilbert series of X_P are totally determined by the singularity content. See [1, 3].

Definition 18 Let $p, q \in \mathbb{Z}_0$ be coprime. Then the *Hirzebruch–Jung continued fraction* of p/q is given by the continued fraction of the form:

$$\frac{p}{q} = a_1 - \cfrac{1}{a_2 - \cfrac{1}{a_3 - \cfrac{1}{\ddots}}}$$

$$= [a_1, \ldots, a_k].$$

Given a cyclic quotient singularity $\sigma = \frac{1}{R}(1, a-1)$, construct the variety $Z = \text{Spec}(\mathbb{C}[x, y]^{\mu_R})$ as in the definition of quotient singularity. Information about a minimal resolution of Z can be calculated from the Hirzebruch–Jung continued fraction of $R/(a-1)$. Consider the minimal resolution $\pi : Y \to Z$ with

$$K_Y = \pi^* K_Z + \sum_{i=1}^{k} d_i E_i.$$

Let $[a_1, \ldots, a_{k_\sigma}]$ be the Hirzebruch–Jung continued fraction of $R/(a-1)$. Then the values $-a_i$ are the self-intersection numbers of the exceptional divisors E_i. Additionally define α_i, β_i for $i \in \{1, \ldots, k_\sigma\}$ by:

$$\alpha_1 = \beta_{k_\sigma} = 1,$$

$$\frac{\alpha_i}{\alpha_{i-1}} = [a_{i-1}, \ldots, a_1], \qquad \text{for } i \in \{2, \ldots, k_\sigma\},$$

$$\frac{\beta_i}{\beta_{i+1}} = [a_{i+1}, \ldots, a_{k_\sigma}], \qquad \text{for } i \in \{1, \ldots, k_\sigma - 1\}.$$

Note that the α_i are in increasing order, and the β_i are in decreasing order. The discrepancy of E_i is given by $d_i = -1 + (\alpha_i + \beta_i)/R$. For further reading on minimal resolutions, see Reid [15].

Proposition 19 ([3, Proposition 3.3, Corollary 3.5]) *Let P be a Fano polygon and let X_P be the corresponding toric surface. Suppose P has singularity content $(n, \mathcal{B})$. Then*

$$(-K_{X_P})^2 = 12 - n - \sum_{\sigma \in \mathcal{B}} A_\sigma,$$

where $A_\sigma = k_\sigma + 1 - \sum_{i=1}^{k_\sigma} d_i^2 a_i + 2\sum_{i=1}^{k_\sigma - 1} d_i d_{i+1}$. *Furthermore the anticanonical Hilbert series of* X_P *admits a decomposition*

$$Hilb(X_P, -K_{X_P}) = \frac{1 + \left((-K_{X_P})^2 - 2\right)t + t^2}{(1-t)^3} + \sum_{\sigma \in \mathcal{B}} Q_\sigma(t),$$

where $Q_{\frac{1}{R}(1,a-1)}(t) = \frac{1}{1-t^R}\sum_{i=1}^{R-1}(\delta_{ai} - \delta_0)t^{i-1}$ *is the Riemann–Roch contribution coming from the singularity* $\frac{1}{R}(1, a-1)$ *and* $\delta_j = \frac{1}{R}\sum_{\epsilon \in \mu_R, \epsilon \neq 1} \frac{\epsilon^j}{(1-\epsilon)(1-\epsilon^{a-1})}$ *are the Dedekind sums.*

Example 20 Recall from Example 16 that $P = \text{conv}\left\{(0, 1), (1, 0), (-5, -1)\right\}$ has singularity content $\left(2, \{\frac{1}{5}(1, 1)\}\right)$. The Hirzebruch–Jung continued fraction of the cyclic quotient singularity $\frac{1}{5}(1, 1)$ is simply [5], so $d_1 = -\frac{3}{5}$ and $A_{\frac{1}{5}(1,1)} = \frac{1}{5}$. Also $Q_{\frac{1}{5}(1,1)} = \frac{t-2t^2+t^3}{5(1-t^5)}$. Therefore the anticanonical degree and Hilbert series of $X = \mathbb{P}(1, 1, 5)$ are given by

$$(-K_X)^2 = \frac{49}{5},$$

$$\text{Hilb}\big(X, -K_X\big) = \frac{1 + 8t + 2t^3 - 2t^4 - 8t^6 - t^7}{(1-t^5)(1-t)^3}.$$

More generally for a polygon P with n primitive T-singularities and basket of singularities $\mathcal{B} = \{m \times \frac{1}{5}(1, 1)\}$:

$$(-K_{X_P})^2 = 12 - n - \frac{1}{5}m,$$

$$\text{Hilb}(X_P, -K_{X_P}) = \frac{-t^7 + (n-10)t^6 + (m-1)t^5 - 2mt^4 + 2mt^3 + (1-m)t^2 + (10-n)t + 1}{(1-t)^3(1-t^5)}.$$

Hirzebruch–Jung fractions can be further studied in [8, 14].

5.3 Minimal Fano Polygons

Mutation-equivalence classes raise the issue about our choice of representative of a mutation-equivalence class of Fano polygons. This leads to the definition of a minimal polygon from [11]. For a polygon P, the notation ∂P denotes the boundary of P.

Definition 21 Let $P \subset N_{\mathbb{R}}$ be a Fano polygon. The polygon P is *minimal* if

$$|\partial P \cap N| \leq |\partial Q \cap N|, \qquad \text{for all } Q = \text{mut}_{(n,F)}(P).$$

For an edge E of P, let $n_E \in M$ be the primitive inner pointing normal of E. Define $h^E_{\min} = \min\{n_E(v) : v \in P\}$ and $h^E_{\max} = \max\{n_E(v) : v \in P\}$. Then [11] tells us P is minimal if and only if $|E \cap N| - 1 \geq |h^E_{\min}|$ implies $|h^E_{\min}| \leq h^E_{\max}$ for all edges E of P.

Given a Fano polygon P, find a minimal representative of the mutation-equivalence class by calculating all possible mutations of P. If none of the Fano polygons obtained via these mutations have fewer boundary points, then P is minimal. Otherwise choose one of the mutations of P that have fewer boundary points as our new representative. Repeat this inductively, to obtain a minimal representative. The process must terminate since the number of boundary points of P is finite and non-negative.

A minimal representative of a mutation-equivalence class is not necessarily unique. We will always choose the representative of any equivalence class of Fano polygons to be minimal.

Example 22 In Example 9, $|\partial P \cap N| = 3$ and $|\partial Q \cap N| = 8$. We know there is only one other existing mutation of P. It is routine to check that this remaining mutation does not have fewer boundary points than P. It follows that P is minimal.

5.4 Algorithm to Calculate Minimal Polygons with Given Basket

5.4.1 Special Facets

We require the notion of a special facet introduced by Øbro [13].

Definition 23 Let $P \subset N_{\mathbb{R}}$ be a Fano polygon. An edge E of P is a *special facet* if

$$\sum_{v \in \text{vert}(P)} v \in \mathbb{R}_{\geq 0} E.$$

Example 24 Considering $P = \text{conv}\{(0, 1), (1, 0), (-5, -1)\}$, calculate that:

$$\sum_{v \in \text{vert}(P)} v = (0, 1) + (1, 0) + (-5, -1) = (-4, 0).$$

So P has a unique special facet given by $F = \text{conv}\{(0, 1), (-5, -1)\}$.

By the definition of Fano polygon, $\mathbf{0} \in \operatorname{int}(P)$. Therefore the union of all cones obtained from a Fano polygon P is equal to $N_{\mathbb{R}}$, so P has at least one special facet. We use a result from [10] which is derived from a proof in [7].

Lemma 25 *Let P be a Fano polygon. Let F be a special facet of P of height h and with inward pointing normal $n_F \in M$. Then*

$$\operatorname{vert}(P) \subset \left\{(a, b) \in N_{\mathbb{R}} : -h(h+1) \leq \langle n_F, (a, b)\rangle \leq h\right\}.$$

5.4.2 Description of Algorithm

Define the maximal local index of a Fano polygon P by

$$m_P = \max\left(\operatorname{height}(E) : E \in \mathcal{F}(P)\right),$$

where $\mathcal{F}(P)$ is the set of edges of P. Similarly define $m_{\mathcal{B}}$ to be the maximum height among the cones representing the R-singularities of P.

The classification of Fano polygons with a given basket of singularities $\mathcal{B}$ up to mutation-equivalence is split into two cases:

1. $m_P = m_{\mathcal{B}}$
2. $m_P > m_{\mathcal{B}}$

The proof of [11, Theorem 6.3] efficiently tackles Case 2. Note the polygons this proof outputs are not necessarily minimal. It remains to deal with Case 1. An algorithm to compute this classification has been completed in [10]. However tackling classifications beyond the case of polygons with only $\frac{1}{3}(1, 1)$ R-singularities is inefficient.

The main result of this paper is an efficient algorithm to tackle Case 1. The algorithm is as follows: start with only a single vertex $(a, l_F) = v \in F$ where F is the special facet. We can assume that the other endpoint of F is (b, l_F), where $b < a$. Since $m_{\mathcal{B}} = m_P$ we have a bound on the heights on all edges. Given a point v_1 consider the set of points

$$S = \{v_2 : v_2 - v_1 \text{ is primitive and } \operatorname{height}(E_{(v_1,v_2)}) = h\}$$

where $E_{(v_1,v_2)}$ is the line segment from v_1 to v_2. Note S is a subset of a line L_h. Given vertices $v_1, v_2, \ldots v_k$ consider the lines $L_1, \ldots L_{m_{\mathcal{B}}}$ then pick all the valid points u on these lines that give us an edge $E_{(v_k,u)}$ that respects convexity and defines either a T-singularity or a residual singularity of $\mathcal{B}$. Lemma 25 gives a bound on how low these lines can go. This often suffices as bound, however it is possible that either

1. there exists h such that L_h is horizontal, or
2. there exists $u \in L_h$ such that the $E_{(v_k,u)}$ is horizontal.

In Case 1 L_h never touches the lower bound. In a similar fashion in Case 2 we are uncertain how far we can extend the edge $E_{(v,u)}$ by adding T-cones resulting in infinitely many points of the form $v_{k+1} = u + n(u - v_k)$ where n is chosen suitably from $\mathbb{Z}_{\geq 0}$.

We start by showing how to deal with Case 2. Suppose $v_{k+1} = u + n(u - v_k) \in \text{vert}(P)$. Once the construction of P is complete then there must be another vertex $v_{k+2} \in \text{vert}(P)$, where subscripts are considered modulo $\#\,\text{vert}(P)$, defining an edge $E_{(v_{k+1},v_{k+2})}$. Consider replacing n by $n + h$, which has the same effect as a shear transformation parallel to the line L_h. Having done this, all new choices of v_{k+2} equal the image of the old v_{k+2} under the same shearing. In particular as $n \to \infty$, the lines L_h tend towards lines going through the origin. This means that at suitably large values of n, since $l_F > 0$, the choices of v_{k+2} will end up violating convexity. Hence there is only a finite choice along each L_h. Note that while this is quite computationally intensive, this happens at most once per a polygon.

The method to deal with Case 1 is similar. Similarly to the previous case we fix L_h and consider points v_{k+1} on it. From the point v_{k+1} we construct the lines L'_h we then pick the next vertex v_{k+2}. At this point we exclude the cases where v_{k+1} and v_{k+2} lie on a horizontal line as this puts us in Case 2. As before we see that that as the x-coordinate of v_{k+1} tends towards $-\infty$, then the x-coordinate of v_{k+2} tends towards ∞. Hence once this cutoff occurs there are a finite set of possible points for v_{k+1}.

Theorem 26 *The algorithm gives a complete classification for Fano polygons with a specific basket of singularities $\mathcal{B}$.*

Proof It suffices to show that at each stage there are only finitely many choices of vertices, this is clear from the above description. There are also only finitely many choices for inputs as up to a linear transformation $0 < a \leq l_f$ and $l_f \leq m_{\mathcal{B}}$, so there are only finitely many choices of a. □

We check our output up to $\text{GL}(N)$-equivalence and mutation-equivalence. Two polygons can be shown to be mutation-equivalent by explicitly calculating a sequence of mutations between the two. Conversely a polygon P has corresponding to it a *maximally mutable Laurent polynomial* $f \in \mathbb{C}[x_1^{\pm 1}, \ldots, x_n^{\pm 1}]$ as defined in [1]. The *classical period* of f given by

$$\begin{aligned} \pi_f(t) &= \left(\frac{1}{2\pi i}\right)^n \int_{|x_1|=\cdots=|x_n|=1} \frac{1}{1 - tf(x_1, \ldots, x_n)} \frac{dx_1}{x_1} \cdots \frac{dx_n}{x_n} \\ &= \sum_{k \geq 0} \text{coeff}_1(f^k) t^k, \end{aligned}$$

is an invariant under mutation. Hence two polygons with different periods cannot be mutation-equivalent. We have computer code in SAGE [16] that efficiently implements the algorithm.

It is important to compare the LDP-algorithm [10] used to calculate the $\frac{1}{3}(1, 1)$ classification of Fano polygons in Theorem 4 with this new algorithm. The LDP-algorithm takes as input a value for the *index* $l_P = \operatorname{lcm}\{\text{height}(E) : E \in \mathcal{F}(P)\}$, and returns all Fano polygons P with this index. Hence when used to calculate the case $m_P = m_{\mathcal{B}}$ in the $\frac{1}{3}(1, 1)$ classification, edges of height 1, 2 and 3 are permitted, and so the LDP-algorithm needs to be run for $l_P \in \{3, 6\}$. In the $\frac{1}{5}(1, 1)$ classification of Theorem 6, the value of l_F can be up to 60. All polygons P satisfying $l_P \leq 16$ have been classified using the LDP-algorithm but this took approximately three days to obtain and the run time will increase at least quadratically with l_P. A classification of Fano polygons with singularity content $\left(n, \{m \times \frac{1}{r}(1, 1)\}\right)$, where $r \geq 7$ would be impossible to calculate using the LDP-algorithm (l_P would be bounded above by 420 if $r = 7$).

In comparison we had the following run times to calculate the classifications of Theorems 5 and 6 using our algorithm:

Basket in classification	Run time
$(n, \{m \times \frac{1}{3}(1, 1)\})$	40 s
$(n, \{m \times \frac{1}{5}(1, 1)\})$	8 min

Furthermore a classification of Fano polygons whose basket of singularities only contains $\frac{1}{7}(1, 1)$ R-singularities has been informally completed. The main reasons for the difference in speed between the LDP-algorithm and our algorithm are as follows.

1. We only look for minimal representatives for each mutation equivalence class. This is not the case in the LDP-algorithm.
2. The LDP-algorithm is not designed to look for polygons based on the singularity content. For example in the $\frac{1}{3}(1, 1)$ classification running the LDP-algorithm for $l_P \in \{3, 6\}$ will output polygons that do not have singularity content $(n, \{m \times \frac{1}{3}(1, 1)\})$.

5.5 Minimal Fano Polygons with $\mathcal{B} = \{m_1 \times \frac{1}{3}(1, 1), m_2 \times \frac{1}{6}(1, 1)\}$

We apply our algorithm to classify all Fano polygons whose only R-singularities are the cyclic quotient singularities $\frac{1}{3}(1, 1)$ and $\frac{1}{6}(1, 1)$. Set $\mathcal{B} = \{m_1 \times \frac{1}{3}(1, 1), m_2 \times \frac{1}{6}(1, 1)\}$, where $m_1 \in \mathbb{Z}_{\geq 0}$ and $m_2 \in \mathbb{Z}_{>0}$. Here m_2 is non-zero since a classification for Fano polygons with only $\frac{1}{3}(1, 1)$ R-singularities has been completed in [11].

In the $\frac{1}{3}(1, 1)$ classification of [11], a bound on the number of R-singularities is found by substituting the degree contribution $A_{\frac{1}{3}(1,1)} > 0$ into an expression for the anticanonical degree of the corresponding toric Fano variety from Proposition 19 since we know this value to be positive. However the degree contribution $A_{\frac{1}{6}(1,1)}$ is negative and a similar method does not yield a bound. We show that this can be done by a purely combinatorial argument instead.

Lemma 27 *There exist no minimal Fano polygons $P \subset N_{\mathbb{R}}$, with $m_P = 3$ and residual basket given by $\mathcal{B} = \{m \times \frac{1}{6}(1, 1)\}$, where $m \geq 3$.*

Proof The result for $m > 3$ follows from the base case $m = 3$.

Let P be a polygon with $\mathcal{B} = \{3 \times \frac{1}{6}(1, 1)\}$. By a $GL(N)$-translation, assume that one of the R-singularities is given by $E_1 = \text{conv}\{(-1, 3), (1, 3)\}$. By mutating with respect to any T-singularity lying between E_1 and a second R-singularity, assume this second R-singulary is adjacent to E_1, given by an edge E_2 with endpoints $(1, 3)$ and (a, b). The primitive inner pointing normal of E_2 is given by

$$n_{E_2} = \left(\frac{b-3}{g}, \frac{1-a}{g}\right) \in M$$

where $g = \gcd(b - 3, 1 - a)$. The height of E_2 is

$$h = -n_{E_2} \cdot (1, 3) = \frac{3a - b}{g}.$$

Since E_2 represents a $\frac{1}{6}(1, 1)$ singularity, set $h = 3$:

$$\frac{3a - b}{g} = 3, \qquad b = 3a - 3\gcd(b - 3, 1 - a).$$

By convexity $b < 3$. The only remaining integer solution with $a \geq 0$, is given by $(0, -3)$. However this point is not primitive so it can not be chosen as a vertex of a Fano polygon. Hence $a < 0$.

Suppose the second edge from $(-1, 3)$, denoted E_3, is vertical. By convexity $a = -1$ and (a, b) is a vertex of E_3. But then E_3 is of height 1 so cannot represent a $\frac{1}{6}(1, 1)$ singularity and $m < 3$. Suppose E_3 is not vertical. Again convexity demands that the second endpoint of E_3 has first coordinate less than -1. Then $\text{height}(E_3) > 3$ which contradicts $m_P = 3$.

Therefore there can be no minimal Fano polygon with residual basket given by $\mathcal{B} = \{3 \times \frac{1}{6}(1, 1)\}$ with $m_P = 3$. □

A similar argument shows that for a basket $\mathcal{B} = \{m_1 \times \frac{1}{3}(1, 1), m_2 \times \frac{1}{6}(1, 1)\}$ as above, then $m_1 + m_2 < 3$.

Examples in this particular classification demonstrate a notion known as shattering introduced by Wormleighton [17]. Let $C_1 = \langle u, v \rangle$, $C_2 = \langle v, w \rangle$ be two cones in $N_{\mathbb{R}}$. Suppose the vectors $v - u$, $w - v$ are parallel. Then define the hyperplane sum of C_1 and C_2 to be given by $C_1 * C_2 = \langle u, w \rangle$.

Corollary 28 ([17, Corollary 2.2]) *Let $\sigma_1 * \sigma_2 * \cdots * \sigma_n = \tau$ be a T-singularity. Then the Riemann–Roch contributions Q_{σ_i} and the degree contributions A_{σ_i} satisfy*

Table 5.1 The polygons with singularity content of the form $(n, \{m_1 \times \frac{1}{3}(1,1), m_2 \times \frac{1}{6}(1,1)\})$ with $m_2 \neq 0$

#	vert(P)	n	m_1	m_2	$(-K_X)^2$
1.1	$(-1,3),(1,3),(0,-1)$	2	0	1	32/3
1.2	$(-1,3),(1,3),(1,2),(0,-1)$	3	0	1	29/3
1.3	$(-1,3),(1,3),(1,1),(0,-1)$	4	0	1	26/3
1.4	$(-1,3),(1,3),(1,0),(0,-1)$	5	0	1	23/3
1.5	$(-1,3),(1,3),(1,2),(0,-1),(-1,0)$	6	0	1	20/3
1.6	$(-1,3),(1,3),(1,2),(0,-1),(-1,-1)$	7	0	1	17/3
1.7	$(-1,3),(1,3),(1,0),(0,-1),(-1,0)$	8	0	1	14/3
1.8	$(-1,3),(1,3),(1,0),(-1,-1)$	8	0	1	14/3
1.9	$(-1,3),(1,3),(1,0),(0,-1),(-1,-1)$	9	0	1	11/3
1.10	$(-1,3),(1,3),(1,2),(-1,-4)$	10	0	1	8/3
1.11	$(-1,3),(1,3),(1,-1),(-1,-3)$	11	0	1	5/3
1.12	$(-1,3),(1,3),(5,-1),(-5,-1)$	12	0	1	2/3
1.13	$(-1,1),(1,1),(5,-1),(-5,-1)$	12	0	2	4/3
1.14	$(-1,3),(1,3),(1,-1),(-1,-2)$	9	1	1	2

$$Q_{\sigma_1} + \cdots + Q_{\sigma_n} = 0,$$

$$A_{\sigma_1} + \cdots + A_{\sigma_n} = A_\tau = \frac{\textit{lattice length}(\tau)}{\textit{lattice height}(\tau)}.$$

Consider a T-cone $C = \operatorname{cone}\{(-2,3),(1,3)\}$ of height 3. By adding an additional ray given by primitive generating vector $(-1,3)$ decompose C into two sub-cones C_1 and C_2 representing a $\frac{1}{3}(1,1)$ and a $\frac{1}{6}(1,1)$ R-singularity respectively. By Corollary 28

$$Q_{\frac{1}{3}(1,1)} + Q_{\frac{1}{6}(1,1)} = 0, \qquad \text{and} \qquad A_{\frac{1}{3}(1,1)} + A_{\frac{1}{6}(1,1)} = 1.$$

Knowing $A_{\frac{1}{3}(1,1)} = \frac{5}{3}$ and $Q_{\frac{1}{3}(1,1)} = -\frac{t}{3(1-t^3)}$, derive:

$$A_{\frac{1}{6}(1,1)} = -\frac{2}{3}, \qquad \text{and} \qquad Q_{\frac{1}{6}(1,1)} = \frac{t}{3(1-t^3)}.$$

By Proposition 19 and Lemma 27, calculate $(-K_{X_P})^2 = 12 - n - \frac{5}{3}m_1 + \frac{2}{3}m_2$. Since we are interested in Fano polygons, $(-K_{X_P})^2 > 0$, so $n \leq 13$. The algorithm can be run a finite number of times to get the desired classification.

The results for the classification of polygons with singularity content of the form $(n, \{m_1 \times \frac{1}{3}(1,1), m_2 \times \frac{1}{6}(1,1)\})$ with $m_2 \neq 0$, up to mutation-equivalence is given in Table 5.1. All the polygons listed arose in Case 1 with the exception of polygon 1.12. These polygons are illustrated in Fig. 5.3.

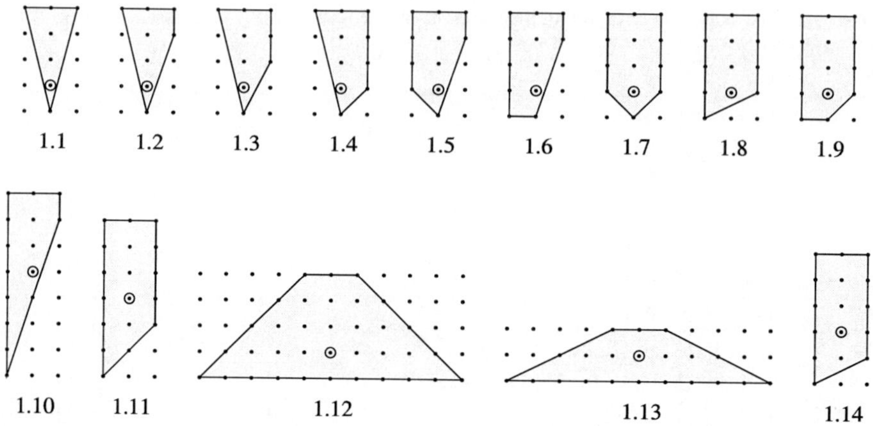

Fig. 5.3 Minimal representatives of mutation-equivalence classes of Fano polygons with singularity content $(n, \{m_1 \times \frac{1}{3}(1,1), m_2 \times \frac{1}{6}(1,1)\})$, where $m_1 \geq 0, m_2 > 0$

Recall that the maximally mutable Laurent polynomial of a polygon P is a polynomial f such that $\mathrm{Newt}(f) = P$, and that the mutations of f remain Laurent polynomials. The periods of maximally mutable Laurent polynomials are mutation invariants by [2, Lemma 1]. The maximally mutable Laurent polynomials of 1.7 and 1.8 are given respectively by:

$$f = xy^3 + 3xy^2 + ay^3 + 3xy + by^2 + \frac{y^3}{x} + x + cy + 3\frac{y^2}{x} + 3\frac{y}{x} + \frac{1}{y} + \frac{1}{x},$$
$$g = xy^3 + 3xy^2 + dy^3 + 3xy + ey^2 + \frac{y^3}{x} + x + fy + 4\frac{y^2}{x} + 6\frac{y}{x} + 4\frac{1}{x} + \frac{1}{xy}.$$

Calculating the corresponding periods of f and g obtain:

$$\pi_f = 1 + (2a+2)x^2 + (3b+36)x^3 + (6a^2 + 24a + 4c + 186)x^4 + (20ab + 360a + 60b + 760)x^5 + \cdots,$$
$$\pi_g = 1 + 14x^2 + 6ax^3 + 546x^4 + (420a + 30b)x^5 + \cdots.$$

It is easy to see that these periods are not equal and hence the polygons cannot be mutation-equivalent. All other Fano polygons in this classification have pairwise distinct singularity content, hence are not mutation equivalent.

5.6 Minimal Fano Polygons with $\mathcal{B} = \{m \times \frac{1}{5}(1, 1)\}$

We find all Fano polygons with singularity content of the form $(n, \{m \times \frac{1}{5}(1, 1)\})$ with $m > 0$. Similarly to Sect. 5.5, bounds on n and m are required to ensure the complete classification.

Lemma 29 *There exist no minimal Fano polygons $P \subset N_{\mathbb{R}}$, with $m_P = 5$ and residual basket given by $\mathcal{B} = \{m \times \frac{1}{5}(1, 1)\}$, where $m \geq 3$.*

Proof Similarly to the proof of Lemma 27, assume the existence of a Fano polygon P with three $\frac{1}{5}(1, 1)$ singularities, and perform a combination of $GL(N)$-translations and mutations so that one of the R-singularities is represented by the edge $E_1 = \text{conv}\{(-3, 5), (-2, 5)\}$, and another by $E_2 = \text{conv}\{(-2, 5), (a, b)\}$, where $(a, b) \neq (-3, 5)$. We show you can always mutate P so that the third R-singularity is represented by $E_3 = \text{conv}\{(-3, 5), (c, d)\}$, where $(c, d) \neq (-3, 5)$, without disrupting the original two $\frac{1}{5}(1, 1)$ singularities sitting adjacently.

Study the possible T-cones that when mutated with respect to would separate the adjacent R-singularities. Calculate the line of points from $(-2, 5)$ that give an edge at height 5, since (a, b) must lie on this line in order for E_2 to define a $\frac{1}{5}(1, 1)$ R-singularity. Unlike Lemma 27, since we are only interested in $\frac{1}{5}(1, 1)$ singularities, assume that the inner pointing normal $n_{E_2} = (b - 5, -2 - a)$ is primitive. This line of points on which (a, b) lies, provides a bound on where (c, d) can lie by convexity. Convexity also determines that $d \leq 5$. Furthermore since P is Fano, the origin $(0, 0)$ must lie in the interior which further bounds the region (c, d) lies in. Finally since we are only interested in the case where the prospective T-singularity would disrupt the adjacent R-cones when mutated with respect to we obtain a final bound on the

Table 5.2 The polygons with singuarity content of the form $(n, \{m \times \frac{1}{5}(1, 1)\})$ with $m > 0$

#	vert(P)	n	m	$(-K_X)^2$
2.1	$(-3, 5), (-2, 5), (1, -2)$	2	1	49/5
2.2	$(-3, 5), (-2, 5), (-1, 3), (1, -2)$	3	1	44/5
2.3	$(-3, 5), (-2, 5), (-1, 3), (1, -2), (-2, 3)$	4	1	39/5
2.4	$(-3, 5), (-2, 5), (-1, 3), (1, -2), (-1, 1)$	5	1	34/5
2.5	$(-3, 5), (-2, 5), (0, 1), (1, -2), (-1, 1)$	6	1	29/5
2.6	$(-3, 5), (-2, 5), (0, 1), (1, -2), (0, -1)$	7	1	24/5
2.7	$(-3, 5), (-2, 5), (1, -1), (0, -1)$	7	1	24/5
2.8	$(-3, 5), (-2, 5), (1, -1), (1, -2), (0, -1)$	8	1	19/5
2.9	$(-3, 5), (-2, 5), (1, -1), (1, -3)$	9	1	14/5
2.10	$(-3, 5), (-2, 5), (2, -3), (2, -5)$	10	1	9/5
2.11	$(-3, 5), (-2, 5), (4, -1), (-3, -1)$	11	1	4/5
2.12	$(-3, 5), (-2, 5), (3, -5), (2, -5)$	10	2	8/5

region in which (c, d) can lie. It is then possible to exhaustively check that none of the primitive lattice points in this region give the second vertex of a T-cone.

Hence assume that the three R-cones lie adjacently. Calculating the points (c, d) so that E_3 is height 5 and comparing with the possible choices of (a, b), note that there are no choices of (a, b) and (c, d) that maintain convexity.

Therefore there can be no minimal Fano polygon with residual basket given by $\mathcal{B} = \left\{3 \times \frac{1}{5}(1, 1)\right\}$ with $m_P = 5$. □

The above method extends very nicely to a combinatorial proof that

Lemma 30 *There exist no Fano polygons* $P \subset N_{\mathbb{R}}$ *with* $m_P = p$ *and residual basket given by* $\mathcal{B} = \{m \times \frac{1}{p}(1, 1)\}$, *where* $m \geq 3$, p *is odd and* $p \geq 3$.

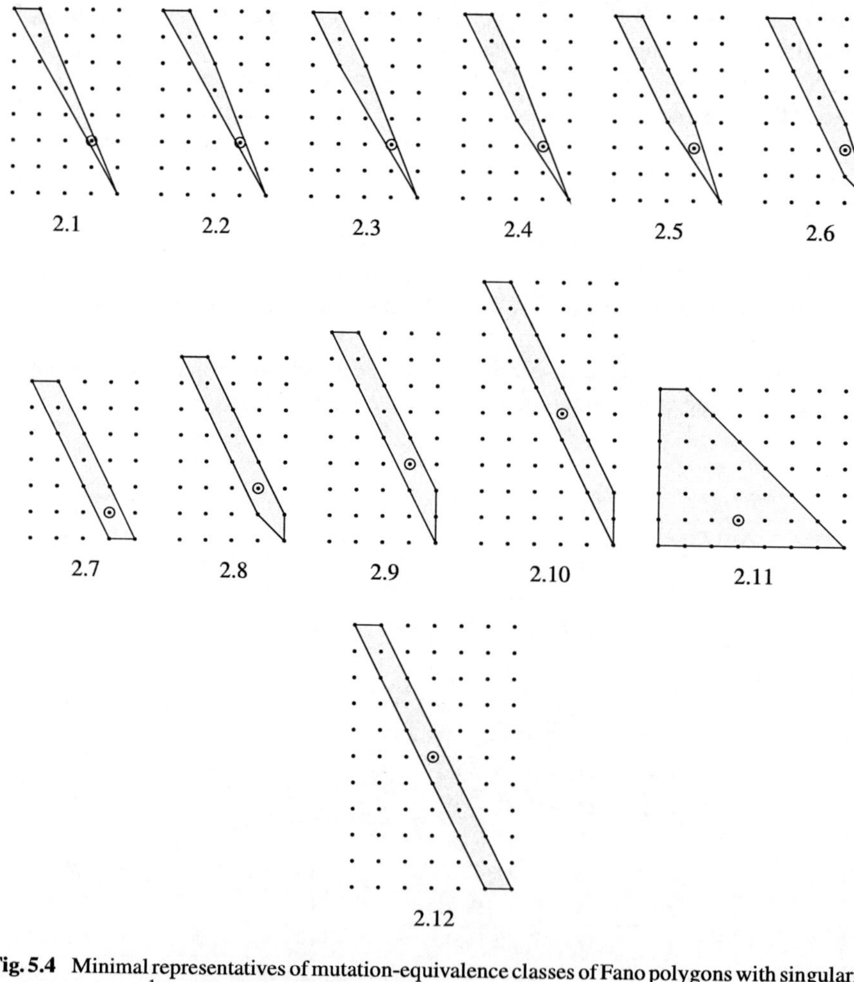

Fig. 5.4 Minimal representatives of mutation-equivalence classes of Fano polygons with singularity content $(n, \{m \times \frac{1}{5}(1, 1)\})$ where $m > 0$

By Example 20, the anticanonical degree of the toric variety corresponding to a Fano polygon with only R-singularities of type $\frac{1}{5}(1,1)$ is given by $(-K_{X_P})^2 = 12 - n - \frac{1}{5}m > 0$. Therefore $n < 12$. We apply the algorithm to complete the classification.

The results for the classification of polygons with singuarity content of the form $(n, \{m \times \frac{1}{5}(1,1)\})$ with $m > 0$ is given in Table 5.2. All the polygons were found in Case 1. None arose in Case 2. These polygons are illustrated in Fig. 5.4.

Similarly to Sect. 5.5 note that polygons 2.6 and 2.7 are not mutation equivalent by looking at the periods π_f, π_g of their respective maximally mutable Laurent polynomials f and g:

$$\pi_f = 1 + 12x^2 + 6ax^3 + 396x^4 + (360a + 30b)x^5 + \cdots,$$

$$\pi_g = 1 + (2c + 12)x^2 + (6c + 3d + 90)x^3 + (6c^2 + 24d + 144c + 636)x^4 + (20cd + 60c^2 + 390d + 1260c + 6900)x^5 + \cdots.$$

All other Fano polygons in the classification have pairwise distinct singularity content and therefore belong to different mutation equivalence classes.

Acknowledgements We thank Alexander Kasprzyk for his guidance and many useful conversations. Much of the paper was written in during a visit of EK to Nottingham supported by EPRSC Fellowship EP/NO22513/1. EK would also like to thank his doctoral advisor, Miles Reid.

References

1. Akhtar, M., Coates, T., Corti, A., Heuberger, L., Kasprzyk, A.M., Oneto, A., Petracci, A., Prince, T., Tveiten, K.: Mirror symmetry and the classification of orbifold del Pezzo surfaces. Proc. Amer. Math. Soc. **144**(2), 513–527 (2016)
2. Akhtar, M., Coates, T., Galkin, S., Kasprzyk, A.M.: Minkowski polynomials and mutations. SIGMA Symmetry Integr. Geom. Methods Appl. **8**, Paper 094, 17 (2012)
3. Akhtar, M., Kasprzyk, A.M.: Singularity content (2014). arXiv:1401.5458 [math.AG]
4. Coates, T., Corti, A., Galkin, S., Golyshev, V., Kasprzyk, A.M.: Mirror symmetry and Fano manifolds. In: European Congress of Mathematics, pp. 285–300. European Mathematical Society, Zürich (2013)
5. Corti, A., Heuberger, L.: Del Pezzo surfaces with $\frac{1}{3}(1,1)$ points. Manuscripta Math. **153**(1–2), 71–118 (2017)
6. Cox, D.A., Little, J.B., Schenck, H.K.: Toric Varieties. Graduate Studies in Mathematics, vol. 124. American Mathematical Society, Providence (2011)
7. Dais, D.I., Nill, B.: A boundedness result for toric log del Pezzo surfaces. Arch. Math. (Basel) **91**(6), 526–535 (2008)
8. Fulton, W.: Introduction to toric varieties. Annals of Mathematics Studies, vol. 131. Princeton University Press, Princeton (1993). The William H. Roever Lectures in Geometry
9. Kasprzyk, A.M.: Bounds on fake weighted projective space. Kodai Math. J. **32**(2), 197–208 (2009)
10. Kasprzyk, A.M., Kreuzer, M., Nill, B.: On the combinatorial classification of toric log del Pezzo surfaces. LMS J. Comput. Math. **13**, 33–46 (2010)
11. Kasprzyk, A.M., Nill, B., Prince, T.: Minimality and mutation-equivalence of polygons. Forum Math. Sigma **5**, e18, 48 (2017)

12. Kollár, J., Shepherd-Barron, N.I.: Threefolds and deformations of surface singularities. Invent. Math. **91**(2), 299–338 (1988)
13. Øbro, M.: An algorithm for the classification of smooth Fano polytopes (2007). arXiv:0704.0049 [math.CO]
14. Reid, M.: Surface cyclic quotient singularities and Hirzebruch–Jung resolutions. http://homepages.warwick.ac.uk/~masda/surf/more/cyclic.pdf
15. Reid, M.: Young person's guide to canonical singularities. In: Algebraic Geometry, Bowdoin, 1985 (Brunswick, Maine, 1985), Proceedings of Symposia in Pure Mathematics, vol. 46, pp. 345–414. American Mathematical Society, Providence, RI (1987)
16. The Sage Developers: SageMath, the Sage Mathematics Software System (Version 8.4.0) (2018). http://www.sagemath.org
17. Wormleighton, B.: Reconstruction of singularities on orbifold del Pezzo surfaces from their Hilbert series. Comm. Algebra **48**(1), 119–140 (2020)

Chapter 6
On the Topology of Fano Smoothings

Tom Coates, Alessio Corti, and Genival da Silva Jr.

Abstract Suppose that X is a Fano manifold that corresponds under Mirror Symmetry to a Laurent polynomial f, and that P is the Newton polytope of f. In this setting it is expected that there is a family of algebraic varieties over the unit disc with general fiber X and special fiber the toric variety defined by the spanning fan of P. Building on recent work and conjectures by Corti–Hacking–Petracci, who construct such families of varieties, we determine the topology of the general fiber from combinatorial data on P. This provides evidence for the Corti–Hacking–Petracci conjectures, and verifies that their construction is compatible with expectations from Mirror Symmetry.

Keywords Fano manifolds · Mirror symmetry · Betti numbers · Vanishing cycles

6.1 Introduction

There has been recent interest in the classification of Fano manifolds via Mirror Symmetry [1, 7, 8]. For us, an n-dimensional Fano manifold X corresponds under Mirror Symmetry to a Laurent polynomial $f \in \mathbb{C}[x_1^{\pm 1}, \ldots, x_n^{\pm 1}]$ if the regularized quantum period of X, which is a generating function for certain genus-zero Gromov–Witten invariants of X, coincides with classical period π_f of f:

$$\pi_f(t) = \frac{1}{(2\pi \mathrm{i})^n} \int_{S^1 \times \cdots \times S^1} \frac{1}{1 - tf} \frac{dx_1}{x_1} \cdots \frac{dx_n}{x_n}.$$

T. Coates (✉) · A. Corti · G. da Silva Jr.
Department of Mathematics, Imperial College London, 180 Queen's Gate, London SW7 2AZ, UK
e-mail: t.coates@imperial.ac.uk

A. Corti
e-mail: a.corti@imperial.ac.uk

G. da Silva Jr.
e-mail: jrriemann@gmail.com

A. M. Kasprzyk and B. Nill (eds.), *Interactions with Lattice Polytopes*, Springer Proceedings in Mathematics & Statistics 386,
https://doi.org/10.1007/978-3-030-98327-7_6

If a Fano manifold X corresponds under Mirror Symmetry to a Laurent polynomial f then it is expected that there is a degeneration $\mathcal{X} \to \Delta$, where $\Delta \subset \mathbb{C}$ is a disc, such that the general fiber is X and the special fiber is the toric variety X_f defined by the spanning fan of the Newton polytope of f. In general, X_f will be singular. It is natural to ask whether one can determine the Betti numbers of the Fano manifold X from its mirror partner f. We will show that, in dimension 3, the answer to this question is 'yes'.

The key new ingredient presented here is work by Corti–Hacking–Petracci [10], in preparation and still partly conjectural, which goes a long way towards establishing the expected picture described above in dimension 3. Corti–Hacking–Petracci construct, given a 3-dimensional reflexive polytope P decorated (as described below) with 'decomposition data', a smoothing $\mathcal{X} \to \Delta$ of the toric variety X_P defined by the spanning fan of P. The general fiber of this smoothing is a three-dimensional Fano manifold X. The decomposition data also determine a Laurent polynomial f with Newton polytope P, so that $X_f = X_P$. In this paper we show that the Betti numbers of X depend on the decomposition data only via f. That is, for each of the (many) choices of decomposition data that give rise to the same Laurent polynomial f, the Betti numbers of the corresponding smoothing are the same. Furthermore, these coincide with the Betti numbers of the Fano manifold that corresponds to f under Mirror Symmetry.

We proceed as follows. Corti–Hacking–Petracci construct, given a choice of decomposition data for a reflexive polytope P, a toric partial resolution $\pi \colon Y \to X_P$ and a family $\mathcal{Y} \to \Delta$ with special fiber Y and fiber over a general point $t \in \Delta$ a weak Fano manifold Y_t. Contracting finitely many $(-1,-1)$ curves in Y_t gives a resolution $\pi_t \colon Y_t \to X_t$, where X_t is a Fano variety with ordinary double points, and this fits into a diagram

$$\begin{array}{ccccc} & & Y_t & \mathrel{\vert\!\!\rightsquigarrow} & Y \\ & & \Big\downarrow{\scriptstyle \pi_t} & & \Big\downarrow{\scriptstyle \pi} \\ X_\eta & \mathrel{\vert\!\!\rightsquigarrow} & X_t & \mathrel{\vert\!\!\rightsquigarrow} & X_P \end{array} \tag{6.1}$$

where the arrow $A \mathrel{\vert\!\!\rightsquigarrow} B$ means that A is the general fiber in a family over Δ with special fiber B. Here $X_\eta \mathrel{\vert\!\!\rightsquigarrow} X_t$ is Namikawa's smoothing of Fano varieties with ordinary double points. The Fano variety X_η is our desired smoothing of X_P, and the diagram above allows us to compute its Betti numbers. The central fiber Y is a toric variety, so we know its cohomology groups explicitly; we can compute the Betti numbers of Y_t by analysing the vanishing cycles of the degeneration $Y_t \mathrel{\vert\!\!\rightsquigarrow} Y$ and, since the left-hand part of the diagram is a conifold transition from Y_t to X_η, this determines the Betti numbers of X_η.

We begin by reviewing the cohomology of toric varieties and the vanishing cycle exact sequence. We then explain in Sect. 6.3 how to compute the Betti numbers of the smoothing X from the decomposition data, and give examples in Sect. 6.4. In Sect. 6.5 we prove that the Betti numbers of X depend on the decomposition data only via the Laurent polynomial f determined by those data, and in Sect. 6.6 we verify that these Betti numbers of X coincide with the Betti numbers of the Fano manifold that corresponds to f under Mirror Symmetry.

6.2 Cohomology and Vanishing Cycles

6.2.1 *The Cohomology of Toric Varieties*

We will compute the Betti numbers of the fiber Y in diagram (6.1) using the fact that it is a toric variety.[1]

Theorem 1 (cf. [16, Proposition 3.5.3]) *Let Σ be a complete fan in a three-dimensional lattice and let X_Σ be the toric threefold defined by Σ. Let d_i denote the number of i-dimensional cones in Σ, and let b_i denote the ith Betti number of X_Σ. Then:*

$$\begin{aligned} d_1 - d_2 + d_3 &= 2 \qquad & b_2 &= \operatorname{rk}\operatorname{Pic}(X_\Sigma) \\ b_0 = b_6 &= 1 & b_3 &= \operatorname{rk}\operatorname{Pic}(X_\Sigma) - d_2 + 2d_1 - 3 \\ b_1 = b_5 &= 0 & b_4 &= d_1 - 3 \end{aligned}$$

and the Euler characteristic of X is d_3.

6.2.2 *The Vanishing Cycle Exact Sequence*

We will compute the Betti numbers of the fiber Y_t in diagram (6.1) by analysing the vanishing cycles for the degeneration $Y_t \rightsquigarrow Y$. Consider a complex variety $\mathcal{Y}$, a disc $\Delta \subset \mathbb{C}$, and a projective morphism $f : \mathcal{Y} \to \Delta$. Let $\Delta^* = \Delta \setminus \{0\}$ be the punctured disc, and

$$\{0\} \xrightarrow{i_0} \Delta \xleftarrow{j_0} \Delta^*$$

be the natural inclusions. Denote the fiber over $t \in \Delta^*$ by Y_t, and the fiber over $0 \in \Delta$ by Y. Choose a universal covering map $p_0 : \widetilde{\Delta^*} \to \Delta^*$, and consider the diagram

$$\begin{array}{ccccccc} Y & \xrightarrow{\ i\ } & \mathcal{Y} & \xleftarrow{\ j\ } & \mathcal{Y} \setminus Y & \xleftarrow{\ p\ } & \widetilde{\mathcal{Y} \setminus Y} \\ \downarrow & & \downarrow{\scriptstyle f} & & \downarrow & & \downarrow \\ \{0\} & \xrightarrow{\ i_0\ } & \Delta & \xleftarrow{\ j_0\ } & \Delta^* & \xleftarrow{\ p_0\ } & \widetilde{\Delta^*}. \end{array}$$

[1] We learned the statement of Theorem 1 from Andrea Petracci.

Let $\mathcal{S}$ be a stratification for $\mathcal{Y}$ and suppose that $\mathbf{F}^\bullet \in \mathbf{D}^b_{\mathcal{S}}(\mathcal{Y})$. The *nearby sheaf* is defined [14, Exposé I] to be the complex

$$\psi_f \mathbf{F}^\bullet = i^* \mathbf{R}(j \circ p)_* (j \circ p)^* \mathbf{F}^\bullet.$$

By adjunction, there is a natural map

$$i^* \mathbf{F}^\bullet \to \psi_f \mathbf{F}^\bullet.$$

The *sheaf of vanishing cycles* $\phi_f \mathbf{F}^\bullet$ is the cone on this map (ibid.), and there is a distinguished triangle:

$$i^* \mathbf{F}^\bullet \to \psi_f \mathbf{F}^\bullet \to \phi_f \mathbf{F}^\bullet \xrightarrow{+1} \tag{6.2}$$

Consider now the cohomology sheaves $\mathbf{H}^i(\psi_f \mathbf{F}^\bullet)$ and $\mathbf{H}^i(\phi_f \mathbf{F}^\bullet)$—these are complexes of sheaves on Y_0—and their stalks at $y \in Y_0$. Embed $\mathcal{Y}$ into an affine space and let $B(y, \epsilon)$ be the open ball of radius ϵ around y. Then for sufficiently small $\epsilon > 0$ and for all $t \in \Delta^*$ such that $|t| < \epsilon$, we have

$$\mathbf{H}^i(\psi_f \mathbf{F}^\bullet)_y \cong \mathbb{H}^i(F_{f,y}, \mathbf{F}^\bullet)$$

where $F_{f,y}$ is the Milnor fiber of f at y, $F_{f,y} := B(y, \epsilon) \cap \mathcal{Y} \cap f^{-1}(t)$. Similarly,

$$\mathbf{H}^i(\phi_f \mathbf{F}^\bullet)_y \cong \mathbb{H}^{i+1}(B(y, \epsilon) \cap \mathcal{Y}, F_{f,y}; \mathbf{F}^\bullet).$$

Now consider the constant sheaf $\mathbb{Q}_{\mathcal{Y}}$ as a complex concentrated in degree 0. Taking stalks of the hypercohomology of the distinguished triangle in (6.2), we get

$$\cdots \to H^i(Y, \mathbb{Q}) \to H^i(Y_t, \mathbb{Q}) \to H^i_v(Y_t, \mathbb{Q}) \to H^{i+2}(Y, \mathbb{Q}) \to \cdots \tag{6.3}$$

where $H^i_v(Y_t, \mathbb{Q})$ is the subspace in $H^i(Y_t, \mathbb{Q})$ generated by vanishing cycles, that is, cycles in the kernel of the natural map $H^i(Y_t, \mathbb{Q}) \to H^i(Y_0, \mathbb{Q})$.

6.3 Smoothing Toric Fano Threefolds

Every three-dimensional Gorenstein toric Fano variety is the toric variety X_P defined by the spanning fan of a three-dimensional reflexive polytope P. This gives a one-to-one correspondence between three-dimensional Gorenstein toric Fano varieties up to isomorphism and three-dimensional reflexive polytopes up to $GL(3, \mathbb{Z})$-equivalence. In general such a toric variety X_P is singular. As mentioned in the Introduction, Corti–Hacking–Petracci construct, starting from a three-dimensional reflexive polytope P decorated with some additional data, a smoothing $\mathcal{X} \to \Delta$ of X_P. In this section we describe their construction and the additional data required.

Definition 2 Let n be an integer such that $n \geq -1$. An A_n *triangle* is a lattice polygon that is $\mathbb{Z}^2 \rtimes \mathrm{GL}(\mathbb{Z}^2)$-equivalent to the polygon in $\mathbb{Q}^2$ with vertices $(0, 0)$, $(0, 1)$, and $(n + 1, 0)$

Example 3 The standard two-dimensional simplex is an A_0-triangle.

Example 4 An A_{-1}-triangle is a line segment of unit length.

Definition 5 Let F be a lattice polygon. An *admissible Minkowski decomposition* of F is a Minkowski decomposition $F = F_1 + \cdots + F_k$ of F as a sum of lattice polygons F_j such that each F_j is an A_n-triangle for some $n \geq -1$. (Here n can depend on j.) We consider admissible Minkowski decompositions of F that differ by reordering and translation of the summands to be equivalent.

We now introduce certain polyhedral decompositions of lattice polygons. Suppose that $F = F_1 + \cdots + F_k$ is an admissible Minkowski decomposition of the lattice polygon F. Recall that the *Cayley polytope* $C_{F_1,\ldots,F_k}$ is the convex hull of

$$F_1 + e_1, F_2 + e_2, \ldots, F_k + e_k$$

in $L \oplus \mathbb{Z}e_1 \oplus \cdots \oplus \mathbb{Z}e_k$, where L is the lattice containing F. The map

$$\begin{aligned} F &\longrightarrow C_{F_1,\ldots,F_k} \cap \left(L_{\mathbb{R}} \times \left\{\left(\tfrac{1}{k}, \cdots, \tfrac{1}{k}\right)\right\}\right) \\ v &\longmapsto \tfrac{1}{k}(v + e_1 + \cdots + e_k) \end{aligned} \tag{6.4}$$

is bijective, and this allows us to obtain polyhedral decompositions of F from polyhedral decompositions of the Cayley polytope $C_{F_1,\ldots,F_k}$.

Definition 6 Let $F = F_1 + \cdots + F_k$ be an admissible lattice Minkowski decomposition as above. A *regular fine mixed subdivision* of F is a subdivision of F induced, via (6.4), by a regular unimodular triangulation of the Cayley polytope $C_{F_1,\ldots,F_k}$. We say that such a subdivision is *subordinate to the Minkowski decomposition* $F = F_1 + \cdots + F_k$.

There is a whole body of theory here, which we will not discuss, concerning polyhedral subdivisions which may be neither fine nor mixed: an introduction to these topics can be found in the extremely beautiful book by De Loera et al. [11]. We will only consider regular fine mixed subdivisions.

For the rest of this section, fix a three-dimensional reflexive polytope P. Corti–Hacking–Petracci consider the polytope P together with *decomposition data*. This is, for each facet F of P:

A. a choice of admissible Minkowski decomposition $F = F_1 + \cdots + F_k$;
B. a choice of regular fine mixed subdivision of F subordinate to (A);

satisfying a condition that we now describe. Note that by taking cones over the regular fine mixed subdivisions of each facet, we obtain a complete fan Σ that refines the

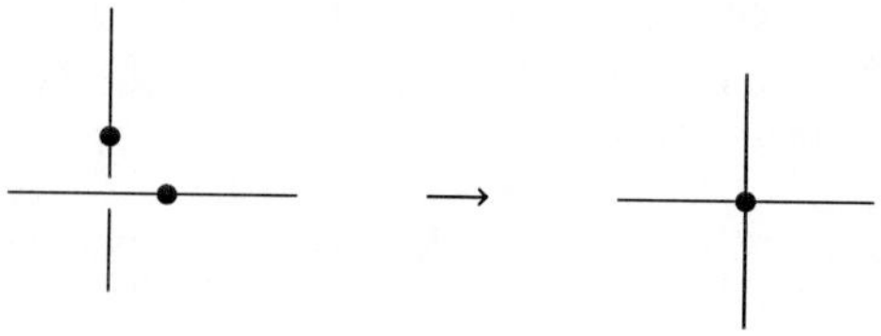

Fig. 6.1 The normalisation of a 4-valent vertex in the toric 1-skeleton

spanning fan of P, and thus a toric crepant partial resolution $Y \to X_P$. Recall first that irreducible toric curves in X_P (respectively in Y) correspond to two-dimensional cones in the spanning fan of P (respectively in the fan Σ). Thus irreducible toric curves in X_P correspond to edges of P, and irreducible toric curves in Y correspond to edges in the polyhedral subdivision of the boundary of P determined by the decomposition data.

Theorem 7 (Corti–Hacking–Petracci) *The singularities of the toric partial resolution Y of X_P are at worst quasi-ordinary double points (qODPs).*

This amounts to the statement that each polygon in the regular fine mixed subdivision (B) above is either a standard 2-simplex or a quadrilateral with sides of unit length. The curves in the toric 1-skeleton of Y, therefore, either meet at 3-valent vertices—which are the torus-fixed points on Y corresponding to the cones over the 2-simplices—or at 4-valent vertices, which are the torus-fixed points on Y corresponding to the cones over the quadrilaterals. The 3-valent vertices are smooth points on Y, and the 4-valent vertices are the qODPs.

Let Γ_e denote the irreducible toric curve in X_P determined by the edge e of P, and let $\widetilde{\Gamma}_e$ denote the set of irreducible toric curves in Y that map dominantly to Γ_e under the partial resolution $Y \to X_P$. The fact that each polygon in the subdivision (B) above is either a standard 2-simplex or a quadrilateral with sides of unit length implies that $|\widetilde{\Gamma}_e| = \ell(e)$, the lattice length of the edge e.

Let Δ denote the toric 1-skeleton of Y. Consider the partial normalisation $\Delta' \to \Delta$ constructed by normalising each 4-valent vertex as shown in Fig. 6.1.

The partial normalisation Δ' consists of rational curves that meet at either bivalent or trivalent vertices. Let e denote an edge of P, and Γ_e denote the corresponding toric curve in X_P. We define two partitions of the set $\widetilde{\Gamma}_e$, as follows. Let $p \in X_P$ be one of the endpoints of Γ_e, and $F \subset P$ the corresponding face of P. Consider the part Δ'_p of the partially-normalised toric 1-skeleton Δ' that lies over p; this consists of the dual graph to the polyhedral subdivision (B) of F, partially normalised as described in Fig. 6.1. The components of the partially-normalised dual graph define a partition of $\widetilde{\Gamma}_e$. There is one such partition for each of the two endpoints p of Γ_e: let us denote them by Π_e and Π'_e (the order will not matter). The condition that Corti–Hacking–Petracci impose on their decomposition data is: for each edge e of P such that the dual edge $e^\star$ has lattice length $\ell(e^\star)$ equal to 1 and for each pair $T \in \Pi$, $T' \in \Pi'$, we have $|T \cap T'| \leq 1$.

6.3.1 Computing the Betti Numbers of the Smoothing

Corti–Hacking–Petracci prove:

Theorem 8 *Let P be a 3-dimensional reflexive polytope and let X_P be the toric Fano threefold associated to the spanning fan of P. Fix decomposition data for P as above, and let $\pi : Y \to X_P$ be the associated crepant toric partial resolution. Then:*

1. *Y is unobstructed and smoothable;*
2. *if Y_t is a general smoothing of Y, then Y_t is a weak Fano threefold and the anticanonical morphism $\pi_t \colon Y_t \to X_t$, where $X_t = \operatorname{Proj} R(Y_t, -K_{Y_t})$, contracts a finite number of disjoint nonsingular rational curves, each with normal bundle $\mathcal{O}(-1) \oplus \mathcal{O}(-1)$;*
3. *X_t is a Fano threefold with ordinary nodes as singularities and it is a deformation of X_P.*

A theorem of Namikawa [18, Theorem 11] now implies that X_t is smoothable. It follows that X_P is smoothable. Let X_η denote a generic smoothing of X_t. Our goal is to compute the Betti numbers of X_η.

We begin by analysing the topology of Y_t. The vanishing cycles for the degeneration $Y_t \rightsquigarrow Y$ are three-dimensional spheres. The sheaf of vanishing cycles ϕ_f from (6.2), which is concentrated at the nodes of Y, therefore has stalk at each node equal to $\mathbb{Q}$ concentrated in degree three, and the vanishing cycle exact sequence (6.3) gives

$$\begin{aligned} b_0(Y_t) &= b_0(Y) & b_4(Y_t) &= b_2(Y) \\ b_1(Y_t) &= b_1(Y) & b_5(Y_t) &= b_1(Y) \\ b_2(Y_t) &= b_2(Y) & b_6(Y_t) &= b_0(Y) \\ b_3(Y_t) &= b_3(Y) - b_4(Y) + b_2(Y) + k \end{aligned} \tag{6.5}$$

where k is the number of nodes on Y. Note that $H_2(Y_t)$ is canonically identified with $H_2(Y)$; also $b_5(Y_t) = b_1(Y_t) = 0$ by Poincaré duality and Kodaira vanishing. Passing from Y_t to X_η is an example of a conifold transition [5], and therefore

$$\begin{aligned} b_0(X_\eta) &= b_0(Y_t) = 1 & b_4(X_\eta) &= b_2(Y_t) - l \\ b_1(X_\eta) &= b_1(Y_t) = 0 & b_5(X_\eta) &= b_1(Y_t) = 0 \\ b_2(X_\eta) &= b_2(Y_t) - l & b_6(X_\eta) &= b_0(Y_t) = 1 \\ b_3(X_\eta) &= b_3(Y_t) + 2m - 2l \end{aligned} \tag{6.6}$$

where l is the dimension of the subspace L of $H_2(Y_t)$ spanned by the classes of curves that are contracted by $\pi_t \colon Y_t \to X_t$, and m is the number of nodes on X_t.

In view of Theorem 1, computing the Betti numbers of the smoothing X_η comes down to computing the integers k, l, and m. We have seen that k is the number of quadrilaterals in the polyhedral decomposition of the boundary of P determined by the decomposition data. Corti–Hacking–Petracci conjecture the values of l and m, as follows. Let us identify $L \subset H_2(Y_t)$ as a subspace of $H_2(Y)$ via the canonical isomorphism $H_2(Y_t) \cong H_2(Y)$ just discussed. Recall that an edge e of P determines a toric rational curve Γ_e in X_P: this is the toric subvariety of X_P defined by the cone over e. The set $\widetilde{\Gamma}_e$ of toric curves that map dominantly to Γ_e under the map $Y \to X_P$ is indexed by the $\ell(e)$ line segments that subdivide e. If $\ell(e) \geq 2$ then e contains interior lattice points: such a lattice point v then corresponds to a toric surface $S_v \subset Y$ that projects to Γ_e. Let us denote the homology class[2] of a fiber of $S_v \to \Gamma_e$ by F_v. For a pair of toric curves $C, C' \in \widetilde{\Gamma}_e$, let c, c' denote the corresponding line segments and $(c, c') \subset P$ denote the relative interior of the convex hull of c and c'. Recall the two partitions of $\widetilde{\Gamma}_e$ defined in the discussion around Fig. 6.1, and define $n_{c,c'}$ by starting with $\ell(e)$ and subtracting one for each partition that has c and c' in the same part. The set of exceptional curves for $\pi_t : Y_t \to X_t$ is conjecturally indexed by edges e of P such that $\ell(e) \geq 2$ and pairs of distinct elements $c, c' \in \widetilde{\Gamma}_e$: it contains precisely $n_{c,c'}$ curves in the homology class

$$\sum_{v \in (c,c')} F_v$$

and no others. Note that $n_{c,c'}$ here is non-negative; it can be zero. This conjecture determines the subspace $L \subset H_2(Y_t)$ spanned by exceptional curves, and thus determines $l = \dim L$. For m, suppose that the elements of the two partitions of $\widetilde{\Gamma}_e$ have sizes $a_1, a_2, \ldots$ and $b_1, b_2, \ldots$ respectively. Set

$$N_e = \ell(e^\star)\binom{\ell(e)}{2} - \sum_i \binom{a_i}{2} - \sum_i \binom{b_i}{2}. \tag{6.7}$$

Then

$$m = \sum_{e\,:\,\ell(e)\geq 2} N_e. \tag{6.8}$$

6.3.2 *Minkowski Polynomials and Smoothings*

An admissible Minkowski decomposition of a three-dimensional reflexive polytope P determines a Minkowski polynomial [2]. This is a Laurent polynomial with Newton polytope P. In the notation of the discussion in Sect. 6.3, it depends on the choices (A) of admissible Minkowski decomposition of each facet of P, but not on the choices (B) of regular fine mixed subdivision.

---

[2] The fiber F_v is homologous in Y to the sum of toric curves corresponding to 2-dimensional cones in Σ that contain the ray spanned by v and that lie entirely on one side of the hyperplane defined by the edge e. The choice of side does not matter here, as the resulting sums are homologous.

It is known that Minkowski polynomials provide mirrors for three-dimensional Fano manifolds [8]. As mentioned in the Introduction, if a Fano manifold X is mirror to a Laurent polynomial f with Newton polytope P, it is expected that there is a degeneration $\mathcal{X} \to \Delta$ with general fiber X and special fiber X_P. For this expectation to be compatible with the results and conjectures of Corti–Hacking–Petracci, therefore, the Betti numbers of the Corti–Hacking–Petracci smoothings X_η must depend only on the Minkowski polynomial. That is, the Betti numbers must depend only on decomposition data only through the choices (A) of admissible Minkowski decompositions of facets: they must be independent of the choices (B) of regular fine mixed subdivision. This is not obvious from the construction; we prove it in Sect. 6.5.

6.4 Examples

6.4.1 Cube

Consider the cube P centered at the origin with vertices $(\pm 1, \pm 1, \pm 1)$. This has six non-simplicial facets and twelve edges of length two; thus the toric variety X_P defined by the spanning fan of P has six singular points and twelve curves of transverse A_1 singularities. These are arranged as on the right-hand side of Fig. 6.2, with the singular points at the vertices of the octahedron and the singular curves as the edges.

Each facet F of P is a square with side-length two; this has a unique admissible Minkowski decomposition, as a Minkowski sum of four line segments, which in turn leads to the unique fine mixed subdivision of F shown in Fig. 6.3.

The fan Σ that defines the partial resolution Y of X_P is therefore obtained by taking cones over the polyhedral decomposition of the boundary of P shown in Fig. 6.4. The variety Y contains 24 ordinary double points and 48 toric curves, arranged as on the right-hand side of Fig. 6.4, with the singular points at the vertices and the toric

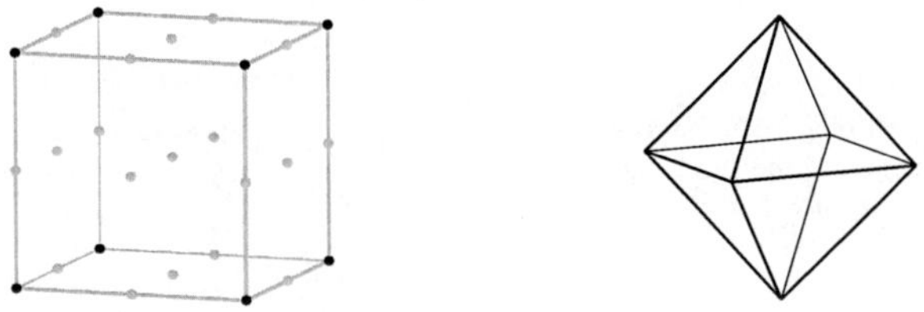

Fig. 6.2 The cube P and a schematic picture of the toric variety X_P

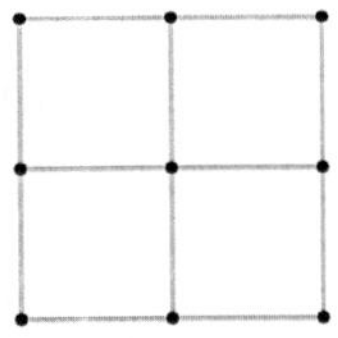

Fig. 6.3 The unique fine mixed subdivision of the facet F

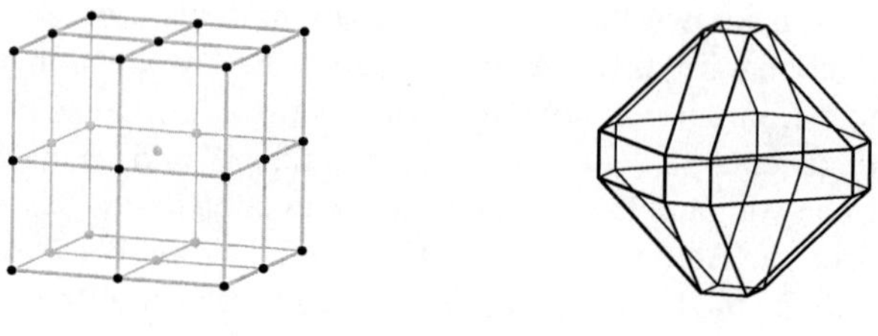

Fig. 6.4 The polyhedral subdivision and a schematic picture of Y

curves (along which Y is non-singular) as the edges. Furthermore each edge in the dual polygon $P^\star$ has length 1, and each partition Π_e of $\widetilde{\Gamma}_e$ is into singleton sets, so the polyhedral decomposition satisfies the condition to be decomposition data.

Applying Theorem 1 gives

$$\begin{aligned} b_0(Y) &= 1 & \qquad b_4(Y) &= 23 \\ b_1(Y) &= 0 & b_5(Y) &= 0 \\ b_2(Y) &= 4 & b_6(Y) &= 1 \\ b_3(Y) &= 5 \end{aligned}$$

There are $k = 24$ quadrilaterals in the polyhedral subdivision of the boundary of P, and the discussion in Sect. 6.3.2 yields

$$\begin{aligned} b_0(Y_t) &= 1 & \qquad b_4(Y_t) &= 4 \\ b_1(Y_t) &= 0 & b_5(Y_t) &= 0 \\ b_2(Y_t) &= 4 & b_6(Y_t) &= 1. \\ b_3(Y_t) &= 10 \end{aligned}$$

From the description of the homology classes F_v in footnote 2 and [12, Proposition 2.1], we see that $l = 3$, with generators for the subspace $L \subset H_2(Y_t) \cong H_2(Y)$ of classes of exceptional curves as shown in Fig. 6.5.

Equation (6.8) gives $m = 12$, and we find:

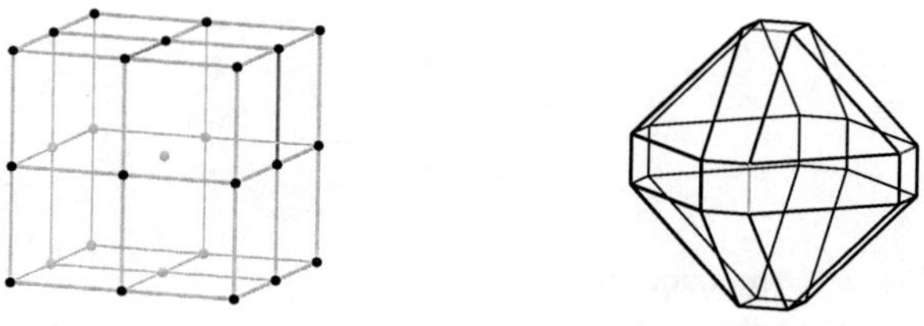

Fig. 6.5 Toric curves that generate the subspace of exceptional curves

$$b_0(X_\eta) = 1 \qquad b_4(X_\eta) = 1$$
$$b_1(X_\eta) = 0 \qquad b_5(X_\eta) = 0$$
$$b_2(X_\eta) = 1 \qquad b_6(X_\eta) = 1$$
$$b_3(X_\eta) = 28$$

The only smooth Fano 3-fold with these Betti numbers is V_8. Thus X_η is isomorphic to V_8, which is consistent with the fact that the Minkowski polynomial

$$f = \frac{(1+x)^2(1+y)^2(1+z)^2}{xyz} - 8$$

defined by our decomposition data for P is a mirror to V_8.

6.4.2 A Singular Toric Variety with Two Different Smoothings

Consider the three-dimensional polytope, pictured in Fig. 6.6, with vertices

$$(0,0,1),\ (0,1,-1),\ (1,1,-1),\ (1,0,-1),\ (0,-1,-1),\ (-1,-1,-1),\ (-1,0,-1).$$

This polytope P is reflexive. It has six facets that are standard simplices, one non-simplicial facet (a hexagon), and 12 edges of length 1. Thus the toric variety X_P defined by the spanning fan of P contains six non-singular toric points, a unique singular point (at which the singularity is a cone over the del Pezzo surface of degree 6), and 12 non-singular toric curves. These are arranged as on the right-hand side of Fig. 6.6, with the non-singular toric points at the 3-valent vertices, the singular point at the 6-valent vertex, and the toric curves as the edges.

The hexagonal facet F of P admits two Minkowski decompositions: as the sum of three line segments, and as the sum of two triangles. Up to automorphism, these each give rise to a unique fine mixed subdivision of F, as shown in Fig. 6.7. Consider first the left-most fine mixed subdivision in Fig. 6.7. This leads to the polyhedral subdivision of the boundary of P shown in Fig. 6.8. The fan Σ given by taking cones over

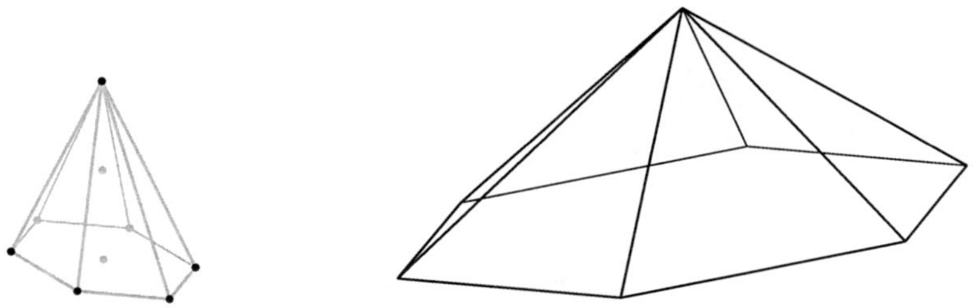

Fig. 6.6 The polytope P and a schematic picture of the toric variety X_P

Fig. 6.7 Two fine mixed subdivisions of the facet F

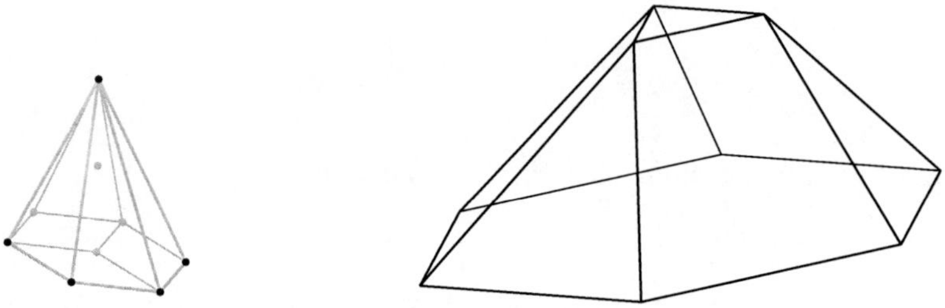

Fig. 6.8 The first polyhedral subdivision and a schematic picture of Y

this polyhedral subdivision defines a toric partial resolution Y of X_P. This variety Y has three ordinary double points, six non-singular toric points, and 15 non-singular toric curves, arranged as on the right-hand side of Fig. 6.8: the ordinary double points are the 4-valent vertices, the non-singular points are the 3-valent vertices, and the toric curves are the edges. The polyhedral decomposition satisfies the condition to be decomposition data.

Applying Theorem 1 gives

$$b_0(Y) = 1 \qquad b_4(Y) = 5$$
$$b_1(Y) = 0 \qquad b_5(Y) = 0$$
$$b_2(Y) = 2 \qquad b_6(Y) = 1$$
$$b_3(Y) = 0$$

and since there are $k = 3$ quadrilaterals in the polyhedral subdivision of the boundary of P, we have

$$b_0(Y_t) = 1 \qquad b_4(Y_t) = 2$$
$$b_1(Y_t) = 0 \qquad b_5(Y_t) = 0$$
$$b_2(Y_t) = 2 \qquad b_6(Y_t) = 1.$$
$$b_3(Y_t) = 0$$

The discussion in Sect. 6.3.1 implies that, assuming the conjectures of Corti–Hacking–Petracci, there are no exceptional curves in Y_t and so the smoothings X_η of X_P and Y_t of Y are isomorphic. Thus

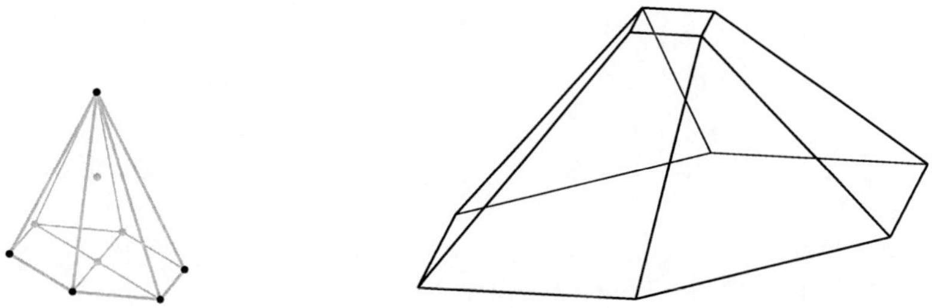

Fig. 6.9 The second polyhedral subdivision and a schematic picture of Y

$$b_0(X_\eta) = 1 \qquad b_4(X_\eta) = 2$$
$$b_1(X_\eta) = 0 \qquad b_5(X_\eta) = 0$$
$$b_2(X_\eta) = 2 \qquad b_6(X_\eta) = 1.$$
$$b_3(X_\eta) = 0$$

These are the Betti numbers of the hypersurface $W_{1,1}$ of bidegree $(1, 1)$ in $\mathbb{P}^2 \times \mathbb{P}^2$, which is consistent with the fact that the Minkowski polynomial

$$f = \frac{(1+x)(1+y)(1+xy)}{xyz} + z$$

defined by our decomposition data for P is a mirror to $W_{1,1}$.

One could instead consider the right-most fine mixed subdivision in Fig. 6.7. This leads to the polyhedral subdivision of the boundary of P shown in Fig. 6.9; again this satisfies the condition to be decomposition data. This time the toric partial resolution Y has two ordinary double points, eight non-singular toric points, and 16 non-singular toric curves. These are arranged as on the right-hand side of Fig. 6.8, with the ordinary double points as the 4-valent vertices, the non-singular points as the 3-valent vertices, and the toric curves as the edges.

Theorem 1 yields

$$b_0(Y) = 1 \qquad b_4(Y) = 5$$
$$b_1(Y) = 0 \qquad b_5(Y) = 0$$
$$b_2(Y) = 3 \qquad b_6(Y) = 1$$
$$b_3(Y) = 0$$

and since there are $k = 2$ quadrilaterals in the polyhedral subdivision of the boundary of P, we have

$$b_0(Y_t) = 1 \qquad b_4(Y_t) = 3$$
$$b_1(Y_t) = 0 \qquad b_5(Y_t) = 0$$
$$b_2(Y_t) = 3 \qquad b_6(Y_t) = 1.$$
$$b_3(Y_t) = 0$$

Once again, there are (conjecturally) no exceptional curves in Y_t and so the smoothings X_η of X_P and Y_t of Y are isomorphic. Thus

$$b_0(X_\eta) = 1 \qquad b_4(X_\eta) = 3$$
$$b_1(X_\eta) = 0 \qquad b_5(X_\eta) = 0$$
$$b_2(X_\eta) = 3 \qquad b_6(X_\eta) = 1$$
$$b_3(X_\eta) = 0$$

These are the Betti numbers of $\mathbb{P}^1 \times \mathbb{P}^1 \times \mathbb{P}^1$, which is consistent with the fact that the Minkowski polynomial

$$f = \frac{(1 + x + xy)(1 + y + xy)}{xyz} + z$$

defined by our decomposition data for P is a mirror to $\mathbb{P}^1 \times \mathbb{P}^1 \times \mathbb{P}^1$.

6.4.3 An Example with Transverse A_2 Singularities

Consider the three-dimensional reflexive polytope P, pictured in Fig. 6.10, with vertices

$$(1, 0, 0),\ (0, 1, 0),\ (0, 0, 1),\ (-1, -1, -1),\ (-1, -1, 2).$$

This polytope has four facets that are standard simplices, two non-standard simplicial facets, eight edges of length 1, and one edge of length 3. Thus the toric variety X_P

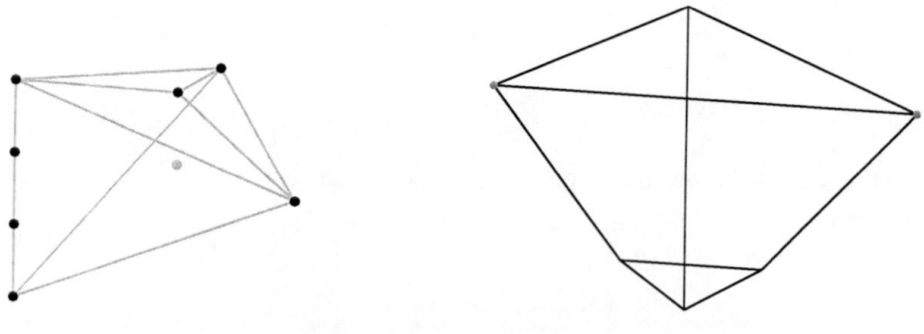

Fig. 6.10 The polytope P and a schematic picture of the toric variety X_P

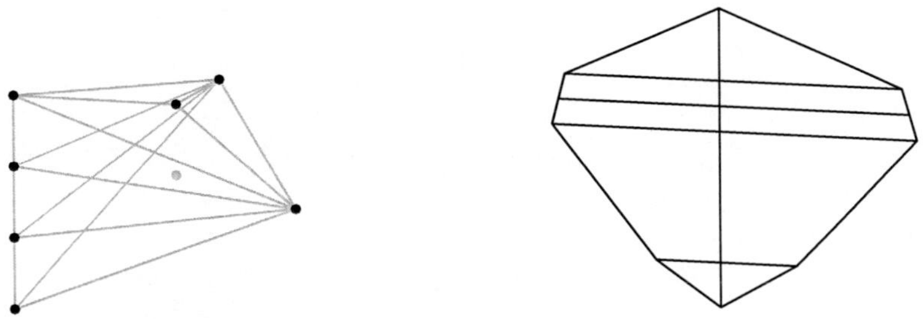

Fig. 6.11 The polyhedral subdivision and a schematic picture of Y

defined by the spanning fan of P contains four non-singular toric points, two orbifold points, eight non-singular toric curves, and one toric curve with transverse A_2 singularities. These are arranged as on the right-hand side of Fig. 6.10, with the toric points at the vertices, the orbifold points indicated in red, and the toric curves as the edges. The curve of A_2 singularities is the edge between the two orbifold points.

Figure 6.11 shows the unique fine mixed subdivision of the boundary of P. This defines decomposition data for P. Taking cones over the polyhedra in this decomposition gives a fan Σ that defines a toric resolution Y of X_P. The variety Y is smooth, with ten toric points and fifteen toric curves arranged as on the right-hand side of Fig. 6.11: the toric points are the vertices and the toric curves are the edges. Note the two toric surfaces in Y that map to the curve of singularities under the resolution $Y \to X_P$.

Applying Theorem 1 gives

$$
\begin{aligned}
b_0(Y) &= 1 & \qquad b_4(Y) &= 4\\
b_1(Y) &= 0 & \qquad b_5(Y) &= 0\\
b_2(Y) &= 4 & \qquad b_6(Y) &= 1\\
b_3(Y) &= 0 & &
\end{aligned}
$$

and since there are no quadrilaterals in the polyhedral subdivision, we find that the Betti numbers of Y_t coincide with those of Y. Computing the subspace $L \subset H_2(Y_t) \cong H_2(Y)$ of classes of exceptional curves, as in the cube example, we find that $l = 2$ and that generators for L are as shown in Fig. 6.12. These generators are the fibers of the toric surfaces in Y that resolve the curve of transverse A_2 singularities.

From (6.7) and (6.8) we find that (conjecturally) there are $m = 3$ exceptional curves in total, and therefore:

$$
\begin{aligned}
b_0(X_\eta) &= 1 & \qquad b_4(X_\eta) &= 2\\
b_1(X_\eta) &= 0 & \qquad b_5(X_\eta) &= 0\\
b_2(X_\eta) &= 2 & \qquad b_6(X_\eta) &= 1.\\
b_3(X_\eta) &= 2 & &
\end{aligned}
$$

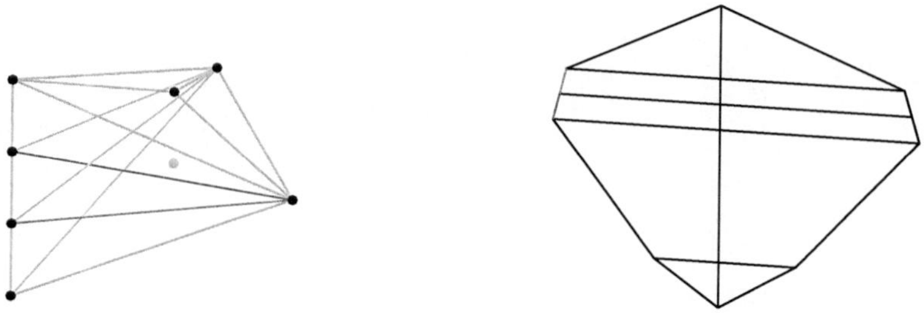

Fig. 6.12 Toric curves that generate the subspace of exceptional curves

These are the Betti numbers of the blow-up X of $\mathbb{P}^3$ in a plane cubic, which is consistent with the fact that the Minkowski polynomial

$$f = \frac{(1+z)^3}{xyz} + x + y + z$$

defined by our decomposition data for P is a mirror to X.

6.5 Betti Numbers Depend Only on the Mirror Laurent Polynomial

We have seen that decomposition data for a three-dimensional reflexive polytope P are, for each facet F of P:

A. a choice of admissible Minkowski decomposition $F = F_1 + \cdots + F_k$;
B. a choice of regular fine mixed subdivision of F subordinate to (A);

and that decomposition data determine a Minkowski polynomial f. In this section we will show that the Betti numbers of the smoothing X of X_P determined by the decomposition data are independent of the choice (B). This implies, in view of the discussion in Sect. 6.3.2, that the Betti numbers of X depend on the decomposition data only via f. We will prove:

Theorem 9 *Let P be a three-dimensional reflexive polytope. Consider two sets of decomposition data for P, where the choices (A) are the same in each case but the choices (B) are different. Let X denote the Corti–Hacking–Petracci smoothing of X_P determined by the first set of decomposition data, with $Y \to X_P$ the toric partial resolution and Y_t the smoothing of Y; let X' denote the smoothing of X_P determined by the second set of decomposition data, with $Y' \to X_P$ the toric partial resolution and Y'_t the smoothing of Y'. Assume the Corti–Hacking–Petracci conjectures described on Sect. 6.3.2 hold. Then:*

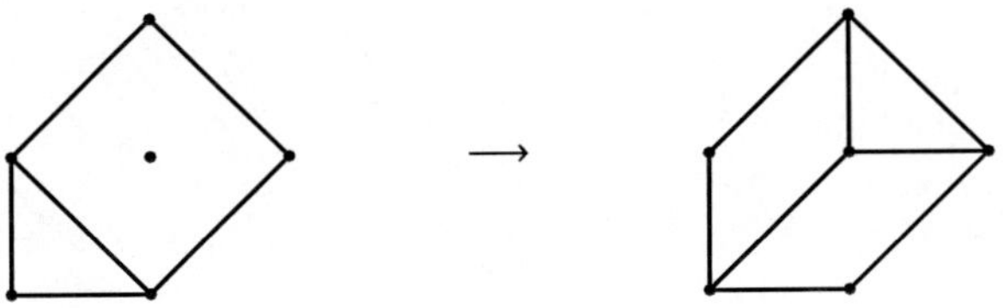

Fig. 6.13 A move of Type I

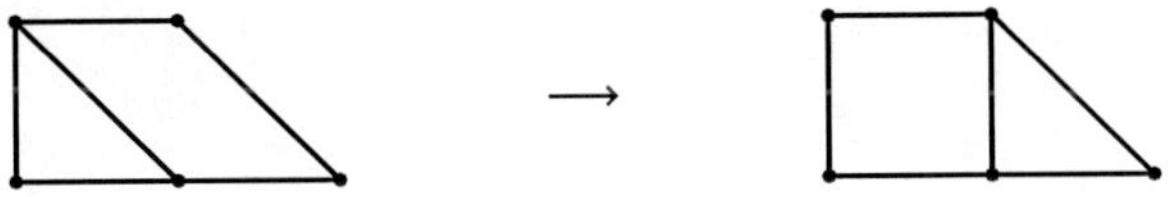

Fig. 6.14 A move of Type II

a. $b_2(Y) = b_2(Y')$;
b. *the Betti numbers of* Y_t *and* Y'_t *coincide;*
c. *the Betti numbers of* X *and* X' *coincide.*

The key point is that the polyhedral decompositions of the boundary of P that define Y and Y' differ by a sequence of the two types of moves depicted in Figs. 6.13 and 6.14, or their inverses. The vertices of the outer pentagon in Fig. 6.13 are $(0, 0)$, $(1, 0)$, $(\alpha + 1, \beta)$, $(\beta, \alpha + 1)$, $(0, 1)$, where α and β are positive coprime integers, and the interior lattice point pictured is at (α, β); unless $\alpha = \beta = 1$ then there are other interior lattice points which are not pictured. The precise values of α and β will not affect the analysis. The vertices of the outer quadrilateral in Fig. 6.14 are at $(0, 0)$, $(2, 0)$, $(1, 1)$, and $(0, 1)$.

6.5.1 Type I Moves

Let us analyse how b_2 changes under a Type I move. Suppose first that Y_1, Z, and Y_2 are three-dimensional toric varieties defined by polyhedral decompositions of the boundary of P that differ only as shown in Fig. 6.15.[3] Then Y_1 and Y_2 are toric partial resolutions of X_P which differ by a Type I move, and there is a diagram

$$\begin{array}{ccc} Y_1 & & Y_2 \\ & {\scriptstyle f_1}\searrow \quad \swarrow{\scriptstyle f_2} & \\ & Z & \end{array}$$

Let $z \in Z$ be the torus-fixed point corresponding to the pentagon pictured, and write $Z^0 = Z \setminus \{z\}$. The maps f_1 and f_2 induce isomorphisms $f_1^{-1}(Z^0) \to Z^0$

[3] The co-ordinates of the vertices and interior lattice point pictured in Fig. 6.15 are as in Fig. 6.13.

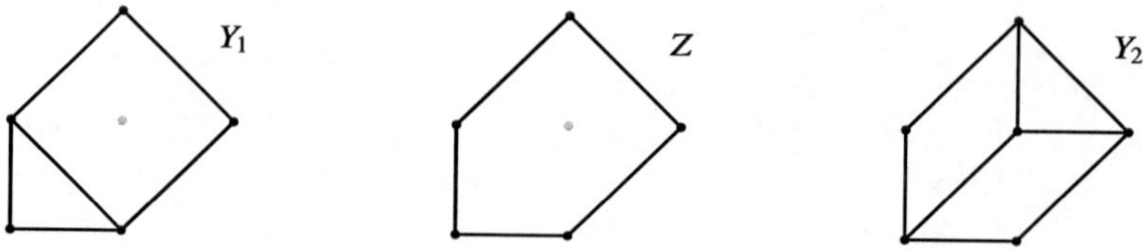

Fig. 6.15 The fans for Y_1, Z, and Y_2 differ only at the cones over these polygons

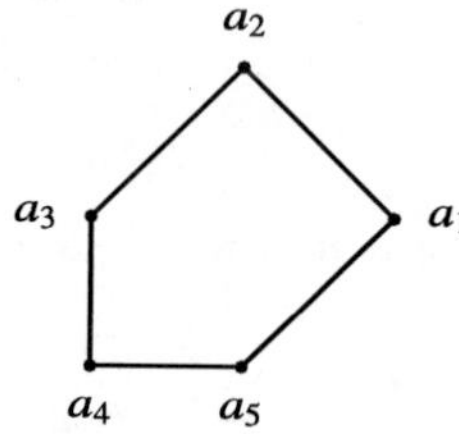

Fig. 6.16 The values of a piecewise-linear function on the fan for Z_0

and $f_2^{-1}(Z^0) \to Z^0$, and the resulting inclusions $j_1 \colon Z^0 \to Y_1$ and $j_2 \colon Z^0 \to Y_2$ define a diagram

$$\begin{array}{ccc} \mathrm{Pic}(Y_1) & & \mathrm{Pic}(Y_2) \\ & {\scriptstyle j_1^*} \searrow \quad \swarrow {\scriptstyle j_2^*} & \\ & \mathrm{Pic}(Z^0) & \end{array}$$

We will identify $\mathrm{Pic}(Y_1)$ and $\mathrm{Pic}(Y_2)$ as subspaces of $\mathrm{Pic}(Z^0)$. To give a line bundle on Z^0 is to give a piecewise-linear function on each maximal cone in the fan for Z^0, subject to the constraint that these piecewise-linear functions agree along faces. Let us write the values of such a piecewise linear function at the vertices of the polyhedral decomposition that we are considering as in Fig. 6.16.

Since the fan for Z^0 does not include the cone over the pentagon pictured in Fig. 6.16, that cone does not impose any relation between the values $a_1, \ldots, a_5$. (There may be other relations from the part of the fan not pictured, but these will be the same for Z, Y_1, and Y_2.) This piecewise-linear function defines a line bundle on Y_1 if and only if it is piecewise-linear on the two cones pictured on the left-hand side of Fig. 6.15, that is, if and only if $a_1 + a_3 = a_2 + a_5$. So

$$\mathrm{Pic}(Y_1) = \{a_1 + a_3 = a_2 + a_5\} \subset \mathrm{Pic}(Z^0).$$

Similarly, the piecewise-linear function on Z_0 defines a line bundle on Y_2 if and only if it is piecewise-linear on the two cones pictured on the right-hand side of Fig. 6.15, that is, if and only if $a_1 + a_4 - a_5 = a_2 + a_4 - a_3$. So

$$\mathrm{Pic}(Y_2) = \{a_1 + a_4 - a_5 = a_2 + a_4 - a_3\} \subset \mathrm{Pic}(Z^0).$$

These are the same subspace of $\mathrm{Pic}(Z_0)$; therefore $\mathrm{Pic}(Y_1)$ and $\mathrm{Pic}(Y_2)$ are canonically isomorphic. Since all maximal cones in the fan for Y_1 are full-dimensional, the Picard group of Y_1 is isomorphic to $H^2(Y_1)$; the same statement is true for Z and for Y_2. Thus b_2 is invariant under Type I moves.

A result of Fulton and Sturmfels [12, Proposition 1.1] implies that $b_4(Y_i)$ is equal to three less than the number of vertices in the polyhedral decomposition that defines Y_i, and so $b_4(Y_2) = b_4(Y_1) + 1$. Furthermore $\chi(Y_2) = \chi(Y_1) + 1$—here we used Theorem 1—and so $b_3(Y_1) = b_3(Y_2)$. In summary:

$$\begin{aligned} b_0(Y_2) &= b_0(Y_1) = 1 & \qquad b_4(Y_2) &= b_4(Y_1) + 1 \\ b_1(Y_2) &= b_1(Y_1) = 0 & b_5(Y_2) &= b_5(Y_1) = 0 \\ b_2(Y_2) &= b_2(Y_1) & b_6(Y_2) &= b_6(Y_1) = 1 \\ b_3(Y_2) &= b_3(Y_1) & & \end{aligned}$$

6.5.2 *Type II Moves*

An essentially identical argument shows that the Betti numbers of Y are invariant under Type II moves. We are now in a position to prove Theorem 9.

6.5.3 *Proof of Theorem 9*

Since Y and Y' differ by a sequence of moves of Type I and II, and their inverses, we have that $b_2(Y') = b_2(Y)$. This is part (a) of the Theorem. Furthermore $b_3(Y') = b_3(Y)$, and if the sequence of moves connecting Y to Y' contains M moves of Type I and N moves of (Type I)$^{-1}$ then $b_4(Y') = b_4(Y) + M - N$, and the numbers k and k' of quadrilaterals in the polyhedral decompositions defining Y and Y' satisfy $k' = k + M - N$. The vanishing cycle analysis (6.5) now implies (b).

To prove (c), it suffices to show that the quantities l and m occurring in Eq. (6.6) are the same for Y and Y'. This is obvious for the number of nodes m: the conjectural formula (6.8) for m depends only on the sizes of the partitions of $\widetilde{\Gamma}_e$, which in turn depends on the choices of Minkowski decomposition (A) but not on the fine mixed subdivisions (B). It remains to show that the dimension l of the subspace L of $H_2(Y)$ spanned by the classes of exceptional curves in the smoothing Y_t is the same as the dimension l' of the subspace L' of $H_2(Y')$ spanned by the classes of exceptional curves in the smoothing Y'_t. Let us return to the situation considered in Sect. 6.5.1, where Y_1 and Y_2 are three-dimensional toric varieties that differ by a Type I move. We showed there that $H^2(Y_1)$ and $H^2(Y_2)$ are isomorphic, via the inclusions

$$H^2(Y_1) \xrightarrow{\;j_1^*\;} H^2(Z^0) \xleftarrow{\;j_2^*\;} H^2(Y_2).$$

Dualising gives

$$H_2(Y_1) \xleftarrow{\;j_{1*}\;} H_2(Z^0) \xrightarrow{\;j_{2*}\;} H_2(Y_2).$$

and since the subspace of $H_2(Y_i)$ spanned by exceptional curves is pushed forward from $H_2(Z^0)$ via j_{i*} it follows that the dimension l of this subspace is also invariant under Type I moves. Repeating this analysis for Type II moves shows that $l = l'$, and proves Theorem 9.

6.6 Systematic Analysis

The computation of Betti numbers described in Sect. 6.3 can be automated. The key ingredients are as follows.

1. Algorithms for computing with lattice polyhedra and their duals. There are several robust and well-tested implementations here, including those in MAGMA [4], SAGE [20], and polymake [13].
2. The Kreuzer–Skarke classification [17] of three-dimensional reflexive polytopes.
3. Altmann's determination [3] of all Minkowski summands of a given polytope.
4. The computation of fine mixed subdivisions (Definition 6), that is, the determination of all regular triangulations of a Cayley polytope. For this we use Jörg Rambau's TOPCOM package [19].
5. An HPC cluster. Some of the computations involved are quite challenging.

Full source code for these computations, written in MAGMA, can be found at [6]. This relies in an essential way on code from the Fanosearch project [9].

There are 4319 three-dimensional reflexive polytopes, which in total admit more than a billion decomposition data. These decomposition data give rise to 3857 distinct Minkowski polynomials,[4] which together give mirrors to the 98 three-dimensional

[4] The number of Minkowski polynomials here differs slightly from the count in [2], because there the authors required Minkowski decompositions of facets to satisfy an additional lattice condition (ibid., Definition 7) and here we do not regard GL(3, $\mathbb{Z}$)-equivalent Minkowski polynomials as the same.

Fano manifolds with very ample[5] anticanonical bundle. We analysed several million decomposition data, including at least one decomposition for each of the 3857 Minkowski polynomials. In each case we found that

> the Betti numbers of the smoothing X_η determined by the decomposition data depend only on the mirror Fano manifold X, and coincide with those of X. (*)

This provides evidence for the conjectural picture described in the introduction[6]: that if a Fano manifold X corresponds under Mirror Symmetry to a Laurent polynomial f then there is a degeneration $\mathcal{X} \to \Delta$ with general fiber X and special fiber the toric variety defined by the spanning fan of the Newton polytope of f. It also provides evidence for the conjectures of Corti–Hacking–Petracci described in Sect. 6.3, on the number and homology class of the exceptional curves in their resolution $\pi_t\colon Y_y \to X_t$. If these conjectures are correct then, in view of Theorem 9, 3857 of these calculations give a computer-assisted rigorous proof of (*).

Acknowledgements This project has received funding from the European Research Council (ERC) under the European Union's Horizon 2020 research and innovation programme (grant agreement No. 682603), and from the EPSRC Programme Grant EP/N03189X/1, *Classification, Computation, and Construction: New Methods in Geometry*. We thank Paul Hacking and Andrea Petracci for many extremely useful conversations, and thank Andy Thomas, Matt Harvey, and the Imperial College Research Computing Service team for invaluable technical assistance. TC thanks Alexander Kasprzyk for a number of very useful conversations about computations.

References

1. Akhtar, M., Coates, T., Corti, A., Heuberger, L., Kasprzyk, A.M., Oneto, A., Petracci, A., Prince, T., Tveiten, K.: Mirror symmetry and the classification of orbifold del Pezzo surfaces. Proc. Am. Math. Soc. **144**(2), 513–527 (2016)
2. Akhtar, M., Coates, T., Galkin, S., Kasprzyk, A.M.: Minkowski polynomials and mutations. SIGMA Symmetry Integrability Geom. Methods Appl. **8**, Paper 094, 17 (2012)
3. Altmann, K.: The versal deformation of an isolated toric Gorenstein singularity. Invent. Math. **128**(3), 443–479 (1997)
4. Bosma, W., Cannon, J., Playoust, C.: The Magma algebra system. I. The user language, pp. 235–265 (1997). Computational algebra and number theory (London, 1993)
5. Candelas, P., Green, P.S., Hübsch, T.: Rolling among Calabi-Yau vacua. Nucl. Phys. B **330**(1), 49–102 (1990)
6. Coates, T., Corti, A., Da Silva Jr., G.: Code Repository (2019). https://bitbucket.org/fanosearch/3d_fano_smoothing
7. Coates, T., Corti, A., Galkin, S., Golyshev, V., Kasprzyk, A.M.: Mirror symmetry and Fano manifolds. In: European Congress of Mathematics, pp. 285–300. European Mathematical Society, Zürich (2013)

[5] The seven three-dimensional Fano manifolds without very ample canonical bundle are not expected to admit Laurent polynomial mirrors with reflexive Newton polytopes, and so fall outside the range of the Corti–Hacking–Petracci construction.

[6] Betti numbers for three-dimensional Fano manifolds can be found in [15].

8. Coates, T., Corti, A., Galkin, S., Kasprzyk, A.M.: Quantum periods for 3-dimensional Fano manifolds. Geom. Topol. **20**(1), 103–256 (2016)
9. Coates, T., Kasprzyk, A.M.: Code repository (2019). https://bitbucket.org/fanosearch/magma-core
10. Corti, A., Hacking, P., Petracci, A.: Smoothing Toric Fano Threefolds (2021). In preparation
11. De Loera, J.A., Rambau, J., Santos, F.: Triangulations. Structures for algorithms and applications. Algorithms and Computation in Mathematics, vol. 25. Springer-verlag, Berlin (2010)
12. Fulton, W., Sturmfels, B.: Intersection theory on toric varieties. Topology **36**(2), 335–353 (1997)
13. Gawrilow, E., Joswig, M.: Polymake: a framework for analyzing convex polytopes. In: Polytopes—Combinatorics and Computation (Oberwolfach, 1997), DMV SEM, vol. 29, pp. 43–73. Birkhäuser, Basel (2000)
14. Grothendieck, A., Raynaud, M., Rim, D. (eds.): Groupes de monodromie en géométrie algébrique. I. Lecture Notes in Mathematics, Vol. 288. Springer, Berlin, New York (1972). Séminaire de Géométrie Algébrique du Bois-Marie 1967–1969 (SGA 7 I)
15. Iskovskikh, V.A., Prokhorov, Y.G.: Fano varieties. In: Algebraic geometry, V. Encylopaedia of Mathematical Sciences, vol. 47, pp. 1–247. Springer, Berlin (1999)
16. Jordan, A.: Homology and cohomology of toric varieties. Ph.D. thesis, Universität Konstanz (1997)
17. Kreuzer, M., Skarke, H.: Classification of reflexive polyhedra in three dimensions. Adv. Theor. Math. Phys. **2**(4), 853–871 (1998)
18. Namikawa, Y.: Smoothing Fano 3-folds. J. Algebr. Geom. **6**(2), 307–324 (1997)
19. Rambau, J.: TOPCOM: triangulations of point configurations and oriented matroids. In: Mathematical Software (Beijing, 2002), pp. 330–340. World Scientific Publishing, River Edge, NJ (2002)
20. The Sage Developers: SageMath, the Sage Mathematics Software System (Version 8.4.0) (2018). http://www.sagemath.org

Chapter 7
Computing Seshadri Constants on Smooth Toric Surfaces

Sandra Di Rocco and Anders Lundman

Abstract In this paper we consider the problem of computing Seshadri constants at a general point on a smooth polarized toric surface. We consider the case when the degree of jet separation is small or the core of the associated polygon is a line segment. Our main result is that under these hypothesis the Seshadri constant at the general point can often be determined in terms of easily computable invariants of the surfaces at hand. When the core of the associated polygon is a point we show that the surface can be constructed via consecutive equivariant blow-ups of either $\mathbb{P}^2$ or $\mathbb{P}^1 \times \mathbb{P}^1$.

Keywords Toric geometry · Polytopes · Surface classification · Seshadri constants

7.1 Introduction

Let X be a complex projective variety and $\mathcal{L}$ be a nef line bundle on X.

Definition 1 For a point $x \in X$ the *Seshadri constant* at x is defined as the number

$$\epsilon(X, \mathcal{L}; x) = \inf_{C} \frac{\mathcal{L} \cdot C}{\mathrm{mult}_x(C)},$$

where the infimum is taken over all irreducible curves C passing through x.

The motivation for studying Seshadri constants is that they measure the local positivity of $\mathcal{L}$ at x. This can be seen from the Seshadri criterion which says that $\mathcal{L}$ is ample if and only if $\epsilon(X, \mathcal{L}; x) > 0$ for all points $x \in X$ [11, Theorem 1.4.13]. These constants were first introduced by Demailly in relation to the Nagata conjecture [1, 4].

S. Di Rocco (✉) · A. Lundman
Department of Mathematics, KTH Royal Institute of Technology, SE-100 44 Stockholm, Sweden
e-mail: dirocco@kth.se

A. Lundman
e-mail: alundman@kth.se

A. M. Kasprzyk and B. Nill (eds.), *Interactions with Lattice Polytopes*,
Springer Proceedings in Mathematics & Statistics 386,
https://doi.org/10.1007/978-3-030-98327-7_7

In most cases the interest lies on the value of the Seshadri constant at a general point, which is harder to compute with respect to specific points, like torus invariant points for toric varieties. By semi-continuity the value of these constants at special points can drop below that of general points. As a consequence the known results are often either bounds or only valid for certain special points of certain classes of varieties, see for example [1, 10, 15]. In this paper we approach the problem of computing Seshadri constants at the general point on smooth polarized toric surfaces. We do so by relating these constants to other invariants from local positivity and adjunction theory that are easy to compute on smooth toric surfaces.

The methods used in this paper rely on the fact that a polarized toric variety $(X, \mathcal{L})$ corresponds to a convex lattice polytope $P_{\mathcal{L}}$. Under this correspondence the fixpoints of the torus action on X correspond to the vertices of P. The Seshadri constant at these fixpoints is equal to the lattice length of the shortest edge through the corresponding vertex of $P_{\mathcal{L}}$, [1, 9], and it is thus easy to compute. It is known that the values of the Seshadri constant at the general point and at the general point of all torus-invariant subvarieties of X determine the Seshadri constant at every point of X [9, Proposition 3.2]. For a general point there is however no exact description, even though lately some very useful bound has been proven [9, 14].

In this note we related the Seshadri constant to the degree of jet separation, $s(\mathcal{L}, 1)$, and the unnormalized spectral value, $\mu(\mathcal{L})$. This allows us to compute the Seshardi constant at the general point for a large class of smooth toric surfaces. Unfortunately not for all. For high values of either the degree of jet separation or unnormalized spectral value the combinatorics quickly becomes intractable, see Example 40. We will therefore concentrate on low values of these constants.

Definition 2 Let x be a point of a projective variety X with maximal ideal $\mathfrak{m}_x$. The degree of jet separation $s(\mathcal{L}, x)$ of a line bundle $\mathcal{L}$ at the point $x \in X$ is the largest integer k for which the natural map $j_x^k \colon H^0(X, \mathcal{L}) \to H^0(X, \mathcal{L} \otimes \mathcal{O}_X/\mathfrak{m}_x^{k+1})$ is onto.

Like Seshadri constants the degree of jet separation is a semi-continuous invariant. In fact for smooth toric varieties $s(\mathcal{L}, x)$ obtains its minimum at some torus fixpoint and its maximum at the general point [14]. We will use the notation $s(\mathcal{L}, 1)$ to denote the degree of jet separation at a general point. Our first main result says that if $s(\mathcal{L}, 1)$ is small on a smooth polarized toric surface, then it equals the Seshadri constant.

Theorem 3 *Let $(X, \mathcal{L})$ be smooth polarized toric surface with $s(\mathcal{L}, 1) \leq 2$, then*

$$\epsilon(X, \mathcal{L}; 1) = s(\mathcal{L}, 1).$$

We remark that we do not know if the bound $s(\mathcal{L}, 1) \leq 2$ in Theorem 3 is sharp. Unfortunately the combinatorics involved in the proof gets out of hand already in the case $s(\mathcal{L}, 1) = 3$. On the one hand it was shown in [12] that if $s(\mathcal{L}, x)$ is constant at all points of a smooth polarized toric variety $(X, \mathcal{L})$, then $\epsilon(X, \mathcal{L}; x) = s(\mathcal{L}, x)$ for all points $x \in X$. Moreover the results in [12] suggest that there might be an integer $k > 2$ such that if $s(\mathcal{L}, 1) \leq k$ then $\epsilon(X_P, \mathcal{L}_P; 1) = s(\mathcal{L}, 1)$.

Example 40, due to Atsushi Ito [8], provides a smooth toric surfaces with $s(\mathcal{L}, 1) = 7$ and $\epsilon(X, \mathcal{L}; 1) \neq 7$. Thus the constant k has to be lower than 7.

Let $\mathcal{K}_X$ denote the canonical divisor on X. A linear system of the form $|\mathcal{K}_X + s\mathcal{L}|$ is called an adjoint linear system. Such systems play a prominent role in the classification of varieties and define important invariants, see [2] for more details. One such invariant is the *unnormalized spectral value*.

Definition 4

$$\mu(\mathcal{L}) = \sup \left\{ s \in \mathbb{Q} \;\middle|\; \begin{array}{l} \dim(H^0(X, m(\mathcal{K}_X + s\mathcal{L})) = 0 \text{ for all integers } m > 0 \\ \text{such that } m(\mathcal{K}_X + s\mathcal{L}) \text{ is an integral Cartier divisor} \end{array} \right\}.$$

Observe that since any ample line bundle is big it holds that $\mu(\mathcal{L}) < \infty$. Rationality of the nef and ample cone implies that $\mu(\mathcal{L}) \in \mathbb{Q}$ for toric varieties. By the toric dictionary any big $\mathbb{Q}$-Weil divisor $\mathcal{L}$ on a toric variety X of dimension n corresponds to a n-dimensional polytope $P_{\mathcal{L}} \subset \mathbb{R}^n$, whose vertices have rational coordinates. Any such polytope has a unique minimal description as

$$P_{\mathcal{L}} = \{x \in \mathbb{R}^n : Ax \geq b\}$$

where $b \in \mathbb{Q}^n$ and A is a matrix with integer coefficient with the entries in every row being relatively prime. If $P_{\mathcal{L}}$ is as above, then $t\mathcal{K}_X + \mathcal{L}$ corresponds to the polytope

$$P_{t\mathcal{K}_X+\mathcal{L}} = \{x \in \mathbb{R}^n : Ax \geq b + t\mathbf{1}\},$$

where $\mathbf{1} = (1, \ldots, 1)^T$. Here the polytope $P_{\frac{a}{b}\mathcal{K}_X+\mathcal{L}}$ is the polytope obtained by first dilating the polytope $P_{\mathcal{L}}$ by the factor b and then moving all supporting hyperplanes a steps inwards. It follows that $\mu(\mathcal{L})$ is the maximum over all t such that $P_{t\mathcal{K}_X+\mathcal{L}}$ is non-empty. Following [5] we will call the polytope $P_{\mu(\mathcal{L})^{-1}\mathcal{K}_X+\mathcal{L}}$ the *core of* $P_{\mathcal{L}}$ and denote it by $\text{core}(P_{\mathcal{L}})$. Observe that, unless P is a point, the dimension of $P_{\mathcal{L}}$ is strictly larger than the dimension of its core. Thus if X is a toric surface, $P_{\mathcal{L}}$ is a polygon and $\text{core}(P_{\mathcal{L}})$ is either a line segment or a point. Our second main result is the following

Theorem 5 *Let $(X, \mathcal{L})$ be a smooth polarized toric surface associated to the polytope $P_{\mathcal{L}}$. If $\text{core}(P_{\mathcal{L}})$ is a line segment and $\mu(\mathcal{L})^{-1} < 3$, then $\epsilon(X, \mathcal{L}; 1) = 2\mu(\mathcal{L})^{-1}$.*

We remark that, unlike the Seshadri constant at the general point, $\mu(\mathcal{L})$ is easily computable in the toric setting, see [5, Proposition 1.14]. It is worth pointing out that for $(\mathbb{P}^2, \mathcal{O}_{\mathbb{P}^2}(1))$ it holds that $\text{core}(P_{\mathcal{O}(1)})$ is a point, $\mu(\mathcal{O}(1))^{-1} = 1/3$, while $\epsilon(\mathbb{P}^2, \mathcal{O}_{\mathbb{P}^2}(1); 1) = 1$. Thus the assumption that $\text{core}(P)$ is a line segment is necessary in Theorem 5. We do not however have an example of a smooth polarized toric surface $(X, \mathcal{L})$ such that $\text{core}(P_{\mathcal{L}})$ is a line segment, while $\epsilon(X, \mathcal{L}; 1) \neq 2\mu(\mathcal{L})^{-1}$. Unfortunately it is apparent from the proof of Theorem 5 that the combinatorics involved quickly grows, with increasing values of $\mu(\mathcal{L})^{-1}$. This is further supported by Example 34 which shows that further complications will appear for higher values of $\mu(\mathcal{L})^{-1}$. Moreover Example 33 indicates that already the case $\mu(\mathcal{L})^{-1} = 3$

will be more involved than the cases we consider. We leave it as an open question to find an optimal bound on $\mu(\mathcal{L})^{-1}$ in Theorem 5.

Our next result makes use of Lemma 31, which says that if $P_{\mathcal{L}}$ is a polygon and $\text{core}(P)$ is a line segment, then there exist two edges e and e' of $P_{\mathcal{L}}$ which are parallel to $\text{core}(P)$. To state our next theorem we let $K(P)$ be the linear space parallel to the affine hull of $\text{core}(P)$. Moreover we will let $\pi : \mathbb{R}^2 \to \mathbb{R}^2/K(P)$ and $\pi^{\perp} : \mathbb{R}^2 \to \mathbb{R}^2/K(P)^{\perp}$ be the natural projections, where $K(P)^{\perp}$ is the orthogonal complement of $K(P)$.

Theorem 6 *Let $(X, \mathcal{L})$ be a smooth polarized toric surface associated to the polytope $P_{\mathcal{L}}$ and assume that* $\text{core}(P)$ *is a line segment. If π has a fiber of lattice length at least $2\mu(\mathcal{L})^{-1}$ that is a rational polytope or if there is a fiber of $\pi^{\perp}$ intersecting both e and e', then $\epsilon(X, \mathcal{L}; 1) = 2\mu(\mathcal{L})^{-1}$.*

For our purposes Theorem 6 serves as an important tool in proving Theorem 5. However as the assumptions of Theorem 6 are easily checked it provides a quick method to compute the Seshadri constant in many cases, even when $\mu(\mathcal{L})^{-1} \geq 3$. Therefore we believe that it can be of independent interest.

In the final part of the paper we focus, in contrast to Theorems 5 and 6, on the case when $\text{core}(P)$ is a point. We show the following characterization.

Theorem 7 *Let $(X, \mathcal{L})$ be a smooth polarized toric surface such that* $\text{core}(P)$ *is a point, then X is given by a sequence of consecutive equivariant blow-ups of $\mathbb{P}^2$ or $\mathbb{P}^1 \times \mathbb{P}^1$.*

7.1.1 Open Questions and Future Directions

This paper offers encouraging results for a class of smooth toric embeddings. Results that open a series of future directions and related questions.

One natural question to ask is if the bounds appearing in Theorems 3 and 5 are sharp. In the case of Theorem 3 this means asking for a minimal integer k such that if $s(\mathcal{L}_P, 1) \leq k$, then $\epsilon(X_P, \mathcal{L}_P; 1) = s(\mathcal{L}_P, 1)$. By Example 40 we know that $k < 7$, but it would be interesting to find a sharp bound.

In relation to Theorem 5, we do not believe that the equality $\epsilon(X_P, \mathcal{L}_P; 1) = 2\mu(\mathcal{L}_P)^{-1}$ will hold in more generality, whenever $\text{core}(P)$ is a line segment. However we are not aware of any counter example. Finding such an example would be enlightening and give further insight into the relation between $\epsilon(X_P, \mathcal{L}_P; 1)$ and $\mu(\mathcal{L}_P)$, for such varieties.

Regarding Theorem 7, notice that as $\mathbb{P}^2$ and $\mathbb{P}^1 \times \mathbb{P}^1$ are projective bundles Theorem 3 implies that $s(\mathcal{L}, 1) = \epsilon(X, \mathcal{L}; 1)$ for these surfaces. However when considering equivariant blow-ups of these surfaces it is not clear how the relationship between the Seshadri constant and degree of jet separation will change. It is worth pointing out that, as shown in Example 40, it can happen that $s(\mathcal{L}, 1)$, $\epsilon(X, \mathcal{L}; 1)$ and $2\mu(\mathcal{L})^{-1}$ are all distinct if $\text{core}(P)$ is a point. The surface appearing in Example 40 can be described both via six consecutive equivariant blow-ups of $\mathbb{P}^2$ and via five consecutive equivariant blow-ups of $\mathbb{P}^1 \times \mathbb{P}^1$. If instead X is obtained by a small number of blow-ups of either $\mathbb{P}^2$ or $\mathbb{P}^1 \times \mathbb{P}^1$ it seems reasonable to believe that $\epsilon(X, \mathcal{L}; 1)$ could be computed in terms of $s(\mathcal{L}, 1)$ or $\mu(\mathcal{L})^{-1}$. For example the blow-up of $\mathbb{P}^2$ at one fixpoint is the projective bundle $\mathbb{F}_1$ so $s(\mathcal{L}, 1) = \epsilon(\mathbb{F}_1, \mathcal{L}; 1)$ for any ample line bundle $\mathcal{L}$ on $\mathbb{F}_1$. The main difficulty in proving such statements lies in the number of different varieties achieved through consecutive equivariant blow-ups grows very fast in the number of blow-ups performed. We leave the computation of the Seshadri constant at the general point for these varieties as an open problem.

Another possible direction for future research is trying to achieve similar results for higher dimensional smooth polarized toric varieties.

7.2 Background

We start with a basic example on Seshadri constants.

Example 8 It holds that $\epsilon(\mathbb{P}^n, \mathcal{O}_{\mathbb{P}^n}(d); x) = d$ for all $x \in \mathbb{P}^n$. This follows from the fact that for any irreducible curve C and point $x \in X$, we have that $\mathcal{O}(d) \cdot C \geq d\,\mathrm{mult}_x(C)$ with equality when C is a line.

Another measure of local positivity is given by considering osculating spaces.

Definition 9 Let X be a projective variety and let $\mathcal{L}$ be a line bundle on X. For a non-negative integer k and a smooth point $x \in X$ consider the natural map

$$j_x^k\colon H^0(X, \mathcal{L}) \to H^0(X, \mathcal{L} \otimes \mathcal{O}_X/\mathfrak{m}_X^{k+1})$$

The projectivization of the image $\mathbb{P}(\mathrm{Im}(j_x^k))$ is called the kth *osculating space*, $\mathbb{T}_x^k(X, \mathcal{L})$ of $(X, \mathcal{L})$ at x. Moreover we will say that $\mathcal{L}$ is *k-jet spanned at x* if j_x^k is onto and call $\mathcal{L}$ *k-jet spanned* if j_x^k is onto for all points $x \in X$.

We remark that the map j_x^k is given by sending a section s to its Taylor expansion around x. Thus if $H^0(X, \mathcal{L}) = \bigoplus_{i=1}^m \mathbb{C}s_i$, then $\dim(\mathbb{T}_x^k(X, \mathcal{L})) = \mathrm{rk}(J_x^k) - 1$, where J_x^k is the matrix whose rows are given by the partial derivatives of $(s_0, \ldots, s_m)$ of order at most k evaluated at x. We will call J_x^k the *matrix of k-jets at x*. Thus given sufficient knowledge about the global sections of a line bundle $\mathcal{L}$, it is straight forward to check if $\mathcal{L}$ is k-jet spanned at a given point x. As in the introduction we will let $s(\mathcal{L}, x)$ denote the *degree of jet separation of* $\mathcal{L}$ at x, which is defined as the maximal k such that $\mathcal{L}$ is k-jet spanned at x. By counting the number of partial derivatives of order at most k it follows from the above that $s(\mathcal{L}, x)$ satisfies the relation

$$\binom{n + s(\mathcal{L}, x)}{s(\mathcal{L}, x)} \leq \dim(H^0(X, \mathcal{L})) < \infty. \tag{7.1}$$

Thus the the degree of jet separation, $s(\mathcal{L}, x)$ can be computed in finite time.

As $s(\mathcal{L}, 1)$ is considerably easier to compute than $\epsilon(X, \mathcal{L}; 1)$ it is of interest to relate the two. An important theorem in this direction is the following

Theorem 10 ([4]) *Let X be a projective variety and $\mathcal{L}$ be a nef line bundle on X. For any point $x \in X$ it holds that*

$$\epsilon(X, \mathcal{L}; x) = \lim_{n \to \infty} \frac{s(n\mathcal{L}, x)}{n}$$

We have the following well-known corollary of Theorem 10, which we will use repeatedly in this paper.

Corollary 11 *Let X be a projective variety and $\mathcal{L}$ be a nef line bundle on X. For any point $x \in X$ it holds that*

$$s(\mathcal{L}, x) \leq \epsilon(X, \mathcal{L}; x).$$

Proof If $\mathcal{L}$ is k-jet spanned line bundle, then $n\mathcal{L}$ is nk-jet spanned. Therefore it holds that $s(n\mathcal{L}, x) \geq ns(\mathcal{L}; x)$ and the corollary follows by taking the limit in Theorem 10. □

7.2.1 Toric Geometry

In this paper we assume that the reader is familiar with the basic properties of toric varieties. We recommend [3] as a good general introduction. A key fact in toric geometry is that any normal toric variety X_Σ corresponds to a polyhedral fan Σ. Moreover every torus invariant subvariety of codimension d on a toric variety X_Σ corresponds to a d-dimensional cone in the fan Σ. In particular any torus invariant prime divisor D_ρ corresponds to a 1-dimensional cone ρ, which is typically called a *ray* of Σ, and fixed points m_σ correspond to the cones σ.

This interplay between toric and convex geometry is stronger when considering polarized toric varieties $(X, \mathcal{L})$. In fact any such variety $(X, \mathcal{L})$ corresponds to a convex lattice polytope $P_\mathcal{L}$. To see how this correspondence is defined let D be a divisor having the property that $\mathcal{O}(D) = \mathcal{L}$. As the class group of a projective toric variety X is generated by the torus invariant prime divisors of X, it holds that $D = \sum_{\rho \in \Sigma(1)} a_\rho D_\rho$, where $\Sigma(1)$ denotes the set of rays of Σ and $a_\rho \in \mathbb{Z}$. We then define the polytope $P_\mathcal{L}$ associated to $(X, \mathcal{L})$ as

$$P_\mathcal{L} = \{x \in \mathbb{R}^n \colon \langle x, \rho \rangle \geq -a_\rho, \forall \rho \in \Sigma(1)\}.$$

Via this connection many concepts in algebraic geometry can be understood in convex geometric terms and vice versa. One example is the following theorem. After choosing a basis of the lattice $\mathbb{Z}^n$ and hence choosing an affine patch U_σ corresponding to the vertex v_σ placed at 0, then one has a convenient monomial basis for the global sections:

Theorem 12 ([3, Theorem 4.3.3]) *Let X be a projective toric variety and $\mathcal{L}$ a line bundle on X, then*

$$H^0(X, \mathcal{L}) \cong \bigoplus_{(m_1,\ldots,m_n)\in P_{\mathcal{L}}\cap\mathbb{Z}^n} \mathbb{C} < t_1^{m_1} \cdots t_n^{m_n} > .$$

where the monomial basis is express in terms of torus coordinates $(t_1, \cdots, t_n)$ in the corresponding affine patch U_σ.

In particular every entry in the matrix of k-jets, J_x^k is the evaluation of a monomial. This leads to the following proposition.

Proposition 13 ([14]) *Let X be a smooth polarized toric variety. Then $s(\mathcal{L}, 1) = k$ if and only if there exist a polynomial of degree $k + 1$ vanishing on the lattice points of $P_{\mathcal{L}}$ and any other polynomial with the same property has at least degree $k + 1$.*

We recall that a polarized toric variety $(X, \mathcal{L})$ of dimension n is smooth if and only if for every vertex v of $P_{\mathcal{L}}$ the shortest lattice vectors along the edges through v form a basis for $\mathbb{Z}^n$. We will call these vectors the *edge-directions at v*.

Next we introduce maps of fans which are the combinatorial equivalent of equivariant maps of toric varieties.

Definition 14 Let Σ and Σ' be fans in $\mathbb{R}^n$. A linear map $\phi : \mathbb{R}^n \to \mathbb{R}^n$ is called a *map of fans* if for every $\sigma' \in \Sigma'$ there exist a $\sigma \in \Sigma$ such that $\phi(\sigma' \cap \mathbb{Z}^n) \subset \sigma \cap \mathbb{Z}^n$. If ϕ is a map of fans, then we often write $\phi : \Sigma' \to \Sigma$ in place of $\phi : \mathbb{R}^n \to \mathbb{R}^n$.

We remark that a map of fans $\Sigma' \to \Sigma$ induces an equivariant morphism $X_{\Sigma'} \to X_\Sigma$ of toric varieties [6, Theorem VI.6.1]. The most important examples of maps of fans are obtained by letting Σ' be a refinement of Σ and ϕ be the identity. Recall that Σ' is a *refinement* of Σ if for every $\sigma' \in \Sigma$ it holds that $\sigma' \subset \sigma$ for some cone $\sigma \in \Sigma$. When Σ' is a refinement of Σ, we will call Σ a *coarsening* of Σ'.

Example 15 Let X_Σ be a smooth toric surface, then blowing up a torus fixpoint of X_Σ can be realised as a map of fans. To see this recall that the fixpoints of X_Σ correspond to the maximal dimensional cones in Σ. Choosing a fixpoint x_σ to blow up corresponds to a choice of such a cone $\sigma = \text{cone}(e_1, e_2)$, where the e_i:s can be taken to be a basis of $\mathbb{Z}^2$ since X_Σ is smooth. The fan Σ' given by replacing σ with the cones $\text{cone}(e_1, e_1 + e_2)$ and $\text{cone}(e_1 + e_2, e_2)$ while keeping all other cones in Σ, defines a new fan Σ'. The identity map $\text{id} : \mathbb{R}^n \to \mathbb{R}^n$ is a map of fans $\text{id}: \Sigma' \to \Sigma$ which induces a toric morphism $\pi : X_{\Sigma'} \to X_\Sigma$. By [6, Theorem VI.7.2], $\pi : X_{\Sigma'} \to X_\Sigma$ is exactly the blow-up of X_Σ at x_σ. A corresponding construction works when blowing-up any torus invariant subvariety of a smooth toric variety of any dimension (Fig. 7.1).

Smooth toric surfaces are characterized by the following theorem.

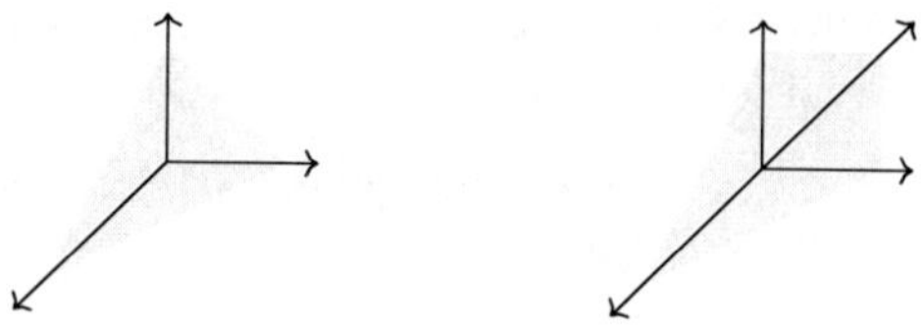

Fig. 7.1 The fan of $\mathbb{P}^2$ and the blow-up of $\mathbb{P}^2$ in a torus fixpoint

Theorem 16 ([6, Theorem VI.7.5]) *Let X be a smooth toric surface. Then there exist a chain of equivariant blow-downs of X to either a Hirzebruch surface or* $\mathbb{P}^2$.

We next focus on Seshadri constants on toric varieties.

Definition 17 Let P be a full-dimensional polytope. Then P is called a *Cayley polytope* if the vertices of P are contained in two parallel hyperplanes H_0 and H_1. Moreover if the lattice distance between H_0 and H_1 is k, then P is said to be *of type* $[P_0 * P_1]^k$, where $P_0 = P \cap H_0$ and $P_1 = P \cap H_1$. Lastly if P is a Cayley polytope, then we call the projection onto the linear space perpendicular to H_0 and H_1 the *defining projection of* P.

In particular P is a Cayley polytope of type $[P_0 * P_1]^1$ if and only if P has lattice width 1. In this case the defining projection projects P onto a line segment of lattice length 1.

The two following theorems relate Seshadri constants and Cayley polytopes.

Theorem 18 ([10, Theorem 1.3]) *Let P be a lattice polytope associated to the polarized toric variety $(X_P, \mathcal{L}_P)$, then the following are equivalent:*

1. $\epsilon(X_P, \mathcal{L}_P; 1) = 1$;
2. $P = [P_0 * P_1]^1$;
3. *for every point $p \in X_P$ there exist a curve C containing p, having the property that* $(C, \mathcal{L}|_C) \cong (\mathbb{P}^1, \mathcal{O}_{\mathbb{P}^1}(1))$.

Theorem 19 ([12]) *Let P be a smooth lattice polytope associated to the polarized toric variety $(X_P, \mathcal{L}_P)$ and let k be a positive integer. Then the following are equivalent:*

1. $\epsilon(X_P, \mathcal{L}_P; x) = k$ *for all points* $x \in X_P$;
2. $s(\mathcal{L}_P, x) = k$ *for all points* $x \in X_P$;
3. $P = [P_0 * P_1]^k$ *and every edge of P has lattice length at least k.*

The following theorem gives bounds on the Seshadri constant at a general point and is the most important result in establishing the new results of this paper.

Theorem 20 ([9, Theorem 3.6]) *Let M be a lattice and $M_{\mathbb{R}} = M \otimes_{\mathbb{R}} \mathbb{R}$. Let further $P \subset M_{\mathbb{R}}$ be a smooth polygon and $\pi : M \to \mathbb{Z}$ be a lattice projection. Then*

$$\min\{|\pi_{\mathbb{R}}(P)|_{\pi(M)}, |F|_M\} \leq \epsilon(X_P, \mathcal{L}_P; 1) \leq |\pi_{\mathbb{R}}(P)|_{\pi(M)}$$

where $\pi_{\mathbb{R}}$ is the map $\pi_{\mathbb{R}} : M_{\mathbb{R}} \to \mathbb{R}$ induced by π, F is any fiber which is also a rational polytope, and $|\cdot|_M$ is the lattice distance with respect to the lattice M.

7.2.2 Adjunction Theory for Toric Varieties

Given a linear system $|\mathcal{L}|$ on projective variety X, a linear system of the form $|\mathcal{K}_X + s\mathcal{L}|$ is called an *adjoint system* to $\mathcal{L}$. In this paper we are interested in two invariants associated to $|\mathcal{K}_X + s\mathcal{L}|$, namely the nef value and the unnormalized spectral value.

Definition 21 Let $(X, \mathcal{L})$ be a polarized variety we call

$$\tau(\mathcal{L}) = \inf \{s \in \mathbb{R} \colon \mathcal{K}_X + s\mathcal{L} \text{ is nef}\}.$$

the *nef value* of $\mathcal{L}$.

We remark that $\tau(\mathcal{L})$ equals the largest s such that $K + s\mathcal{L}$ is nef but not ample [2]. The second invariant of interest to us is the unnormalized spectral value and we recall its definition from the introduction.

Definition 22 Let $(X, \mathcal{L})$ be a polarized variety, then the *unnormalized spectral value of* $\mathcal{L}$ is

$$\mu(\mathcal{L}) = \sup \left\{ s \in \mathbb{Q} \;\middle|\; \begin{array}{l} \dim(H^0(X, m(\mathcal{K}_X + s\mathcal{L})) = 0 \text{ for all integers } m > 0 \\ \text{such that } m(\mathcal{K}_X + s\mathcal{L}) \text{ is an integral Cartier divisor} \end{array} \right\}.$$

Both the nef value and the unnormalized spectral value have convex geometric interpretations for toric varieties. We briefly recall these interpretations.

Remark 23 Let P be a lattice polytope associated to the polarized toric variety $(X_P, \mathcal{L}_P)$. For any $s \in \mathbb{Q}$ the polytope $P_{s\mathcal{K}_{X_P}+\mathcal{L}_P}$ is denoted by $P^{(s)}$. The nef value is the minimum over all s such that the inner-normal fan of P is a refinement of the inner-normal fan of $P^{(1/s)}$. The unnormalized spectral value $\mu(\mathcal{L}_P)$ of $(X_P, \mathcal{L}_P)$ is the minimum over all s such that $P^{(1/s)}$ is non-empty. The polytope $P^{(1/\mu(\mathcal{L}_P))}$, called the core of P, is a polytope of strictly smaller dimension than P and is denoted by core(P). We remark that the core of a polytope and $\mu(\mathcal{L}_P)^{-1}$ are easily computable using [5, Proposition 1.14]. These definitions are illustrated in Fig. 7.2, for further details on these correspondences we refer to [5].

Definition 24 Let $P \subset \mathbb{R}^n$ be a polytope and let $K(P)$ be the linear space parallel to the affine span of the core of P. The projection $\pi : \mathbb{R}^n \to \mathbb{R}^n/K(P)$ is called the *natural projection* associated to P.

Remark 25 If one restricts the attention to polarized toric surfaces, then there are two possibilities, either core(P) is a point or a line segment. When the core of P is a point the natural projection is the identity and when it is a line segment the natural projection is the projection onto a line.

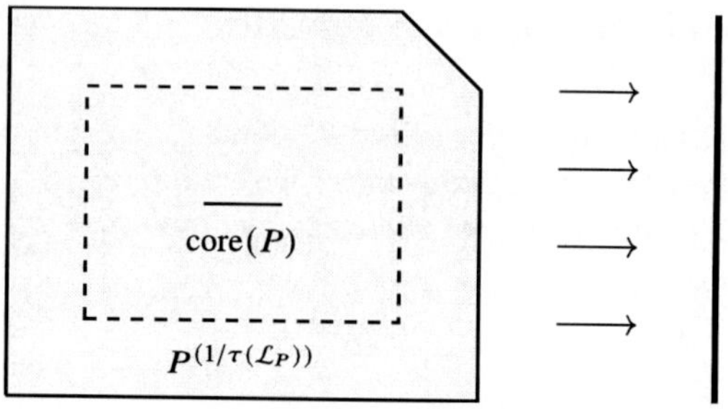

Fig. 7.2 A polytope P, $P^{(1/\tau(\mathcal{L}_P))}$, $\mathrm{core}(P) = P^{(1/\mu(\mathcal{L}_P))}$ and the natural projection

7.3 Seshadri Constants and Jet Separation

In the following section we prove Theorem 3. To make the exposition more accessible we first prove Lemmas 27 and 28 which are key-steps in the proof of the theorem. We further make the following notational definition.

Definition 26 Let P be a smooth polytope, then we say that P is *canonically positioned* if P has a vertex at the origin and an edge along each coordinate axis in the positive direction.

Lemma 27 *Let P be a smooth Cayley polygon associated to the smooth polarized toric surface $(X_P, \mathcal{L}_P)$, then $s(\mathcal{L}_P, 1) = \epsilon(X_P, \mathcal{L}_P; 1)$.*

Proof Since P has dimension 2 the assumption that P is a Cayley polytope means that either $P \cong k\Delta_2 = \mathrm{conv}((0,0), (k,0), (0,k))$ or P is the convex hull of two line segments. If $P \cong k\Delta_2$ then $s(\mathcal{L}_P, 1) = \epsilon(X_P, \mathcal{L}_P; 1) = k$. If instead $P = [l_0, l_1]^k$, where l_0 and l_1 are line segments, then there are two cases: either $k = s(\mathcal{L}_P, 1)$ or $k \neq s(\mathcal{L}_P, 1)$. If $k = s(\mathcal{L}_P, 1)$, then projecting onto the defining projection of P shows that $\epsilon(X_P, \mathcal{L}_P; 1) \leq s(\mathcal{L}_P, 1)$ which by Corollary 11 implies $s(\mathcal{L}_P, 1) = \epsilon(X_P, \mathcal{L}_P; 1)$. If instead $k \neq s(\mathcal{L}_P, 1)$, we first observe that it must hold that $k > s(\mathcal{L}_P, 1)$, otherwise the defining projection of P would give that $\epsilon(X_P, \mathcal{L}_P; 1) < s(\mathcal{L}_P, 1)$ by Theorem 20, which contradicts Corollary 11. Thus without loss of generality we can assume that $k > s(\mathcal{L}_P, 1)$ and that P is canonically positioned with the longest defining line segment along the x-axis. The assumption $k > s(\mathcal{L}_P, 1)$ implies that $(0, s(\mathcal{L}_P, 1) + 1) \in P$. There are now two cases:

1. $(s(\mathcal{L}_P, 1) + 1, 0) \notin P$, in which case the projection onto the first coordinate axis shows that $\epsilon(X_P, \mathcal{L}_P; 1) = s(\mathcal{L}_P, 1)$ by Theorem 20 and Corollary 11.
2. $(s(\mathcal{L}_P, 1) + 1, 0) \in P$, in which case $(s(\mathcal{L}_P, 1) + 1)\Delta_2 \subseteq P$ by convexity. This implies that P is $(s(\mathcal{L}_P, 1) + 1)$-jet spanned at the general point contradicting the definition of $s(\mathcal{L}_P, 1)$.

From the above we conclude that $s(\mathcal{L}_P, 1) = \epsilon(X_P, \mathcal{L}_P; 1)$. □

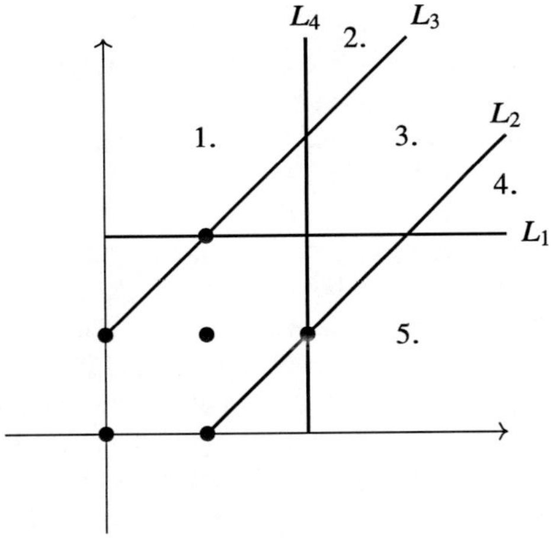

Fig. 7.3 Five lattice points in P, a configuration of four lines and open regions

Lemma 28 *Let P be a smooth polygon associated to the smooth toric surface $(X_P, \mathcal{L}_P)$. If $s(\mathcal{L}_P, 1) = 2$, then $\epsilon(X_P, \mathcal{L}_P; 1) = 2$.*

Proof By Lemma 27 we can without loss of generality assume that P is not a Cayley polytope. Assume that P is canonically positioned, then $(0, 0), (1, 0), (0, 1) \in P$ by the smoothness assumption. Furthermore we claim that the points $(2, 1)$ and $(1, 2)$ must lie in P. To see why this is the case note that if either of these points fail to be in P, then by smoothness and convexity either $P = 2\Delta_2$ or all lattice points of P lie on two parallel lines at lattice distance 1 apart. In the former case $s(\mathcal{L}_P, 1) = \epsilon(X_P, \mathcal{L}_P; 1) = 2$ and the latter is a contradiction to $s(\mathcal{L}_P, 1) = 2$. From convexity it now follows that $(1, 1) \in P$. Thus the indicated lattice points in Fig. 7.3 lie in P. In Fig. 7.3 we also introduce four lines and enumerate some open regions delimited by these lines.

We now observe that if either the horizontal, vertical or both diagonal lines are supporting lines of P, then there exists a projection showing that $\epsilon(X_P, \mathcal{L}_P; 1) \leq 2$ by Theorem 20. Thus without loss of generality we can assume that P contains at least one point on the other side of the horizontal and vertical line and one point outside the diagonal strip. Those constrains can be satisfied in a few different ways and we will end the proof by considering all possibilities.

(1) If P contains a lattice point in region 4, then convexity implies that the points $(2, 2)$ and $(3, 2)$ lie in P. Furthermore if there is a point in region 4, then the line L_2 cannot be a supporting line of P. By the smoothness assumption at the vertex of P on the x-axis, different from the origin, it then follows that $(3, 1) \in P$. The matrix of 3-jets evaluated at 1, for the 9 lattice points we now know lie in P, consists of 9 independent rows and a row of zeros corresponding to the derivative $\frac{\partial^3}{\partial y^3}$. However any monomial corresponding to a point in region 4 will yield a non-zero result when taking the derivative $\frac{\partial^3}{\partial y^3}$ and evaluating at 1. Thus we can conclude that $\mathcal{L}_P$ is 3-jet spanned at the general point if it contains a point in region 4, which is a contradiction. We remark that this argument also takes care of the case when

there is a point of P on the line L_1 to the right of $(3, 2)$. Since in that case either L_1 is a supporting line of P or P contains a point whose y-coordinate is at least 3. In the former case $\epsilon(X_P, \mathcal{L}_P; 1) \leq 2$ and there is nothing to prove while in the latter case $\mathcal{L}_P$ is generically 3-jet spanned. Moreover by symmetry P cannot have a point in region 2 and can be assumed not to have a point on L_2 above $(2, 3)$.

(2) Assume that P contains a lattice point in region 5. We claim first that without loss of generality we can assume that $(3, 1) \in P$. To see this note simply that otherwise $(3, 0) \in P$ but $(3, 1) \notin P$, which implies that P must be a Cayley polytope. By smoothness and convexity it then follows that $(3, 0) \in P$ or $(3, 2) \in P$. If $(3, 2) \in P$ we can apply the same argument as in the case when P contains a point in region 4. Thus we can without loss of generality assume that $(3, 0) \in P$ but that $(3, 2) \notin P$. Next we note that we can assume that $(2, 2) \in P$. To see this note that otherwise one readily checks that, $(1, 2)$ has to be a vertex of P. Thus L_1 is a supporting line of P and the projection onto the y-axis shows that $\epsilon(X_P, \mathcal{L}_P; 1) = 2$. Thus we can assume that $(2, 2) \in P$. Let J_1^3 be the matrix of 3-jets at the general point for the convex hull of the points we now can assume lie in P. Then J_1^3 has nine linearly independent rows and one row of zeros corresponding to the derivative $\frac{\partial^3}{\partial y^3}$. Thus there are two possibilities: either P contains a point above L_1 and $\mathcal{L}_P$ is 3-jet spanned at the general point or the projection onto the y-axis shows that $\epsilon(X_P, \mathcal{L}_P; 1) = 2$. In conclusion if P contains a point in region 5 and $s(\mathcal{L}_P, 1) = 2$, then $\epsilon(X_P, \mathcal{L}_P; 1) = 2$. The case when P contains a point in region 1, follows by symmetry.

By the above we can assume that all lattice points of P lie in the convex hull of L_2 and L_3, except possibly the points $(2, 0)$ and $(0, 2)$. Thus if $\epsilon(X_P, \mathcal{L}_P; 1) \neq 2$, then P must contain a point in the region 3 or on one of the lines L_2 and L_3 further away from the origin than $(2, 1)$ or $(2, 1)$ respectively. However convexity and smoothness then show that neither $(2, 0)$ nor $(0, 2)$ can be a point in P, since if for example $(2, 0) \in P$ then $(3, 1) \in P$, which is a contradiction. Thus P is contained in the convex hull of L_2 and L_3 and the diagonal projection shows that $\epsilon(X_P, \mathcal{L}_P; 1) = 2$. □

We are now ready to prove Theorem 3.

Proof (Proof of Theorem 3) We can assume $s(\mathcal{L}_P, 1) \geq 1$, since $\mathcal{L}_P$ is very ample. Thus there are two cases either $s(\mathcal{L}_P, 1) = 1$ or $s(\mathcal{L}_P, 1) = 2$. In the first case we have that $s(\mathcal{L}_P, x) = 1$ al all points in X and thus Theorem 19 gives the conclusion. In the latter case Lemma 28 gives $\epsilon(X_P, \mathcal{L}_P; 1) = s(\mathcal{L}_P, 1)$. □

Corollary 29 *Let P be a smooth polygon such that $|P \cap \mathbb{Z}^n| \leq 20$, then it holds that $\epsilon(X_P, \mathcal{L}_P; 1) = s(\mathcal{L}_P, 1)$.*

Proof If $|P \cap \mathbb{Z}^n| < 20$, then $s(\mathcal{L}_P, 1) \leq 2$, by the bound (7.1) and Theorem 3 implies the claim. If instead $|P \cap \mathbb{Z}^n| = 20$, then either $s(\mathcal{L}_P, 1) \leq 2$ or $s(\mathcal{L}_P, 1) = 3$ and $P = 3\Delta_2$ by the main theorem in [7]. However $3\Delta_2$ corresponds to $(\mathbb{P}^2, \mathcal{O}(3))$, so the corollary follows since $s(\mathcal{O}(3), 1) = \epsilon(\mathbb{P}^2, \mathcal{O}(3); 1) = 3$. □

The following example shows that if one considers non-complete embeddings, then it may happen that $s(\mathcal{L}_P, 1_P) < \epsilon(X_P, \mathcal{L}_P, 1_P)$.

Example 30 Let P be the following smooth polygon corresponding to a non-complete embedding. (The white point is not included.)

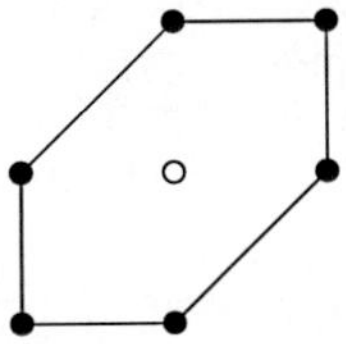

We claim that $s(\mathcal{L}_P, 1) = 1$ while $\epsilon(X_P, \mathcal{L}_P, 1) = 2$. To prove the claim note first that $s(\mathcal{L}_P, 1) = 1$ since $f(x, y) = x(x-1) + y(y-1) - xy$ is a degree 2 polynomial vanishing at the black points. To show that $\epsilon(X_P, \mathcal{L}_P, 1) \geq 2$ it is by Theorem 10 enough to show that $s(2\mathcal{L}_P, 1) = s(2\mathcal{L}_{P'}, 1) = 4$, where P' is the *complete* embedding corresponding to the polytope conv$(P \cap M)$. This follows from the fact that $2P' \cap M = 2P \cap M$. Which in turn follows from the following set of inclusions of sets.

$$2P \cap M \supseteq P \cap M + P \cap M = P' \cap M + P' \cap M = 2P' \cap M \supseteq 2P \cap M$$

Here the first and last inclusion follows by definition, while the second equality follows by the projective normality of P'. Checking the first equality can be done directly since a priori a lattice point $p = (x, y) + (z, w) \in P' \cap M + P' \cap M$ can fail to be in $P \cap M + P \cap M$ if and only if $(x, y) = (1, 1)$ or $(z, w) = (1, 1)$. By symmetry assume $(x, y) = (1, 1)$. If $(z, w) \in \{(0, 0), (1, 1), (2, 1)\}$, then $(x, y) - (0, 1) \in P \cap M$ and $(z, w) + (0, 1) \in P \cap M$ hence

$$p = (x, y) - (0, 1) + (z, w) + (0, 1) \in P \cap M + P \cap M.$$

If $(z, w) \in \{(0, 1), (1, 0)\}$, then $(x, y) - (1, 1) \in P \cap M$ and $(z, w) + (1, 1) \in P \cap M$. If instead $(z, w) = (1, 2)$, then $(x, y) - (1, 0) \in P \cap M$ and $(z, w) + (1, 0) \in P \cap M$. Finally if $(z, w) = (2, 2)$, then $(x, y) + (1, 0) \in P \cap M$ and $(z, w) - (1, 0) \in P \cap M$. We conclude on the one hand that $2P \cap M = 2P' \cap M$, i.e. that $\epsilon(X_P, \mathcal{L}_P, 1) \geq 2$. On the other hand it is clear that $s(n\mathcal{L}_P, 1) \leq s(n\mathcal{L}_{P'}, 1)$, since the global sections of $\mathcal{L}_P$ is a subspace of the global sections of $\mathcal{L}_{P'}$. Thus $\epsilon(X_P, \mathcal{L}_P, 1) = \epsilon(X_P, L_{P'}, 1) = 2$ by Theorem 10.

7.4 Seshadri Constants and Unnormalized Spectral Values

In the current section we prove Theorems 5 and 6. We start by showing the following lemma.

Lemma 31 *Let P be a polygon. If* core(P) *is a line segment, then*

$$\epsilon(X_P, \mathcal{L}_P; 1) \leq 2\mu(\mathcal{L}_P)^{-1}$$

and there exist two edges e and e' which are parallel to core(P).

Proof Let aff(core(P)) be the affine hull of core(P) and $K(P)$ be the parallel linear space. Furthermore we let $d_{\mathbf{n}}(\text{aff}(\text{core}(P)))$ denote the distance from the supporting hyperplane of P with normal $\mathbf{n}$ to aff(core(P)). By [5, Lemma 2.2] the normals of the supporting hyperplanes with $d_{\mathbf{n}}(\text{aff}(\text{core}(P))) = \mu(\mathcal{L}_P)^{-1}$ positively span $K(P)^{\perp}$. Since aff(core(P)) is one dimensional and P is convex it then follows that on either side of aff(core(P)) there is an edge which is parallel and at lattice distance $\mu(\mathcal{L}_P)^{-1}$ from aff(core(P)). Denote these edges by e and e'. Theorem 20 implies that $\epsilon(X_P, \mathcal{L}_P; 1) \leq |\pi(P)|$ for any projection π induced by a lattice projection. Thus projecting onto $K(P)^{\perp}$ shows that $\epsilon(X_P, \mathcal{L}_P; 1) \leq 2\mu(\mathcal{L}_P)^{-1}$. □

Remark 32 We remark that the assumption that the core of P is a line segment is a necessary condition even in the smooth setting. To see this it is enough to consider $(\mathbb{P}^2, \mathcal{O}(1))$. It is well-known that $s(\mathcal{O}(1), 1) = \epsilon(\mathbb{P}^2, \mathcal{O}(1); 1) = 1$, which can also be seen from Theorem 19. However it is straight forward to check that $P^{(1/3)}$ is a point, so that core(P) is a point and $2\mu(\mathcal{L}_P)^{-1} = 2/3$.

We will prove Theorem 6 before proving Theorem 5.

Proof (Proof of Theorem 6) By Lemma 31 it is sufficient to show that

$$\epsilon(X_P, \mathcal{L}_P; 1) \geq 2\mu(\mathcal{L}_P)^{-1}.$$

Assume that P is canonically positioned with the edge e of Lemma 31 along the x-axis. By Theorem 20 it is enough to show the existence of a projection π such that $|\pi(P)| \geq 2\mu(\mathcal{L}_P)^{-1}$ and a fiber F of π that satisfies $|F| \geq 2\mu(\mathcal{L}_P)^{-1}$ and corresponds to a rational polytope. Choosing π to be the natural projection proves the first part of the theorem since e and e' are at distance $2\mu(P)^{-1}$ apart.

It remains to consider the case when there is a fiber of the projection onto the x-axis containing both a point in e and a point in e'. Let π be the projection onto the x-axis. The assumption is equivalent to requiring that there exists a fiber of π with length exactly $2\mu(P)^{-1}$. Since e and e' are lattice line segments, the fiber can be taken to be a lattice polytope. It remains to show that $|\pi(P)| \geq 2\mu(P)^{-1}$.

Define now e $= (e_1, e_2, \ldots, e_p)$ to be the sequence of distinct edges of maximal length such that:

1. e_1 is the edge sharing a vertex with e different from the origin;
2. e_i shares a vertex with e_{i-1};
3. $e_p \neq e'$.

Let furthermore $\mathcal{K}$ be the sequence defined by letting k_i be the slope of the edge e_i and $k_i = \infty$ if e_i is a vertical edge. From convexity we first observe that the sequence $\mathcal{K}$ can be:

1. positive and strictly monotonically increasing, not including ∞;
2. positive and strictly monotonically increasing and then negative and strictly monotonically increasing.

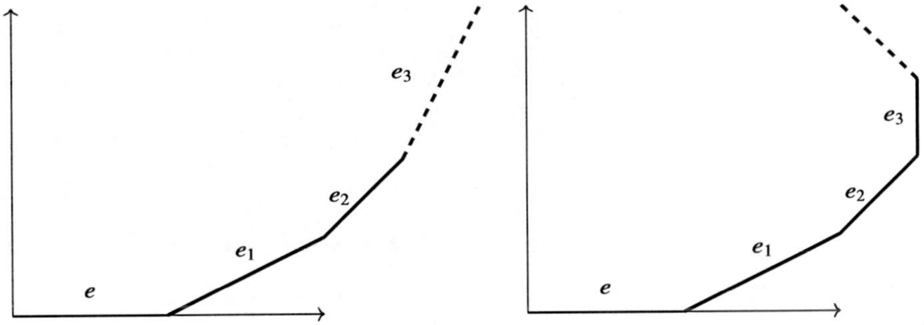

Fig. 7.4 Illustration of e when $\mathcal{K}$ has one and two phases

In the former case we will say that $\mathcal{K}$ has one phase and in the latter that $\mathcal{K}$ has two phases (see Fig. 7.4). We will call the subsequence of $\mathcal{K}$ that is strictly positive and monotonically increasing the first phase of $\mathcal{K}$ and the subsequence which is negative and strictly monotonically increasing the second phase of $\mathcal{K}$.

Consider first the case when $\mathcal{K}$ has one phase and let e_p be the last edge in e. By the smoothness assumption e_p must pass through a lattice point on the line $y = 2\mu(\mathcal{L}_P)^{-1} - 1$ and end at a vertex of the edge e'. As e' lies on the line $y = \mu(\mathcal{L}_P)^{-1}$ we can therefore conclude that e_p has slope at most 1. Because $\mathcal{K}$ is strictly increasing the same holds for all edges in e. Thus $|\pi(\text{e})|$ is larger or equal to the distance between the x-axis and the line $y = 2\mu(\mathcal{L}_P)^{-1}$, i.e. $|\pi(P)| \geq 2\mu(\mathcal{L}_P)^{-1}$.

We now turn to the case when $\mathcal{K}$ has two phases. We claim that by the smoothness of P these two phases of the sequence $\mathcal{K}$ are separated by ∞, corresponding to the slope of a vertical line. To see this let $(-a, -b)$ and (x, y) be the edge-direction of two adjacent edges in e, where $a, b, y > 0$ while $x \leq 0$. Thus $(-a, -b)$ is the direction vector of an edge of the first phase of $\mathcal{K}$ and (x, y) is a connecting edge which does not belong to this phase. Then the smoothness condition at the vertex of intersection between these edges says that

$$\begin{vmatrix} -a & x \\ -b & y \end{vmatrix} = bx - ay = \pm 1.$$

As $a, b, y > 0$ while $x \leq 0$ it follows that $x = 0$ and $y = a = 1$. Thus the edge having direction (x, y) is vertical. Furthermore there is a vertical edge of P along the y-axis. Let these edges lie on the lines $x = 0$ and $x = c$. Then the lines $x = \mu(\mathcal{L}_P)^{-1}$ and $x = c - \mu(\mathcal{L}_P)^{-1}$ give halfspaces containing core(P). As core(P) is a line segment it then follows that $c > 2\mu(\mathcal{L}_P)^{-1}$. As a consequence $|\pi(P)| > 2\mu(\mathcal{L}_P)^{-1}$ and $\epsilon(X_P, \mathcal{L}_P; 1) = 2\mu(\mathcal{L}_P)^{-1}$. □

Example 33 Consider the smooth polygon P depicted in Fig. 7.5, which has the property that core(P) is a line segment and $\mu(\mathcal{L}_P)^{-1} = 3$.

Here the natural projection is the projection on the y-axis. The longest fibers of that projection are the fibers between the two dashed lines. The length of any such

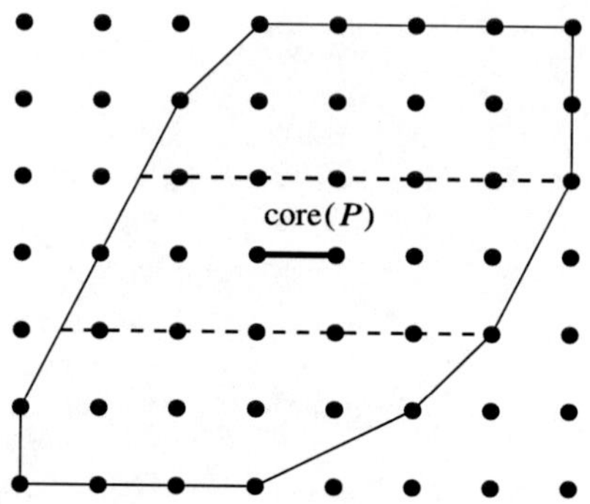

Fig. 7.5 A polytope P and its core

fiber is $11/2$. However the point 3 under the projection onto the x-axis has a fiber containing a point in e and a point in e'. Thus Theorem 6 shows that $\epsilon(X_P, \mathcal{L}_P; 1) = 2\mu(\mathcal{L}_P)^{-1} = 6$.

We now prove Theorem 5.

Proof (Proof of Theorem 5) We will proceed as in the proof of Theorem 6. Therefore we assume without loss of generality that P is canonically positioned with the edge e of Lemma 31 along the x-axis. As P is a lattice polytope $\mu(\mathcal{L}_P)^{-1} \in \frac{1}{2}\mathbb{Z}$, since e and e' must contain lattice points. We will proceed by considering all possible values of $\mu(\mathcal{L}_P)^{-1}$ less than three.

If $\mu(\mathcal{L}_P)^{-1} = 1/2$, then $\epsilon(X_P, \mathcal{L}_P; 1) = 2\mu(\mathcal{L}_P)^{-1} = 1$ by Theorem 6, since both e and e' are fibers of the projection onto the y-axis.

Consider next the case $\mu(\mathcal{L}_P)^{-1} = 1$. In this case the edge of P along the y-axis either contains a vertex of both e and e' or ends at $(0, 1)$. In the latter case the smoothness assumption at $(0, 1)$ implies that $(1, 2) \in P$. In either case there is a fiber of the projection onto the x-axis with lattice length $2\mu(\mathcal{L}_P)^{-1}$, which shows that $\epsilon(X_P, \mathcal{L}_P; 1) = 2\mu(\mathcal{L}_P)^{-1} = 2$.

Consider next the case $\mu(\mathcal{L}_P)^{-1} = 3/2$. If either $(3, 0)$ or $(3, 1)$ lies in P, then by projecting onto the y-axis it follows that $\epsilon(X_P, \mathcal{L}_P; 1) = 2\mu(\mathcal{L}_P)^{-1}$ by Theorem 6. Thus the vertex v of e different from the origin lies at a point $(a, 1)$ with $a < 3$. Consider now the edge f sharing the vertex v with e. By smoothness f must have an edge-direction of the form $(b, 1)$, with $b \in \mathbb{Z}$. Assume now that v is at $(2, 0)$. If f passes through the point $(0, 1)$ or $(1, 1)$, then the projection onto the y-axis has length less than 3, which is a contradiction. Hence without loss of generality we can assume that f passes through the point $(2, 1)$. Thus in this case $x = 2$ is a supporting line of P, but then $P^{(3/2)}$ is empty which is a contradiction. Therefore we can assume that f has a vertex at $(1, 0)$ and that $a = 1$, but then again $P^{(3/2)}$ is again empty.

Consider the case $\mu(\mathcal{L}_P)^{-1} = 2$. Arguing in the same way as for $\mu(\mathcal{L}_P)^{-1} = 3/2$ we see that the edge f, defined as above, can be assumed to have a vertex at either $(1, 0)$ or $(2, 0)$. If f has a vertex at $(2, 0)$, then we can assume that f passes through the point $(3, 1)$. This is because otherwise either the fiber of 1 under the projection onto the y-axis has length at least $2\mu(\mathcal{L}_P)^{-1}$ or core(P) is empty.

However it then follows that f has slope one and that core(P) is a point (or empty) which is a contradiction.

Thus we can assume that f has a vertex at $(1, 0)$. Then f can be assumed to pass through the point (3, 1) since otherwise either the fiber of 1 has length at least $2\mu(\mathcal{L}_P)^{-1}$ or we get a contradiction to that core(P) is a line segment. In this case let g be the edge sharing a vertex with f, other than e. By convexity the slope of g is greater than that of f. However there must be an edge passing through a lattice point on the line $y = 2\mu(\mathcal{L}_P)^{-1} - 1$ that also shares a vertex with e'. Thus by smoothness the only possibilities are $g = e'$ or that g has slope 1 or 2/3. As a consequence either f has length at least 2 of g has slope $2/3$, since otherwise core(P) would be a point (or empty). If f has length 2 the point (5, 2) lies in P. Therefore the fiber of 2 under the projection onto the y-axis has length at least $2\mu(\mathcal{L}_P)^{-1}$, since $(1, 2) \in P$. If instead g has slope 2/3, then $(6, 3) \in P$. However by convexity and smoothness either $(2, 3) \in P$ or there is an edge h of slope 1 passing through the lattice points (0, 1) and (1, 2). In the former case the fiber of 3 under the projection onto the y-axis has length at least 4. In the latter case by moving the edges e, e', f and h we see that $P^{(2)}$ is a point which is a contradiction. This shows that $\epsilon(X_P, \mathcal{L}_P; 1) = 2\mu(\mathcal{L}_P)^{-1} = 4$ in this case.

It remains to consider the case $\mu(\mathcal{L}_P)^{-1} = 5/2$. Arguing in the same way as above we see that we can assume that f has a vertex at either (1, 0) or (2, 0) or (3, 0). If f has a vertex at (3, 0), then we can assume that f pass through the point (4, 1). Since otherwise the fiber of 1 under the projection onto the y-axis has length at least $2\mu(\mathcal{L}_P)^{-1}$ or $P^{(5/2)}$ would be empty. Thus f can be assumed to have slope 1. Furthermore f can be assumed to have length 1, since otherwise either $(0, 2) \in P$ and the fiber of 2 under the projection onto the y-axis has length $2\mu(\mathcal{L})^{-1}$ or $(0, 2) \notin P$ and $P^{\mu(\mathcal{L}_P)^{-1}}$ is empty. Let g be the edge different than e sharing an edge with f. By smoothness and convexity the edge-direction of g is (0, 1), (1, 2) or (2, 3), since there must be an edge through a lattice point on the line $y = 2\mu(\mathcal{L}_P)^{-1} - 1$ sharing a vertex with e'. If g has edge-direction (1, 2), then the edge sharing a vertex with g different from f must be vertical. Therefore if g has edge-direction (0, 1) or (1, 2), we get that $P^{(5/2)}$ is a point or empty by moving e, e' and the vertical edges, yielding a contradiction. Assume instead that g has edge-direction (2, 3). Then the edge sharing the vertex (6, 4) with g must have slope 1 by smoothness and convexity, i.e. the same slope as f which is a contradiction.

If instead f has a vertex at (2, 0), then f can be assume to pass through the point (4, 1). Again because otherwise $P^{(5/2)}$ would be empty or the fiber of 1 would have length at least $2\mu(\mathcal{L}_P)^{-1}$. If f has length at least 2, then $(6, 2) \in P$ and the fiber of 2 under the projection on the y-axis has length at least $2\mu(\mathcal{L}_P)^{-1}$. Thus we can assume that f ends at (4, 1). Let g be the edge sharing an edge with f at (4, 1) and let (a, b) be the edge-direction of g. Note first that $b = 1, 2$ since there must be an edge of P passing through a lattice point on the line $y = 2\mu(\mathcal{L}_P)^{-1}$ that shares a vertex with e' having a different slope than f.

If $b = 1$ then g has slope 1. In this case $(1, 3) \in P$, since otherwise $P^{(5/2)}$ is a point or empty. Now either g has length at least 2 or g ends at (5, 2). In the former case the fiber of 3 under the projection onto the y-axis has length $2\mu(\mathcal{L}_P)^{-1}$. In the latter case let h be the edges sharing a vertex with g different from f. As there must be an edge i passing through a lattice point on the line $y = 2\mu(\mathcal{L}_P)^{-1} - 1$ sharing a vertex with e', it follows from the smoothness assumption that h has edge-direction $(1, 2)$ and i is vertical. In this case the point $(1, 4) \in P$, since otherwise there is an edge of P passing through the lattice points $(1, 3)$ and $(2, 4)$ and $P^{(5/2)}$ is empty. But as $(6, 4) \in P$ the fiber of 4 under the projection onto the y-axis has a length $2\mu(\mathcal{L}_P)^{-1}$ proving that $\epsilon(X_P, \mathcal{L}_P; 1) = 2\mu(\mathcal{L}_P)^{-1}$.

If instead $b = 2$, then smoothness and convexity implies that $a = 3$, i.e. $(7, 3) \in P$. Consider now the edge h different than f which shares a vertex with g. As there must be an edge i through a lattice point on the line $y = 2\mu(\mathcal{L}_P)^{-1} - 1$ sharing a vertex with e', the edge-directions of h are on the form $(a, 1)$, with $a \in \mathbb{Z}$. By convexity and smoothness it then follows that $a = 1$ so that h has slope 1. We now claim that $(2, 3) \in P$ which proves that the fiber of 3 under the projection onto the y-axis has length at least $2\mu(\mathcal{L}_P)^{-1}$. To see that $(2, 3) \in P$ note first that by smoothness the edge j different from e sharing a vertex with the edge along the y-axis must contain a point on the line $x = 1$. If $(2, 3) \notin P$, then j must have slope 1 and end at $(1, 2)$. However that leads to a contradiction since then $P^{(5/2)}$ is a point (or empty), as can be seen by moving the edges e, e', h and j.

Lastly we consider the case when $(1, 0)$ is a vertex of P. In this final case we can assume that the edge f must pass through either $(3, 1)$ or $(4, 1)$. Like before this is because otherwise we either get a contradiction to that core(P) is a line segment or the fiber of 1 under the projection onto the y-axis has length at least $2\mu(\mathcal{L}_P)^{-1}$. If f passes through $(4, 1)$, then either f has length at least 2 or f ends at $(4, 1)$. By the same argument as above the next edge g has edge-direction $(2, 1)$, $(5, 2)$ or $(8, 3)$. Thus regardless if f ends at $(4, 1)$ or not it holds that $(6, 2) \in P$. Since $(1, 2) \in P$ it then follows that the fiber of 2 under the projection onto the y-axis has length at least $2\mu(\mathcal{L}_P)^{-1}$.

It remains to check the case when f passes through $(3, 1)$. In this case the point $(1, 3) \in P$, since otherwise $P^{(\mu(\mathcal{L}_P)^{-1})}$ is empty. There are then two possibilities either f has length at least 2 or it ends at $(3, 1)$. It is easy to check that, like before, the edge-directions of the next edge g is $(5, 3)$, $(3, 2)$ or $(1, 1)$ by smoothness and convexity. If f has length 1 and g has direction $(1, 1)$ then one gets a contradiction to that core(P) is a line segment. In all other cases $(6, 3) \in P$ so the fiber of 3 under the projection onto the y-axis has length at least $2\mu(\mathcal{L}_P)^{-1}$. This concludes the case $\mu(\mathcal{L}_P)^{-1} = 5/2$ (Figs. 7.6, 7.7, 7.8 and 7.9). □

Example 34 Let P be the smooth polygon depicted in Fig. 7.10. It can readily be checked that core(P) is a line segment and that $\mu(\mathcal{L}_P)^{-1} = 4$. By direct computation one can also show that $s(\mathcal{L}, 1) = 8$, which implies that $\epsilon(X_P, \mathcal{L}_P; 1) = 8$. However the lattice length of a fiber is bounded from above by the maximal lattice length of a line segment contained in P. It is tedious but straight forward to check that the lattice length of any line segment in P is strictly less than 8. Thus there exists no projection of P such that the lower bound in Theorem 20 suffices to compute $\epsilon(X_P, \mathcal{L}_P; 1)$.

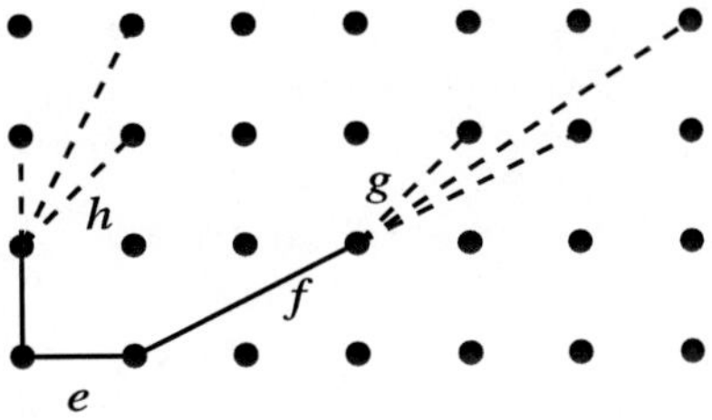

Fig. 7.6 Illustration of the case $\mu(\mathcal{L}_P)^{-1} = 2$ when $(1, 0)$ is a vertex of P

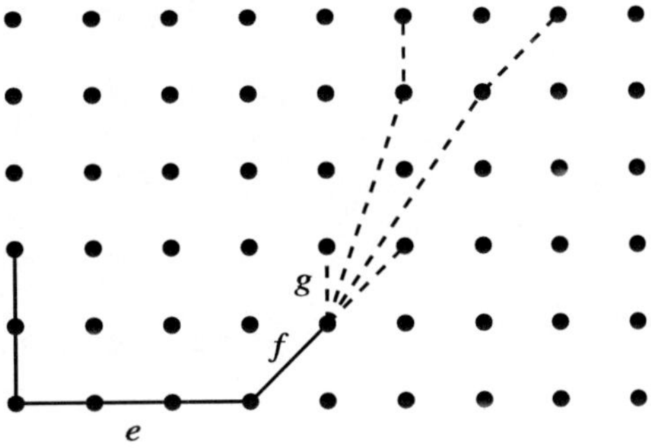

Fig. 7.7 Illustration for the case $\mu(P)^{-1} = 5/2$ when $(3, 0)$ is a vertex of P

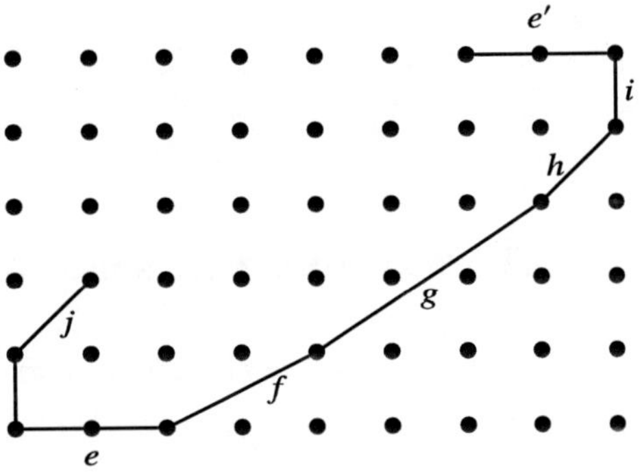

Fig. 7.8 Illustration for a special case when $\mu(P)^{-1} = 5/2$ and $(2, 0)$ is a vertex of P

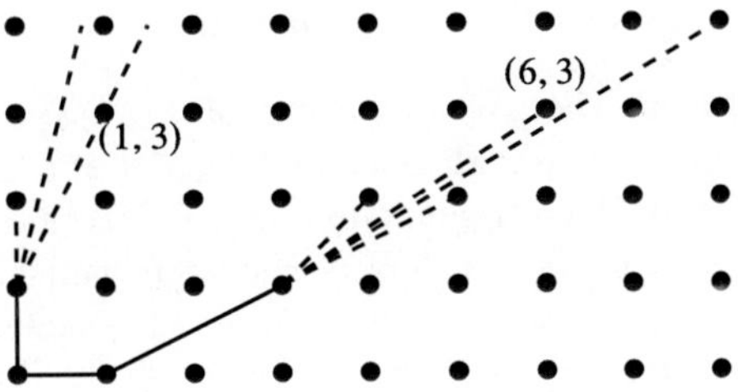

Fig. 7.9 Illustration for the case when $\mu(P)^{-1} = 5/2$ and $(1, 0)$ is a vertex of P

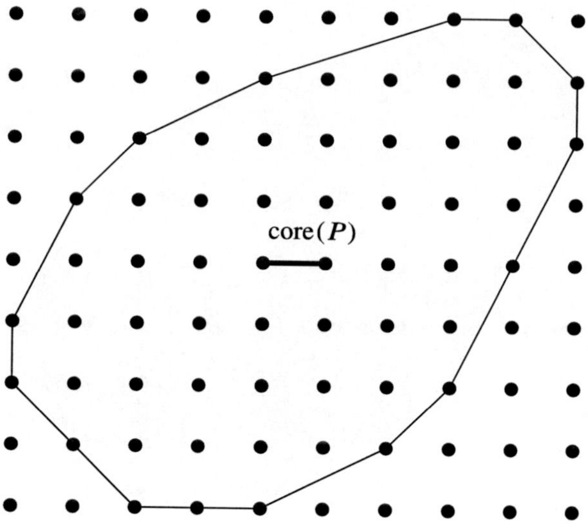

Fig. 7.10 The smooth polygon from Example 34

Remark 35 We are not aware of any smooth polygon P such that the core of P is a line segment but $2\mu(P)^{-1} \neq \epsilon(X_P, \mathcal{L}_P; 1)$. However Example 34 illustrates that if $\mu(\mathcal{L}_P)^{-1} \geq 4$, then the methods used to prove Theorem 5 will not suffice to show that $\epsilon(X_P, \mathcal{L}_P; 1) = 2\mu(\mathcal{L}_P)^{-1}$.

7.5 Characterizing Polygons Whose Core is a Point

In the current section we characterize smooth polygons P such that core(P) is a point.

Proposition 36 *Assume P is a polytope such that* core(P) *is a point. If $\mu(\mathcal{L}_P) = \tau(\mathcal{L}_P)$, then X_P is a Fano variety.*

Proof The assumption that core(P) is a point is equivalent to $H^0(X_P, \mu(\mathcal{L}_P)^{-1}\mathcal{K}_X + \mathcal{L}_P) = \mathbb{C}$. Recall that for toric varieties $\mu(\mathcal{L}_P)^{-1}\mathcal{K}_{X_P} + \mathcal{L}_P$ being nef is equivalent to being globally generated, see [3, Theorem 6.3.12]. Thus we can assume that at every point $x \in X$ there exists a nowhere vanishing global section of $\mu(\mathcal{L}_P)^{-1}\mathcal{K}_{X_P} + \mathcal{L}_P$. But as all such sections are linearly dependent it follows that $\mu(\mathcal{L}_P)^{-1}\mathcal{K}_{X_P} + \mathcal{L}_P$ has a global section s such that $s(x) \neq 0$ for all $x \in X_P$. Thus $\mu(\mathcal{L}_P)^{-1}\mathcal{K}_{X_P} + \mathcal{L}_P$ is free of rank 1, i.e.

$$\mu(\mathcal{L}_P)^{-1}\mathcal{K}_{X_P} + \mathcal{L}_P = \mathcal{O}_{X_P} \iff -\mathcal{K}_{X_P} = \mu(\mathcal{L}_P)\mathcal{L}_P.$$

Thus X_P is a Fano variety since $\mathcal{L}_P$ was assumed to be ample and $\mu(\mathcal{L}_P) > 0$. $\square$

Corollary 37 *Assume that P is a smooth polygon such that* core(P) *is a point and $\mu(\mathcal{L}_P) = \tau(\mathcal{L}_P)$. Then X_P is a Del Pezzo surface.*

Proposition 38 *Assume P is a polygon such that* core(P) *is a point. If $\mu(\mathcal{L}_P) \neq \tau(\mathcal{L}_P)$ then X_P is constructed by a sequence of toric morphisms onto a Fano variety. Moreover this sequence of morphisms can be taken to only correspond to refinements of the associated fans.*

Proof By definition $\tau(\mathcal{L}_P)^{-1}\mathcal{K}_X + \mathcal{L}_P$ is a line bundle which is nef but not ample. It follows that $(X_P, \tau(\mathcal{L}_P)^{-1}\mathcal{K}_{X_P} + \mathcal{L}_P)$ corresponds to a polytope P_1 whose inner-normal fan is a coarsening of the inner-normal fan of X, since every normal of an edge in P_1 also is a normal of an edge in P. Let $(X_1, \mathcal{L}_1)$ be the polarized toric variety and ample line bundle associated to P_1. Because the fan of X_1 is a coarsening of the fan of X we get a toric morphism $\phi : X_P \to X_1$ which contracts some torus invariant curves on X_P. Observe now that if $a < \tau(\mathcal{L}_P)$, then $(X, a^{-1}\mathcal{K}_{X_P} + \mathcal{L}_P)$ will yield the same polygon as $(X_1, (a^{-1} - \tau(\mathcal{L}_P)^{-1})\mathcal{K}_{X_1} + \mathcal{L}_1)$. To see this let E be a torus invariant divisor contracted by ϕ, then $(\tau(\mathcal{L}_P)^{-1}\mathcal{K}_{X_P} + \mathcal{L}_P) \cdot E = 0$ and $\mathcal{K}_{X_P} \cdot E < 0$. Thus

$$\left(\frac{1}{a}\mathcal{K}_{X_P} + \mathcal{L}_P\right) \cdot E = \left(\frac{1}{a} - \frac{1}{\tau(\mathcal{L}_P)}\right)\mathcal{K}_{X_P} \cdot E + \left(\frac{1}{\tau(\mathcal{L}_P)}\mathcal{K}_{X_P} + \mathcal{L}_P\right) \cdot E < 0.$$

Therefore for any $a < \tau(\mathcal{L}_P)$, there is no edge of the rational polygon P' associated to the linear system $(X_P, a^{-1}\mathcal{K}_{X_P} + \mathcal{L}_P)$ which corresponds to E. Thus by the main theorem of polytopes P' can be defined without the inequality corresponding to E. We can apply the same procedure to P_1 using the nef value of $(X_1, \mathcal{L}_1)$. Continuing in this fashion, we obtain a sequence of toric morphisms:

$$X_P \longrightarrow X_1 \longrightarrow \dots \longrightarrow X_r \tag{7.2}$$

Here X_r has the property that $\mu(\mathcal{L}_r) = \tau(\mathcal{L}_r)$. The argument for replacing $\mathcal{K}_{X_P}$ with $\mathcal{K}_{X_1}$ shows that once the edge corresponding to a prime divisor has disappeared it will not reappear later in the sequence. Thus all maps corresponds to refinements of the associated fans. Furthermore the sequence is finite since its length is bounded from above by the Picard number of X minus one. By the argument in the proof of Proposition 36 it now follows that X_r is a Fano variety. □

Corollary 39 *Assume that P is a smooth polygon such that* core(P) *is a point, then X_P can be constructed by taking consecutive blow-ups of $\mathbb{P}^2$ or $\mathbb{P}^1 \times \mathbb{P}^1$.*

Proof We start by showing that if X_P is smooth, then the variety X_r of Proposition 38 is smooth. Assume that this is not the case, then the sequence (7.2) is a resolution of the singularities of X_r, since all maps are birational and proper. Let f be the composition of these maps, then $\mathcal{K}_{X_P} = f^*\mathcal{K}_{X_r} + \sum E_i$ where the E_i are the torus invariant curves contracted by the chain of maps. Thus X_r is a Fano variety with at worse terminal singularities. Furthermore by [13, Lemma 1.17] a toric Fano variety

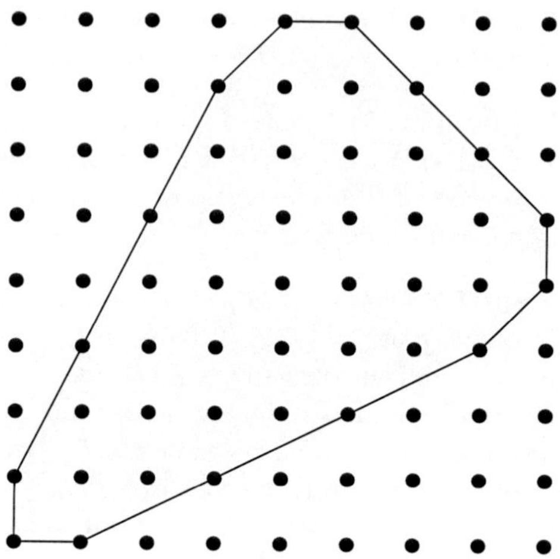

Fig. 7.11 The smooth polygon from Example 40

of dimension $d \leq 3$ with terminal singularities is smooth. Hence X_r is a Del Pezzo surface, i.e. it can be further blown-down to $\mathbb{P}^2$ or $\mathbb{P}^1 \times \mathbb{P}^1$. Furthermore as X_P and X_r are both smooth, all maps in (7.2) must be blow-ups, by [6, Theorem V.6.2]. □

The following example due to Atsushi Ito shows that $\epsilon(X_P, \mathcal{L}_P; 1)$ may fail to be an integer for smooth polarized toric surfaces. When this happens no combination of Corollary 11 and Theorem 20 can be used to compute it.

Example 40 Let P be the smooth polygon depicted in Fig. 7.11. It is straight forward to check that the core of P is a point, $s(\mathcal{L}_P, 1) = 7$ and $\mu(\mathcal{L}_P)^{-1} = 4$. However as shown by Ito [8], $\epsilon(X_P, \mathcal{L}_P; 1) = 15/2$. Moreover this example is part of an infinite family of smooth polarized toric surfaces having the property that the Seshadri constant is an odd multiple of $1/2$. Furthermore for any n Ito has constructed an infinite family of smooth polarized toric n-folds such that $\epsilon(X_P, \mathcal{L}_P; 1)$ is an odd multiple of $1/2$.

Acknowledgements We are grateful to Atsushi Ito for sharing Example 40 with us and letting us use it in the paper. The authors' research was supported by the Swedish Research Council, grant number 2014-4763. The research for this paper was supported by a VR grant [NT:2014-4763].

References

1. Bauer, T., Di Rocco, S., Harbourne, B., Kapustka, M., Knutsen, A., Syzdek, W., Szemberg, T.: A primer on Seshadri constants. In: Interactions of Classical and Numerical Algebraic Geometry. Contemporary Mathematics, vol. 496, pp. 33–70. American Mathematical Society, Providence (2009)
2. Beltrametti, M.C., Sommese, A.J.: The Adjunction Theory of Complex Projective Varieties. De Gruyter Expositions in Mathematics, vol. 16. Walter de Gruyter & Co., Berlin (1995)

3. Cox, D.A., Little, J.B., Schenck, H.K.: Toric Varieties. Graduate Studies in Mathematics, vol. 124. American Mathematical Society, Providence (2011)
4. Demailly, J.P.: Singular Hermitian metrics on positive line bundles. In: Complex Algebraic Varieties (Bayreuth, 1990). Lecture Notes in Mathematics, vol. 1507, pp. 87–104. Springer, Berlin (1992)
5. Di Rocco, S., Haase, C., Nill, B., Paffenholz, A.: Polyhedral adjunction theory. Algebra Number Theory **7**(10), 2417–2446 (2013)
6. Ewald, G.: Combinatorial Convexity and Algebraic Geometry. Graduate Texts in Mathematics, vol. 168. Springer, New York (1996)
7. Fulton, W., Kleiman, S., Piene, R., Tai, H.: Some intrinsic and extrinsic characterizations of the projective space. Bull. Soc. Math. France **113**(2), 205–210 (1985)
8. Ito, A.: Personal communication
9. Ito, A.: Seshadri constants via toric degenerations. J. Reine Angew. Math. **695**, 151–174 (2014)
10. Ito, A.: Algebro-geometric characterization of Cayley polytopes. Adv. Math. **270**, 598–608 (2015)
11. Lazarsfeld, R.: Positivity in algebraic geometry. I. Ergebnisse der Mathematik und ihrer Grenzgebiete. 3. Folge. A Series of Modern Surveys in Mathematics [Results in Mathematics and Related Areas. 3rd Series. A Series of Modern Surveys in Mathematics], vol. 48. Springer, Berlin (2004). Classical setting: line bundles and linear series
12. Lundman, A.: Local positivity of line bundles on smooth toric varieties and Cayley polytopes. J. Symbolic Comput. **74**, 109–124 (2016)
13. Nill, B.: Gorenstein toric Fano varieties. Manuscripta Math. **116**(2), 183–210 (2005)
14. Perkinson, D.: Inflections of toric varieties. pp. 483–515 (2000). Dedicated to William Fulton on the occasion of his 60th birthday
15. Szemberg, T.: An effective and sharp lower bound on Seshadri constants on surfaces with Picard number 1. J. Algebra **319**(8), 3112–3119 (2008)

Chapter 8
The Characterisation Problem of Ehrhart Polynomials of Lattice Polytopes

Akihiro Higashitani

Abstract One of the most important invariants of a lattice polytope is the Ehrhart polynomial. The problem of which polynomials can be Ehrhart polynomials of lattice polytopes is now well-studied. In this survey paper, after recalling the fundamental properties of the Ehrhart polynomials of lattice polytopes, we survey the results on this problem including recent developments. We discuss the characterisation of Ehrhart polynomials in several particular cases: small dimensions; small volumes; palindromic; small degrees. We also suggest some possible further questions.

Keywords Lattice polytopes · Ehrhart polynomials · h^*-vector · h^*-polynomial

8.1 Introduction

We say that a convex polytope $P \subset \mathbb{R}^d$ is a *lattice polytope* if all of its vertices belong to the standard lattice $\mathbb{Z}^d$. Given a lattice polytope $P \subset \mathbb{R}^d$ of dimension d, we can associate the enumerative function $n \mapsto \sharp(nP \cap \mathbb{Z}^d)$, which counts the number of lattice points contained in the nth dilation of P. Ehrhart [11] showed that there exists a polynomial $E_P(n)$ in n of degree d, where d is the dimension of P, such that $E_P(n) = \sharp(nP \cap \mathbb{Z}^d)$ for any positive integer n. We call the polynomial $E_P(n)$ the *Ehrhart polynomial* of P. The Ehrhart polynomial $E_P(n)$ has the following properties:

1. the constant term of the polynomial $E_P(n)$ is always equal to 1;
2. the leading term of the polynomial is equal to the Euclidean volume of P.

Moreover, $E_P(n)$ also satisfies the *Ehrhart-Macdonald reciprocity* [27]:

$$\sharp(nP^\circ \cap \mathbb{Z}^d) = (-1)^d E_P(-n) \quad \text{for } n \in \mathbb{Z}_{>0}, \tag{8.1}$$

A. Higashitani (✉)
Department of Pure and Applied Mathematics, Graduate School of Information Science and Technology, Osaka University, Osaka 565-0871, Japan
e-mail: higashitani@ist.osaka-u.ac.jp

A. M. Kasprzyk and B. Nill (eds.), *Interactions with Lattice Polytopes*, Springer Proceedings in Mathematics & Statistics 386,
https://doi.org/10.1007/978-3-030-98327-7_8

where P° denotes the relative interior of P. Namely, the number of lattice points contained in the nth dilation of the interior of P can be expressed by using the Ehrhart polynomial of P.

When formally expressed as a rational function, the generating function of $E_P(n)$ is known to be

$$\sum_{n=0}^{\infty} E_P(n)t^n = \frac{\sum_{i\geq 0} h_i^* t^i}{(1-t)^{d+1}},$$

and it is known that $h_i^* = 0$ for $i \geq d+1$ and that each h_i^* is an integer [30]. We call the integer sequence

$$h^*(P) = (h_0^*, h_1^*, \ldots, h_d^*)$$

the *h^*-vector* of P, denoted by $h^*(P)$, and the polynomial

$$h_P^*(t) = \sum_{i=0}^{d} h_i^* t^i$$

the *h^*-polynomial* of P, denoted by $h_P^*(t)$. In some literature, the h^*-vector or the h^*-polynomial is called the *δ-vector* or the *δ-polynomial* (see, e.g., [13, 16, 17, 19]).

For the study of Ehrhart polynomials of lattice polytopes, it is often preferable to use h^*-vectors (h^*-polynomials) instead of Ehrhart polynomials since the h^*-vectors behave better than the coefficients of Ehrhart polynomials. For example, it is known that h^*-vectors are always nonnegative integers [30], while the coefficients of Ehrhart polynomials are not necessarily integers (just rational numbers) and not necessarily nonnegative except for the leading, the second and the constant terms (see, e.g., [18]). Given a lattice polytope, knowing its h^*-vector is equivalent to knowing its Ehrhart polynomial. In fact, when $h^*(P) = (h_0^*, h_1^*, \ldots, h_d^*)$, we see that

$$E_P(n) = \sum_{i=0}^{d} h_i^* \binom{n+d-i}{d}. \tag{8.2}$$

We refer the reader to [7, 13] for an introduction to Ehrhart theory.

It follows from (8.1) that for a lattice polytope P with its h^*-vector $(h_0^*, h_1^*, \ldots, h_d^*)$, we have

$$\sum_{n=1}^{\infty} \sharp(nP^\circ \cap \mathbb{Z}^d)t^n = \frac{\sum_{i=0}^{d} h_{d-i}^* t^{i+1}}{(1-t)^{d+1}}. \tag{8.3}$$

The h^*-vectors of lattice polytopes have the following properties:

1. $h_0^* = 1$, $h_1^* = \sharp(P \cap \mathbb{Z}^d) - (d+1)$ and $h_d^* = \sharp(P^\circ \cap \mathbb{Z}^d)$;

2. if $P^\circ \cap \mathbb{Z}^d$ is non-empty, i.e., $h_d^* > 0$, then $h_i^* \geq h_1^*$ holds for $i = 1, \ldots, d - 1$ [15];
3. the leading coefficient of $E_P(n)$ is equal to $(\sum_{i=0}^d h_i^*)/d!$ by (8.2).

We call $\sum_{i=0}^d h_i^*$ the *normalized volume* of P, denoted by $\mathrm{vol}(P)$.

Note that $h_1^* \geq h_d^*$ holds. In fact, since P is of dimension d, we have $\sharp(\partial P \cap \mathbb{Z}^d) \geq d + 1$, which implies that

$$h_1^* = \sharp(P^\circ \cap \mathbb{Z}^d) + \sharp(\partial P \cap \mathbb{Z}^d) - d + 1 \geq \sharp(P^\circ \cap \mathbb{Z}^d) = h_d^*.$$

Hence, we see that $h_1^* = h_d^*$ if and only if $\sharp(\partial P \cap \mathbb{Z}^d) = d + 1$, which implies that P is a simplex. In particular, when $h_1^* = h_d^* = 0$, P is a simplex which contains no lattice point except for its vertices, called an *empty simplex*.

Let $\deg(P)$ denote the degree of the polynomial $h_P^*(t)$, i.e., $\deg(P) = \max\{i : h_i^* \neq 0\}$. The invariant $\deg(P)$ is called the *degree* of P. It is follows from (8.3) that

$$\deg(P) = d + 1 - \min\{m : mP^\circ \cap \mathbb{Z}^d \neq \varnothing\}. \tag{8.4}$$

In particular, $\deg(P) \leq d$.

The following is one of the most important unsolved problems in Ehrhart theory:

Problem 1 Characterise the sequences (or polynomials) that are the h^*-vectors (or h^*-polynomials) of lattice polytopes.

This problem has two steps: to prove the necessity and the sufficiency. More precisely, one step is to show some conditions for an integer sequence $(h_0^*, h_1^*, \ldots, h_d^*)$ to be the h^*-vector of some lattice polytope (e.g., $h_i^* \geq h_1^*$ for each $1 \leq i \leq d - 1$ if $h_d^* > 0$, $h_1^* \geq h_d^*$ and so on), and the other step is to construct a concrete lattice polytope whose h^*-vector coincides with a desired integer sequence.

We see that it is quite difficult to solve this problem in general. Furthermore, it might be believed that there exists *no* nice solution to this problem. Thus, in this survey paper, we will restrict to some particular cases. More concretely, we discuss the following cases:

1. small dimensions, i.e., d is small—see Sect. 8.3;
2. small volumes, i.e., $\sum_{i=0}^d h_i^*$ is small—see Sect. 8.4;
3. palindromic, i.e., $h_i^* = h_{\deg(P)-i}^*$ for any i—see Sect. 8.5;
4. small degrees, i.e., $\deg(P)$ is small—see Sect. 8.6.

8.2 Preliminaries

In this section, we present the basic concepts that we will use in this paper.

1. For two lattice polytopes $P, P' \subset \mathbb{R}^d$, we say that P and P' are *unimodularly equivalent* if there exist $f \in \mathrm{GL}_d(\mathbb{Z})$ and $\mathbf{u} \in \mathbb{Z}^d$ such that $P' = f(P) + \mathbf{u}$.

2. For a lattice polytope $P \subset \mathbb{R}^d$ of dimension d, let

$$\mathrm{Pyr}(P) = \mathrm{conv}(\{(\alpha, 0) \in \mathbb{R}^{d+1} : \alpha \in P\} \cup \{(0, \ldots, 0, 1)\}) \subset \mathbb{R}^{d+1}.$$

This new lattice polytope is said to be a *lattice pyramid* over P. It is not so hard to see that for a lattice polytope P, we have [7, Theorem 2.4]

$$\sum_{n=0}^{\infty} E_{\mathrm{Pyr}(P)}(n)t^n = \frac{h_P^*(t)}{(1-t)^{d+2}}.$$

In particular, we have

$$h_P^*(t) = h_{\mathrm{Pyr}(P)}^*(t) \qquad \text{and} \qquad \deg(P) = \deg(\mathrm{Pyr}(P)).$$

We recall the finite abelian group associated with a lattice simplex and discuss some properties on a lattice simplex in terms of this group. Let $\Delta \subset \mathbb{R}^d$ be a lattice simplex of dimension d with its vertices $\mathbf{v}_1, \mathbf{v}_2, \ldots, \mathbf{v}_{d+1} \in \mathbb{Z}^d$. Let

$$\Lambda_\Delta = \left\{ (x_1, x_2, \ldots, x_{d+1}) \in [0,1)^{d+1} : \sum_{i=1}^{d+1} x_i \mathbf{v}_i \in \mathbb{Z}^d, \ \sum_{i=1}^{d+1} x_i \in \mathbb{Z} \right\}$$

equipped with its addition defined by $\mathbf{x} + \mathbf{y} = (\{x_1 + y_1\}, \ldots, \{x_{d+1} + y_{d+1}\}) \in [0,1)^{d+1}$ for $\mathbf{x} = (x_1, \ldots, x_{d+1}) \in [0,1)^{d+1}$ and $\mathbf{y} = (y_1, \ldots, y_{d+1}) \in [0,1)^{d+1}$, where $\{r\} = r - \lfloor r \rfloor$ denotes the fractional part of $r \in \mathbb{R}$. We see that Λ_Δ is a finite abelian group. In fact, one can see that $\mathbf{0} = (0, \ldots, 0) \in \Lambda_\Delta$ and $-\mathbf{x} := (\{1 - x_1\}, \ldots, \{1 - x_{d+1}\}) \in \Lambda_\Delta$ for any $\mathbf{x} \in \Lambda_\Delta$.

Let $\mathcal{F}(d)$ denote the set of unimodular equivalence classes of lattice simplices of dimension d with a fixed vertex order, and let $\mathcal{A}(d)$ denote the set of finite abelian subgroups Λ of $[0,1)^{d+1}$ satisfying that the sum of all entries of each element in Λ is an integer. Actually, [6, Theorem 2.3] says that the correspondence

$$\mathcal{F}(d) \to \mathcal{A}(d); \quad \Delta \mapsto \Lambda_\Delta$$

is a bijection. In particular, a unimodular equivalence class of lattice simplices Δ is uniquely determined by the finite abelian group Λ_Δ up to permutation of coordinates.

We can discuss $h_\Delta^*(t)$, $\deg(\Delta)$, $\mathrm{vol}(\Delta)$ and whether Δ is a lattice pyramid in terms of Λ_Δ. We fix some notation. For $\mathbf{x} = (x_1, \ldots, x_{d+1}) \in \Lambda_\Delta$

1. let $\mathrm{ht}(\mathbf{x}) = \sum_{i=1}^{d+1} x_i \in \mathbb{Z}_{\geq 0}$;
2. let $\mathrm{supp}(\mathbf{x}) = \{i \in [d+1] : x_i \neq 0\}$, where $[n] = \{1, \ldots, n\}$ for $n \in \mathbb{Z}_{>0}$;
3. let $\mathrm{wt}(\mathbf{x}) = \sharp(\mathrm{supp}(\mathbf{x}))$.

Then we have [7, Corollary 3.11]

$$h^*_\Delta(t) = \sum_{\mathbf{x} \in \Lambda_\Delta} t^{\mathrm{ht}(\mathbf{x})}.$$

In particular,

$$\deg(\Delta) = \max\{\mathrm{ht}(\mathbf{x}) : \mathbf{x} \in \Lambda_\Delta\} \text{ and } \mathrm{vol}(\Delta) = |\Lambda_\Delta|.$$

We see that a lattice simplex Δ of dimension d is not a lattice pyramid if and only if for each $i \in [d+1]$, there is $(x_1, \ldots, x_{d+1}) \in \Lambda_\Delta$ such that $x_i \neq 0$.

Furthermore, we have

$$\mathrm{wt}(\alpha) = \mathrm{ht}(\alpha) + \mathrm{ht}(-\alpha) \tag{8.5}$$

for any $\alpha \in \Lambda_\Delta$ since

$$\mathrm{wt}(\alpha) = \sum_{i \in \mathrm{supp}(\alpha)} 1 = \sum_{i \in \mathrm{supp}(\alpha)} (\alpha_i + 1 - \alpha_i) = \mathrm{ht}(\alpha) + \mathrm{ht}(-\alpha).$$

8.3 Small Dimensions

We first consider the classification of h^*-polynomials of lattice polytopes in low dimensions. Namely, we discuss whether a given integer sequence $(h^*_0, h^*_1, \ldots, h^*_d)$ is a possible h^*-vector of some lattice polytope of dimension d for the case d is small.

In the case $d = 1$, the following is easy to prove.

Proposition 2 *For any nonnegative integer a, the integer sequence $(1, a)$ is the h^*-vector of some lattice polytope of dimension 1.*

Proof Take the lattice segment $\{x \in \mathbb{R} : 0 \leq x \leq a + 1\}$ of length $a + 1$. □

The case $d = 2$ is highly non-trivial.

Theorem 3 ([29]) *Given nonnegative integers a and b, the integer sequence $(1, a, b)$ is the h^*-vector of a lattice polytope of dimension 2 if and only if (a, b) satisfies one of the following conditions:*

1. $b = 0$;
2. $0 < b \leq a \leq 3b + 3$;
3. $(a, b) = (7, 1)$.

The "only if" part is due to [29]. The "if" part is easy to see. In fact, we can construct lattice polygons for any (a, b) satisfying 1, 2 or 3: see Fig. 8.1. Note that lattice polygons in the Case 1 can be given by Proposition 2 and lattice pyramid construction.

For the case $d = 3$, the characterisation of h^*-vectors is wide open, with the exception of a few partial results: the result in the case of palindromic h^*-polynomials

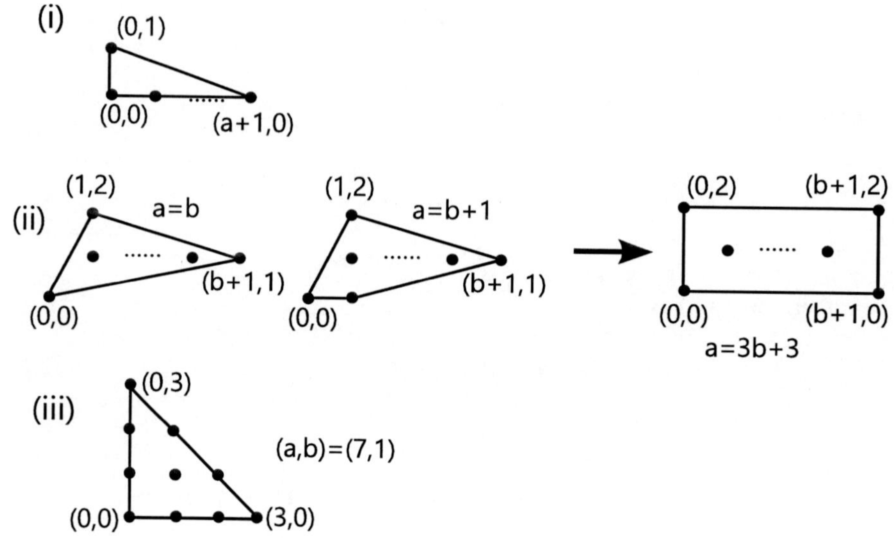

Fig. 8.1 Construction of lattice polygons whose h^*-vectors are one of 1, 2 or 3 in Theorem 3

will appear in Sect. 8.5 and the result in the case $h_3^* = 0$ will appear in Sect. 8.6. The characterisation in the case $h_3^* = 1$ can be essentially obtained from [23] and the case $h_3^* = 2$ is from [3].

Moreover, by the series of the papers [8–10] by Blanco–Santos, an algorithm which provides a complete list of all lattice polytopes of dimension 3 up to unimodular equivalence has been given. This enables us to get the complete list of h^*-vectors of lattice polytopes of dimension 3 containing only a few lattice points. In [10], the list of lattice polytopes $P \subset \mathbb{R}^3$ of dimension 3 with $\sharp(P \cap \mathbb{Z}^3) \leq 11$ is given up to unimodular equivalence. As a by-product, we also know the characterisation of h^*-vectors $(h_0^*, h_1^*, h_2^*, h_3^*)$ with $h_1^* \leq 7$.

The contribution of the papers [8–10] naturally suggests the following problem:

Problem 4 Characterise the possible h^*-vectors of lattice polytopes of dimension 3.

In fact, Balletti [1, Conjecture 8.7] predicts some necessary inequalities for h^*-vectors $(h_0^*, h_1^*, h_2^*, h_3^*)$ of lattice polytopes of dimension 3.

8.4 Small Volumes

Next, we consider the case of small normalized volumes. Namely, we discuss whether a given integer sequence $(h_0^*, h_1^*, \dots, h_d^*)$ is a possible h^*-vector of some lattice polytope of dimension d for the case $\sum_{i=0}^{d} h_i^*$ is small.

In the case $\sum_{i=0}^{d} h_i^* = 1$, since each h_i^* is nonnegative and $h_0^* = 1$, we conclude that $h_0^* = 1$ and $h_1^* = \cdots = h_d^* = 0$. This is the h^*-vector of a *unit simplex* of dimension d, where a unit simplex of dimension d is the convex hull of the origin of $\mathbb{R}^d$ and the unit vectors of $\mathbb{R}^d$.

For the case $\sum_{i=0}^{d} h_i^* \geq 2$, we recall two well-known inequalities on h^*-vectors. Let $P \subset \mathbb{R}^d$ be a lattice polytope of dimension d with degree s. One inequality is

$$h_0^* + h_1^* + \cdots + h_i^* \leq h_s^* + h_{s-1}^* + \cdots + h_{s-i}^* \quad \text{for } i = 0, 1, \ldots, s. \tag{8.6}$$

This is proved by Stanley [31]. Another one is

$$h_d^* + h_{d-1}^* + \cdots + h_{d-i}^* \leq h_1^* + h_2^* + \cdots + h_{i+1}^* \quad \text{for } i = 0, 1, \ldots, d-1. \tag{8.7}$$

This appears in the work of Hibi [15]. Note that these inequalities (8.6) and (8.7) are vastly generalized by Stapledon [32, 33]. All inequalities in (8.6) hold with equality for any i if and only if $h_i^* = h_{s-i}^*$ for $i = 0, 1, \ldots, s$. This is equivalent to the lattice polytope being Gorenstein. See Sect. 8.5. All inequalities in (8.7) hold with equality for any i if and only if $h_{i+1}^* = h_{d-i}^*$ for $i = 0, 1, \ldots, d-1$. This is called *shifted symmetric*, introduced in [19]. A lattice polytope P has a shifted symmetric h^*-vector if and only if P is a lattice simplex all of whose facets have normalized volume 1 (see [19, Theorem 2.1]).

A characterisation in the cases of $\sum_{i=0}^{d} h_i^* = 2$ and 3 is given as follows:

Theorem 5 ([17, Theorem 0.1]) *Given a sequence $(h_0^*, h_1^*, \ldots, h_d^*)$ of nonnegative integers, where $h_0^* = 1$ and $\sum_{i=0}^{d} h_i^* \leq 3$, there exists a lattice polytope of dimension d whose h^*-vector coincides with $(h_0^*, h_1^*, \ldots, h_d^*)$ if and only if $(h_0^*, h_1^*, \ldots, h_d^*)$ satisfies the inequalities* (8.6) *and* (8.7).

The “only if” part directly follows from the works by Stanley and Hibi. The essential part of this theorem is the “if” part.

The following example shows that Theorem 5 is no longer true for the case $\sum_{i=0}^{d} h_i^* = 4$.

Example 6 ([17, Example 1.2]) We consider the sequence $(1, 0, 1, 0, 1, 1, 0, 0)$. Then we see that this cannot be the h^*-vector of any lattice polytope of dimension 7. Suppose, on the contrary, that there exists a lattice polytope P with its h^*-vector $(1, 0, 1, 0, 1, 1, 0, 0)$. Since $h_1^* = 0$, we know that P should be a simplex. Let $\Lambda_P = \{\mathbf{0}, \alpha, \beta, \gamma\}$ be as in Sect. 8.2, where $\mathrm{ht}(\alpha) = 2$, $\mathrm{ht}(\beta) = 4$ and $\mathrm{ht}(\gamma) = 5$. Since $-\gamma \in \Lambda_P$ and $\mathrm{wt}(\gamma) \leq 8$ together with $\mathrm{ht}(-\gamma) = \mathrm{wt}(\gamma) - 5 \in \{2, 4, 5\}$ by (8.5), we see that $\mathrm{wt}(\gamma) = 7$ and $-\gamma = \alpha$. Thus, we also see that $-\beta = \beta$. This implies that $\mathrm{wt}(\beta) = 8$ by (8.5). Hence, $\mathrm{supp}(\gamma) = \mathrm{supp}(\alpha) = [8] \setminus \{i\}$ for some $i \in [8]$ and $\mathrm{supp}(\beta) = [8]$. Then one has $i \in \mathrm{supp}(\alpha + \beta)$. This means that $\alpha + \beta = \beta$, which means that $\alpha = \mathbf{0}$, a contradiction.

Before considering the statement in the case $\sum_{i=0}^{d} h_i^* = 4$, we introduce the following notation. For a sequence $(h_0^*, h_1^*, \ldots, h_d^*)$ of nonnegative integers with $h_0^* = 1$ and $\sum_{i=0}^{d} h_i^* = m$, let $i_1, \ldots, i_{m-1}$ be the positive integers such that

$$\sum_{i=0}^{d} h_i^* t^i = 1 + t^{i_1} + \cdots + t^{i_{m-1}},$$

where $1 \leq i_1 \leq \cdots \leq i_{m-1} \leq d$. For example, for a sequence $(1, 0, 2, 3, 0, 1, 0)$, we have $i_1 = i_2 = 2$, $i_3 = i_4 = i_5 = 3$ and $i_6 = 5$. Notice that the sequence $(h_0^*, h_1^*, \ldots, h_d^*)$ and the positive integers $i_1, \ldots, i_{m-1}$ have the same information. Actually, we can rephrase the inequalities (8.6) and (8.7) as follows:

Proposition 7 ([20, Proposition 2.2]) *With notation as above:*

1. *the inequalities* $i_j + i_{m-j-1} \geq i_{m-1}$, *where* $1 \leq j \leq m-2$, *are equivalent to the inequalities* (8.6)*;*
2. *the inequalities* $i_j + i_{m-j} \leq d+1$, *where* $1 \leq j \leq m-1$, *are equivalent to the inequalities* (8.7).

A characterisation in the case of $\sum_{i=0}^{d} h_i^* = 4$ is given as follows:

Theorem 8 ([16, Theorem 5.1]) *Let* $1 + t^{i_1} + t^{i_2} + t^{i_3}$ *be a polynomial with* $1 \leq i_1 \leq i_2 \leq i_3 \leq d$. *Then there exists a lattice polytope of dimension* d *whose* h^*-*polynomial equals* $1 + t^{i_1} + t^{i_2} + t^{i_3}$ *if and only if the triple* (i_1, i_2, i_3) *satisfies:*

1. $i_1 + i_2 \geq i_3$, $i_1 + i_3 \leq d+1$ *and* $2i_2 \leq d+1$ *hold;*
2. *either* $2i_2 \leq i_1 + i_3$ *or* $i_2 + i_3 \leq d+1$ *holds.*

We remark that as mentioned in Proposition 7, the first inequality in Condition 1 is equivalent to (8.6) and the second and third inequalities are equivalent to (8.7). Condition 2 is the special one for the case $\sum_{i=0}^{d} h_i^* = 4$.

Example 9 (*Example 6 continued*) Consider the sequence $(1, 0, 1, 0, 1, 1, 0, 0)$ again. For the above notation, we have $i_1 = 2$, $i_2 = 4$, $i_3 = 5$ and $d+1 = 8$. Then Condition 1 in Theorem 8 are satisfied, while both $2i_2 > i_1 + i_3$ and $i_2 + i_3 > d+1$ hold. Hence Condition 2 is not satisfied.

Remark 10 ([16, Remark 5.3]) For the case of $\sum_{i=0}^{d} h_i^* \leq 4$, we see from the proofs of Theorems 5 and 8 that all the possible h^*-vectors can be obtained by lattice *simplices*. However, when $\sum_{i=0}^{d} h_i^* = 5$, the h^*-vector $(1, 3, 1)$ cannot be obtained from any simplex, while it is a possible h^*-vector of a lattice polygon (see Theorem 3).

This remark implies that for the complete characterisations with larger normalized volumes, we have to see the h^*-vectors of not only lattice simplices but also non-simplices. Hence, for the further investigation, it is natural to think of the h^*-vectors of lattice *simplices* as a first step.

Theorem 11 ([20, Theorem 1.1]) *Let* P *be a lattice simplex of dimension* d *and let* $h^*(P) = (h_0^*, h_1^*, \ldots, h_d^*)$ *be its* h^*-*vector. Suppose that* $\sum_{i=0}^{d} h_i^* = p$ *is an odd prime. Let* $i_1, \ldots, i_{p-1}$ *be the positive integers such that* $\sum_{i=0}^{d} h_i^* t^i = 1 + t^{i_1} + \cdots + t^{i_{p-1}}$, *where* $1 \leq i_1 \leq \cdots \leq i_{p-1} \leq d$.*Then:*

1. $i_1 + i_{p-1} = i_2 + i_{p-2} = \cdots = i_{(p-1)/2} + i_{(p+1)/2} \leq d + 1$; *and*
2. $i_k + i_\ell \geq i_{k+\ell}$ *for* $k, \ell \geq 1$ *with* $k + \ell \leq p - 1$.

By using these necessary conditions for integer sequences to be the h^*-vector of lattice simplices, we can prove the following:

Theorem 12 ([20, Theorem 1.2]) *Let* $1 + t^{i_1} + t^{i_2} + t^{i_3} + t^{i_4}$ *be a polynomial with* $1 \leq i_1 \leq \cdots \leq i_4 \leq d$. *Then there exists a lattice simplex of dimension d whose* h^*-*polynomial equals* $1 + t^{i_1} + t^{i_2} + t^{i_3} + t^{i_4}$ *if and only if* (i_1, i_2, i_3, i_4) *satisfies* $i_1 + i_4 = i_2 + i_3 \leq d + 1$ *and* $i_k + i_\ell \geq i_{k+\ell}$ *for* $k, \ell \geq 1$ *with* $k + \ell \leq 4$.

Theorem 13 ([20, Theorem 1.3]) *Let* $1 + t^{i_1} + t^{i_2} + t^{i_3} + t^{i_4} + t^{i_5} + t^{i_6}$ *be a polynomial with* $1 \leq i_1 \leq \cdots \leq i_6 \leq d$. *Then there exists a lattice simplex of dimension d whose* h^*-*polynomial equals* $1 + t^{i_1} + t^{i_2} + t^{i_3} + t^{i_4} + t^{i_5} + t^{i_6}$ *if and only if* $(i_1, i_2, i_3, i_4, i_5, i_6)$ *satisfies* $i_1 + i_6 = i_2 + i_5 = i_3 + i_4 \leq d + 1$ *and* $i_k + i_\ell \geq i_{k+\ell}$ *for* $k, \ell \geq 1$ *with* $k + \ell \leq 6$.

Remark 14 (*see* [20, Sect. 5.1]) Theorems 12 and 13 say that when $\sum_{i=0}^{d} h_i^*$ is 5 or 7, the necessary conditions in Theorem 11 are also sufficient for lattice simplices. However, this is not true for all prime numbers. In fact, since the volume of a lattice polytope containing a unique lattice point in its interior has an upper bound (see, e.g., [26]), if p is a sufficiently large prime number, then the integer sequence $(1, 1, p - 3, 1)$ cannot be the h^*-vector of any lattice simplex of dimension 3, although $(1, 1, p - 3, 1)$ satisfies all the conditions in Theorem 11.

Recently, a complete characterisation of the possible h^*-vectors $(h_0^*, h_1^*, \ldots, h_d^*)$ of lattice *polytopes* with $\sum_{i=0}^{d} h_i^* = 5$ has been given.

Theorem 15 ([35, Theorem 0.4]) *Let* $1 + t^{i_1} + t^{i_2} + t^{i_3} + t^{i_4}$ *be a polynomial with* $1 \leq i_1 \leq \cdots \leq i_4 \leq d$. *Then there exists a lattice polytope of dimension d whose* h^*-*polynomial equals* $1 + t^{i_1} + t^{i_2} + t^{i_3} + t^{i_4}$ *if and only if*

1. (i_1, i_2, i_3, i_4) *satisfies the conditions in Theorem 12; or*
2. $1 + t^{i_1} + t^{i_2} + t^{i_3} + t^{i_4} = 1 + 3t + t^2$ *or* $1 + t + 3t^2$ *or* $1 + t + t^2 + 2t^3$.

Note that the polynomials $1 + 3t + t^2$ and $1 + t + 3t^2$ and $1 + t + t^2 + 2t^3$ are the h^*-polynomials of some lattice polytopes respectively (see [35, Example 2.4] for a more detailed construction of those lattice polytopes), but these are never the h^*-polynomials of lattice simplices.

The proof of Theorem 15 relies on the following theorem, which is a recent, major contribution to Ehrhart theory:

Theorem 16 ([22, Theorem 1.3]) *If a lattice polytope P is spanning then the* h^*-*vector of P has no gap, i.e.,* $h_i^* > 0$ *for any* $i = 0, 1, \ldots, \deg(P)$.

Here, we say that a lattice polytope $P \subset \mathbb{R}^d$ is *spanning* if the affine lattice generated by $P \cap \mathbb{Z}^d$, denoted by Γ_P, is equal to $\mathbb{Z}^d$, Note that the converse of Theorem 16 is not true in general. In fact, the lattice simplex $P = \text{conv}(\{\mathbf{e}_1, \mathbf{e}_2, \pm(\mathbf{e}_1 + \mathbf{e}_2 + 2\mathbf{e}_3)\}) \subset \mathbb{R}^3$ has the h^*-vector $(1, 1, 2, 0)$, while this simplex is not spanning since $\mathbf{e}_3 \in \mathbb{Z}^3$ is not contained in Γ_P.

The key lemma for the proof of Theorem 15 is the following:

Lemma 17 ([35, Theorem 1.1]) *Let P be a lattice polytope which is not a simplex. Suppose that* $\mathrm{vol}(P)$ *is prime. Then P is spanning.*

Note that this lemma is not true for non-prime case. For example, let

$$P = \mathrm{conv}(\{\mathbf{0}, \mathbf{e}_1, \mathbf{e}_2, \mathbf{e}_1 + \mathbf{e}_2, \mathbf{e}_3, \mathbf{e}_1 + \mathbf{e}_2 + \mathbf{e}_3 + 2\mathbf{e}_4\}) \subset \mathbb{R}^4.$$

Then P is a lattice polytope with $\mathrm{vol}(P) = 4$. Moreover, P has 6 vertices, i.e., P is not a simplex, and $P \cap \mathbb{Z}^4$ equals the set of vertices. Hence, we observe that $\mathbf{e}_4 \notin \Gamma_P$.

Proof *(Sketch of proof of Theorem* 15*)* For the "if" part, we may construct the lattice polytopes whose h^*-polynomials are $1 + 3t + t^2$, $1 + t + 3t^2$ and $1 + t + t^2 + 2t^3$, respectively.

For the "only if" part, we consider the h^*-polynomial $1 + t^{i_1} + \cdots + t^{i_4}$ of the lattice polytope P which is not a simplex. Then it follows from Theorem 16 and Lemma 17 that the h^*-polynomial has no gap. Thus, the possible h^*-polynomials are the following polynomials:

$$1+4t,\ \ 1+3t+t^2,\ \ 1+t+3t^2,\ \ 1+2t+2t^2,\ \ 1+t+2t^2+t^3,\ \ 1+t+t^2+2t^3,\ \ 1+t+t^2+t^3+t^4.$$

Note that $1 + 2t + t^2 + t^3$ violates the inequality (8.6). Since $1 + 4t$, $1 + 2t + 2t^2$, $1 + t + 2t^2 + t^3$ and $1 + t + t^2 + t^3 + t^4$ are the possible h^*-polynomials of lattice simplices by Theorem 12, we conclude the desired assertion. □

As a future problem, we suggest the following:

Problem 18 Characterise the h^*-polynomials of lattice polytopes with $\mathrm{vol}(P) = 7$.

Towards this characterisation, the similar idea to the proof of Theorem 15 enables us to deduce that we must consider the following polynomials:

$$1 + 2t + 3t^2 + t^3,\ \ 1 + t + 3t^2 + 2t^3,\ \ 1 + t + 2t^2 + 3t^3,\ \ 1 + t + t^2 + 4t^3,$$
$$1 + t + t^2 + 3t^3 + t^4,\ \ 1 + t + 3t^2 + t^3 + t^4,\ \ 1 + t + t^2 + t^3 + 2t^4 + t^5.$$

More precisely, we may determine the existence or non-existence of a lattice polytope whose h^*-polynomial is one of them.

The following is also a natural problem to try as a next step:

Problem 19 Characterise the h^*-polynomials of lattice simplices with $\mathrm{vol}(P) = 6$.

8.5 Palindromic

We say that a polynomial $\sum_{i=0}^{s} a_i t^i$ of degree s is *palindromic* if $a_i = a_{s-i}$ for $i = 0, 1, \ldots, s$. Next, we consider the case of palindromic h^*-polynomials. Namely, we discuss whether a given palindromic polynomial $\sum_{i=0}^{s} h_i^* t^i$ of degree s is a possible h^*-polynomial of some lattice polytope with degree s.

Recall that a lattice polytope $P \subset \mathbb{R}^d$ is *reflexive* if the origin of $\mathbb{R}^d$ is contained in the interior of P and the polar polytope

$$P^\vee = \{\mathbf{x} \in \mathbb{R}^d : \langle \mathbf{x}, \mathbf{y} \rangle \leq 1 \text{ for any } \mathbf{y} \in P\}$$

is also a lattice polytope, where $\langle \cdot, \cdot \rangle$ denotes the usual inner product of $\mathbb{R}^d$. Reflexivity of lattice polytopes is characterised as follows:

Proposition 20 (cf. [5, 14]) *Let P be a lattice polytope of dimension d, let $(h_0^*, h_1^*, \ldots, h_d^*)$ be its h^*-vector and let $E_P(n) = a_d n^d + a_{d-1} n^{d-1} + \cdots + 1$ be its Ehrhart polynomial. Then the following four conditions are equivalent:*

1. *P is a reflexive polytope;*
2. *$\deg(P) = d$ and $h_P^*(t)$ is palindromic, i.e., $h_i^* = h_{d-i}^*$ for $i = 0, 1, \ldots, d$;*
3. *the functional equation $E_P(n) = (-1)^d E_P(-n-1)$ holds;*
4. *we have $da_d = 2a_{d-1}$.*

We also recall that a lattice polytope $P \subset \mathbb{R}^d$ is called *Gorenstein of index ℓ* if ℓP is unimodularly equivalent to a reflexive polytope. Since $\sharp(\ell P \cap \mathbb{Z}^d) = 1$, we see that $\deg(P) = \dim P + 1 - \ell$ by (8.4). The following easily follows from Proposition 20.

Proposition 21 *Let P be a lattice polytope of dimension d with its h^*-vector $(h_0^*, h_1^*, \ldots, h_d^*)$. Then the following three conditions are equivalent:*

1. *P is a Gorenstein polytope of index ℓ;*
2. *$\deg(P) = s$ and $h_P^*(t)$ is palindromic, i.e., $h_i^* = h_{s-i}^*$ for $i = 0, 1, \ldots, s$, where $s = d + 1 - \ell$;*
3. *the functional equation $E_P(n) = (-1)^d E_P(-n-\ell)$ holds.*

Reflexive polytopes together with Gorenstein polytopes are particularly important in many areas, e.g., commutative algebra, toric geometry and mirror symmetry (e.g., [4, 5, 31] and so on). Hence, many researchers are quite interested in reflexive polytopes and they have been intensively studied. On the characterisation of palindromic h^*-polynomials, i.e., the h^*-polynomials of Gorenstein polytopes, the case of 2-dimensional reflexive polytopes is well known:

Proposition 22 (See e.g. [28]) *Given an integer a, the sequence $(1, a, 1)$ is the h^*-vector of a reflexive polytope of dimension 2 if and only if $a \in \{1, 2, \ldots, 7\}$.*

The case of Gorenstein polytopes of degree 2 is also known, which is highly non-trivial:

Theorem 23 ([4, Theorem 2.10]) *Given an integer a, the polynomial $1 + at + t^2$ is the h^*-polynomial of a Gorenstein polytope of degree 2 if and only if $a \in \{0, 1, \ldots, 7\}$.*

The following is one natural analogue of Theorem 23.

Theorem 24 ([21, Corollary 1.1]) *Given an integer a, the polynomial $1 + at^k + t^{2k}$ is the h^*-polynomial of a Gorenstein polytope of degree $2k$ with $k \geq 2$ if and only if one of the following conditions holds:*

1. $a \in \{0, 1, 2, 4, 6\}$;
2. *there exist* $\ell \geq 4$ *and* $m \geq 1$ *such that* $a = 2^\ell - 2$ *and* $k = 2^{\ell-3}m$;
3. *there exist* $\ell \geq 3$ *and* $m \geq 1$ *such that* $a = 3^\ell - 2$ *and* $k = 3^{\ell-2}m$.

The case of 3-dimensional reflexive polytopes is also known:

Proposition 25 ([24]) *Given an integer* a*, the integer sequence* $(1, a, a, 1)$ *is the* h^**-vector of a reflexive polytope of dimension* 3 *if and only if* $a \in \{1, 2, \dots, 35\} \setminus \{33, 34\}$.

Similar to this theorem, the necessary and sufficient condition for the integer sequence $(1, a, b, a, 1)$ to be the h^*-vector of a reflexive polytope of dimension 4 is also known by [25]. (There are 14, 373 possible h^*-vectors.)

Since we already know the possible h^*-vectors of reflexive polytopes of dimension 3, the following is a natural problem to try as a next step:

Problem 26 Characterise the h^*-polynomials of Gorenstein polytopes of degree 3.

8.6 Small Degrees

Next, we consider the case of small degrees. Namely, we discuss whether a given polynomial $\sum_{i=0}^s h_i^* t^i$ is a possible h^*-polynomial of some lattice polytope of degree s when s is small.

In general, we see the following:

Proposition 27 *If a given polynomial* $\sum_{i=0}^d h_i^* t^i$ *is the* h^**-polynomial of some lattice polytope of dimension* d*, then it is also the* h^**-polynomial of some lattice polytope of dimension* $d + m$ *for any* $m \in \mathbb{Z}_{\geq 0}$.

Proof Take the m times iterated lattice pyramids. □

Thus, we have the inclusion

$$\begin{aligned}&\{h^*\text{-polynomials of lattice polytopes of dimension } d\}\\ &\qquad \subset \{h^*\text{-polynomials of lattice polytopes of degree at most } d\}.\end{aligned}$$

In the case degree at most 2, we see that there is a difference between these two sets. More precisely, the following is known. It looks quite similar to Theorem 3 but slightly different.

Theorem 28 ([34, Theorem 2]) *Let* P *be a lattice polytope of degree at most* 2*. Then its* h^**-polynomial* $1 + h_1^* t + h_2^* t^2$ *satisfies one of the following conditions:*

1. $h_2^* = 0$;
2. $h_1^* \leq 3h_2^* + 3$ *and* $h_2^* > 0$; *or*
3. $(h_1^*, h_2^*) = (7, 1)$.

Note that the conditions in Theorem 28 are also sufficient. Namely, it is proved in [12, Proposition 1.10] that given nonnegative integers (h_1^*, h_2^*) satisfying one of the conditions in Theorem 28, there exists a lattice polytope P such that $h_P^*(t) = 1 + h_1^*t + h_2^*t^2$. Those polytopes can be constructed by lattice polytopes of dimension at most 3. Hence, the possible h^*-polynomials of degree at most 2 are completely characterised.

In Theorem 28, the condition "$h_2^* \leq h_1^*$" is missing, while $b \leq a$ appears in Theorem 3. The inequality $b \leq a$ comes from $h_1^* \geq h_d^*$ (see Introduction) and $b \leq a$ is a particular condition for $d = 2$. Hence, Theorem 3 essentially characterises the possible h^*-polynomials of lattice polytopes of not only dimension 2 but also degree at most 2.

The following is natural to try as a further problem in addition to Problem 4.

Problem 29 Characterise the h^*-polynomials of lattice polytopes of degree ≤ 3.

8.7 Universal Inequalities

Finally, we introduce a new kind of condition on h^*-vectors of lattice polytopes. Recently, the following theorem has been proved:

Theorem 30 ([2, Theorem 1.4]) *Let P be a lattice polytope and let $h_P^*(t) = \sum_{i \geq 0} h_i^* t^i$. Assume that $h_3^* = 0$. Then (h_1^*, h_2^*) satisfies one of the following conditions:*

1. $h_2^* = 0$;
2. $h_1^* \leq 3h_2^* + 3$ *and* $h_2^* > 0$; *or*
3. $(h_1^*, h_2^*) = (7, 1)$.

Theorem 30 is a generalization of Theorem 28 since $\deg(P) \leq 2$ is equivalent to $h_3^* = h_4^* = \cdots = 0$. Theorem 30 says that the conditions in Theorem 28 is valid independently of both the dimension and the degree of lattice polytopes. An inequality is therefore called *universal* if it does not depend on the dimension or the degree of the lattice polytopes. (This terminology is suggested in [2].)

Example 31 ([2, Examples 1.5, 1.6])

1. Theorem 30 is no longer true if $h_3^* \neq 0$. In fact, there is a lattice polytope of dimension 5 whose h^*-polynomial equals $1 + 8t + t^2 + 8t^3$, while Theorem 28 says that there is no lattice polytope of degree 2 whose h^*-polynomial equals $1 + 8t + t^2$.
2. Theorem 30 cannot be generalized into the case $h_3^* < h_1^*$. In fact, there is a lattice polytope of dimension 9 whose h^*-polynomial equals $1 + 17t + 4t^2 + 14t^3 + 8t^4 + 10t^5 + 12t^6 + 6t^7$, while Theorem 28 says that there is no lattice polytope of degree 2 whose h^*-polynomial equals $1 + 17t + 4t^2$.

As a further investigation for this direction, we suggest the following:

Problem 32 Find other universal inequalities for h^*-vectors.

Acknowledgements The author would like to be grateful to Christopher Borger, Johannes Hofscheier, Alexander Kasprzyk and Benjamin Nill for the organization of and the invitation to the wonderful conference "Interactions with Lattice Polytopes", held in September 14–16 2017 at Magdeburg. The author would also like to thank the anonymous referee for a lot of helpful comments.

References

1. Balletti, G.: Enumeration of lattice polytopes by their volume. Discrete Comput. Geom. **65**(4), 1087–1122 (2021)
2. Balletti, G., Higashitani, A.: Universal inequalities in Ehrhart theory. Israel J. Math. **227**(2), 843–859 (2018)
3. Balletti, G., Kasprzyk, A.M.: Three-dimensional lattice polytopes with two interior lattice points (2016). arXiv:1612.08918 [math.CO]
4. Batyrev, V., Juny, D.: Classification of Gorenstein toric del Pezzo varieties in arbitrary dimension. Mosc. Math. J. **10**(2), 285–316, 478 (2010)
5. Batyrev, V.V.: Dual polyhedra and mirror symmetry for Calabi-Yau hypersurfaces in toric varieties. J. Algebraic Geom. **3**(3), 493–535 (1994)
6. Batyrev, V.V., Hofscheier, J.: Lattice polytopes, finite abelian subgroups in $\mathrm{SL}(n, \mathbb{C})$ and coding theory (2013). arXiv:1309.5312 [math.CO]
7. Beck, M., Robins, S.: Computing the continuous discretely, 2nd edn. Undergraduate Texts in Mathematics. Springer, New York (2015). Integer-point enumeration in polyhedra, With illustrations by David Austin
8. Blanco, M., Santos, F.: Lattice 3-polytopes with few lattice points. SIAM J. Discrete Math. **30**(2), 669–686 (2016)
9. Blanco, M., Santos, F.: Lattice 3-polytopes with six lattice points. SIAM J. Discrete Math. **30**(2), 687–717 (2016)
10. Blanco, M., Santos, F.: Enumeration of lattice 3-polytopes by their number of lattice points. Discrete Comput. Geom. **60**(3), 756–800 (2018)
11. Ehrhart, E.: Sur les polyèdres rationnels homothétiques à n dimensions. C. R. Acad. Sci. Paris **254**, 616–618 (1962)
12. Henk, M., Tagami, M.: Lower bounds on the coefficients of Ehrhart polynomials. Eur. J. Combin. **30**(1), 70–83 (2009)
13. Hibi, T.: Algebraic Combinatorics on Convex Polytopes. Carslaw Publications, Glebe (1992)
14. Hibi, T.: Dual polytopes of rational convex polytopes. Combinatorica **12**(2), 237–240 (1992)
15. Hibi, T.: A lower bound theorem for Ehrhart polynomials of convex polytopes. Adv. Math. **105**(2), 162–165 (1994)
16. Hibi, T., Higashitani, A., Li, N.: Hermite normal forms and δ-vectors. J. Combin. Theory Ser. A **119**(6), 1158–1173 (2012)
17. Hibi, T., Higashitani, A., Nagazawa, Y.: Ehrhart polynomials of convex polytopes with small volumes. Eur. J. Combin. **32**(2), 226–232 (2011)
18. Hibi, T., Higashitani, A., Tsuchiya, A., Yoshida, K.: Ehrhart polynomials with negative coefficients. Graphs Combin. **35**(1), 363–371 (2019)
19. Higashitani, A.: Shifted symmetric δ-vectors of convex polytopes. Discrete Math. **310**(21), 2925–2934 (2010)
20. Higashitani, A.: Ehrhart polynomials of integral simplices with prime volumes. Integers **14**, Paper No. A45, 15 (2014)
21. Higashitani, A., Nill, B., Tsuchiya, A.: Gorenstein polytopes with trinomial h^*-polynomials. Beitr. Algebra Geom. **62**(3), 667–685 (2021)

22. Hofscheier, J., Katthän, L., Nill, B.: Ehrhart theory of spanning lattice polytopes. Int. Math. Res. Not. IMRN **19**, 5947–5973 (2018)
23. Kasprzyk, A.M.: Canonical toric Fano threefolds. Canad. J. Math. **62**(6), 1293–1309 (2010)
24. Kreuzer, M., Skarke, H.: Classification of reflexive polyhedra in three dimensions. Adv. Theor. Math. Phys. **2**(4), 853–871 (1998)
25. Kreuzer, M., Skarke, H.: Complete classification of reflexive polyhedra in four dimensions. Adv. Theor. Math. Phys. **4**(6), 1209–1230 (2000)
26. Lagarias, J.C., Ziegler, G.M.: Bounds for lattice polytopes containing a fixed number of interior points in a sublattice. Canad. J. Math. **43**(5), 1022–1035 (1991)
27. Macdonald, I.G.: Polynomials associated with finite cell-complexes. J. Lond. Math. Soc. **2**(4), 181–192 (1971)
28. Poonen, B., Rodriguez-Villegas, F.: Lattice polygons and the number 12. Amer. Math. Monthly **107**(3), 238–250 (2000)
29. Scott, P.R.: On convex lattice polygons. Bull. Austral. Math. Soc. **15**(3), 395–399 (1976)
30. Stanley, R.P.: Decompositions of rational convex polytopes. Ann. Discrete Math. **6**, 333–342 (1980)
31. Stanley, R.P.: On the Hilbert function of a graded Cohen-Macaulay domain. J. Pure Appl. Algebra **73**(3), 307–314 (1991)
32. Stapledon, A.: Inequalities and Ehrhart δ-vectors. Trans. Amer. Math. Soc. **361**(10), 5615–5626 (2009)
33. Stapledon, A.: Additive number theory and inequalities in Ehrhart theory. Int. Math. Res. Not. IMRN **5**, 1497–1540 (2016)
34. Treutlein, J.: Lattice polytopes of degree 2. J. Combin. Theory Ser. A **117**(3), 354–360 (2010)
35. Tsuchiya, A.: Ehrhart polynomials of lattice polytopes with normalized volumes 5. J. Comb. **10**(2), 283–290 (2019)

Chapter 9
The Ring of Conditions for Horospherical Homogeneous Spaces

Johannes Hofscheier

Abstract These are notes of a five talks lecture series during the "Graduate Summer School in Algebraic Group Actions", at McMaster University, June 11th–15th, 2018. The aim of this lecture series is to introduce the ring of conditions of a spherical homogeneous space with a special emphasis on the horospherical case, i.e., homogeneous spaces with respect to a connected complex reductive group which are torus bundles over a flag variety. In these notes, we start with an example from enumerative geometry which naturally yields first instances of spherical varieties. We continue with a recollection of the necessary background on reductive groups needed throughout the rest of the manuscript. After that we introduce spherical varieties: we discuss the Luna–Vust theory of spherical embeddings and explain the complete combinatorial description of horospherical varieties (an important subfamily of spherical varieties). We conclude with the definition of the ring of conditions of spherical homogeneous spaces and give an explicit description for the horospherical case.

Keywords Spherical variety · Linear algebraic group · Enumerative geometry · Ring of conditions

9.1 Motivation

In the following, some elementary knowledge of algebraic geometry is expected from the reader. Introductory texts which cover the required topics are, e.g., [9, 23, 24, 27]. Parts of this manuscript follow the lecture notes [11] by Kiritchenko.

First examples of spherical varieties emerged from enumerative geometry such as Grassmannians. It turns out that many enumerative problems reduce to intersection theoretic questions on algebraic varieties equipped with a "good" transitive (or almost transitive) action of an algebraic group. Here is a classical example:

J. Hofscheier (✉)
School of Mathematical Sciences, University of Nottingham, Nottingham NG7 2RD, UK
e-mail: johannes.hofscheier@nottingham.ac.uk

A. M. Kasprzyk and B. Nill (eds.), *Interactions with Lattice Polytopes*, Springer Proceedings in Mathematics & Statistics 386,
https://doi.org/10.1007/978-3-030-98327-7_9

Problem 1 How many lines in 3-space $\mathbb{C}^3$ intersect 4 given lines in general position?

Recall the general trick how to rephrase an affine geometric question into a linear one: Suppose X is an affine geometric object in $\mathbb{C}^n$. Introduce one further dimension and consider the linear span of X regarded as a subset of the affine hyperplane $\{x_{n+1} = 1\}$ where x_{n+1} denotes the additional coordinate.

Hence, the above question reduces to a problem in the Grassmannian $\mathrm{Gr}(2, 4)$ (2-planes in $\mathbb{C}^4$). This algebraic variety admits a transitive action by GL_4 (the general linear group of invertible 4×4-matrices with complex entries). Indeed, let $e_1, \ldots, e_4$ be the standard basis in $\mathbb{C}^4$ and consider the natural action of GL_4 on $\mathbb{C}^4$. Clearly GL_4 acts transitively on planes in $\mathbb{C}^4$ and the stabilizer P of the coordinate plane $\mathrm{span}\{e_1, e_2\}$ is given by

$$P = \left\{ \begin{pmatrix} A & C \\ 0 & B \end{pmatrix} : A, B \in \mathrm{GL}_2, C \in \mathrm{Mat}(2 \times 2) \right\}.$$

Hence $\mathrm{Gr}(2, 4) \cong \mathrm{GL}_4 / P$ is a homogeneous space under GL_4 and P is an example of a parabolic subgroup (see definition below).

Note that by the transition to $\mathrm{Gr}(2, 4)$, we implicitly consider the lines as sitting in the projective 3-space $\mathbb{P}^3$ and intersections are taken in the projective sense. Indeed, two parallel lines do not intersect in the affine 3-space, but their corresponding 2-planes do. This corresponds to the fact that two parallel lines intersect at infinity in $\mathbb{P}^3$.

Let us recall the crucial ideas of Schubert's solution to Problem 1. To solve it, he developed the calculus of "conditions" (see [22]), which has since become known as *Schubert Calculus*. Examples of conditions are:

1. for a given point a, denote by σ_a the condition that a line contains a;
2. for a given line ℓ, denote by σ_ℓ the condition that a line intersects ℓ; or
3. for a plane Π, denote by σ_Π the condition that a line is contained in Π.

Schubert's brilliant idea was that conditions can be added and multiplied and this corresponds to logical "or" and "and" operations on the conditions, e.g., $\sigma_{\ell_1} + \sigma_{\ell_2}$ is the condition that a line intersects line ℓ_1 or line ℓ_2 while $\sigma_{\ell_1} \cdot \sigma_{\ell_2}$ is the condition that a line intersects both lines ℓ_1 and ℓ_2. So, for instance, Problem 1 can be reformulated to: What is $\sigma_{\ell_1} \cdots \sigma_{\ell_4}$ where ℓ_i are four lines in general position? In particular, we can reformulate the problem in an algebraic equation and obtain

$$\sigma_{\ell_1} \cdot \sigma_{\ell_2} \cdot \sigma_{\ell_3} \cdot \sigma_{\ell_4} = (\sigma_{\ell_1} \cdot \sigma_{\ell_2}) \cdot (\sigma_{\ell_3} \cdot \sigma_{\ell_4}) = ?.$$

So we have to understand the conditions $\sigma_{\ell_1} \cdot \sigma_{\ell_2}$ and $\sigma_{\ell_3} \cdot \sigma_{\ell_4}$. By using some heuristics, Schubert came to the conclusion that "perturbations" of the condition $\sigma_{\ell_1} \cdots \sigma_{\ell_4}$ do not change the answer (the conservation of number principle), i.e., we are allowed to move the lines ℓ_i. In particular, we may assume that ℓ_1, ℓ_2 lie on a plane, and so do ℓ_3 and ℓ_4 (see Fig. 9.1). From that it straightforwardly follows that a line intersects both ℓ_1 and ℓ_2 if and only if it is either contained in the

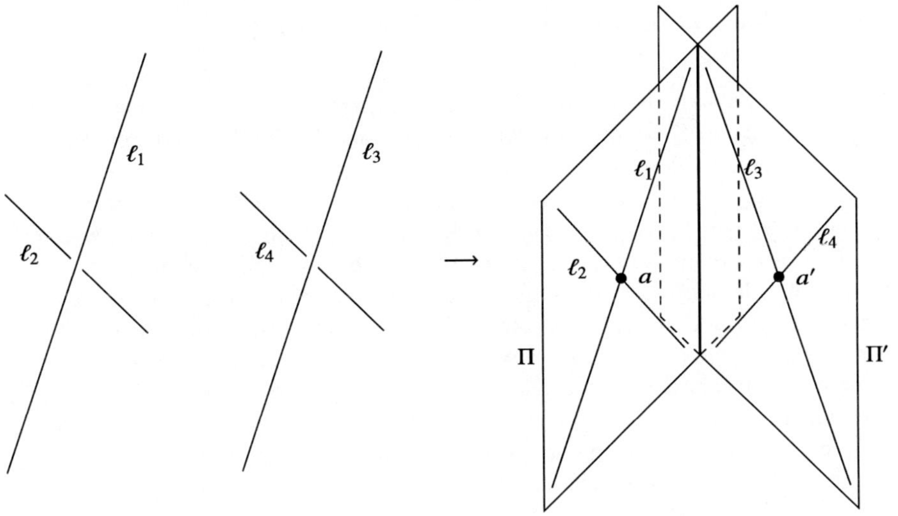

Fig. 9.1 The conservation of number principle implies that we are allowed to move the lines ℓ_i

plane Π spanned by ℓ_1 and ℓ_2 (recall that we take intersections in $\mathbb{P}^3$) or it contains the intersection point of ℓ_1 and ℓ_2. Using Schubert calculus this means

$$\sigma_{\ell_1} \cdot \sigma_{\ell_2} = \sigma_a + \sigma_\Pi \qquad \text{and} \qquad \sigma_{\ell_3} \cdot \sigma_{\ell_4} = \sigma_{a'} + \sigma_{\Pi'}$$

where a is the intersection point of ℓ_1 and ℓ_2 and Π is the plane spanned by ℓ_1 and ℓ_2 and similarly for ℓ_3, ℓ_4.

Thus, we get

$$\sigma_{\ell_1} \cdot \sigma_{\ell_2} \cdot \sigma_{\ell_3} \cdot \sigma_{\ell_4} = (\sigma_a + \sigma_\Pi) \cdot (\sigma_{a'} + \sigma_{\Pi'}) = \sigma_a \cdot \sigma_{a'} + \sigma_a \cdot \sigma_{\Pi'} + \sigma_{a'} \cdot \sigma_\Pi + \sigma_\Pi \cdot \sigma_{\Pi'}.$$

Clearly there is exactly one line passing through both a and a' and there is exactly one line contained in both Π and Π'. On the other hand, as a is not contained in Π', the condition $\sigma_a \cdot \sigma_{\Pi'}$ is not satisfied by any line, and similarly for $\sigma_{a'} \cdot \sigma_\Pi$.

We obtain

$$\sigma_{\ell_1} \cdot \sigma_{\ell_2} \cdot \sigma_{\ell_3} \cdot \sigma_{\ell_4} = \underbrace{\sigma_a \cdot \sigma_{a'}}_{=1} + \underbrace{\sigma_a \cdot \sigma_{\Pi'}}_{=0} + \underbrace{\sigma_{a'} \cdot \sigma_\Pi}_{=0} + \underbrace{\sigma_\Pi \cdot \sigma_{\Pi'}}_{=1} = 2.$$

Of course, we haven't given a precise explanation yet and in his fifteenth problem Hilbert asked for a rigorous foundation of Schubert Calculus. Our goal will be to understand De Concini's and Procesi's solution to Hilbert's problem. For that, we also need to understand spherical geometry, a topic which is exciting in its own right.

9.2 Linear Algebraic Groups: A Crash Course

The classical books by Borel, Humphreys and Springer [2, 10, 25] are excellent references for what follows. A more modern and accessible book is [18]. For convenience, we work over the field of complex numbers $\mathbb{C}$.

An algebraic variety G is called an *algebraic group* if G is a group and the maps $G \times G \to G$, $(g, h) \mapsto gh$ and $G \to G$, $g \mapsto g^{-1}$ are morphisms of algebraic varieties. The *Lie algebra* of G, usually denoted by $\mathfrak{g}$, is the tangent space T_eG at the identity element $e \in G$ equipped with a binary operation $[\cdot, \cdot]$ called the *Lie bracket*. Important examples of algebraic groups are GL_n (=the general linear group of invertible $n \times n$-matrices with complex entries), SL_n (=the special linear group of $n \times n$-matrices with complex entries and determinant 1), abelian varieties (=complete connected algebraic groups) or elliptic curves (=1-dimensional abelian varieties). We will work with *linear algebraic groups*, i.e., Zariski closed subgroups of GL_n. If $G \subseteq \mathrm{GL}_n$ is a linear algebraic group, then $T_eG = \mathfrak{g} \subseteq \mathfrak{gl}_n = T_e\,\mathrm{GL}_n = \{(n \times n) - \text{matrices}\}$, and the Lie bracket can be defined as the commutator of matrices

$$[A, B] := AB - BA.$$

Remark 2 If one replaces "algebraic varieties" and "morphisms of algebraic varieties" by "smooth manifolds" and "smooth maps", one obtains the definition of a Lie group.

Exercise 3 Let G be an algebraic group.

1. Show that only one irreducible component of G can pass through e. This component is called the *identity component* of G, usually denoted by G°.
2. Show that G° is a normal subgroup of finite index in G, whose cosets are the connected as well as irreducible components of G.

Exercise 4 Which of the following algebraic groups are linear?

1. $(\mathbb{C}^n, +)$,
2. An elliptic curve,
3. PGL_n.

From now on all algebraic groups are assumed to be linear, unless stated otherwise.

Definition 5 An element $g \in G \subseteq \mathrm{GL}_n$ is called *semisimple* if the matrix g is diagonalizable. It is called *unipotent* if all eigenvalues of g are equal to 1. (This definition is independent of the choice of the embedding $G \subseteq \mathrm{GL}_n$.)

Exercise 6 Let $\pi : G \to \mathrm{GL}_n$ be a (rational) representation of an algebraic group G, i.e., π is a morphism of algebraic groups. Show that:

1. If $G = (\mathbb{C}^*)^n$, then any matrix in $\pi(G)$ is diagonalizable.
2. If $G = \mathbb{C}^n$, then any matrix in $\pi(G)$ is unipotent.

Exercise 7 (*Jordan decomposition*) Show that every element $g \in G$ has a unique decomposition $g = g_s g_u$, where $g_s \in G$ is semisimple, $g_u \in G$ is unipotent, and g_s and g_u commute.

The *radical*, denoted by $R(G)$, of an algebraic group G is the identity component of its maximal normal solvable subgroup. The *unipotent radical*, denoted by $R_u(G)$, is the set of unipotent elements in the radical of G.

Definition 8 An algebraic group G is *reductive* if $R_u(G) = \{e\}$. It is *semisimple* if $R(G) = \{e\}$.

Theorem 9 (Characterization of reductive groups) *Let G be an algebraic group. The following conditions are equivalent:*

1. *G is reductive;*
2. *$R(G)$ is a torus;*
3. *$G^\circ = T \cdot S$, where T is a torus and S is a connected semisimple subgroup;*
4. *any finite-dimensional rational representation of G is completely reducible (recall: a* rational representation *of G in a vector space V is a homomorphism $G \to \mathrm{GL}(V)$ of algebraic groups);*
5. *G admits a faithful finite-dimensional completely reducible rational representation;*
6. *the Lie algebra of G admits a direct sum decomposition $\mathfrak{g} = \mathfrak{h} \oplus i\mathfrak{h}$ where $\mathfrak{h}$ is the Lie algebra of a maximal compact real Lie subgroup of G.*

Exercise 10 Which of the following groups are reductive?

1. $\mathbb{C}^n$,
2. GL_n,
3. An elliptic curve.

Exercise 11 Show that an algebraic group G is reductive if and only if G does not contain a normal subgroup isomorphic to $\mathbb{C}^n$.

A *character* of an algebraic group G is a homomorphism of algebraic groups $\chi : G \to \mathbb{C}^*$ and the set of all characters gives the *character group* of G, i.e., $\mathfrak{X}(G) := \{\chi : G \to \mathbb{C}^* \text{ character}\}$.

An *algebraic torus* is an algebraic group T that is isomorphic to

$$(\mathbb{C}^*)^n = \{(z_1, \ldots, z_n) \in \mathbb{C}^n : z_i \neq 0\}.$$

If G is an algebraic group, then a maximal element of the set

$$\{H \subseteq G \text{ closed subgroup, } H \text{ an algebraic torus}\}$$

(which is ordered by inclusion) is called a *maximal torus* of G.

Theorem 12 *In an algebraic group, any two maximal tori are conjugated.*

The dimension of T is called the *rank* of G. The character lattice $\mathfrak{X}(T)$ of T is also called the *weight lattice* of G, and its elements are called *weights* of G.

Exercise 13 Find a maximal torus of the following groups:

1. GL_n,
2. SL_n,
3. SO_n.

The set $\{H \subseteq G$ closed connected solvable subgroup$\}$ is partially ordered by inclusion. A maximal element of this set is called a *Borel subgroup*.

Exercise 14 Show that the upper-triangular invertible matrices form a Borel subgroup in GL_n and that any two Borel subgroups are conjugated. (Hint: Use the Lie–Kolchin theorem [10, Theorem 17.6].)

Theorem 15 *In an algebraic group, any two Borel subgroups are conjugated.*

Definition 16 A (Zariski) closed subgroup $P \subseteq G$ is called *parabolic* if P contains a Borel subgroup of G.

Exercise 17 Let G be a linear algebraic group, $B \subseteq G$ a Borel subgroup and $T \subseteq G$ a maximal torus.

1. Show that up to conjugation $T \subseteq B$.
2. Show that restricting characters from B to T induces an isomorphism of character lattices $\mathfrak{X}(B) \cong \mathfrak{X}(T)$.

The *Weyl group* W of G is defined as $N_G(T)/C_G(T)$, where $N_G(T)$ and $C_G(T)$ denote the normalizer and centralizer, respectively, of a maximal torus $T \subseteq G$. The Weyl group acts on T by conjugation: if $w = nC_G(T)$ for $n \in N_G(T)$, then $w(t) := (ntn^{-1})$ for $t \in T$.

Theorem 18 *If G is a connected reductive group, then $C_G(T) = T$ for any maximal torus $T \subseteq G$. In particular, the Weyl group is given by $W = N_G(T)/T$.*

Exercise 19 If G is a connected reductive group, $T \subseteq G$ a maximal torus, and $B \subseteq G$ a Borel subgroup with $T \subseteq B$, show that for any $w \in W$ the double coset $B\dot{w}B$ is independent of the choice of a representative $\dot{w} \in N_G(T)$. Thus, by abuse of notation, we will denote this double coset by BwB.

Theorem 20 (Bruhat decomposition) *If G is a connected reductive group, $T \subseteq G$ a maximal torus, and $B \subseteq G$ a Borel subgroup with $T \subseteq B$, then there is a disjoint union, i.e., $BwB = Bw'B$ if and only if $w = w'$ in W,*

$$G = \bigsqcup_{w \in W} BwB$$

In particular,

$$G/B = \bigsqcup_{w \in W} BwB/B.$$

Exercise 21 You may want to start with $n = 3$ or $n = 4$ in the following problems.

1. Explicitly compute the Bruhat decomposition of GL_n (take T to be the maximal torus of diagonal matrices and B the Borel subgroup of upper triangular matrices).
2. Classify all parabolic subgroups in GL_n (up to conjugation). (Hint: There is a relationship between parabolic subgroups in GL_n and flags in $\mathbb{C}^n$. Recall that a *flag* is an increasing sequence of subspaces of $\mathbb{C}^n$, i.e., $\{0\} = V_0 \subsetneq V_1 \subsetneq \dots \subsetneq V_k = \mathbb{C}^n$. The dimensions $d_i := \dim V_i$ yield an increasing sequence of integers $0 = d_0 < d_1 < \dots < d_k = n$, called the *signature* of the flag.)

Exercise 22 Let G be a connected reductive group, $T \subseteq G$ a maximal torus, $B \subseteq G$ a Borel subgroup with $T \subseteq B$ and $P \subseteq G$ a parabolic subgroup with $B \subseteq P$. Show that

$$G/P = \bigsqcup_{w \in W/W_P} BwP/P$$

where $W_P = N_P(T)/T$ is the Weyl group of P.

The closure of B-orbits in G/P are the *Schubert varieties* (denoted by $X(w)$). They play an important role in the study of G/P. The dimension of $X(w)$ equals the *length* $l(w)$ of w (i.e., the minimal number of factors needed to write w as a product of simple reflections). In particular, there exists a unique element w_0 of maximal length in W/W_P.

Example 23 Let $G = \mathrm{SL}_n$ and T be the maximal torus of diagonal matrices contained in the Borel subgroup B of upper-triangular matrices. The Lie algebra $\mathfrak{g} = \mathfrak{sl}_n$ (i.e., the tangent space $T_e \mathrm{SL}_n$ equipped with the Lie bracket $[\cdot, \cdot]$) is the set of traceless matrices in $\mathrm{Mat}(n \times n)$ equipped with the commutator bracket $[A, B] = AB - BA$. Furthermore, the Lie algebra $\mathfrak{t}$ of T coincides with the subspace of diagonal matrices in $\mathfrak{sl}_n$. Observe that the Lie bracket induces a map $\mathrm{ad}\colon \mathfrak{t} \to \mathrm{End}(\mathfrak{g})$; $A \mapsto [A, \cdot]$ which is a *representation of Lie algebras*, i.e., $\mathrm{ad}([A, B]) = \mathrm{ad}(A)\,\mathrm{ad}(B) - \mathrm{ad}(B)\,\mathrm{ad}(A)$ for any $A, B \in \mathfrak{t}$ (check this!). Let $\varepsilon_1, \dots, \varepsilon_n$ be the linear forms in $\mathfrak{t}^*$ induced by the diagonal entries, i.e., $\varepsilon_i(\mathrm{diag}(t_1, \dots, t_n)) = t_i$ and set $\varepsilon_{ij} = \varepsilon_i - \varepsilon_j$. It is straightforward to show that the Lie algebra decomposes into eigenspaces as follows

$$\mathfrak{g} = \mathfrak{t} \oplus \bigoplus_{\varepsilon_{ij} \in R} \mathfrak{g}_{\varepsilon_{ij}}, \quad \text{where } \mathfrak{g}_{\varepsilon_{ij}} = \mathbb{C} \begin{pmatrix} 0 & & 0 \\ & 1 & \\ 0 & & 0 \end{pmatrix} \begin{matrix} \\ i\text{th} \\ \\ \end{matrix} ,$$

(with the entry 1 in the ith row and jth column)

and $R = \{\varepsilon_{ij} \colon 1 \leq i, j \leq n, i \neq j\}$. If $\mathfrak{b}$ is the Lie algebra of B, then

$$\mathfrak{b} = \mathfrak{t} \oplus \bigoplus_{\alpha \in R^+} \mathfrak{g}_\alpha$$

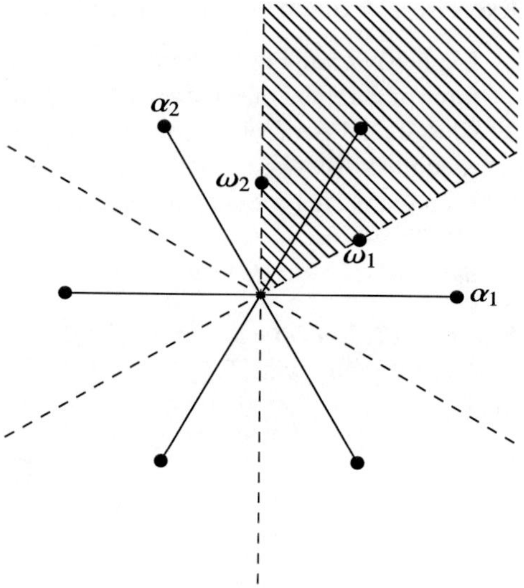

Fig. 9.2 Illustration of the root system A_2. We identify the hyperplane $\{x + y + z = 0\} \subseteq \mathbb{R}^3$ with $\mathbb{R}^2$ via the basis obtained by applying the Gram–Schmidt algorithm to the basis (α_1, α_2) where $\mathbb{R}^3$ and $\mathbb{R}^2$ are equipped with the usual euclidian scalar products

where $R^+ = \{\varepsilon_{ij} : i < j\}$ and this set is called the set of *positive roots*. The set of *simple roots* $S = \{\alpha_i := \varepsilon_{i,i+1} : i = 1, \ldots, n-1\}$ (cf. Fig. 9.2) forms a basis of $\mathfrak{t}^*$ (check this!) and induces an isomorphism $\mathfrak{t}^* \cong \{(x_1, \ldots, x_n) \in \mathbb{R}^n : x_1 + \ldots + x_n = 0\} \subseteq \mathbb{R}^n$. To any simple root α_i one associates a reflection s_i, namely the linear transformation on $\mathbb{R}^n$ which swaps the coordinates with index i and $(i+1)$. We identify s_i with an element in $W = N_G(T)/T$:

$$\begin{pmatrix} I_{i-1} & & \\ & \begin{matrix} 0 & 1 \\ -1 & 0 \end{matrix} & \\ & & I_{n-i-1} \end{pmatrix} T \in W.$$

Then W is generated by the s_i, i.e., $W = \langle s_i : i = 1, \ldots, n-1 \rangle$ (check this!). It straightforwardly follows that W is isomorphic to the group S_n of permutations on the coordinates of $\mathbb{R}^n$ via $s_i \mapsto (i, i+1)$ (transposition swapping i with $i+1$).

In general, the Lie bracket induces a natural representation $\mathrm{ad}\colon \mathfrak{t} \to \mathrm{End}(\mathfrak{g});\ x \mapsto [x, \cdot]$. There is a set of linear forms, called *roots*, $R \subseteq \mathfrak{t}^*$ such that

$$\mathfrak{g} = \mathfrak{t} \oplus \bigoplus_{\alpha \in R} \mathfrak{g}_\alpha$$

where $\mathfrak{g}_\alpha$ denotes the linear subspace of eigenvectors of weight α, i.e., the set of vectors $x \in \mathfrak{g}$ such that $[h, x] = \alpha(h)x$ for all $h \in \mathfrak{t}$. The Lie algebra $\mathfrak{b}$ of B can be written as

$$\mathfrak{b} = \mathfrak{t} \oplus \bigoplus_{\alpha \in R_+} \mathfrak{g}_\alpha$$

for some subset $R^+ \subseteq R$, called the set of *positive roots*. There exists a unique basis S contained in R^+ such that all positive roots are linear combinations of elements in S with nonnegative integer coefficients. The elements of S are called *simple roots*.

The following fundamental theorem on parabolic subgroups can be found in any introductory text on algebraic groups. Recall the definition of the Weyl group: $W = N_G(T)/T$. Let R be the set of roots and let S be the set of simple roots induced by the choice of the Borel subgroup B.

Theorem 24 *The assignment* $I \mapsto P_I = \bigsqcup_{w \in W_I} BwB$ *induces a bijection between subsets of the set of simple roots* S *and parabolic subgroups of* G *which contain* B *(here* W_I *denotes the group generated by the simple reflections* s_α *for* $\alpha \in I$*).*

9.3 Spherical Varieties

Recall the definition of a toric variety:

Definition 25 Let T be an algebraic torus. A normal irreducible T-variety is called *toric variety* if it contains an open dense T-orbit.

Spherical varieties can be thought of as a generalization of toric varieties where one allows also non-abelian group actions. Unfortunately, the straightforward generalization does not work:

Definition 26 (*Incorrect definition*) Let G be a connected linear algebraic group. A normal irreducible G-variety is called *spherical* if it contains an open dense G-orbit.

Exercise 27 Show that the "incorrect definition" of spherical varieties does not imply finiteness of the number of orbits, a property one would expect from a generalization of toric varieties. (Hint: Consider the action of GL_n on the space of $(n \times n)$-matrices by left multiplication. Show that the GL_n-orbits are classified by matrices in reduced row echelon form. If $n \geq 2$, deduce that, although there is an open GL_n orbit, the number of GL_n-orbits is infinite.)

So the definition of spherical varieties is more subtle: Let G be a connected reductive complex linear algebraic group (this assumption has several implications which make this choice important: finite generation properties, good representation theory, cf. Theorem 9).

Definition 28 A normal irreducible G-variety is said to be *spherical* if it contains an open orbit under the action of a Borel subgroup of G. (In particular, it contains an open G-orbit.)

Example 29 Examples of spherical varieties are toric varieties (a Borel subgroup of $(\mathbb{C}^*)^n$ is $(\mathbb{C}^*)^n$ itself).

Another point of view on spherical varieties, important to the theory, is as follows: First consider the open G-orbit which is a homogeneous space G/H for some subgroup H of G. Then consider the embedding of G/H in X. So we make the following definitions:

Definition 30 A closed subgroup $H \subseteq G$ is called *spherical* if G/H has a dense open orbit for a Borel subgroup of G. In this case, G/H is called a *spherical homogeneous space*.

Recall that in Exercise 27, we have seen that an open G-orbit does not guarantee the finiteness of G-orbits, a property one would expect from a generalization of toric varieties. It is interesting that one can use this property as a characterization of spherical homogeneous spaces:

Theorem 31 ([1]) *A homogeneous G-space O is spherical if and only if any G-variety X with an open orbit equivariantly isomorphic to O has finitely many G-orbits.*

A morphism $\varphi\colon X \to Y$ of G-varieties is called *equivariant* if $\varphi(g \cdot x) = g \cdot \varphi(x)$ for any $g \in G$ and all $x \in X$.

Definition 32 Suppose G/H is a spherical homogeneous space. An equivariant open embedding of G/H into a normal irreducible G-variety X is called a *spherical embedding*, and X is called a *spherical variety*.

In particular, the description of spherical varieties splits into two parts:

1. Classify all spherical homogeneous spaces G/H.
2. For a fixed spherical homogeneous space, classify all G-equivariant open embeddings $G/H \hookrightarrow X$ into normal irreducible G-varieties.

Historically, the second problem has been answered first by the work of Luna and Vust [17]. Only recently, the first problem has been answered by work of several researchers (see [3, 6, 15, 16]).

Exercise 33 Show the following statements:

1. A closed subgroup $H \subseteq \mathrm{SL}_2$ is spherical if and only if $\dim H > 0$.
2. Table 9.1 shows a list of all spherical subgroups of SL_2 (up to conjugation). (Although spherical varieties with an SL_2-action seem to be special, they actually play a crucial role in the development of spherical varieties. See, for example, [17] or [8, 14].) Hints: If this is too hard, then verify explicitly that the subgroups given in Table 9.1 are spherical:

   a. $\mathrm{SL}_2/B \cong \mathbb{P}^1$ where SL_2 naturally acts on $\mathbb{P}^1$,
   b. it is enough to show that U_1 is spherical (why?) and to do that consider the natural action of SL_2 on $\mathbb{A}^2$,
   c. consider the natural action of SL_2 on $\mathbb{P}^1 \times \mathbb{P}^1$ to show that T is a spherical subgroup of SL_2,
   d. consider the natural action of SL_2 on the symmetric (2×2)-matrices to show that N is a spherical subgroup of SL_2.

Exercise 34 A closed subgroup $H \subseteq G$ is called *horospherical* if it contains a maximal unipotent subgroup U of G. Show that horospherical subgroups are spherical. In particular, flag varieties are spherical. (Hint: Use the Bruhat decomposition.)

Table 9.1 Classification of spherical subgroups of SL_2 (up to conjugation)

H	Details
SL_2	
B	Borel subgroup
$U_k = \left\{ \begin{pmatrix} \xi & * \\ 0 & \xi^{-1} \end{pmatrix} : \xi \in \mu_k \right\}$	$k \in \mathbb{N}$, μ_k group of kth roots of unity
T	Maximal torus
N	Normalizer of a maximal torus

9.3.1 The Luna–Vust Theory of Spherical Embeddings

Recall that for a fixed algebraic torus T, all toric embeddings $T \hookrightarrow X$ into a normal irreducible T-variety can be combinatorially described by polyhedral fans in the vector space $\mathrm{Hom}(\mathfrak{X}(T), \mathbb{Q})$. A similar description exists for spherical embeddings which we now explain. This is called the *Luna–Vust theory of spherical embeddings*. Good references for this theory are [13, 17]:

Let G be a connected reductive complex algebraic group and fix a spherical subgroup $H \subseteq G$. Let B be a Borel subgroup of G and T a maximal torus of G contained in B. We now explain how all spherical embeddings $G/H \hookrightarrow X$ can be described combinatorially.

Definition 35 The combinatorial objects needed in the Luna–Vust theory are listed in Table 9.2. The rank of $\mathcal{M}$ is also called the *rank of the spherical homogeneous space* G/H, i.e., $\mathrm{rk}(G/H) = \mathrm{rk}(\mathcal{M})$. Let $\mathcal{N} := \mathrm{Hom}(\mathcal{M}, \mathbb{Z})$ be the dual lattice of $\mathcal{M}$ and note that we have a dual pairing $\langle \cdot, \cdot \rangle : \mathcal{N} \times \mathcal{M} \to \mathbb{Z}$. Furthermore, set $\mathcal{M}_{\mathbb{Q}} := \mathcal{M} \otimes \mathbb{Q}$ and $\mathcal{N}_{\mathbb{Q}} = \mathrm{Hom}(\mathcal{M}, \mathbb{Q})$. Recall, that in our context a *valuation* is a map $\nu : \mathbb{C}(G/H)^* = \mathbb{C}(G/H) \setminus \{0\} \to \mathbb{Q}$ which satisfies:

1. $\nu(f_1 + f_2) \geq \min\{\nu(f_1), \nu(f_2)\}$ whenever $f_1, f_2, f_1 + f_2 \in \mathbb{C}(G/H)^*$;
2. $\nu(f_1 f_2) = \nu(f_1) + \nu(f_2)$ for all $f_1, f_2 \in \mathbb{C}(G/H)^*$; and
3. $\nu(\mathbb{C}^*) = 0$.

A valuation ν is called G-invariant if $\nu(g \cdot f) = \nu(f)$ for all $g \in G$ and $f \in \mathbb{C}(G/H)^*$.

As the set of B-semi-invariant rational functions on G/H will appear frequently below, we introduce the notation $\mathbb{C}(G/H)^{(B)}$ for it.

Lemma 36 *As G/H has an open B-orbit, a B-semi-invariant rational function f is determined by its weight χ_f up to a scalar multiple. Said in other words: For any $\chi \in \mathcal{M}$, there is $f_\chi \in \mathbb{C}(G/H)^{(B)}$ (unique up to a scalar multiple) such that $b \cdot f_\chi = \chi(b) f_\chi$.*

Table 9.2 The *Luna–Vust data*

Object	Definition
Weight lattice	$\mathcal{M} = \{\chi \in \mathfrak{X}(B)\colon$ $\exists f \in \mathbb{C}(G/H)^*$, B-semi-invariant with $b \cdot f = \chi(b) f$ for $b \in B\}$
Colors	$\mathcal{D} = \{B\text{-invariant prime divisors in } G/H\}$
Valuation cone	$\mathcal{V} = \{\nu\colon \mathbb{C}(G/H)^* \to \mathbb{Q},\ G\text{-invariant valuation}\}$

Table 9.3 The "Luna–Vust data" for the toric case

Object	Toric case
Weight lattice	$M = \mathfrak{X}(T)$
Colors	$\mathcal{D} = \varnothing$
Valuation cone	$\mathcal{V} = N_{\mathbb{Q}}$

The following interpretation of a valuation $\nu\colon \mathbb{C}(G/H) \to \mathbb{Q}$ (invariant or not) will be useful:

$$\rho\colon \{\nu\colon \mathbb{C}(G/H) \to \mathbb{Q} \text{ valuation}\} \to \mathcal{N}_{\mathbb{Q}};\ \nu \mapsto [\chi \mapsto \nu(f_\chi)].$$

Theorem 37 ([4]) *The map $\rho|_{\mathcal{V}}\colon \mathcal{V} \hookrightarrow \mathcal{N}_{\mathbb{Q}}$ is injective and its image is a polyhedral cone whose dual cone is simplicial and not necessarily full-dimensional.*

Any color $D \in \mathcal{D}$ induces a valuation ν_D and by abuse of notation, we will write $\rho(D) := \rho(\nu_D)$. In general, the map $\rho|_{\mathcal{D}}\colon \mathcal{D} \to \mathcal{N}_{\mathbb{Q}}$ is not that well-behaved (see Exercise 38).

Exercise 38 Find the "Luna–Vust data" for the spherical homogeneous spaces from Exercise 33. In particular, you should see the following phenomena:

1. SL_2/T: the map $\rho\colon \mathcal{D} \to \mathcal{N}$ might not be injective;
2. SL_2/N: the image of a color $\rho(D)$ might not be primitive in $\mathcal{N}$;
3. SL_2/B: the image of a color might even be zero, i.e., $\rho(D) = 0$.

Example 39 The "Luna–Vust data" of the toric case is listed in Table 9.3.

Example 40 Consider the natural action of SL_2 on $\mathbb{C}^2$. Let B be the Borel subgroup of upper triangular matrices, T the maximal torus of diagonal matrices and U the unipotent radical of B. Denote the standard basis of $\mathbb{C}^2$ by e_1, e_2. Then $\mathrm{SL}_2/U \cong \mathrm{SL}_2 \cdot e_1 = \mathbb{C}^2 \setminus \{0\}$ and $B \cdot e_2 = \mathbb{C} \times \mathbb{C}^*$ is the open B-orbit. The rational functions on SL_2/U are given by $\mathbb{C}(x, y) = \mathbb{C}(\mathbb{A}^2)$. It is straightforward to verify that

$$\mathbb{C}(\mathrm{SL}_2/U)^{(B)} \cong \{cy^k\colon c \in \mathbb{C}, k \in \mathbb{Z}\}.$$

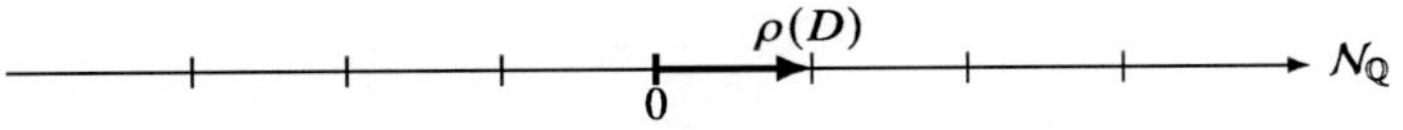

Fig. 9.3 Illustrating the "Luna–Vust data" of SL_2/U

Hence, $\mathcal{M} = \mathbb{Z}\omega$ where ω is the fundamental weight of SL_2 induced by the diagonal elements of T. Clearly $D = \mathrm{div}(y)$ is the only B-stable prime divisor in SL_2/U, i.e., we only have one color $\mathcal{D} = \{D\}$ and $\rho(D) = \check{\alpha}|_{\mathcal{M}}$ where $\check{\alpha}$ denotes the coroot associated to the simple root α of SL_2 (with respect to our choice of Borel). If we consider the blowup of $\mathbb{A}^2$ at the origin, we obtain an exceptional SL_2-invariant divisor E which induces an SL_2-invariant valuation ν_E such that $\rho(\nu_E) = \check{\alpha}|_{\mathcal{M}}$. In particular, the ray $\mathbb{Q}_{\geq 0}\check{\alpha}$ is contained in the valuation cone $\mathcal{V}$. If we consider the spherical embedding $\mathrm{SL}_2/U \hookrightarrow \mathbb{P}^2$, we see that the line at infinity induces a G-invariant valuation ν with $\rho(\nu) = -\check{\alpha}|_{\mathcal{M}}$, and thus $\mathcal{V} = \mathcal{N}_{\mathbb{Q}}$.

One usually illustrates the combinatorial data in a picture (see Fig. 9.3).

Definition 41 We introduce the following "colored" extensions of the notions of polyhedral cone, face and fan from the toric case.

1. A *colored cone* is a pair $(C, \mathcal{F})$ with
   a. $\mathcal{F} \subseteq \mathcal{D}$,
   b. $C \subseteq \mathcal{N}_{\mathbb{Q}}$ convex cone generated by $\rho(\mathcal{F})$ and finitely many elements of $\mathcal{V} \cap \mathcal{N}$,
   c. the relative interior of C intersects $\mathcal{V}$,
   d. C contains no lines and $0 \notin \rho(\mathcal{F})$.

   Such colored cones are usually called strictly convex colored cones, but as we are only interested in strictly convex cones, we will omit the specifier and just speak of colored cones.
2. A *colored face* of a colored cone $(C, \mathcal{F})$ is a pair $(C', \mathcal{F}')$ such that
   a. C' is a face of C (in the usual sense),
   b. the relative interior of C' intersects $\mathcal{V}$,
   c. $\mathcal{F}' = \{D \in \mathcal{F} : \rho(D) \in C'\}$.
3. A *colored fan* is a finite set Σ of colored cones with the following properties:
   a. every colored face of a colored cone of Σ is in Σ,
   b. for all $\nu \in \mathcal{V}$, there exists at most one colored cone $(C, \mathcal{F}) \in \Sigma$ such that ν is in the relative interior of C.
4. The *support* of a color fan Σ is the set $\bigcup_{(C,\mathcal{F})}(C \cap \mathcal{V}) \subseteq \mathcal{V}$ where $(C, \mathcal{F})$ runs through all elements in Σ.

Let us explain how to associate a colored fan Σ_X to a spherical embedding $G/H \hookrightarrow X$.

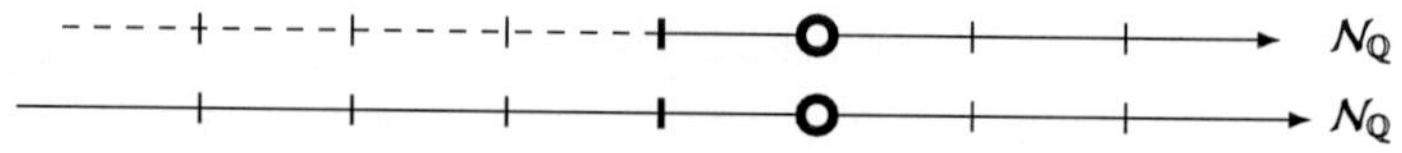

Fig. 9.4 The colored fans of $\mathrm{SL}_2/U \hookrightarrow \mathbb{A}^2$ and $\mathrm{SL}_2/U \hookrightarrow \mathbb{P}^2$

Theorem 42 *X is covered by finitely many G-invariant open subvarieties of X containing a unique closed G-orbit (such varieties are called* simple embeddings*).*

Let $X' \subseteq X$ be an open G-invariant subvariety which is a simple embedding and denote the G-invariant divisors of X' by $X_1, \ldots, X_{m'}$. Let $\mathcal{F}'$ be the set of colors $D \in \mathcal{D}$ whose closure in X contain the closed orbit of X'. We define C' to be the cone in $\mathcal{N}_\mathbb{Q}$ generated by $\rho(D)$ for $D \in \mathcal{F}'$ and $\rho(X_i) := \rho(\nu_{X_i})$ for $i = 1, \ldots, m'$. Then $(C', \mathcal{F}')$ is a colored cone in $\mathcal{N}_\mathbb{Q}$. Moreover the set of colored cones constructed this way (together with their colored faces) forms a colored fan, which we denote by Σ_X.

Theorem 43 (Luna–Vust) *The map $X \mapsto \Sigma_X$ is a bijection from the isomorphism classes of spherical G/H-embeddings and the set of colored fans.*

Example 44 (*Example* 40 *continued*) Clearly $\mathrm{SL}_2/U \hookrightarrow \mathbb{A}^2$ is a simple spherical embedding (the origin is the only closed SL_2-orbit). On the other hand the spherical embedding $\mathrm{SL}_2/U \hookrightarrow \mathbb{P}^2$ is not simple (indeed we can cover it with an affine chart $\mathbb{A}^2$ and the complement of the unique SL_2 fixed point). The corresponding colored fans are illustrated in Fig. 9.4 (understand how to get them and which colored fan corresponds to which spherical embedding). Note that the circle means that the cone $\mathbb{Q}_{\geq 0}$ is "colored" by the unique color of SL_2/U, i.e., $(\mathbb{Q}_{\geq 0}, \{D\})$.

Exercise 45 Use the Luna–Vust theory to classify all spherical embeddings of SL_2/T and SL_2/N. Draw the corresponding colored fans. Hint: You should find 2 in both cases.

Many results about spherical varieties are known. Unfortunately, due to lack of time, we won't be able to dig any deeper.

Theorem 46 *A list of selected results:*

1. *The number of B-orbits is finite;*
2. *X is complete if and only if any $\nu \in \mathcal{V}$ is contained in some colored cone of Σ_X;*
3. *there is a bijective correspondence between G-orbits in X and colored cones in Σ_X;*
4. *a combinatorial smoothness criterion;*
5. *combinatorial descriptions of the Picard group and the divisor class group;*
6. *ampleness criterion for divisors;*

and many more . . .

To learn more about the features of spherical varieties, the interested reader is encouraged to consult [26] for further reading.

9.3.2 The Classification of Spherical Homogeneous Spaces

The classification of spherical homogeneous spaces G/H turns out to be quite hard. Luna's brilliant insight in spherical varieties made it possible to find such a description from Wasserman's list of certain spherical varieties of rank 2 [28]. Inspired by it, Luna [16] formulated a conjectural description and proved it for spherical varieties of type A. Only recently Luna's Conjecture was proven in general with the combined efforts of several researchers [3, 6, 15]. Unfortunately, time does not permit to give more details on this exciting topic, instead we will see a complete answer for an interesting special case.

9.3.3 The Complete Picture in the Horospherical Case

Recall from Exercise 34 that a closed subgroup $H \subseteq G$ is called *horospherical* if it contains a maximal unipotent subgroup of G. An exceedingly well written introduction to horospherical varieties can be found in [19, 20] by Pasquier.

Fix a maximal unipotent subgroup $U \subseteq G$, a Borel subgroup $U \subseteq B$ of G and a maximal torus $T \subseteq B$.

Let us list some fundamental properties of horospherical subgroups. We refer to [19] for further details and references.

Proposition 47 *For a horospherical subgroup $H \subseteq G$ with $U \subseteq H$, the following statements hold:*

1. *the normalizer $P := N_G(H)$ is a parabolic subgroup containing B. Let $I \subseteq S$ be the unique set of simple roots such that $P = P_I$ (see Theorem 24);*
2. $\mathcal{M} = \{\chi \in \mathfrak{X}(P) \colon \chi|_H = 1\} \subseteq \{\chi \in \mathfrak{X}(T) \colon \langle \check{\alpha}, \chi\rangle = 0$ *for all* $\alpha \in I\}$*;*
3. $H = \bigcap_{\chi \in \mathcal{M}} \ker(\chi)$*;*
4. $\mathcal{D} = \{D_\alpha := \overline{Bw_0 s_\alpha P/H} \colon \alpha \in S \setminus I\}$ *where $w_0 \in W$ is the longest element in the Weyl group $W = N_G(T)/T$ and s_α denotes the simple reflection associated to the simple root α;*
5. *$P_{-w_0(I)}$ coincides with the stabilizer of the open B-orbit in G/H;*
6. $\mathcal{V} = \mathcal{N}_{\mathbb{Q}}$.

Theorem 48 ([13, Theorem 6.1]) *If $H \subseteq G$ is a spherical subgroup, then $N_G(H)/H$ is diagonalizable. In particular, if H contains U, then P/H is a torus where $P := N_G(H)$.*

Now, we come to an important geometric characterization of horospherical homogeneous spaces.

Recall that a continuous surjective map $p \colon E \to X$ of topological spaces is called a *fiber bundle* with fiber F (another topological space) if X can be covered with open subsets U such that there are homeomorphisms $\varphi \colon p^{-1}(U) \to U \times F$ in such a way that p agrees with the projection onto the first factor (see Fig. 9.5). It is said to be

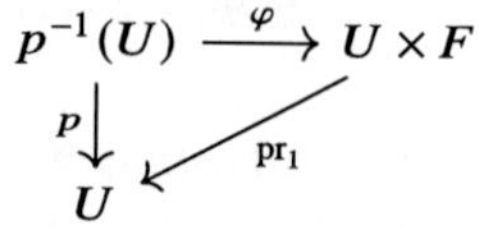

Fig. 9.5 Local trivialization of fiber bundles

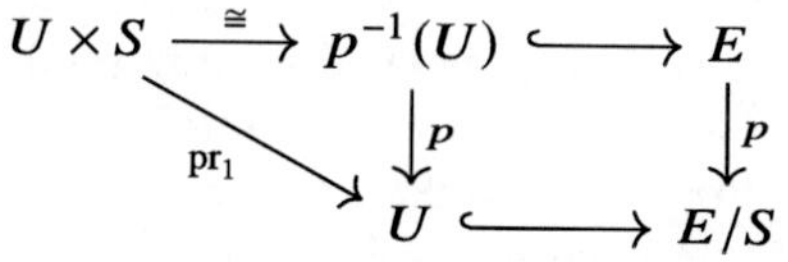

Fig. 9.6 Quotient by free torus action is locally trivial in Zariski topology

a *principal* Γ*-bundle*, for Γ a topological group, if in addition E is equipped with a continuous Γ-action $\Gamma \times E \to E$ preserving the fibers of p, i.e., if $y \in p^{-1}(x)$ for some $x \in X$, then $\gamma \cdot y \in p^{-1}(x)$ for any $\gamma \in \Gamma$, and acts freely and transitively on them.

Here is a crucial observation on principal torus bundles in algebraic geometry.

Lemma 49 *Let the torus S act freely on the normal irreducible variety E with good geometric quotient $p\colon E \to E/S$. Then for each $y \in E/S$ there exists an affine open neighbourhood $U \subseteq E/S$ of x such that the diagram in Fig. 9.6 commutes and the upper left isomorphism is S-equivariant.*

Exercise 50 Show that if an algebraic torus S acts freely on a normal irreducible variety E with good geometric quotient $p\colon E \to E/S$, then p admits Zariski open trivializations (i.e., prove Lemma 49). What if we replace S by a connected reductive G? Hint: For the second part of the question, you may want to consider the morphism $\phi\colon X \to Y;\ (A, B) \mapsto (\det(A), \operatorname{tr}(AB), \det(B))$ where $X = \{(A, B) \in \mathrm{Mat}(2 \times 2, \mathbb{C})^2\colon\ \det(AB - BA) \neq 0, \operatorname{tr}(A) = \operatorname{tr}(B) = 0\}$ and $Y = \{y \in \mathbb{C}^3\colon 4y_1y_3 - y_2^2 \neq 0\}$.

Let $p\colon E \to X$ be a Γ-principal bundle. Suppose that E and X are G-spaces for another topological group G. Then $p\colon E \to X$ is an *equivariant principal* Γ*-bundle* if p is equivariant (i.e., $p(g \cdot y) = g \cdot p(y)$ for any $g \in G$ and $y \in E$) and the two actions by Γ and G commute (usually one assumes that Γ acts from the right while G acts from the left).

Proposition 51 *If $H \subseteq G$ is a closed subgroup, then the following statements are equivalent:*

1. *H is horospherical, i.e., contains the unipotent radical of a Borel subgroup;*
2. *G/H is an (algebraic) equivariant principal torus bundle over a flag variety G/P where G naturally acts on G/H resp. G/P by left translations (the dimension of the torus fiber coincides with the* rank *of G/H);*
3. *$H = \bigcap_{\chi \in M} \ker \chi$ for some parabolic subgroup P of G and some sublattice $M \subseteq \mathfrak{X}(P)$.*

Furthermore, $P = N_G(H)$ *and* $P = TH = BH$ *for all maximal tori* T *of* B *contained in* P *and all Borel subgroups* B *of* G *contained in* P.

Proof (1) $\Rightarrow$ (2): By Theorem 48, $S := P/H = N_G(H)/H$ is an algebraic torus. It acts on G/H by right-translations, i.e.,

$$S \times G/H \to G/H;\ (pH, xH) \mapsto xp^{-1}H.$$

It is straightforward to check that S acts freely on G/H, and thus the result follows by Lemma 49.

(2) $\Rightarrow$ (1): Suppose that the fibers of the torus bundle $p\colon G/H \to G/P$ are isomorphic to the algebraic torus T. As the two actions by G and T commute, the morphisms $\varphi_t\colon G/H \to G/H; xH \mapsto t \cdot xH$ for $t \in T$ are G-equivariant automorphisms. It follows by [26, Proposition 1.8] that we may consider T as a subgroup of $N_G(H)/H$. Set $N := N_G(H)$. Let $\widetilde{T}$ be the preimage of T under the natural projection map $N \to N/H$. Note that $N \to N/H$ is a morphism of algebraic groups and that $\widetilde{T}$ is a closed subgroup of G. Since $p^{-1}(xP) \cong T$ for any $xP \in G/P$, it follows that a conjugate of P is contained in $\widetilde{T}$, and thus it contains a maximal unipotent subgroup U. As the natural projection morphism of algebraic groups $\widetilde{T} \to T$ maps unipotent elements on unipotent elements, it follows that U is in its kernel which implies that $U \subseteq H$.

(1) $\Leftrightarrow$ (3) straightforwardly follows from Proposition 47 (3). □

Exercise 52 Show that SL_2/U_k is indeed a torus bundle over SL_2/B.

Exercise 53 Use the Luna–Vust theory to classify all spherical embeddings of SL_2/U where $U \subseteq \mathrm{SL}_2$ is a maximal unipotent subgroup. Draw the corresponding colored fans. Hint: You should find 6.

Proposition 54 ([20, Proposition 1.6]) *The assignment which associates to a horospherical subgroup* $H \subseteq G$ *the pair* $(\mathcal{M}, I)$ *(see Proposition 47) induces a bijection between horospherical subgroups of* G *and pairs* (M, J) *where* $J \subseteq S$ *and* $M \subseteq \mathfrak{X}(T)$ *is a sublattice such that* $\langle \check{\alpha}, \chi \rangle = 0$ *for any* $\alpha \in J$ *and all* $\chi \in M$.

The horospherical subgroup associated to a pair (M, J) as in Proposition 54 is given by $H = \bigcap_{\chi \in M} \ker \chi$ where $M \subseteq \mathfrak{X}(P_J)$.

Exercise 55 Use the combinatorial description of horospherical subgroups to classify those contained in SL_2.

A colored fan Σ is called *toroidal* if $\mathcal{F} = \varnothing$ for any $(C, \mathcal{F}) \in \Sigma$. Observe that in the horospherical case toroidal fans coincide with fans in the toric sense. In this special case, we have the following explicit construction of horospherical toroidal varieties:

Proposition 56 ([20, Examples 1.13 (2)]) *If* $H \subseteq G$ *is a horospherical subgroup containing* U *and* Σ *is a toroidal fan, then the corresponding spherical embedding is* G*-equivariantly isomorphic to* $G \times_P X_\Sigma$ *where* X_Σ *denotes the toric variety corresponding to the fan* Σ *(with acting torus* P/H *where* $P = N_G(H)$*).*

In the situation of Proposition 56, recall that P acts on $G \times X_\Sigma$ by $p \cdot (g, x) := (gp^{-1}, pH \cdot x)$ with good geometric quotient $G \times_P X_\Sigma = (G \times X_\Sigma)/P$.

9.4 The Ring of Conditions of a Horospherical Variety

A good reference for the ring of conditions is the classical paper by De Concini and Procesi [7].

Let G be a connected complex algebraic group and $H \subseteq G$ a closed subgroup (not necessarily spherical). Consider the homogeneous space G/H.

Recall that two subvarieties $X, Y \subseteq G/H$ are said to *intersect properly* if either $X \cap Y = \varnothing$ or each irreducible component of the intersection $X \cap Y$ has dimension $\dim(X) + \dim(Y) - \dim(G/H)$. They are said to *intersect transversally* if the intersection $X \cap Y$ is smooth and has pure dimension $\dim(X) + \dim(Y) - \dim(G/H)$.

Theorem 57 (Kleiman's transversality theorem [12, Corollary 4]) *Let $X, Y \subseteq G/H$ be two irreducible subvarieties. The left translate of X by $g \in G$ we denote by gX.*

1. *There exists a dense open subset $U \subseteq G$ such that gX and Y intersect properly for each $g \in U$.*
2. *If X, Y are smooth, then there exists a dense open subset $U \subseteq G$ such that gX and Y intersect transversally for any $g \in U$.*

In particular, if X and Y have complementary dimensions (but are not necessarily smooth), the intersection $gX \cap Y$ consists of finitely many points and this number is constant for generic $g \in G$.

Remark 58 There is a slight strengthening of Kleiman's transversality theorem in [7, Sect. 6.1].

Recall that the free abelian group $\mathcal{Z}^k(G/H) = \bigoplus_{X \subseteq G/H} \mathbb{Z}X$, where the sum is over closed irreducible subvarieties of codimension k, is said to be the group of algebraic cycles of codimension k. Theorem 57 makes it possible to introduce an intersection pairing between groups of algebraic cycles of complementary codimensions

$$\mathcal{Z}^k(G/H) \quad \text{and} \quad \mathcal{Z}^{\dim(G/H)-k}(G/H).$$

It is enough to define it for irreducible cycles and then extend bilinearly:

$$\mathcal{Z}^k(G/H) \times \mathcal{Z}^{\dim(G/H)-k}(G/H) \to \mathbb{Z};$$
$$(X, Y) \mapsto (X \cdot Y) := \#(gX \cap Y) \quad \text{(for generic } g \in G).$$

Here $X, Y \subseteq G/H$ are assumed to be irreducible subvarieties.

Definition 59 Two algebraic cycles $X, Y \in \mathcal{Z}^k(G/H)$ are said to be equivalent, i.e., $X \sim Y$, if for any algebraic cycle of complementary codimension $Z \in \mathcal{Z}^{\dim(G/H)-k}(G/H)$ the intersection products are the same $(X \cdot Z) = (Y \cdot Z)$. We denote the group of equivalence classes by $C^k(G/H) := \mathcal{Z}^k(G/H)/\sim$ and consider it as the "group of conditions of dimension $\dim(G/H) - k$".

Clearly the intersection pairing factors through the equivalence relation, so that we obtain an intersection pairing on the groups of conditions: $C^k(G/H) \times C^{\dim(G/H)-k}(G/H) \to \mathbb{Z}$. So far $C^*(G/H) := \bigoplus_{k=0}^{\dim(G/H)} C^k(G/H)$ is only a group, but we want to introduce a product on it, so that it becomes a ring. Again, it is enough to define a product structure for classes of irreducible cycles $X \in \mathcal{Z}^k(G/H)$ and $Y \in \mathcal{Z}^l(G/H)$ and then extend bilinearly. Here is a naive approach:

Definition 60 Define the intersection product of $[X]$ and $[Y]$ where $X, Y \subseteq G/H$ are two irreducible subvarieties by $[X] \cdot [Y] := [gX \cap Y]$ for generic $g \in G$.

Unfortunately this definition of an intersection product may not be well-defined in general (see Exercise 61).

Exercise 61 Show that the naive definition of an intersection product of two irreducible subvarieties $X, Y \subseteq G/H$ is not well-defined in general. (Hint: Consider $G = (\mathbb{C}^3, +)$ acting on $\mathbb{A}^3$ by translations. Let $H = \{0\}$ and compute the intersection product of $X = \{y = 0\}$, $Y = \{x = yz\} \subseteq \mathbb{C}^3$.)

Proposition 62 ([7]) *For a flag variety G/P, the intersection product in Definition 60 is well-defined and the ring $C^*(G/P)$ can be identified with the Chow ring $A^*(G/P)$ or with the cohomology ring $H^*(G/P, \mathbb{Z})$.*

Led by this observation, De Concini and Procesi showed the remarkable fact that the intersection product of Definition 60 is well-defined on spherical homogeneous spaces. Let C be the set of smooth (or complete) spherical embeddings $G/H \hookrightarrow X$. This set C admits the partial ordering defined such that a spherical embedding $G/H \hookrightarrow X_1$ is greater than $G/H \hookrightarrow X_2$ if there exists an equivariant morphism $X_1 \to X_2$. De Concini's and Procesi's idea is to show that for any $X \in C$ and any algebraic cycle $Y \subseteq G/H$, there is an $X' \in C$ with $X \leq X'$ such that the closure $\overline{Y}$ of Y in X' intersects the boundary of the open G-orbit in X' properly. The existence of such a "good compactification" ensures that if one considers the embedding X' then we may always assume (up to generic translations by G) that the intersection with $\overline{Y}$ takes place in the open G-orbit G/H. To get an isomorphism of rings, we have to consider "good compactifications" of all algebraic cycles at once.

Theorem 63 ([7, Sect. 6.3]) *The intersection product from Definition 60 is well-defined on a spherical homogeneous space G/H and there is a canonical isomorphism of graded rings*

$$C^*(G/H) = \varinjlim_{X'} A^*(X') = \varinjlim_{X'} H^*(X', \mathbb{Z})$$

where the limit is taken over complete (or equivalently smooth) spherical embeddings $G/H \hookrightarrow X'$.

Remark 64 Any complete spherical embedding is dominated by a smooth projective toroidal one, and thus they form a cofinal set.

Exercise 65 Explicitly compute the ring of conditions for some spherical homogeneous spaces SL_2/H where $H \subseteq \mathrm{SL}_2$ is a spherical subgroup. (Hint: In this case, the rank of the spherical homogeneous space is bounded by 1, and thus there are only finitely many spherical embeddings, so that we can straightforwardly compute the direct limit of cohomology rings.)

Exercise 66 Use the ring of conditions of $\mathrm{Gr}(2, 4)$ to solve the "4-lines problem".

9.4.1 The Horospherical Case

From now on let $H \subseteq G$ be a horospherical subgroup containing the unipotent radical U of a Borel subgroup B. Set $P := N_G(H)$ which is a parabolic subgroup containing B.

Any character $\alpha \in \mathfrak{X}(P)$ induces an action of P on the affine line $\mathbb{C}_\alpha$ by $p \cdot x = \alpha(p)x$. We obtain an action of P on $G \times \mathbb{C}_\alpha$ by $p \cdot (g, x) = (gp^{-1}, \alpha(p)x)$. The geometric quotient by this action exists and is denoted by $G \times_P \mathbb{C}_\alpha$, i.e., $G \times_P \mathbb{C}_\alpha = (G \times \mathbb{C}_\alpha)/P$. For the equivalence classes in $G \times_P \mathbb{C}_\alpha$ we write $g \star x$. The projection map $G \times_P \mathbb{C}_\alpha \to G/P$; $g \star x \mapsto gP$ yields an equivariant line bundle on G/P where G acts on the left, i.e., $g' \cdot (g \star x) = (g'g) \star x$. We write $\delta(\alpha) := G \times_P \mathbb{C}_\alpha$ and note that these bundles are usually called homogeneous fiber bundles (see [26, Section 2.1]). If we compose the map $\mathfrak{X}(P) \to \mathrm{Pic}(G/P)$ with the inclusion $\mathcal{M} \subseteq \mathfrak{X}(P)$, we obtain a map $\delta \colon \mathcal{M} \to \mathrm{Pic}(G/P)$.

The following statement combinatorially describes the cohomology ring of smooth projective toroidal horospherical varieties. It is a special case of a more general result.

Theorem 67 ([21, Theorem 1.2]) *Let X_Σ be a smooth projective toroidal horospherical variety defined by an (uncolored) fan Σ with rays $\rho_1, \ldots, \rho_n$. Let $v_1, \ldots, v_n \in \mathcal{N}$ be the primitive vectors along the rays ρ_i. Then the cohomology ring $H^*(X_\Sigma, \mathbb{Q})$ is isomorphic as an $H^*(G/P, \mathbb{Q})$-algebra to the quotient of $H^*(G/P, \mathbb{Q})[x_1, \ldots, x_n]$ by the sum of ideals*

$$\langle x_{j_1} \cdots x_{j_k} : \rho_{j_1}, \ldots, \rho_{j_k} \text{ do not span a cone of } \Sigma \rangle +$$
$$\left\langle c_1(\delta(m)) - \sum_{i=1}^{n} \langle v_i, m \rangle x_i : m \in \mathcal{M} \right\rangle,$$

where $c_1(\delta(m)) \in H^2(G/P, \mathbb{Z})$ denotes the first Chern class of the line bundle $\delta(m)$.

Note that the first ideal in the sum of ideals in Theorem 67 corresponds to the Stanley-Reisner ideal of the toric variety. The challenge is to find a good description of the ring in Theorem 67 as we want to take the direct limit over all smooth projective toroidal fans Σ.

The following approach is inspired by [5]. To keep notation simple, set $\mathcal{M}_{\mathbb{Q}} = \mathcal{M} \otimes \mathbb{Q}$ and $\mathcal{N}_{\mathbb{Q}} = \mathrm{Hom}_{\mathbb{Z}}(\mathcal{M}, \mathbb{Q})$. Let Σ be a smooth projective toroidal fan in $\mathcal{N}_{\mathbb{Q}}$. A map $f : \mathcal{N}_{\mathbb{Q}} \to \mathbb{Q}$ is *piecewise polynomial* if for any $\sigma \in \Sigma$, the map $f|_{\sigma} : \sigma \to \mathbb{Q}$ extends to a polynomial function on the linear space $\mathrm{span}_{\mathbb{Q}}\{\sigma\}$, i.e., a piecewise polynomial function f on Σ is a collection of compatible polynomial functions $f_{\sigma} : \sigma \to \mathbb{Q}$. In particular, such a function is continuous. We denote by $\mathcal{R}_{\Sigma}$ the set of all piecewise polynomial functions on Σ which is a ring under pointwise addition and multiplication. Let $S^*(\mathcal{M}_{\mathbb{Q}})$ be the symmetric algebra of the $\mathbb{Q}$-vector space $\mathcal{M}_{\mathbb{Q}}$. Recall that $S^*(\mathcal{M}_{\mathbb{Q}})$ can be naturally identified with the polynomial functions on $\mathcal{N}_{\mathbb{Q}}$. Note that $\mathcal{R}_{\Sigma}$ is a positively graded $\mathbb{Q}$-algebra with graded subalgebra $S^*(\mathcal{M}_{\mathbb{Q}})$. Indeed, any piecewise polynomial function $f = (f_{\sigma})_{\sigma \in \Sigma}$ uniquely decomposes into a sum of homogeneous piecewise polynomial functions.

Exercise 68 Let Σ be a smooth projective toroidal fan in $\mathcal{N}_{\mathbb{Q}}$. Show that for any ray ρ there is a piecewise linear function φ_{ρ} on Σ which vanishes on all the other rays and satisfies $\varphi(u_{\rho}) = 1$ where u_{ρ} is the primitive ray generator in $\mathcal{N}$ of the ray ρ.

Let us write $\Sigma(1)$ for the set of rays of a fan Σ and u_{ρ} for the primitive generator in $\mathcal{N}$ of the ray $\rho \in \Sigma(1)$.

Lemma 69 *If Σ is a smooth projective toroidal fan, then $\{\varphi_{\rho} : \rho \in \Sigma(1)\}$ (where φ_{ρ} is defined in Exercise 68) forms a basis of $\mathcal{R}^1_{\Sigma}$ the space of piecewise linear functions on Σ.*

Exercise 70 Let Σ be a smooth projective toroidal fan. Show that $\mathcal{R}_{\Sigma}$ is isomorphic to the *Stanley-Reisner algebra* R_{Σ}, i.e., the quotient ring of $\mathbb{Q}[T_{\rho} : \rho \in \Sigma(1)]$ by the relations $\prod_{i=1}^{k} T_{\rho_i} = 0$ whenever $\rho_1, \ldots, \rho_k$ are distinct rays which do not generate a cone of Σ. (Hint: Clearly $\prod_{i=1}^{k} \varphi_{\rho_i} = 0$ whenever $\rho_1, \ldots, \rho_k$ do not generate a cone of Σ. Therefore, there is a unique algebra homomorphism from R_{Σ} to $\mathcal{R}_{\Sigma}$, which sends T_{ρ} to φ_{ρ}. Show that this map is an isomorphism.)

We can now reformulate Theorem 67:

Proposition 71 *Let X_{Σ} be a smooth projective toroidal horospherical variety defined by an (uncolored) fan Σ. Then the cohomology ring $H^*(X_{\Sigma}, \mathbb{Q})$ is isomorphic as an $H^*(G/P, \mathbb{Q})$-algebra to the quotient of $H^*(G/P, \mathbb{Q}) \otimes \mathcal{R}_{\Sigma}$ by the ideal*

$$\left\langle c_1(\delta(m)) \otimes 1 - \sum_{\rho \in \Sigma(1)} \langle u_{\rho}, m \rangle 1 \otimes \varphi_{\rho} : m \in \mathcal{M} \right\rangle = \langle c_1(\delta(m)) \otimes 1 - 1 \otimes \langle \cdot, m \rangle : m \in \mathcal{M} \rangle,$$

where $\langle \cdot, m \rangle \in S^*(\mathcal{M}_{\mathbb{Q}})$ is a (piecewise) linear function on Σ.

Proof The statement is a reformulation of Theorem 67 except the equality of the two ideals which remains to be shown. Recall from Lemma 69 that the set of piecewise linear functions $\{\varphi_\rho \colon \rho \in \Sigma(1)\}$ (where φ_ρ are defined in Exercise 68) forms a basis of $\mathcal{R}^1_\Sigma$. Then the (piecewise) linear function $\langle \cdot, m\rangle$ for $m \in \mathcal{M}$ can be expressed as a linear combination in this basis, namely $\langle \cdot, m\rangle = \sum_{\rho\in\Sigma(1)} \langle u_\rho, m\rangle \varphi_\rho$. □

We denote by $\mathcal{R}$ the set of all piecewise polynomial functions on smooth projective toroidal fans in $\mathcal{N}_\mathbb{Q}$, i.e., $\mathcal{R} = \bigcup_\Sigma \mathcal{R}_\Sigma$ where the union is taken over all smooth projective toroidal fans Σ.

Theorem 72 *We have that*

$$C^*(G/H) \otimes \mathbb{Q} \cong \left(H^*(G/P, \mathbb{Q}) \otimes \mathcal{R}\right) / \left\langle c_1(\delta(m)) \otimes 1 - 1 \otimes \langle \cdot, m\rangle \colon m \in \mathcal{M}\right\rangle,$$

where $\langle \cdot, m\rangle \in S^*(\mathcal{M}_\mathbb{Q})$ is a piecewise linear function on any smooth projective toroidal fan.

Proof For convenience, let us write $A := H^*(G/P, \mathbb{Q})$.

By Theorem 63, we have

$$C^*(G/H) \otimes \mathbb{Q} = \left(\varinjlim_{X'} H^*(X', \mathbb{Z})\right) \otimes \mathbb{Q} = \varinjlim_{X'} H^*(X', \mathbb{Q})$$

where the limit is taken over all smooth projective toroidal embeddings of G/H which is a directed set. Indeed, for any two smooth projective toroidal embeddings with corresponding fans Σ_1, Σ_2, we can find a third smooth projective toroidal fan Σ which refines both fans Σ_1 and Σ_2. We introduce the relation $\Sigma \preceq \Sigma'$ whenever Σ' refines Σ. Suppose $\Sigma \preceq \Sigma'$, so that we obtain an equivariant map $X_{\Sigma'} \to X_\Sigma$. Our goal is to understand how the representation of cohomology rings given in Proposition 71 behaves under this map. By Proposition 71, the cohomology rings (as A-algebras) are generated by classes of divisors, so that the map corresponding to $X_{\Sigma'} \to X_\Sigma$ is given by pulling back divisors which in turn induces the natural inclusion $\mathcal{R}_\Sigma \subseteq \mathcal{R}_{\Sigma'}$. Let $I_\Sigma := \langle c_1(\delta(m)) \otimes 1 - 1 \otimes \langle \cdot, m\rangle \colon m \in \mathcal{M}\rangle \subseteq A \otimes \mathcal{R}_\Sigma$. Similarly define $I_{\Sigma'}$ in $A \otimes \mathcal{R}_{\Sigma'}$. As $I_\Sigma \subseteq I_{\Sigma'}$, we obtain the natural map $\mu_{\Sigma,\Sigma'} \colon (A \otimes \mathcal{R}_\Sigma)/I_\Sigma \to (A \otimes \mathcal{R}_{\Sigma'})/I_{\Sigma'}$. Then $((A \otimes \mathcal{R}_\Sigma)/I_\Sigma, \mu_{\Sigma,\Sigma'})$ is the direct system yielding the direct limit $\varinjlim H^*(X', \mathbb{Q})$. Moreover, we obtain two more direct systems, namely $(I_\Sigma, I_\Sigma \subseteq I_{\Sigma'})$ and $(A \otimes \mathcal{R}_\Sigma, A \otimes \mathcal{R}_\Sigma \subseteq A \otimes \mathcal{R}_{\Sigma'})$ (for $\Sigma \preceq \Sigma'$). Indeed, we obtain a direct system of exact sequences:

$$0 \to I_\Sigma \to A \otimes \mathcal{R}_\Sigma \to (A \otimes \mathcal{R}_\Sigma)/I_\Sigma \to 0.$$

The statement follows by the fact that taking direct limits in the category of modules is an exact functor, $\varinjlim A \otimes \mathcal{R}_\Sigma = A \otimes \mathcal{R}$, and $\varinjlim I_\Sigma = I$, where I denotes the ideal in the statement. □

Acknowledgements We want to express our gratitude to Megumi Harada and Adam Van Tuyl for organizing the exciting "Graduate Summer School in Algebraic Group Actions" at McMaster University and their great support. We also want to thank the participants of the summer school for their constructive feedback. Finally, we are grateful to the anonymous referee for several valuable suggestions and remarks.

References

1. Akhiezer, D.N.: Actions with a finite number of orbits. Funktsional. Anal. i Prilozhen. **19**(1), 1–5, 96 (1985)
2. Borel, A.: Linear algebraic groups. Graduate Texts in Mathematics, vol. 126, 2nd edn. Springer, New York (1991)
3. Bravi, P., Pezzini, G.: Primitive wonderful varieties. Math. Z. **282**(3–4), 1067–1096 (2016)
4. Brion, M.: Vers une généralisation des espaces symétriques. J. Algebra **134**(1), 115–143 (1990)
5. Brion, M.: Piecewise polynomial functions, convex polytopes and enumerative geometry. In: Parameter spaces (Warsaw, 1994), Banach Center Publications, vol. 36, pp. 25–44. Polish Academy of Sciences Institute of Mathematics, Warsaw (1996)
6. Cupit-Foutou, S.: Wonderful varieties: a geometrical realization (2009). arXiv:0907.2852
7. De Concini, C., Procesi, C.: Complete symmetric varieties. II. Intersection theory. In: Algebraic Groups and Related Topics (Kyoto/Nagoya, 1983), Advanced Studies in Pure Mathematics, vol. 6, pp. 481–513. North-Holland, Amsterdam (1985)
8. Gagliardi, G., Hofscheier, J.: Gorenstein spherical Fano varieties. Geom. Dedicata **178**, 111–133 (2015)
9. Hartshorne, R.: Algebraic Geometry. Springer, New York, Heidelberg (1977). Graduate Texts in Mathematics, No. 52
10. Humphreys, J.E.: Linear Algebraic Groups. Springer, New York, Heidelberg (1975). Graduate Texts in Mathematics, No. 21
11. Kiritchenko, V.: Geometry of spherical varieties (2009). Course notes. https://users.mccme.ru/valya/sph1.pdf
12. Kleiman, S.L.: The transversality of a general translate. Compositio Math. **28**, 287–297 (1974)
13. Knop, F.: The Luna-Vust theory of spherical embeddings. In: Proceedings of the Hyderabad Conference on Algebraic Groups (Hyderabad, 1989), pp. 225–249. Manoj Prakashan, Madras (1991)
14. Knop, F.: Localization of spherical varieties. Algebra Number Theory **8**(3), 703–728 (2014)
15. Losev, I.V.: Uniqueness property for spherical homogeneous spaces. Duke Math. J. **147**(2), 315–343 (2009)
16. Luna, D.: Variétés sphériques de type A. Publ. Math. Inst. Hautes Études Sci. **94**, 161–226 (2001)
17. Luna, D., Vust, T.: Plongements d'espaces homogènes. Comment. Math. Helv. **58**(2), 186–245 (1983)
18. Malle, G., Testerman, D.: Linear algebraic groups and finite groups of Lie type. In: Cambridge Studies in Advanced Mathematics, vol. 133. Cambridge University Press, Cambridge (2011)
19. Pasquier, B.: Fano horospherical varieties. Ph.D. thesis, Université Joseph-Fourier-Grenoble I (2006)
20. Pasquier, B.: Variétés horosphériques de Fano. Bull. Soc. Math. France **136**(2), 195–225 (2008)
21. Sankaran, P., Uma, V.: Cohomology of toric bundles. Comment. Math. Helv. **78**(3), 540–554 (2003)
22. Schubert, H.: Kalkül der abzählenden Geometrie. Springer, Berlin-New York: Reprint of the 1879 original. With an introduction by Steven L, Kleiman (1979)
23. Shafarevich, I.R.: Basic algebraic geometry. 1, Russian edn. Springer, Heidelberg (2013). Varieties in projective space

24. Shafarevich, I.R.: Basic Algebraic Geometry. 2, 3rd edn. Springer, Heidelberg (2013). Schemes and complex manifolds, Translated from the 2007 third Russian edition by Miles Reid
25. Springer, T.A.: Linear Algebraic Groups, 2nd edn. Modern Birkhäuser Classics. Birkhäuser Boston Inc, Boston, MA (2009)
26. Timashev, D.A.: Homogeneous spaces and equivariant embeddings. Encyclopaedia of Mathematical Sciences, vol. 138. Springer, Heidelberg (2011). Invariant Theory and Algebraic Transformation Groups, 8
27. Vakil, R.: The rising sea: foundations of algebraic geometry (2017). Course notes. http://math.stanford.edu/~vakil/216blog/FOAGnov1817public.pdf
28. Wasserman, B.: Wonderful varieties of rank two. Transform. Groups **1**(4), 375–403 (1996)

Chapter 10
Linear Recursions for Integer Point Transforms

Katharina Jochemko

Abstract We consider the integer point transform

$$\sigma_P(\mathbf{x}) = \sum_{\mathbf{m} \in P \cap \mathbb{Z}^n} \mathbf{x}^{\mathbf{m}} \in \mathbb{C}[x_1^{\pm 1}, \ldots, x_n^{\pm 1}]$$

of a polytope $P \subset \mathbb{R}^n$. We show that if P is a lattice polytope then for any polytope Q the sequence $\{\sigma_{kP+Q}(\mathbf{x})\}_{k \geq 0}$ satisfies a multivariate linear recursion that only depends on the vertices of P. We recover Brion's Theorem and by applying our results to Schur polynomials we disprove a conjecture of Alexandersson (2014).

Keywords Lattice Polytops · Integer point transforms · Valuations · Brison's Theorem · Schur polynomials

10.1 Introduction

A **polytope** is the convex hull of finitely many points in $\mathbb{R}^n$. A polytope is a **lattice polytope** if all its vertices lie in the integer lattice $\mathbb{Z}^n$. The **integer point transform** of a polytope P is defined by

$$\sigma_P(x) = \sum_{\mathbf{m} \in P \cap \mathbb{Z}^n} \mathbf{x}^{\mathbf{m}} \in \mathbb{C}[x_1^{\pm 1}, \ldots, x_n^{\pm 1}],$$

where $\mathbf{x}^{\mathbf{m}}$ denotes $x_1^{m_1} \cdots x_n^{m_n}$ for all $\mathbf{m} \in \mathbb{Z}^n$. In this note we study sequences $\{\sigma_{kP}(\mathbf{x})\}_{k \geq 0}$ of integer point transforms of integer dilates of polytopes P and relatives. We prove the following linear recursion.

Theorem 1 *Let Q be a polytope in $\mathbb{R}^n$ and let P be a lattice polytope with vertex set* $\mathrm{vert}(P) = \{v_1, \ldots, v_r\}$. *Then the sequence $\{\sigma_{kP+Q}(\mathbf{x})\}_{k \geq 0}$ satisfies the linear*

K. Jochemko (✉)
Department of Mathematics, KTH Royal Institute of Technology, SE–100 44 Stockholm, Sweden
e-mail: jochemko@kth.se

A. M. Kasprzyk and B. Nill (eds.), *Interactions with Lattice Polytopes*,
Springer Proceedings in Mathematics & Statistics 386,
https://doi.org/10.1007/978-3-030-98327-7_10

recursion

$$\sigma_{(k+r)P+Q}(\mathbf{x}) = \sum_{\emptyset \neq I \subseteq [r]} (-1)^{1+|I|} \mathbf{x}^{\sum_{i\in I} v_i} \sigma_{(k+r-|I|)P+Q}(\mathbf{x})$$

with characteristic polynomial

$$\chi_{P;Q}(X) := \prod_{\mathbf{v} \in \mathrm{vert}(P)} (X - \mathbf{x}^{\mathbf{v}}) .$$

If Q is a lattice polytope, then $\chi_{P;Q}$ is minimal.

In particular, the recursion only depends on the vertices of P. This improves results by Alexandersson [2] where it was assumed that P has the integer decomposition property and $Q = \{0\}$.

Employing classical results from valuation theory, in Sect. 10.2 we first prove a recursion for indicator functions of dilated polytopes. Then, in Sect. 10.3, we apply these results to integer point transforms and prove Theorem 1. We recover Brion's Theorem in Sect. 10.4 and by applying our results to Schur polynomials we disprove a conjecture of Alexandersson [1] in Sect. 10.5.

10.2 Characteristic Functions and Valuations

In this section we prove a linear recursion for indicator functions of integer dilates of a polytope P. Let $\mathcal{P}$ denote the set of polytopes in $\mathbb{R}^n$ and let G be an abelian group. A **valuation** is a map $\varphi\colon \mathcal{P} \to G$ such that $\varphi(\emptyset) = 0$ and

$$\varphi(P \cup Q) = \varphi(P) + \varphi(Q) - \varphi(P \cap Q) ,$$

for all $P, Q \in \mathcal{P}$ such that also $P \cup Q \in \mathcal{P}$. The volume, the number of lattice points inside a polytope and the integer point transform are examples of valuations. It was shown by Volland [19] that every valuation satisfies the inclusion-exclusion property. That is, for polytopes $P, P_1, \ldots, P_r$ such that $P = P_1 \cup \cdots \cup P_r$

$$\varphi(P) = \sum_{\emptyset \neq I \subseteq [r]} (-1)^{|I|+1} \varphi(P_I) ,$$

where $P_I := \bigcap_{i \in I} P_i$. Stronger even, it follows from a result of Groemer [9], that if $\sum \alpha_i \mathbf{1}_{P_i} = 0$ for polytopes $P_1, \ldots, P_m$ and some $\alpha_1, \ldots, \alpha_m \in \mathbb{Z}$ then $\sum_i \alpha_i \varphi(P_i) = 0$ where $\mathbf{1}_P$ denotes the indicator function for every polytope P. A function of the form $\sum \alpha_i \mathbf{1}_{P_i}$ is called a **polytopal simple function**. By Groemer's result, every valuation uniquely defines a homomorphism from the abelian group of polytopal simple functions to G, that is, every polytope can be identified with its indicator function. For valuations on lattice polytopes this was proved by McMullen [12]. It is well-known that for every affine linear map $T\colon \mathbb{R}^n \to \mathbb{R}^m$

$$\mathbf{1}_P \mapsto \mathbf{1}_{T(P)} \tag{10.1}$$

defines a valuation (see, e.g., [4, Chap. 8]). Using this push forward map we obtain the following recursion on indicator functions.

Theorem 2 *Let P be a polytope in $\mathbb{R}^n$ with vertex set* $\mathrm{vert}(P) = \{v_1, \ldots, v_r\}$. *Then*

$$\mathbf{1}_{(k+r)P} = \sum_{\emptyset \neq I \subseteq [r]} (-1)^{1+|I|} \mathbf{1}_{Q_I^{k,r}}$$

for all $k \geq 0$ where $Q_I^{k,r} = (k + r - |I|)P + \sum_{i \in I} v_i$.

Proof We first assume that P is the $(d-1)$-dimensional standard simplex $\Delta_{d-1} = \{x \in \mathbb{R}^d : x_1 + \cdots + x_d = 1, x_1, \ldots, x_d \geq 0\}$. Its $(k+d)$th dilate is given by

$$(k+d)\Delta_{d-1} = \{x \in \mathbb{R}^d : x_1 + \cdots + x_d = d + k, x_1, \ldots, x_d \geq 0\}$$

For all $I \subseteq [d]$, let

$$P_I = (k+d)\Delta_{d-1} \cap \{x \in \mathbb{R}^d : x_i \geq 1 \text{ for all } i \in I\}.$$

Then $P_I = \bigcap_{i \in I} P_{\{i\}}$ for all $\emptyset \neq I \subseteq [r]$. As in [15] we observe that $(k+d)\Delta_{d-1} = P_\emptyset = \bigcup_{i \in [d]} P_{\{i\}}$ for all $k \geq 0$. Therefore, by inclusion-exclusion,

$$\mathbf{1}_{(k+d)\Delta_{d-1}} = \sum_{\emptyset \neq I \subseteq [d]} (-1)^{1+|I|} \mathbf{1}_{P_I}$$

and we finish the proof of this case by observing that $P_I = (k + d - |I|)\Delta_{d-1} + \sum_{i \in I} e_i$.

For the general case, we recall that every polytope is an affine linear projection of a standard simplex and, thus, the claim follows by applying the push forward map (10.1). □

For fixed $Q \in \mathcal{P}$, $\mathbf{1}_P \mapsto \mathbf{1}_{P+Q}$ defines a valuation (see, e.g., [16]) where $P + Q = \{p + q : p \in P, q \in Q\}$ is the **Minkowski sum**. The family of all polytopal simple functions forms an algebra where the multiplicative structure is given by the Minkowski sum of polytopes: $\mathbf{1}_P \star \mathbf{1}_Q := \mathbf{1}_{P+Q}$ for all polytopes P and Q. Another proof of Theorem 2 can be obtained from the following result which was proved in [11]. See also [14] for related material.

Theorem 3 ([11, Lemma 5]) *Let P be a polytope with vertex set* $\mathrm{vert}(P) = \{v_1, \ldots, v_r\}$. *Then*

$$\left(\mathbf{1}_P - \mathbf{1}_{v_1}\right) \star \cdots \star \left(\mathbf{1}_P - \mathbf{1}_{v_r}\right) = 0. \tag{10.2}$$

Proof (2nd proof of Theorem 2) The proof follows from Theorem 3 by expanding Eq. (10.2) and multiplying both sides with $\mathbf{1}_{kP}$. □

By the discussion above, Theorem 2 is equivalent to the following.

Theorem 4 *Let P be a polytope in $\mathbb{R}^n$ with vertex set* $\mathrm{vert}(P) = \{v_1, \ldots, v_r\}$, *and let $\varphi\colon \mathcal{P} \to G$ be a valuation. Then*

$$\varphi((k+r)P) = \sum_{\emptyset \neq I \subseteq [r]} (-1)^{1+|I|} \varphi(Q_I^{k,r})$$

for all $k \geq 0$ where $Q_I^{k,r} = (k + r - |I|)P + \sum_{i \in I} v_i$.

10.3 A Multivariate Recursion

A sequence $\mathbf{a} = \{a_k\}_{k \geq 0}$ of elements in $\mathbb{C}(x_1, \ldots, x_n)$ satisfies a **linear recursion** of order $d \geq 1$ if there are $c_1, \ldots, c_d \in \mathbb{C}(x_1, \ldots, x_n)$, $c_d \neq 0$, such that

$$a_k = \sum_{j=1}^{d} c_j a_{k-j}$$

for all $k \geq d$. The corresponding **characteristic polynomial** $\chi_{\mathbf{c}}$ is defined as $X^d - \sum_{j=1}^{d} c_j X^{d-j} \in \mathbb{C}(x_1, \ldots, x_n)[X]$. The polynomial $\chi_{\mathbf{c}}$ is called **minimal** if for every vector $\mathbf{c}' = (c_1', \ldots, c_{d'}')$ corresponding to a linear recursion of $\mathbf{a}$ we have $\chi_{\mathbf{c}} | \chi_{\mathbf{c}'}$. Since $\mathbb{C}(x_1, \ldots, x_n)[X]$ is a principal ideal domain a uniquely determined minimal polynomial exists.

We are now ready to proof Theorem 1.

Proof (Proof of Theorem 1) Let $r = |\mathrm{vert}(P)|$ be the number of vertices of P. Since the maps $P \mapsto P + Q$ and also $P \mapsto \sigma_P(x)$ define valuations, by Theorem 4

$$\begin{aligned}\sigma_{(k+r)P+Q}(\mathbf{x}) &= \sum_{\emptyset \neq I \subseteq [r]} (-1)^{1+|I|} \sigma_{(k+r-|I|)P + \sum_{i \in I} v_i + Q}(\mathbf{x}) \\ &= \sum_{\emptyset \neq I \subseteq [r]} (-1)^{1+|I|} \mathbf{x}^{\sum_{i \in I} v_i} \sigma_{(k+r-|I|)P+Q}(\mathbf{x}),\end{aligned}$$

where the last equation follows by observing that $\sigma_{P+v}(\mathbf{x}) = \mathbf{x}^v \sigma_P(\mathbf{x})$ for all $v \in \mathbb{Z}^n$. We observe that $\chi_{P;Q}$ is the characteristic polynomial of this linear recursion.

Now let Q be a lattice polytope and suppose that $\chi_{P;Q}$ is not minimal. Then, for some vertex $\mathbf{u}$ of P, $\{\sigma_{kP+Q}(\mathbf{x})\}_{k \geq 0}$ satisfies a linear recursion with characteristic polynomial $\prod_{\mathbf{v} \in \mathrm{vert}(P) \setminus \{\mathbf{u}\}} (X - \mathbf{x}^{\mathbf{v}})$. That is

$$\sigma_{(k+r)P+Q}(\mathbf{x}) + \sum_{j=1}^{|\text{vert}(P)|-1} (-1)^j e_j(\{\mathbf{x}^{\mathbf{v}} : \mathbf{v} \in \text{vert}(P) \setminus \{\mathbf{u}\}\})\sigma_{(k+r-j)P+Q}(\mathbf{x}) = 0$$

where e_j denotes the jth elementary symmetric polynomial in $|\text{vert}(P)| - 1$ variables. Now let $\mathbf{v}$ be a vertex of Q such that $\mathbf{u} + \mathbf{v}$ is a vertex of $P + Q$. Then $(k + r)\mathbf{u} + \mathbf{v}$ is a vertex of $(k + r)P + Q$ and thus $x^{(k+r)\mathbf{u}+\mathbf{v}}$ appears as a summand in $\sigma_{(k+r)P+Q}(x)$. However, it does not appear in $e_j(\{\mathbf{x}^{\mathbf{v}} : \mathbf{v} \in \text{vert}(P) \setminus \{\mathbf{u}\}\})\sigma_{(k+r-j)P+Q}(x)$ for any $1 \leq j \leq |\text{vert}(P)| - 1$. To see that, it suffices to argue that $(k + r)\mathbf{u} + \mathbf{v}$ is not contained in $(k + r - j)P + Q + \sum_{l=1}^{j} v_l$ for any choice of $v_1, \ldots, v_j \in \text{vert}(P) \setminus \{\mathbf{u}\}$. For that, let $\ell : \mathbb{R}^n \to \mathbb{R}$ be a linear functional such that $\ell(\mathbf{u}) > \ell(p)$ for all $p \neq \mathbf{u}$ in P and $\ell(\mathbf{v}) > \ell(q)$ for all $q \neq \mathbf{v}$ in Q. Then $\ell((k + r - j)p + q + \sum_{l=1}^{j} v_l) = (k + r - j)\ell(p) + \ell(q) + \sum_{l=1}^{j} \ell(v_l) < (k + r - j)\ell(\mathbf{u})$
$+ \ell(\mathbf{v}) + j\ell(\mathbf{u}) = \ell((k + r)\mathbf{u} + \mathbf{v})$ for all $p \in P$ and $q \in Q$ and the conclusion follows. □

Every linear map $f : \mathbb{R}^n \to \mathbb{R}^l$ with the property that $f(\mathbb{Z}^n) \subseteq \mathbb{Z}^l$ induces an algebra homomorphism

$$\begin{aligned} \bar{f} : \mathbb{C}\left[x_1^{\pm 1}, \ldots, x_n^{\pm 1}\right] &\to \mathbb{C}\left[x_1^{\pm 1}, \ldots, x_l^{\pm 1}\right] \\ \mathbf{x}^{\mathbf{m}} &\mapsto \mathbf{x}^{f(\mathbf{m})} \end{aligned}$$

As a consequence of Theorem 1 we therefore obtain the following.

Proposition 5 *Let Q be a polytope in $\mathbb{R}^n$ and P be a lattice polytope with vertex set* $\text{vert}(P)$, *and let $f : \mathbb{R}^n \to \mathbb{R}^l$ a linear map such that $f(\mathbb{Z}^n) \subseteq \mathbb{Z}^l$. Then $\{\bar{f}(\sigma_{kP+Q}(\mathbf{x}))\}_{k \geq 0}$ satisfies a linear recursion with characteristic polynomial*

$$\chi_{P;Q}^{f}(X) := \prod_{\mathbf{v} \in \text{vert}(P)} (X - x^{f(\mathbf{v})}).$$

The following two examples show that the minimality of a characteristic polynomial is not necessarily preserved under affine transformations or taking Minkowski sums.

Example 6 If Q in Theorem 1 is not a lattice polytope then $\chi_{P;Q}$ is not necessarily minimal. A counterexample is given by the lattice segment $P = [0, 1]$ and the point $Q = \{(0.5, 0.5)\}$ in $\mathbb{R}^2$. In that case $\sigma_{kP+Q} \equiv 0$ is constant.

Example 7 (*Ehrhart polynomials*) For $f : \mathbb{R}^n \to \mathbb{R}^0$ and $f \equiv 0$ we obtain $\bar{f}(\sigma_{kP}(\mathbf{x})) = |kP \cap \mathbb{Z}^n|$ and thus recover the Ehrhart function counting lattice points in integer dilates of P. If P is a lattice polytope then this function is known to agree with a polynomial of degree dim P [7]. Therefore the order of the minimal polynomial of the sequence is dim P as was demonstrated in [15] and is thus in general smaller than $|\text{vert}(P)|$.

These examples motivate the following question.

Question 8 What are necessary and sufficient conditions on Q and on f that guarantee that $\chi^f_{P;Q}$ is minimal?

10.4 Brion's Theorem

In this section we provide a proof of Brion's Theorem using the recursion given in Theorem 1. For a polytope $P \subseteq \mathbb{R}^n$ and a vertex $\mathbf{v}$ of P the **tangent cone** $\mathcal{K}_\mathbf{v}$ is defined as $\{\mathbf{v}+\mathbf{w} \colon \mathbf{v}+\varepsilon\mathbf{w} \in P \text{ for } 0<\varepsilon \ll 1\}$. If the polytope P has rational edge directions, in particular, if it is a lattice polytope, then the integer point transform of $\mathcal{K}_\mathbf{v}$ is a rational function.

Theorem 9 (Brion's Theorem [6]) *Let P be a lattice polytope with vertex set* vert(P). *Then*

$$\sigma_P(\mathbf{x}) = \sum_{\mathbf{v}\in \mathrm{vert}(P)} \sigma_{\mathcal{K}_\mathbf{v}}(\mathbf{x})$$

as rational functions.

The following is an immediate consequence of [5, Lemma 13.5.]

Lemma 10 ([5]) *Let $u_1,\ldots,u_k \in \mathbb{Z}^n$ such that the cone $\mathcal{K} :=$* cone$(u_1,\ldots,u_k)$ *generated by $u_1,\ldots,u_k$ is pointed. Then*

$$\sigma_{\mathcal{K}}(\mathbf{x}) = \sum_{\mathbf{m}\in\mathcal{K}\cap\mathbb{Z}^n} \mathbf{x}^\mathbf{m}$$

is a rational function and converges absolutely for all $\mathbf{x}$ in $\{\mathbf{x}\in\mathbb{C}^n \colon |\mathbf{x}^{u_i}|<1 \text{ for } i=1,\ldots,k\}$.

A further ingredient for our proof of Brion's Theorem is the following well-known result (see, e.g., [17, Chap. 5]).

Lemma 11 *Let $\{a_n\}_{n\in\mathbb{N}}$ be a sequence of elements of a field K that satisfy a linear recursion of order d with characteristic polynomial*

$$\prod_{i=1}^{d}(X-r_i).$$

If all roots $r_1,\ldots,r_d$ are distinct then there are $\alpha_1,\ldots,\alpha_d \in K$ such that

$$a_n = \sum_{i=1}^{d} \alpha_i r_i^n, \qquad \text{for all } n\in\mathbb{N}.$$

Proof (Proof of Theorem 9) By Theorem 1 and Lemma 11 there exist $c_{\mathbf{v}} \in \mathbb{C}(x_1, \ldots, x_n)$ for all $\mathbf{v} \in \operatorname{vert}(P)$ such

$$\sigma_{kP}(\mathbf{x}) = \sum_{\mathbf{v} \in \operatorname{vert}(P)} c_{\mathbf{v}} \mathbf{x}^{k\mathbf{v}} \tag{10.3}$$

for all $k \geq 0$. Our goal is to show that $c_{\mathbf{w}} \cdot \mathbf{x}^{\mathbf{w}} = \sigma_{\mathcal{K}_{\mathbf{w}}}(\mathbf{x})$ as rational functions for all $\mathbf{w} \in \operatorname{vert}(P)$, or, equivalently, that $c_{\mathbf{w}}$ equals the integer point transform of the tangent cone $\tilde{\mathcal{K}}_0$ of the vertex 0 of the translated polytope $P - \mathbf{w}$. Equation (10.3) is equivalent to $\sigma_{k(P-\mathbf{w})}(\mathbf{x}) = \sum_{\mathbf{v} \in \operatorname{vert}(P)} c_{\mathbf{v}} \mathbf{x}^{k(\mathbf{v}-\mathbf{w})}$. As k goes to infinity $\sigma_{k(P-\mathbf{w})}(\mathbf{x})$ converges absolutely to $\sigma_{\tilde{\mathcal{K}}_0}(\mathbf{x})$ on $W_{\tilde{\mathcal{K}}_0} = \{\mathbf{x} \in \mathbb{C}^n \colon\ |\mathbf{x}^{\mathbf{v}-\mathbf{w}}| < 1 \text{ for all } \mathbf{v} \in \operatorname{vert}(P) \setminus \{\mathbf{w}\}\}$ by Lemma 10. On the other hand, $\sum_{\mathbf{v} \in \operatorname{vert}(P)} c_{\mathbf{v}} \mathbf{x}^{k(\mathbf{v}-\mathbf{w})}$ converges to $c_{\mathbf{w}}$. Thus $\sigma_{\tilde{\mathcal{K}}_0}(x)$ and $c_{\mathbf{w}}$ coincide on $W_{\tilde{\mathcal{K}}_0}$ and are therefore the same as rational functions. □

10.5 Schur Polynomials

In this section we apply our results to Schur polynomials.

A **partition** is a vector $\boldsymbol{\lambda} = (\lambda_1 \geq \lambda_2 \geq \cdots \geq \lambda_n)$ of weakly decreasing nonnegative integers. The number of strictly positive entries λ_i is called the **length** of $\boldsymbol{\lambda}$. A partition $\boldsymbol{\mu}$ is smaller than a partition $\boldsymbol{\lambda}$ with respect to the **inclusion order** if $\mu_i \leq \lambda_i$ for all i. The partition $\boldsymbol{\mu}$ is smaller than a partition $\boldsymbol{\lambda}$ with respect to the **domination order**, denoted $\boldsymbol{\lambda} \trianglerighteq \boldsymbol{\mu}$, if $\sum_{i=1}^{n} \lambda_i = \sum_{i=1}^{n} \mu_i$ and $\sum_{i=1}^{k} \lambda_i \geq \sum_{i=1}^{k} \mu_i$ for all k. A **skew Young diagram** of shape $\boldsymbol{\lambda}/\boldsymbol{\mu}$ is an axes-parallel arrangement of unit squares in the plane centered at the coordinates $\left\{(i, j) \in \mathbb{Z}^2 \colon \mu_i < j \leq \lambda_i\right\}$. A **semi-standard Young tableau** is a Young diagram together with a filling of the boxes with natural numbers such that the numbers are strictly increasing in each column and weakly increasing in every row. Let $\mathbb{T}^n_{\boldsymbol{\lambda}/\boldsymbol{\mu}}$ denote the set of semi-standard Young tableaux filled with numbers in $[n] = \{1, 2, \ldots, n\}$. For every T in $\mathbb{T}^n_{\boldsymbol{\lambda}/\boldsymbol{\mu}}$ let $w(T)$ be the vector $\mathbf{t} = (t_1, \ldots, t_n)$ where t_i is the number of boxes filled with i. The vector $w(T)$ is called the **weight** of T. The **Kostka coefficient** $K_{\boldsymbol{\lambda}/\boldsymbol{\mu},\mathbf{w}}$ equals the number of tableaux of shape $\boldsymbol{\lambda}/\boldsymbol{\mu}$ with weight $\mathbf{w}$. In particular, $K_{\boldsymbol{\lambda}/\boldsymbol{\mu},\mathbf{w}} > 0$ if and only if there is $T \in \mathbb{T}^n_{\boldsymbol{\lambda}/\boldsymbol{\mu}}$ with $w(T) = \mathbf{w}$. The **skew Schur polynomial** of shape $\boldsymbol{\lambda}/\boldsymbol{\mu}$ is defined as

$$s_{\boldsymbol{\lambda}/\boldsymbol{\mu}}(\mathbf{x}) = \sum_{T \in \mathbb{T}^n_{\boldsymbol{\lambda}/\boldsymbol{\mu}}} \mathbf{x}^{w(T)} \in \mathbb{C}[x_1^{\pm 1}, \ldots, x_n^{\pm 1}].$$

In [1] Alexandersson proved the following recursion for Schur polynomials.

Theorem 12 ([1, Theorem 1]) *Let n be a natural number and let $\boldsymbol{\kappa}, \boldsymbol{\lambda}, \boldsymbol{\mu}, \boldsymbol{\nu}$ be partitions of length at most n such that $\boldsymbol{\lambda} \supseteq \boldsymbol{\mu}$ and $\boldsymbol{\kappa} + k\boldsymbol{\lambda} \supseteq \boldsymbol{\nu} + k\boldsymbol{\mu}$ for some positive integer k. Then there is a natural number r such that the sequence $\{s_{\boldsymbol{\kappa}+l\boldsymbol{\lambda}/\boldsymbol{\nu}+l\boldsymbol{\mu}}(\mathbf{x})\}_{l=r}^{\infty}$ satisfies a linear recursion with characteristic polynomial*

$$\begin{array}{ccccccc} & & & x_{1,1} & & x_{1,2} \;\cdots\; x_{1,n} \\ & & x_{2,1} & & x_{2,2} \;\cdots\; x_{2,n} & \\ & \cdot^{\cdot^{\cdot}} & & \cdot^{\cdot^{\cdot}} & \cdot^{\cdot^{\cdot}} & \\ x_{n+1,1} & & x_{n+1,2} \;\cdots\; x_{n+1,n} & & & \end{array} \qquad \begin{array}{ccccccc} & & & 1 & 3 & 5 \\ & & 1 & 3 & 5 & \\ & 1 & 2 & 4 & & \\ 1 & 1 & 4 & & & \end{array}$$

Fig. 10.1 Gelfand–Tsetlin patterns

$$\chi(X) = \prod_{T \in \mathbb{T}^n_{\lambda/\mu}} \left(X - \mathbf{x}^{w(T)}\right).$$

Furthermore, in [1] the following conjecture concerning the minimal polynomial was stated. For every vector $\mathbf{w}$ let $\overline{\mathbf{w}}$ denote the vector obtained from $\mathbf{w}$ by rearranging its coordinates in non-increasing order.

Conjecture 13 ([1, Conjecture 25]) Let $\boldsymbol{\kappa}, \boldsymbol{\lambda}, \boldsymbol{\mu}, \boldsymbol{\nu}$ be as in Theorem 12 and let

$$W = \{\mathbf{w} \in \mathbb{N}^n : K_{\boldsymbol{\lambda}/\boldsymbol{\mu},\mathbf{w}} > 0 \text{ and } \overline{\mathbf{w}} \trianglerighteq \overline{\boldsymbol{\lambda} - \boldsymbol{\mu}}\}.$$

Then, for sufficiently large r, $\{s_{\kappa+l\mu/\lambda+l\nu}(\mathbf{x})\}_{l=r}^{\infty}$ satisfies a linear recursion with minimal polynomial

$$\chi(X) = \prod_{\mathbf{w} \in W} \left(X - \mathbf{x}^{\mathbf{w}}\right).$$

We use Theorem 1 and a well-known correspondence between elements in $\mathbb{T}^n_{\lambda/\mu}$ and lattice points in the Gelfand–Tsetlin-Polytope $\mathbf{GL}_{\boldsymbol{\lambda}/\boldsymbol{\mu}}$ to improve Theorem 12 and to given an example in which the polynomial in Conjecture 13 is not minimal thus refuting the conjecture.

There is a one-to-one correspondence between semi-standard Young tableaux and Gelfand–Tsetlin patterns. A **Gelfand–Tsetlin pattern** is a rectangular array of nonnegative real numbers $\{x_{i,j}\}_{\substack{i=1,\ldots,n+1 \\ j=1,\ldots,n}}$ arranged as in Fig. 10.1 such that the entries are weakly increasing in north-east and south-east direction, that is $x_{i,j} \leq x_{i+1,j+1}$ for all i, j and $x_{i,j} \leq x_{l,j}$ for all $i > l$. For fixed top and bottom rows the family of Gelfand–Tsetlin patterns forms a polytope, the **Gelfand–Tsetlin polytope**, which belongs to the class of marked order polytopes introduced by Ardila, Bliem and Salazar [3]. There is a well-known one-to-one correspondence between elements of $\mathbb{T}^n_{\lambda/\mu}$ and integer valued Gelfand–Tsetlin patterns with top row $\boldsymbol{\lambda}$ and bottom row $\boldsymbol{\mu}$, that is, lattice points in the corresponding Gelfand–Tsetlin polytope $\mathbf{GL}_{\boldsymbol{\lambda}/\boldsymbol{\mu}}$. Via this correspondence, the weight function can be represented as a linear function on $\mathbf{GL}_{\boldsymbol{\lambda}/\boldsymbol{\mu}}$, namely for every Gelfand–Tsetlin pattern $\mathbf{x} = \{x_{i,j}\}$ the ith coordinate of the weight $w(\mathbf{x})$ equals $\sum_{k=1}^{n}(x_{i,k} - x_{i+1,k})$ for all $1 \leq i \leq n$. Further details may be found in [18]. It follows that

$$s_{\boldsymbol{\lambda}/\boldsymbol{\mu}}(\mathbf{x}) = \sum_{p} \mathbf{x}^{w(p)},$$

where p is over all lattice points in $\mathbf{GL}_{\boldsymbol{\lambda}/\boldsymbol{\mu}}$.

As a corollary of Theorem 1 we obtain the following.

Corollary 14 *Let n be a natural number and let $\boldsymbol{\kappa}, \boldsymbol{\lambda}, \boldsymbol{\mu}, \boldsymbol{\nu}$ be partitions of length at most n such that $\boldsymbol{\lambda} \supseteq \boldsymbol{\mu}$ and $\boldsymbol{\kappa} + k\boldsymbol{\lambda} \supseteq \boldsymbol{\nu} + k\boldsymbol{\mu}$ for some positive integer k. Let V be the set of vertices of $\mathbf{GL}_{\boldsymbol{\lambda}/\boldsymbol{\mu}}$. Then there is an integer $r \gg 0$ such that $\{s_{\boldsymbol{\kappa}+l\boldsymbol{\lambda}/\boldsymbol{\nu}+l\boldsymbol{\mu}}(\mathbf{x})\}_{l=r}^{\infty}$ satisfies a linear recursion with characteristic polynomial*

$$\chi(X) = \prod_{\mathbf{v} \in V} \left(X - \mathbf{x}^{w(\mathbf{v})}\right).$$

Proof Let $\mathbf{f} = (\boldsymbol{\lambda}, \boldsymbol{\mu})$ and $\mathbf{g} = (\boldsymbol{\kappa}, \boldsymbol{\nu})$. Then there is an $r \gg 0$ such that if $f_i < f_j$ then $rf_i + g_i < rf_j + g_j$ for all $i \neq j$. In particular, one can find a permutation $\sigma \in S_{2n}$ such that

$$f_{\sigma(1)} \leq f_{\sigma(2)} \leq \cdots \leq f_{\sigma(2n)} \quad \text{and} \quad rf_{\sigma(1)} + g_{\sigma(1)} \leq \cdots \leq rf_{\sigma(2n)} + g_{\sigma(2n)}.$$

Then, by Theorem [8, Theorem 2.10],

$$\mathbf{GL}_{\boldsymbol{\kappa}+l\boldsymbol{\lambda}/\boldsymbol{\nu}+l\boldsymbol{\mu}} = \mathbf{GL}_{\boldsymbol{\kappa}+r\boldsymbol{\lambda}/\boldsymbol{\nu}+r\boldsymbol{\mu}} + (l-r)\mathbf{GL}_{\boldsymbol{\lambda}/\boldsymbol{\mu}}$$

for all $l \geq r$. The claim now follows from Proposition 5 since the weight function w is linear. □

Since typically there are more lattice points in $\mathbf{GL}_{\boldsymbol{\lambda}/\boldsymbol{\mu}}$ than vertices, Corollary 14 shows that the characteristic polynomial given in Theorem 12 is in general not minimal. The next example shows that also the polynomial given in Conjecture 13 is not minimal in general, thus refuting it.

Example 15 Let $n = 3$, $\boldsymbol{\lambda} = (5, 3, 1)$ and $\boldsymbol{\mu} = (3, 0, 0)$. Consider the skew Young tableau T and its corresponding Gelfand–Tsetlin pattern p depicted in Fig. 10.2. Then

$$w(T) = w(p) = (4, 2, 0) \trianglerighteq (3, 2, 1) = \overline{\boldsymbol{\lambda} - \boldsymbol{\mu}}.$$

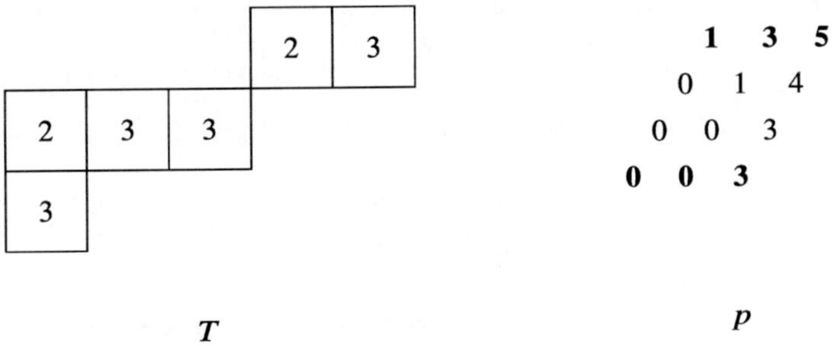

Fig. 10.2 The skew Young tableau T in Example 15 and its corresponding Gelfand–Tsetlin pattern p

From the face structure studied in [10, 13] it follows that the coordinates of any vertex of $\mathbf{GL}_{\lambda/\mu}$ are in the set $\{0, 1, 3, 5\}$. Let $\mathbf{x} = \{x_{i,j}\}$ be a Gelfand–Tsetlin pattern that is a vertex of $\mathbf{GL}_{\lambda/\mu}$. Then $x_{4,1}, x_{4,2}$ and $x_{3,1}$ are 0. Furthermore, $x_{2,1} \in \{0, 1\}$. If $x_{2,1} = 0$, then the sum of entries of the first row of $\mathbf{x}$ is odd and the sum of entries of the second is even, therefore $w(\mathbf{x})_1$ is odd and $w(\mathbf{x}) \neq (4, 2, 0)$. On the other hand, if $x_{2,1} = 1$, then $x_{3,2} \in \{1, 3\}$ and in that case $w(\mathbf{x})_2$ is odd and again $w(\mathbf{x}) \neq (4, 2, 0)$. In summary, $(4, 2, 0) \in W$ is not the weight of a vertex of $\mathbf{GL}_{\lambda/\mu}$ and therefore

$$\prod_{\mathbf{w}\in W} (X - \mathbf{x}^{\mathbf{w}}) \nmid \prod_{\mathbf{v}\in V} (X - \mathbf{x}^{w(\mathbf{v})}) .$$

Therefore, by Corollary 14, $\prod_{\mathbf{w}\in W} (X - \mathbf{x}^{\mathbf{w}})$ cannot be the minimal polynomial.

Remark 16 To verify the counterexample given in Example 15, one can also use [2, Proposition 6].

Acknowledgements The author would like to thank Per Alexandersson and Raman Sanyal for inspiring and fruitful discussions and many helpful comments. The author was partially supported by a Hilda Geiringer Scholarship at the Berlin Mathematical School, the Knut and Alice Wallenberg Foundation and a Microsoft Research Fellowship of the Simons Institute for the Theory of Computing.

References

1. Alexandersson, P.: Stretched skew Schur polynomials are recurrent. J. Combin. Theory Ser. A **122**, 1–8 (2014)
2. Alexandersson, P.: Polynomials defined by tableaux and linear recurrences. Electron. J. Combin. **23**(1), Paper 1.47, 24 (2016)
3. Ardila, F., Bliem, T., Salazar, D.: Gelfand-Tsetlin polytopes and Feigin-Fourier-Littelmann-Vinberg polytopes as marked poset polytopes. J. Combin. Theory Ser. A **118**(8), 2454–2462 (2011)
4. Barvinok, A.: A course in convexity, Graduate Studies in Mathematics, vol. 54. American Mathematical Society, Providence (2002)
5. Barvinok, A.: Integer points in polyhedra. Zurich Lectures in Advanced Mathematics. European Mathematical Society (EMS), Zürich (2008)
6. Brion, M.: Points entiers dans les polyèdres convexes. Ann. Sci. École Norm. Sup. (4) **21**(4), 653–663 (1988)
7. Ehrhart, E.: Sur les polyèdres rationnels homothétiques à n dimensions. C. R. Acad. Sci. Paris **254**, 616–618 (1962)
8. Fang, X., Fourier, G., Pegel, C.: The Minkowski property and reflexivity of marked poset polytopes. Electron. J. Combin. **27**(1), Paper No. 1.27, 19 (2020)
9. Groemer, H.: On the extension of additive functionals on classes of convex sets. Pacific J. Math. **75**(2), 397–410 (1978)
10. Jochemko, K., Sanyal, R.: Arithmetic of marked order polytopes, monotone triangle reciprocity, and partial colorings. SIAM J. Discrete Math. **28**(3), 1540–1558 (2014)
11. Lawrence, J.: Three rings of polyhedral simple functions. J. Res. Natl. Inst. Stand. Technol. **111**(2), 127–134 (2006)
12. McMullen, P.: Valuations on lattice polytopes. Adv. Math. **220**(1), 303–323 (2009)

13. Pegel, C.: The face structure and geometry of marked order polyhedra. Order **35**(3), 467–488 (2018)
14. Pukhlikov, A.V., Khovanskiĭ, A.G.: Finitely additive measures of virtual polyhedra. Algebra i Analiz **4**(2), 161–185 (1992)
15. Sam, S.V.: A bijective proof for a theorem of Ehrhart. Amer. Math. Monthly **116**(8), 688–701 (2009)
16. Schneider, R.: Convex bodies: the Brunn-Minkowski theory. Encyclopedia of Mathematics and its Applications, vol. 151, expanded edn. Cambridge University Press, Cambridge (2014)
17. Spiegel, M.R.: Calculus of Finite Differences and Difference Equations. Schaum's Outline Series. McGraw-Hill, New York (1971)
18. Stanley, R.P.: Enumerative combinatorics, vol. 2. Cambridge Studies in Advanced Mathematics, vol. 62. Cambridge University Press, Cambridge (1999). With a foreword by Gian-Carlo Rota and appendix 1 by Sergey Fomin
19. Volland, W.: Ein Fortsetzungssatz für additive Eipolyederfunktionale im euklidischen Raum. Arch. Math. (Basel) **8**, 144–149 (1957)

Chapter 11
Schubert Calculus on Newton–Okounkov Polytopes

Valentina Kiritchenko and Maria Padalko

Abstract A Newton–Okounkov polytope of a complete flag variety can be turned into a convex geometric model for Schubert calculus. Namely, we can represent Schubert cycles by linear combinations of faces of the polytope so that the intersection product of cycles corresponds to the set-theoretic intersection of faces (whenever the latter are transverse). We explain the general framework and survey particular realizations of this approach in types A, B and C.

Keywords Schubert calculus · Mitosis · Gelfand-Zetlin polytopes

11.1 Introduction

Theory of Newton–Okounkov convex bodies [12, 19] allows us to apply ideas of toric geometry in the non-toric setting. In this paper, we explore non-toric applications of *polytope rings* (see Sect. 11.2 for a definition) introduced by Khovanskii and Pukhlikov [25]. With a convex polytope $P \subset \mathbb{R}^d$, they associated a graded commutative ring (the polytope ring):

$$R_P = \bigoplus_{i=0}^{d} R_P^i$$

that has Poincaré duality. The polytope rings were originally used to give a convenient functorial description of the cohomology rings of smooth toric varieties. In this case, P is always a simple lattice polytope, that is, all vertices of P belong to $\mathbb{Z}^d \subset \mathbb{R}^d$,

V. Kiritchenko (✉)
Laboratory of Algebraic Geometry and Faculty of Mathematics, National Research University Higher School of Economics, Usacheva str. 6, 119048 Moscow, Russia

Institute for Information Transmission Problems, Moscow, Russia
e-mail: vkiritch@hse.ru

M. Padalko
SimilarWeb, Tel Aviv, Israel

A. M. Kasprzyk and B. Nill (eds.), *Interactions with Lattice Polytopes*, Springer Proceedings in Mathematics & Statistics 386,
https://doi.org/10.1007/978-3-030-98327-7_11

and only d edges meet at every vertex of P. In [10], Kaveh noted that polytope rings can also be used for a partial description of the cohomology rings of spherical varieties. In this case, P is still a lattice polytope but not necessarily simple.

For simple polytopes, every face $\Gamma \subset P$ can be naturally identified with an element $x_\Gamma \in R_P$ so that

$$x_\Gamma x_{\Gamma'} = x_{\Gamma \cap \Gamma'}$$

for any two transverse faces Γ and Γ'. This is no longer true for non-simple polytopes, that is, individual faces of P do not have natural counterparts in R_P. However, it is still possible to identify every element of R_P with a linear combination of faces of P so that the product in the polytope ring corresponds to the intersection of faces. In [14], the first author, Smirnov and Timorin developed a general framework for such calculus on polytopes, and studied its applications to Schubert calculus on Gelfand–Zetlin polytopes in type A. In this paper, we mainly consider applications to Schubert calculus in types B and C.

Representation theory of classical groups is a source of several interesting families of lattice convex polytopes. For $SL_n(\mathbb{C})$ (type A), there is a well-known family of Gelfand–Zetlin (GZ) polytopes GZ_λ. Here $\lambda := (\lambda_1, \ldots, \lambda_n) \in \mathbb{Z}^n$ runs through dominant weights of $SL_n(\mathbb{C})$, that is, $\lambda_1 \geq \lambda_2 \geq \cdots \geq \lambda_n$. Originally, GZ polytopes were constructed using representation theory, namely, lattice points in the polytope GZ_λ parameterize the vectors in a special basis in the irreducible representation V_λ of $SL_n(\mathbb{C})$ with the highest weight λ (see [21] for a survey on GZ bases). In convex geometric terms, the GZ polytope $GZ_\lambda \subset \mathbb{R}^d$, where $d := \frac{n(n-1)}{2}$, is defined as the set of all points $(z_1^1, z_2^1, \ldots, z_{n-1}^1; z_1^2, \ldots, z_{n-2}^2; \ldots; z_1^{n-1}) \in \mathbb{R}^d$ that satisfy the following interlacing inequalities:

$$\begin{array}{ccccccccccc}
\lambda_1 & & \lambda_2 & & \lambda_3 & & \ldots & & & & \lambda_n \\
& z_1^1 & & z_2^1 & & \ldots & & & & z_{n-1}^1 & \\
& & z_1^2 & & \ldots & & & & z_{n-2}^2 & & \\
& & & \ddots & & \ddots & & & & & \\
& & & & z_1^{n-2} & & z_2^{n-2} & & & & \\
& & & & & z_1^{n-1} & & & & &
\end{array} \qquad (GZ_A)$$

where the notation

$$\begin{array}{ccc} a & & b \\ & c & \end{array}$$

means $a \geq c \geq b$ (the table encodes $2d$ inequalities). Figure 11.1 shows the 3-dimensional GZ polytope for $n = 3$ and $\lambda = (3, 0, -3)$. Note that GZ polytopes are not simple.

GZ polytopes in types B, C and D were defined in [1] (see Sect. 11.2.2 for definitions in types B and C) and are related to representation theory of $SO_{2n+1}(\mathbb{C})$, Sp_{2n} and $SO_{2n}(\mathbb{C})$, respectively. They are special cases of *string polytopes* introduced by Berenstein–Zelevinsky and Littelmann [20]. There are other families of poly-

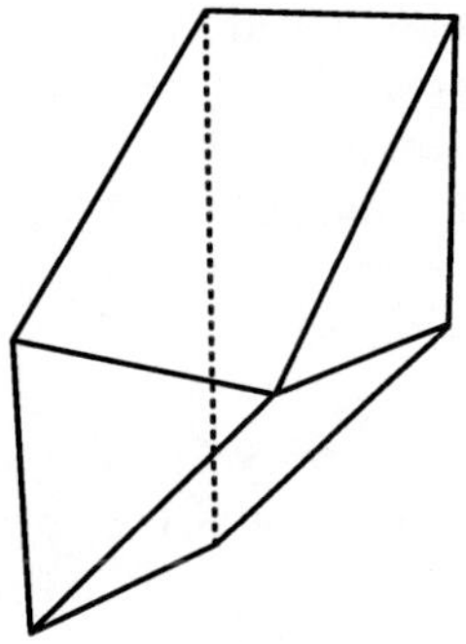

Fig. 11.1 GZ polytope in type A for $n = 3$ and $\lambda = (3, 0, -3)$

topes in representation theory such as Nakashima–Zelevinsky polyhedral realizations of crystal bases and Feigin–Fourier–Littelmann–Vinberg polytopes. They have representation-theoretic meaning similar to that of string polytopes but are not combinatorially equivalent to the latter. All these polytopes were exhibited as Newton–Okounkov polytopes of complete flag varieties for certain geometric valuations [3, 4, 6, 11, 17] (see Sect. 11.2.3 for more details).

For $G = SL_n(\mathbb{C})$, the complete flag variety G/B (here $B \subset G$ denotes the subgroup of upper-triangular matrices) can be thought of as a variety of complete flags of subspaces $(\{0\} \subset V^1 \subset V^2 \subset \cdots \subset V^{n-1} \subset \mathbb{C}^n)$ where $\dim V^i = i$, and there are no gaps. There are similar descriptions of complete flag varieties G/B for other classical groups G (see Sect. 11.2.3). Recall that globally generated line bundles L_λ on G/B are in bijective correspondence with irreducible representations V_λ of G so that $H^0(L_\lambda, G/B) \simeq V_\lambda^*$ [2, Proposition 1.4.5]. Here λ runs through the dominant weights of G. We denote by $\deg_\lambda(G/B)$ the degree of the image of G/B under the map $G/B \to \mathbb{P}(V_\lambda) = \mathbb{P}(H^0(L_\lambda, G/B)^*)$.

In [10], polytope rings of string polytopes were identified with the cohomology rings of complete flag varieties. More generally, string polytope in this description can be replaced with any linear family (in the sense of [13]) of convex polytopes P_λ parameterized by the dominant weights λ whenever the following identity holds:

$$\mathrm{Vol}(P_\lambda) = d!\,\deg_\lambda(G/B) \tag{11.1}$$

where $d := \dim G/B$. We regard both sides of this identity as polynomials in λ. In particular, polytopes P_λ yield an analog of Kushnirenko's theorem for G/B.

Since Newton–Okounkov polytopes of line bundles on G/B by construction satisfy identity (11.1) they can be used to model Schubert calculus. Recall that the cohomology ring $H^*(G/B, \mathbb{Z})$ has a special basis of *Schubert cycles* $[X_w]$ with striking positivity properties. Namely, the structure constants (i.e., the coefficients c_{vw}^u in the decomposition $[X_w][X_v] = \sum_u c_{vw}^u [X_u]$) are always non-negative. However, no enumerative meaning (in the spirit of Littlewood–Richardson rule for Grassmannians) of these coefficients is known. Polytope rings provide a new framework for combinatorial interpretation of structure constants. An important task is to find presentations of Schubert cycles in polytope rings by linear combinations of faces with positive coefficients. Another task is to find Newton–Okounkov polytopes for which these presentations have especially simple combinatorics. It is tempting to use Grossberg–Karshon cubes [7, 8] since they are combinatorial cubes. However, there are might be issues with positivity, that is, some Schubert cycles will be represented by linear combinations of faces with negative coefficients (see Example 10).

There is an algorithm (geometric mitosis) for finding positive presentations of Schubert cycles by faces using convex geometric analogs of Demazure operators from representation theory [15, 16]. In the present paper, we describe geometric mitosis in more combinatorial terms, outline its applications and formulate conjectures. For GZ polytopes in type A, this algorithm reduces to Knutson–Miller mitosis on pipe dreams and was used in [14]. In types B and C, geometric mitosis reduces to a different combinatorial rule that conjecturally yields presentations of Schubert cycles by faces of GZ polytopes in respective types. In particular, 4-dimensional GZ polytope in type C_2 can be used to model Schubert calculus on the variety of isotropic flags in $\mathbb{C}^4$ [24]. Another convex geometric model for the same flag variety was constructed in [9] using a different string polytope in type C_2.

11.2 Preliminaries

In this section, we recall the definitions of polytope rings, GZ polytopes and flag varieties in types B and C. We discuss the relationship between the polytope rings of GZ polytopes and cohomology rings of flag varieties. We also define Newton–Okounkov polytopes of flag varieties.

11.2.1 Polytope Ring

Let $L \subset \mathbb{R}^d$ be a lattice, and $P \subset \mathbb{R}^d$ a convex polytope whose vertices lie in L. We say that P is a *lattice polytope* with respect to L. By the *standard lattice* $\mathbb{Z}^d$ we mean the lattice $\{(x_1, \ldots, x_d) \in \mathbb{R}^d \mid x_i \in \mathbb{Z} \text{ for all } i = 1, \ldots, d\}$. We choose the translation invariant volume form on $\mathbb{R}^d$ so that the covolume of L is 1.

Recall that two convex polytopes P and Q are called *analogous* if they have the same normal fan, i.e. there is a one-to-one correspondence between the faces of P and the faces of Q such that any linear functional, whose restriction to P attains its maximal value at a given face $F \subseteq P$ has the property that its restriction to Q attains its maximal value at the corresponding face of Q.

Denote by S_P the set of all polytopes analogous to P. This set can be endowed with the structure of a commutative semigroup using *Minkowski sum*

$$P_1 + P_2 = \{x_1 + x_2 \in \mathbb{R}^d \mid x_1 \in P_1,\ x_2 \in P_2\}$$

It is not hard to check that this semigroup has cancelation property. We can also multiply polytopes in S_P by positive real numbers using dilation:

$$\lambda P = \{\lambda x \mid x \in P\}, \quad \lambda \geq 0.$$

Hence, we can embed the semigroup of convex polytopes into its Grothendieck group V_P, which is a real vector space. The elements of V_P are called *virtual polytopes* analogous to P.

On the vector space V_P, there is a homogeneous polynomial vol_P of degree d, called the *volume polynomial*. It is uniquely characterized by the property that its value $\mathrm{vol}_P(Q)$ on any convex polytope $Q \in S_P$ is equal to the volume of Q.

Let Λ_P be a lattice in V_P generated by some lattice polytopes (with respect to L) analogous to P (we do not assume that Λ_P contains all lattice polytopes analogous to P). The symmetric algebra $\mathrm{Sym}(\Lambda_P)$ of Λ_P can be thought of as the ring of differential operators with constant integer coefficients acting on $\mathbb{R}[V_P]$, the space of all polynomials on V_P. If $D \in \mathrm{Sym}(\Lambda_P)$ and $\varphi \in \mathbb{R}[V_P]$, then we write $D\varphi \in \mathbb{R}[V_P]$ for the result of this action. Define A_P as the homogeneous ideal in $\mathrm{Sym}(\Lambda_P)$ consisting of all differential operators D such that $D\,\mathrm{vol}_P = 0$. Set $R_P = \mathrm{Sym}(\Lambda_P)/A_P$. This ring is called *the polytope ring* associated with the polytope P and the lattice Λ_P.

Example 1 Let $L = \mathbb{Z}^d$ be the standard lattice, and P an integrally simple lattice polytope (that is, only d edges meet at every vertex of P, and primitive vectors on these edges span L over $\mathbb{Z}$). Let Λ_P be the lattice in V_P generated by all lattice polytopes (with respect to L) analogous to P. Then the ring R_P is isomorphic to the Chow (or cohomology) ring $H^*(X_P, \mathbb{Z})$ of the smooth toric variety X_P associated with the normal fan of P [25].

When P is simple, every facet $\Gamma \subset P$ defines a differential operator $\partial_\Gamma \in R_P$ (see [14, Sect. 2.3] for the details). Recall that the closures of torus orbits in X_P are in bijective correspondence with faces of P. They also give a generating set in the cohomology ring $H^*(X_P, \mathbb{Z})$. Every face $F = \Gamma_1 \cap \cdots \cap \Gamma_k$ can be identified with the operator $[F] = \partial_{\Gamma_1} \cdots \partial_{\Gamma_k} \in R_P$. Using linear relations between ∂_Γ in R_P we can compute products in $H^*(X_P, \mathbb{Z})$ by intersecting faces of P.

For instance, if $P \subset \mathbb{R}^2$ is the trapezoid with vertices $(0, 0)$, $(1, 0)$, $(0, 1)$, $(1, 2)$, then the corresponding toric variety X_P is the blow-up of $\mathbb{CP}^2$ at one point. The edge $\Gamma_1 = \{x = 0\}$ corresponds to the exceptional divisor $E \subset X_P$. The other edges are $\Gamma_2 = \{y - x = 1\}$, $\Gamma_3 = \{x = 1\}$ and $\Gamma_4 = \{y = 0\}$. There are two linear relations between ∂_{Γ_i}. Namely, the parallel translations along x and y axes do not change the area of P, hence, $\partial_{\Gamma_1} + \partial_{\Gamma_2} = \partial_{\Gamma_3}$ and $\partial_{\Gamma_2} = \partial_{\Gamma_4}$. In particular, the identity $[E]^2 = -[pt]$ in $H^*(X_P, \mathbb{Z})$ can be obtained as follows:

$$[\Gamma_1]^2 = [\Gamma_1]([\Gamma_3] - [\Gamma_2]) = [\Gamma_1 \cap \Gamma_3] - [\Gamma_1 \cap \Gamma_2] = -[pt].$$

Example 2 Let $L = \mathbb{Z}^d$, and P the GZ polytope in type A corresponding to a strictly dominant $\lambda = (\lambda_1, \ldots, \lambda_n)$ (that is, $\lambda_1 > \lambda_1 > \cdots > \lambda_n$). Let Λ_P be the lattice in V_P generated by all GZ polytopes P_λ for all dominant λ. Then the ring $R_{GZ} := R_P$ is isomorphic to the cohomology ring $H^*(GL_n(\mathbb{C})/B, \mathbb{Z})$ of the complete flag variety in type A [10].

Since the GZ polytope is not simple, there is no correspondence between individual faces of P and elements of $H^*(GL_n(\mathbb{C})/B, \mathbb{Z})$. However, it is possible to identify every element of $H^*(GL_n(\mathbb{C})/B, \mathbb{Z})$ with a linear combination of faces of P (see [14, Sect. 2] for more details). Again, we can compute all products in $H^*(GL_n(\mathbb{C})/B, \mathbb{Z})$ by intersecting faces of P (see [14, Sect. 2.4] for an example of such computations).

In what follows, L will be a sublattice of $\frac{1}{2}\mathbb{Z}^d := \{(x_1, \ldots, x_d) \mid 2x_i \in \mathbb{Z} \text{ for all } i = 1, \ldots, d\}$. We always compute volumes of faces of P with respect to the lattice L. More precisely, if $F \subset P$ is a face, and $\mathbb{R}F$ is its affine span then the volume of the face is computed using the volume form on $\mathbb{R}F$ normalized so that the covolume of $L \cap \mathbb{R}F$ is 1.

11.2.2 GZ Polytopes in Types B and C

Let $\lambda = (\lambda_1, \ldots, \lambda_n)$ be a non-increasing collection of non-negative integers. Put $d = n^2$. Denote coordinates in $\mathbb{R}^d$ by $(x_1^1, \ldots, x_n^1; y_1^1, \ldots, y_{n-1}^1; \ldots; x_1^{n-1}, x_2^{n-1}, y_1^{n-1}; x_1^n)$. For every λ, define the *symplectic GZ polytope* $SGZ_\lambda \subset \mathbb{R}^d$ for $Sp_{2n}(\mathbb{C})$ by the following interlacing inequalities:

$$
\begin{array}{ccccccccccc}
\lambda_1 & & \lambda_2 & & \lambda_3 & & \ldots & & \lambda_n & & 0 \\
& x_1^1 & & x_2^1 & & \ldots & & & & x_n^1 & \\
& & y_1^1 & & y_2^1 & & \ldots & & y_{n-1}^1 & & 0 \\
& & & x_1^2 & & \ldots & & & & x_{n-1}^2 & \\
& & & & y_1^2 & & \ldots & & y_{n-2}^2 & & 0 \\
& & & & & \ddots & & & \vdots & & \vdots \\
& & & & & & & x_1^{n-1} & & x_2^{n-1} & \\
& & & & & & & & y_1^{n-1} & & 0 \\
& & & & & & & & & x_1^n &
\end{array}
\qquad GZ_C
$$

Again, every coordinate in this table is bounded from above by its upper left neighbor and bounded from below by its upper right neighbor (the table encodes $2d$ inequalities). We regard SGZ_λ as a lattice polytope with respect to the standard lattice $\mathbb{Z}^d$. Roughly speaking, SGZ_λ is the polytope defined using half of the GZ pattern (GZ_A) for $SL_{2n}(\mathbb{C})$.

Example 3 The polytope $SGZ_\lambda \subset \mathbb{R}^4$ for $Sp_4(\mathbb{C})$ is given by 8 inequalities:

$$\lambda_1 \ge x_1^1 \ge \lambda_2; \quad \lambda_2 \ge x_2^1 \ge 0; \quad x_1^1 \ge y_1^1 \ge x_2^1; \quad y_1^1 \ge x_1^2 \ge 0.$$

It is not hard to compute the volume polynomial of SGZ_λ:

$$\mathrm{vol}_{SGZ}(\lambda_1, \lambda_2) = \frac{1}{6}\lambda_1\lambda_2(\lambda_1 - \lambda_2)(\lambda_1 + \lambda_2).$$

This volume times 4! is equal to the degree $\deg_\lambda(Sp_4(\mathbb{C})/B)$ of the isotropic flag variety.

The polytope ring R_{SGZ} defined by the family of symplectic GZ polytopes is isomorphic to the cohomology ring $H^*(Sp_{2n}(\mathbb{C})/B, \mathbb{Z})$. Indeed, by [10] it is isomorphic to the subring of $H^*(Sp_{2n}(\mathbb{C})/B, \mathbb{Z})$ generated by the first Chern classes of line bundles L_λ corresponding to the weights of $Sp_{2n}(\mathbb{C})$. Since the torsion index of $Sp_{2n}(\mathbb{C})$ is 1, this subring coincides with the whole ring (see [26] for the details on torsion indices of classical groups).

The *odd orthogonal GZ polytope* $OGZ_\lambda \subset \mathbb{R}^d$ for $SO_{2n+1}(\mathbb{C})$ is defined using the same pattern (GZ_C) but a different lattice $L_B \subset \mathbb{R}^d$. Namely, L_B consists of all points $(x_1^1, \dots, x_n^1; y_1^1, \dots, y_{n-1}^1; \dots; x_1^{n-1}, x_2^{n-1}, y_1^{n-1}; x_1^n) \in \frac{1}{2}\mathbb{Z}^d$ such that all coordinates except for x_n^1, x_{n-1}^2,..., x_1^n are integer. Lattice points $SGZ_\lambda \cap \mathbb{Z}^d$ and $SGZ_\lambda \cap L_B$ parameterize basis vectors in irreducible representations of $Sp_{2n}(\mathbb{C})$ and $SO_{2n+1}(\mathbb{C})$, respectively (see [20, Sect. 6] for more details).

Remark 4 Family of odd orthogonal GZ polytopes (as defined in [1, 20]) consists of two subfamilies parameterized by integer and half-integer λ. The group $SO_{2n+1}(\mathbb{C})$ is not simply connected, and half-integer weights correspond to the characters of the maximal torus in the universal cover $Spin(2n+1)$. If we define the polytope ring R_{SGZ} using the first subfamily we get a subring of $H^*(SO_{2n+1}/B, \mathbb{Z})$ generated by the first Chern classes of line bundles L_λ corresponding to the characters λ of the maximal torus in $SO_{2n+1}(\mathbb{C})$.

Example 5 The polytope $OGZ_\lambda \subset \mathbb{R}^4$ for $Sp_4(\mathbb{C})$ is given by the same 8 inequalities as in Example 3. However, its volume polynomial is computed using a different volume form chosen so that the covolume of L_B is 1. Since $\mathbb{Z}^4 \subset L_B$ has index 4, we get $\mathrm{vol}_{OGZ} = 4\,\mathrm{vol}_{SGZ}$.

There is an exceptional isomorphism $Sp_4(\mathbb{C})/\pm 1 \simeq SO_5(\mathbb{C})$. In particular, flag varieties in types B_2 and C_2 are the same. This isomorphism takes the dominant weight $\lambda = (\lambda_1, \lambda_2)$ of $Sp_4(\mathbb{C})$ to the dominant weight $\widetilde{\lambda} = (\lambda_1 + \lambda_2)/2, (\lambda_1 - \lambda_2)/2)$ of $SO_5(\mathbb{C})$. This agrees with the identity $\mathrm{vol}(SGZ_\lambda) = \mathrm{vol}(OGZ_{\widetilde{\lambda}})$.

11.2.3 Newton–Okounkov Polytopes of Flag Varieties

We recall a definition of Newton–Okounkov convex bodies in the case of flag varieties. We refer the reader to [12, 19] for definitions in the more general setting.

Recall that the complete flag variety $SL_n(\mathbb{C})/B$ is defined as the variety of complete flags of subspaces $M^\bullet = (\{0\} \subset V^1 \subset V^2 \subset \dots \subset V^{n-1} \subset \mathbb{C}^n)$. We define $SO_n(\mathbb{C})/B$ and Sp_{2n}/B as subvarieties of *orthogonal* and *isotropic* flags in $SL_n(\mathbb{C})/B$ and SL_{2n}/B, respectively. A complete flag $M^\bullet$ in $\mathbb{C}^n$ is *orthogonal* if V^i is orthogonal to to V^{n-i} with respect to a non-degenerate symmetric bilinear form fixed by $SO_n(\mathbb{C})$. Let ω be a non-degenerate skew-symmetric bilinear form fixed by $Sp_{2n}(\mathbb{C})$. A complete flag $M^\bullet$ in $\mathbb{C}^{2n}$ is called *isotropic* if the restriction of ω to V^n is zero, and $V^{2n-i} = \{v \in \mathbb{C}^{2n} \mid \omega(v, u) = 0 \text{ for all } u \in V^i\}$.

Every flag variety X of dimension d has an open dense subset C (*open Schubert cell*) isomorphic to the affine space $\mathbb{C}^d$. It can be constructed as follows. Fix a complete flag $F^\bullet := (F^1 \subset F^2 \subset \cdots \subset F^{n-1} \subset \mathbb{C}^n)$ such that $F^\bullet \in X$ (this amounts to fixing a Borel subgroup $B \subset G$). Also fix a basis $e_1,\ldots, e_n$ in $\mathbb{C}^n$ compatible with $F^\bullet$ (or a maximal torus in B), that is, $F^i = \langle e_1, \ldots, e_i\rangle$. *The open Schubert cell C with respect to $F^\bullet$ is defined as the set of all flags $M^\bullet$ that are in general position with the standard flag $F^\bullet$*, i.e., all intersections $M^i \cap F^j$ are transverse. Let $x_1, \ldots, x_d$ be coordinates on the open Schubert cell C.

Example 6 In type A, we can identify the open Schubert cell C with an affine space $\mathbb{C}^d$ (for $d = n(n-1)/2$) by choosing for every flag $M^\bullet$ a basis $v_1,\ldots, v_n$ in $\mathbb{C}^n$ of the form:

$$v_1 = e_n + x_1^{n-1}e_{n-1} + \cdots + x_1^1 e_1,$$

$$v_2 = e_{n-1} + x_2^{n-2}e_{n-2} + \cdots + x_2^1 e_1, \quad \ldots \quad , v_{n-1} = e_2 + x_{n-1}^1 e_1, \quad v_n = e_n,$$

so that $M^i = \langle v_1, \ldots, v_i\rangle$. Such a basis is unique, hence, the coefficients $(x_j^i)_{i+j<n}$ are coordinates on the open cell. In other words, every flag $M^\bullet \in C$ gets identified with a triangular matrix:

$$\begin{pmatrix} x_1^1 & x_2^1 & \ldots & x_{n-1}^1 & 1 \\ x_1^2 & x_2^2 & \ldots & 1 & 0 \\ \vdots & \vdots & & & 0 \\ x_1^{n-1} & 1 & \ldots & 0 & 0 \\ 1 & 0 & & 0 & 0 \end{pmatrix} \qquad (FFLV).$$

Similar coordinates can be introduced on flag varieties in other types.

Let $V \subset \mathbb{C}(X) = \mathbb{C}(x_1, \ldots, x_d)$ be a finite-dimensional subspace of rational functions on X. Our main examples are spaces of global sections $H^0(L_\lambda, X) \simeq V_\lambda^*$ of line bundles on X. We fix a section $s_0 \in H^0(L_\lambda, X)$, and identify sections $s \in H^0(L_\lambda, X)$ with rational functions $f = \frac{s}{s_0} \in \mathbb{C}(X)$.

Example 7 (*Example* 6 *continued*) If

$$\lambda = (\underbrace{1, \ldots, 1}_{k}, \underbrace{0 \ldots, 0}_{n-k}),$$

then V_λ^* can be identified with the subspace of $\mathbb{C}(x_j^i)_{i+j<n}$ spanned by the minors of the $n \times k$ matrix formed by the first k columns of the matrix $(FFLV)$. These minors are exactly the Plücker coordinates of the Grassmannian $G(k, n)$ in the Plücker embedding. The map $X \to H^0(L_\lambda, X)^*$ is the composition of the projection $X \to$

$G(k,n)$ (obtained by forgetting all subspaces in the flag $M^\bullet$ except for the V^k) and the Plücker embedding of $G(k,n)$.

To assign the *Newton–Okounkov convex body* to V we need an extra ingredient. Choose a translation-invariant total order on the lattice $\mathbb{Z}^d$ (e.g., we can take the lexicographic order). Consider a map

$$v : \mathbb{C}(x_1, \dots, x_d) \setminus \{0\} \to \mathbb{Z}^d,$$

that behaves like the lowest order term of a polynomial, namely: $v(f+g) \geq \min\{v(f), v(g)\}$ and $v(fg) = v(f) + v(g)$ for all nonzero f, g. Recall that maps with such properties are called *valuations*.

Definition 8 The *Newton–Okounkov convex body* $\Delta_v(X, V)$ is the closure of the convex hull of the set

$$\bigcup_{k=1}^{\infty} \left\{ \frac{v(f)}{k} \mid f \in V^k \right\} \subset \mathbb{R}^d.$$

By V^k we denote the subspace spanned by the kth powers of the functions from V.

Example 9 Using coordinates of Example 6 we can define the valuation v as follows. Order the coefficients $(x_j^i)_{i+j<n}$ of the matrix $(FFLV)$ by starting from column $(n-1)$ and going from top to bottom in every column and from right to left along columns. Then $\Delta_v(X, V_\lambda^*)$ coincides with the Feigin–Fourier–Littelmann–Vinberg polytope $FFLV(\lambda)$ [17]. Moreover, the inclusion $FFLV(\lambda) \subset \Delta_v(X, V_\lambda^*)$ follows from a straightforward computation of the valuation v on the minors of the matrix $(FFLV)$ (see [17, Example 2.9] for more details).

Different valuations might yield different Newton–Okounkov convex bodies. In particular, GZ polytopes can also be obtained as Newton–Okounkov polytopes of flag varieties [6, 11]. Okounkov made the first explicit computation of this kind, namely, he exhibited symplectic GZ polytopes as Newton–Okounkov polytopes of the isotropic flag varieties [22].

11.3 Geometric Mitosis

In [15], convex geometric analogs of Demazure (or divided difference) operators are defined on convex polytopes and used to construct *DDO polytopes* that have the same properties as Newton–Okounkov polytopes of flag varieties. In [16], operations on faces of a DDO polytope (geometric mitosis) are defined that yield positive presentations of Schubert cycles by faces. Here we define the same operations in more combinatorial terms using a vertex cone instead of a DDO polytope. We refer the reader to [15, Theorem 3.6], [16, Proposition 2.5] for connections with representation theory and convex geometry.

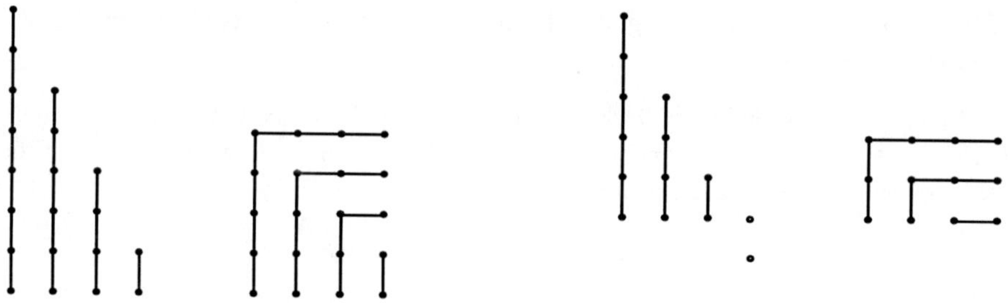

Fig. 11.2 Mitosis as explained in Example 10

Example 10 Figure 11.2 illustrates the idea of mitosis in the simplest example. The trapezoid and rectangle on the left picture have the same number of lattice points with given sum of coordinates. The same is true for the right picture. However, the trapezoid on the right picture becomes a virtual polytope (in particular, lattice points marked with circles have to be counted with the zero coefficient) while the rectangle remains a true polytope. There is a price to pay: the left vertical edge of the trapezoid corresponds to two edges of the rectangle (that is, a single edge of the trapezoid has the same number of lattice points as two edges of the rectangle). In short, mitosis preserves positivity at the cost of more involved combinatorics.

Consider a vector space with the direct sum decomposition

$$\mathbb{R}^d = \mathbb{R}^{d_1} \oplus \cdots \oplus \mathbb{R}^{d_r},$$

and choose coordinates $x = (x_1^1, \ldots, x_{d_1}^1; \ldots; x_1^r, \ldots, x_{d_r}^r)$ with respect to this decomposition. Let $C \subset \mathbb{R}^d$ be a convex polyhedral cone with the vertex at the origin 0. Assume that C is given by inequalities either of type $x_j^i \le a x_{j'}^{i'}$ where $a > 0$ and $i \ne i'$ or of type $0 \le x_j^i$. In what follows, we use the bijective correspondence between facets of C and inequalities, namely, every inequality $x_j^i \le a x_{j'}^{i'}$ defines the facet $H(i, j; i', j')$ given by the equation $x_j^i = a x_{j'}^{i'}$, and every inequality $0 \le x_j^i$ defines the facet $H(0, 0; i, j)$ given by the equation $x_j^i = 0$.

In addition, assume that C does not contain any rays parallel to the x_j^i-axis unless $j = 1$. Then the geometric mitosis of [15, Sect. 5.1] can be defined on faces of C. Below we describe the resulting *mitosis operations* $M_1, \ldots, M_r$ from a combinatorial viewpoint.

Let Γ be a face of the cone C of codimension ℓ. The ith *mitosis operation* M_i applied to Γ will produce a collection $M_i(\Gamma)$ (possibly empty) of faces of C. Choose a minimal subset of facets $H_1, \ldots, H_\ell$ of C such that $\Gamma = H_1 \cap \cdots \cap H_\ell$. If none of these facets coincides with $H(p, q; i, d_i)$ for some p and q, then $M_i(\Gamma) = \emptyset$. Otherwise, let s be the smallest number such that the subset $\{H_1, \ldots, H_\ell\}$ contains facets of type $H(\cdot, \cdot; i, j)$ for all $j = s, s+1, \ldots, d_i$. For brevity, we label these facets by $H^+(i, s)$, $H^+(i, s+1)$, …, $H^+(i, d_i)$. For every $j = s+1, s+2, \ldots, d_i$, we now label by $H_+(i, j)$ the facet of type $H(i, j; \cdot, \cdot)$. If there are two such facets $H(i,j;p,q)$ and $H(i,j;p',q')$, and $x_q^p \le x_{q'}^{p'}$ everywhere on Γ then we put $H_+(i,j) := H(i,j;p,q)$.

Let $J_i(\Gamma) \subset \{s, s+1, \dots, d_i\}$ consist of indices j such that $H_+(i,j) \notin \{H_1, \dots, H_\ell\}$. For every $j \in J_i(\Gamma)$, we define an *offspring* $\Delta_j \in M_i(\Gamma)$ as the intersection of facets

$$\Delta_j = H_1(j) \cap H_2(j) \cap \dots \cap H_{\ell-1}(j),$$

where the set $\{H_1(j), \dots, H_{\ell-1}(j)\}$ is obtained from the set $\{H_1, \dots, H_\ell\}$ by the following rule. First, remove the facet $H^+(i,j)$. Second, for every $k \in J_i(\Gamma)$ such that $k > j$ replace the facet $H^+(i,k)$ by the facet $H_+(i,k)$. Note that $\dim \Delta_j = \dim \Gamma + 1$.

Definition 11 The ith *mitosis operation* M_i sends Γ to

$$M_i(\Gamma) = \{\Delta_j \mid j \in J_i(\Gamma)\}.$$

11.3.1 Type A: GZ Polytopes

Let C be the vertex cone of the GZ polytope in type A for the vertex $a = (\lambda_2, \dots, \lambda_n; \lambda_3, \dots, \lambda_n; \dots; \lambda_n)$ (see table (GZ_A)). After an affine change of coordinates $x = z - a$ the inequalities that define C can be written as follows:

$$0 \le x_1^1; \quad 0 \le x_2^1 \le x_1^2; \quad \dots; \quad 0 \le x_{n-1}^1 \le x_{n-2}^2 \le \dots \le x_1^{n-1}.$$

The cone C has $d = \frac{n(n-1)}{2}$ facets: $H(0,0;1,i)$ for $1 \le i \le (n-1)$ and $H(i-1, j+1; i, j)$ for $2 \le i \le (n-1)$, $1 \le j \le n-i$. In particular, we have the following identifications of facets:

$$H(0,0;1,i) = H^+(1,i), \quad H(i-1,j+1;i,j) = H^+(i,j) = H_+(i-1,j+1).$$

It is convenient to encode a face Γ of C by an $n \times n$ table (*pipe dream*) filled with $+$ as follows. The table contains $+$ in cell $(i, i+j)$ iff $\Gamma \subset H(i-1, j+1; i, j)$ and $i \ge 2$ or $\Gamma \subset H(0,0;i,j)$ and $i = 1$. In particular, only cells above the main diagonal might have $+$. In this notation, mitosis operations M_1, M_2 applied to the vertex 0 produce the following faces (only cells $(1,2)$, $(1,3)$ and $(2,3)$ of 3×3 tables are shown since the other cells never contain $+$):

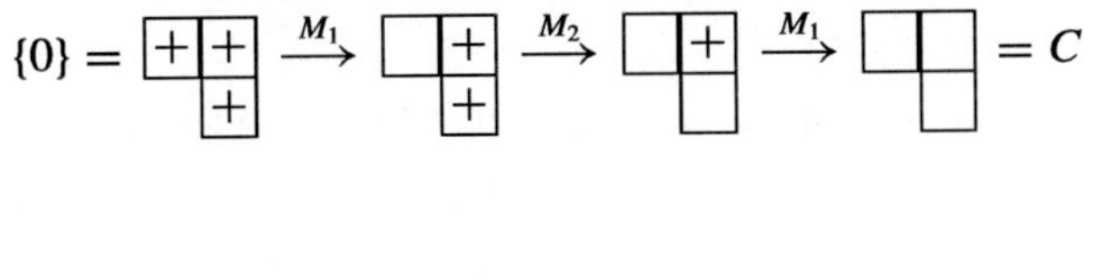

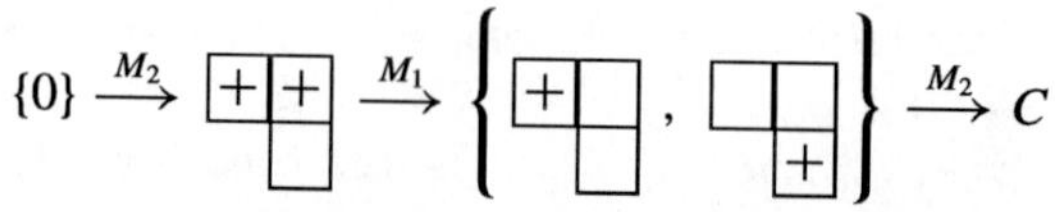

For arbitrary n, the mitosis operations $M_1,\dots, M_{n-1}$ encoded by tables coincide with Knutson–Miller mitosis on pipe dreams [18] after reflecting tables in a vertical line. Instead of the vertex cone C we could take the GZ polytope in type A and consider mitosis operations on faces that contain the vertex a (so called *Kogan faces*). Geometric meaning of the resulting collections of faces is described in [14, Theorem 5.1, Corollary 5.3]. In particular, the following analog of Kushnirenko's theorem holds.

Recall that Schubert subvarieties X_w are labeled by the elements of the Weyl group of G, namely, X_w is the closure of the B-orbit BwB/B, where w is an element of the Weyl group of G. The Weyl group of $G = SL_n(\mathbb{C})$ is the symmetric group S_n. By $s_1,\dots, s_{n-1}$ we denote the elementary transpositions.

Theorem 12 ([14, Theorem 5.4]) *Let $X_w \subset SL_n(\mathbb{C})/B$ be the Schubert subvariety corresponding to a permutation $w \in S_n$. Let $w = s_{j_1} \dots s_{j_\ell}$ be a reduced decomposition of a permutation $w \in S_n$ such that $(j_1, \dots, j_\ell)$ is a subword of* $(1; 2, 1; 3, 2, 1; \dots; n-1, \dots, 1)$. *Let $\mathfrak{S}_w \subset GZ_\lambda$ be the set of all faces produced from the vertex $a \in GZ_\lambda$ by applying successively the operations $M_{n-j_\ell},\dots, M_{n-j_1}$:*

$$\mathfrak{S}_w = M_{n-j_1} \cdots M_{n-j_\ell}(a).$$

Then

$$\deg_\lambda(X_w) = \ell! \sum_{\Gamma\in\mathfrak{S}_w} \mathrm{Vol}(\Gamma).$$

This implies that the Schubert cycle $[X_w]$ (that is, the cohomology class of X_w in $H^*(SL_n(\mathbb{C})/B, \mathbb{Z})$) in the polytope ring $R_{GZ} \simeq H^*(SL_n(\mathbb{C})/B, \mathbb{Z})$ is represented by the sum of faces in $\mathfrak{S}_w$.

Example 13 For $n = 3$, we have $[X_{s_1s_2}] = M_2M_1(a)$ and $[X_{s_2s_1}] = M_1M_2(a)$. Since the faces in these two presentations are transverse and their intersection consists of two edges $M_1(a)$ and $M_2(a)$ we get the identity: $[X_{s_1s_2}] \cdot [X_{s_2s_1}] = [X_{s_1}] + [X_{s_2}]$ (see Fig. 11.3).

11.3.2 Type $C_{2\text{–}3}$: DDO Polytopes

In [16, Example 2.9], the following family of DDO polytopes in $\mathbb{R}^4 = \mathbb{R}^2 \oplus \mathbb{R}^2$ is considered:

$$0 \le x_1^1 \le \lambda_1, \quad x_1^2 \le x_1^1 + \lambda_2, \quad x_2^1 \le 2x_1^2,$$

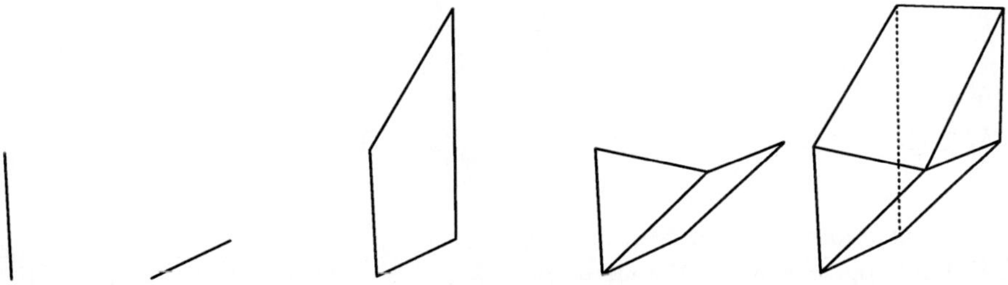

Fig. 11.3 Faces $M_1(a)$, $M_2(a)$, $M_2M_1(a)$, $M_1M_2(a)$, $M_1M_2M_1(a)$ of the GZ polytope in type A, $n = 3$

$$x_2^1 \leq x_1^2 + \lambda_2, \quad 0 \leq x_2^2 \leq \lambda_2, \quad x_2^2 \leq \frac{x_2^1}{2}$$

(these polytopes can also be realized as Newton–Okounkov polytopes of the isotropic flag variety Sp_4/B [16, Proposition 4.1]). The vertex cone C of the vertex 0 is given by 4 homogeneous inequalities: $0 \leq x_1^1, 0 \leq 2x_2^2 \leq x_2^1 \leq 2x_1^2$. It is convenient to encode a face Γ of C by a $(2n-1) \times n$ table (*skew pipe dream*) for $n = 2$ filled with + as follows (see Sect. 11.3.3 for the general definition of skew pipe dreams). The table contains + in cell $(3-i, i)$ (for $i = 1, 2$) iff $\Gamma \subset H(0, 0; i, i)$, + in cell $(2, 2)$ iff $\Gamma \subset H(2, 2; 1, 2)$ and + in cell $(3, 2)$ iff $\Gamma \subset H(1, 2; 2, 1)$. There are two mitosis operations M_1 and M_2.

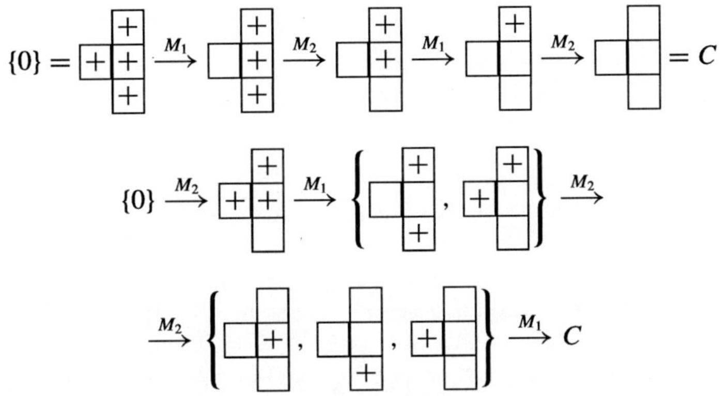

The Weyl group of $G = Sp_4(\mathbb{C})$ is the dihedral group D_4. By s_1, s_2 we denote simple reflections that generate D_4 so that s_2 corresponds to the longer root. By [16, Corollary 3.6] we have

Proposition 14 *Let $X_w \subset Sp_4(\mathbb{C})/B$ be the Schubert subvariety corresponding to a permutation $w \in D_4$. Let $w = s_{j_1} \ldots s_{j_\ell}$ be a reduced decomposition of a permutation $w \in D_4$ such that $(j_1, \ldots, j_\ell)$ is a subword of $(1, 2; 1, 2)$. Let $\mathfrak{S}_w \subset GZ_\lambda$ be the set of all faces produced from the vertex $a \in GZ_\lambda$ by applying successively the operations $M_{j_\ell}, \ldots, M_{j_1}$:*

$$\mathfrak{S}_w = M_{j_1} \cdots M_{j_\ell}(0).$$

Then

$$\deg_\lambda(X_w) = \ell! \sum_{\Gamma \in \mathfrak{S}_w} \mathrm{Vol}(\Gamma).$$

This example can be extended to DDO polytopes for Sp_{2n}. For $n = 3$ and the DDO polytope for $(s_3s_2s_1)^3$ (where s_3 is the simple reflection with respect to the longer root) this was done in [23]. The corresponding family of DDO polytopes in $\mathbb{R}^9 = \mathbb{R}^3 \oplus \mathbb{R}^3 \oplus \mathbb{R}^3$ is given by inequalities:

$$0 \le x_1^1 \le \lambda_1; \quad x_1^2 \le \lambda_2 + x_1^1; \quad x_1^3 \le \lambda_3 + x_1^2;$$

$$0 \le x_2^1 \le \min\{x_1^2, \lambda_2\}; \quad x_2^2 \le \min\{\lambda_3 + x_2^1 + x_1^3, 2x_1^3\};$$

$$x_2^3 \le \min\left\{x_2^1 + \lambda_3, \frac{1}{2}x_2^2\right\}; \quad x_3^1 \le \min\{x_2^2, x_1^3 + \lambda_3, \lambda_3 + x_2^2 - x_2^3\};$$

$$x_3^2 \le \min\{x_3^1, x_2^3 + \lambda_3, 2x_2^3\}; \quad 0 \le x_3^3 \le \min\left\{\frac{1}{2}x_3^2, \lambda_3\right\}.$$

In particular, the vertex cone at 0 is not simplicial. It is defined by 10 inequalities:

$$0 \le x_1^1; \quad 0 \le x_2^1 \le x_1^2; \quad 0 \le x_3^3 \le \frac{1}{2}x_3^2 \le \frac{1}{2}x_3^1 \le \frac{1}{2}x_2^2 \le x_1^3;$$

$$\frac{1}{2}x_3^2 \le x_2^3 \le \frac{1}{2}x_2^2.$$

An analog of Proposition 14 follows easily from [16, Corollary 3.6]. However, combinatorics of mitosis becomes more involved as analogs of pipe dreams in this case have a loop.

Recently, Fujita identified DDO polytopes with certain Nakashima–Zelevinsky polyhedral realizations of crystal bases [5, Theorem 4.1]. In particular, there are explicit inequalities for these polytopes in types A, B, C, D and G_2 [5, Example 4.3]. In type A, they coincide with the GZ polytope and in type $C_{2\text{–}3}$ with the polytopes described in this section. It would be interesting to apply geometric mitosis to these polytopes in the other cases.

11.3.3 *Type C: GZ Polytopes*

The combinatorics of C_2 example from Sect. 11.3.2 can be extended to C_n in a different way by using the string cone C for the reduced decomposition $\overline{w_0} =$

$(s_ns_{n-1}\dots s_2s_1s_2\dots s_{n-1}s_n)\dots(s_2s_1s_2)(s_1)$ of the longest element in the Weyl group (here s_1 corresponds to the longer root). The corresponding string polytope coincides with the symplectic GZ polytope after a unimodular change of coordinates [20, Sect. 6]. The cone C is simplicial and is given by $d = n^2$ inequalities:

$$0 \le x_2^i \le x_4^{i-1} \le x_6^{i-2} \le \dots \le x_{2i-2}^2 \le x_i^1 \le x_{2i-3}^2 \le \dots \le x_5^{i-2} \le x_3^{i-1} \le x_1^i$$

for all $i = 1,\dots, n$. We define *symplectic mitosis* as the geometric mitosis associated with the cone C. Combinatorics of the symplectic mitosis is quite simple and described in detail in [16, Sect. 5.2] using *skew pipe dreams*. However, arguments of [16, Corollary 3.6] do not directly yield presentations for Schubert cycles since the symplectic GZ polytope does not satisfy the necessary conditions. Still computations for $n = 2, 3$ suggest that the collections of faces of the symplectic GZ polytope obtained using symplectic mitosis do represent the corresponding Schubert cycles in the polytope ring R_{SGZ}. Below we describe a bijection between faces of C and faces of SGZ_λ that we used.

Let v be the vertex of SGZ_λ given by equations $\lambda_s = x_j^i = y_l^k$ for all triples λ_s, x_j^i and y_l^k such that $s = i + j - 1 = k + l$. We now define a bijection between those facets of P_λ that contain v and *skew pipe dreams* of size n with exactly one $+$. Recall that a *skew pipe dream* of size n is a $(2n-1)\times n$ table whose cells are either empty or filled with $+$. Only cells (i, j) with $n - j < i < n + j$ are allowed to have $+$ (see [16, Sect. 5.2] for more details on skew pipe dreams). Put $y_i^0 := \lambda_i$ for $i = 1,\dots, n$. The facet given by equation $x_j^i = y_j^{i-1}$ corresponds to the skew pipe dream with $+$ in cell $(i + j - 1, n - i + 1)$. The facet given by equation $y_j^i = x_{j+1}^i$ corresponds to the skew pipe dream with $+$ in cell $(2n - i - j + 1, n - i + 1)$. In what follows, we denote by $H_{(k,l)}$ the facet whose skew pipe dream under this correspondence contains $+$ in cell (k, l).

This correspondence between facets and skew pipe dreams with a single $+$ extends to all faces of the symplectic GZ polytope that contain the vertex v. Namely, the face $H_{k_1,l_1}\cap\dots\cap H_{k_s,l_s}$ obtained as the intersection of s facets corresponds to the skew pipe dream that has $+$ precisely in cells $(k_1, l_1),\dots, (k_s, l_s)$. In particular, the vertex v corresponds to the skew pipe dream D_0 that has $+$ in all (fillable) cells. In what follows, we denote by F_D the face corresponding to a skew pipe dream D.

We now formulate a conjecture. Let w be an element of the Weyl group of $G = Sp_{2n}$. Choose a reduced decomposition $w = s_{j_1}\dots s_{j_\ell}$ such that it is a subword of $(s_ns_{n-1}\dots s_2s_1s_2\dots s_{n-1}s_n)\dots(s_2s_1s_2)(s_1)$.

Conjecture 15 Define the set $\mathfrak{S}_w$ of faces of the symplectic GZ polytope as follows:

$$\mathfrak{S}_w = \{F_D \mid D \in M_{n+1-j_\ell}\cdots M_{n+1-j_1}(D_0)\}$$

where M_i denotes the ith symplectic mitosis operation. Then

$$\deg_\lambda(X_w) = \ell! \sum_{F\in\mathfrak{S}_w} \mathrm{Vol}(F).$$

This conjecture is verified in the case $n = 2$ and for certain w in the case $n = 3$ [24]. Note that the bijection between faces of SGZ_λ that contain the vertex v and faces of the string cone C does not come from the unimodular change of coordinates that identifies the string polytope and the symplectic GZ polytope. There are might be piecewise linear transformations (such as the ones used in [17, Sect. 5.2]) that yield scissor congruence of unions of faces of SGZ_λ and faces of another polytope for which geometric meaning of symplectic mitosis is more transparent.

11.3.4 Type B: GZ Polytopes

Note that the Weyl groups of $Sp_{2n}(\mathbb{C})$ and $SO_{2n+1}(\mathbb{C})$ are the same. Since the GZ polytopes for both groups differ only by lattices symplectic mitosis is also a natural tool for finding presentations of Schubert cycles by faces of OGZ_λ in type B. However, coefficients will be rational rather than integer (with powers of 2 in denominator) because the torsion index of $SO_{2n+1}(\mathbb{C})$ is a power of 2. Note also that the volumes of faces of both SGZ_λ and OGZ_λ should be computed with respect to their lattices. The difference is already visible in the case $n = 2$ (see Example 5).

Acknowledgements The study has been partially funded by the Russian Academic Excellence Project '5–100'.

References

1. Berenstein, A.D., Zelevinsky, A.V.: Tensor product multiplicities and convex polytopes in partition space. J. Geom. Phys. **5**(3), 453–472 (1988)
2. Brion, M.: Lectures on the geometry of flag varieties. In: Topics in Cohomological Studies of Algebraic Varieties, Trends in Mathematics, pp. 33–85. Birkhäuser, Basel (2005)
3. Fang, X., Fourier, G., Littelmann, P.: Essential bases and toric degenerations arising from birational sequences. Adv. Math. **312**, 107–149 (2017)
4. Feigin, E., Fourier, G., Littelmann, P.: Favourable modules: filtrations, polytopes, Newton-Okounkov bodies and flat degenerations. Transform. Groups **22**(2), 321–352 (2017)
5. Fujita, N.: Polyhedral realizations of crystal bases and convex-geometric Demazure operators. Selecta Math. (N.S.) **25**(5), Paper No. 74 (2019)
6. Fujita, N., Oya, H.: A comparison of Newton-Okounkov polytopes of Schubert varieties. J. Lond. Math. Soc. (2) **96**(1), 201–227 (2017)
7. Harada, M., Yang, J.J.: Newton-Okounkov bodies of Bott-Samelson varieties and Grossberg-Karshon twisted cubes. Michigan Math. J. **65**(2), 413–440 (2016)
8. Harada, M., Yang, J.J.: Singular string polytopes and functorial resolutions from Newton-Okounkov bodies. Illinois J. Math. **62**(1–4), 271–292 (2018)
9. Ilyukhina, M.: Schubert Calculus and Geometry of a String Polytope for the Group sp_4. Moscow State University, Diploma (2012)
10. Kaveh, K.: Note on cohomology rings of spherical varieties and volume polynomial. J. Lie Theory **21**(2), 263–283 (2011)
11. Kaveh, K.: Crystal bases and Newton-Okounkov bodies. Duke Math. J. **164**(13), 2461–2506 (2015)

12. Kaveh, K., Khovanskii, A.G.: Newton-Okounkov bodies, semigroups of integral points, graded algebras and intersection theory. Ann. of Math. (2) **176**(2), 925–978 (2012)
13. Kaveh, K., Villella, E.: On a notion of anticanonical class for families of convex polytopes (2018). arXiv:1802.06674 [math.AG]
14. Kirichenko, V.A., Smirnov, E.Y., Timorin, V.A.: Schubert calculus and Gelfand-Tsetlin polytopes. Uspekhi Mat. Nauk **67**(4(406)), 89–128 (2012)
15. Kiritchenko, V.: Divided difference operators on polytopes. In: Schubert calculus—Osaka 2012, Advanced Studies in Pure Mathematics, vol. 71, pp. 161–184. Mathematics Society Japan, [Tokyo] (2016)
16. Kiritchenko, V.: Geometric mitosis. Math. Res. Lett. **23**(4), 1071–1097 (2016)
17. Kiritchenko, V.: Newton-Okounkov polytopes of flag varieties. Transform. Groups **22**(2), 387–402 (2017)
18. Knutson, A., Miller, E.: Gröbner geometry of Schubert polynomials. Ann. of Math. (2) **161**(3), 1245–1318 (2005)
19. Lazarsfeld, R., Mustaţă, M.: Convex bodies associated to linear series. Ann. Sci. Éc. Norm. Supér. (4) **42**(5), 783–835 (2009)
20. Littelmann, P.: Cones, crystals, and patterns. Transform. Groups **3**(2), 145–179 (1998)
21. Molev, A.I.: Gelfand-Tsetlin bases for classical Lie algebras. In: Handbook of Algebra. vol. 4, Handbook of Algebra, vol. 4, pp. 109–170. Elsevier/North-Holland, Amsterdam (2006)
22. Okounkov, A.: Multiplicities and Newton Polytopes. In: Kirillov's seminar on representation theory, American Mathematical Society Translations: Series 2, vol. 181, pp. 231–244. American Mathematical Society, Providence, RI (1998)
23. Padalko, M.: Mitosis for symplectic flag varieties. B.Sc. Thesis, National Research University Higher School of Economics (2015)
24. Padalko, M.: Schubert calculus and symplectic Gelfand–Zetlin polytopes. M.Sc. Thesis, National Research University Higher School of Economics (2017)
25. Pukhlikov, A.V., Khovanskiĭ, A.G.: The Riemann-Roch theorem for integrals and sums of quasipolynomials on virtual polytopes. Algebra i Analiz **4**(4), 188–216 (1992)
26. Totaro, B.: The torsion index of the spin groups. Duke Math. J. **129**(2), 249–290 (2005)

Chapter 12
An Eisenbud–Goto-Type Upper Bound for the Castelnuovo–Mumford Regularity of Fake Weighted Projective Spaces

Bach Le Tran

Abstract We will give an upper bound for the k-normality of very ample lattice simplices, and then give an Eisenbud–Goto-type bound for some special classes of projective toric varieties.

Keywords k-normality · Toric variety · Very ample lattice simplex · Einsenbud-Goto conjecture · Castelnuovo-Mumford regularity

12.1 Introduction

The study of the Castelnuovo–Mumford regularity for projective varieties has been greatly motivated by the Eisenbud–Goto conjecture [7] which asks for any irreducible and reduced variety X, is it always the case that

$$\operatorname{reg}(X) \leq \deg(X) - \operatorname{codim}(X) + 1?$$

The Eisenbud–Goto conjecture is known to be true for some particular cases. For example, it holds for smooth surfaces in characteristic zero [13], connected reduced curves [8], etc. Inspired by the conjecture, there are also many attempts to give an upper bound for the Castelnuovo–Mumford regularity for various types of algebraic and geometric structures [5, 12, 15, 20].

For toric geometry, suppose that (X, L) is a polarized projective toric varieties such that L is very ample. Then there is a corresponding very ample lattice polytope $P := P_L$ associated to L such that $\Gamma(X, L) = \bigoplus_{m \in P \cap M} \mathbb{C} \cdot \chi^m$ [4, Sect. 5.4]. Therefore, by studying the k-normality of P (cf. Definition 2), we can obtain the k-normality and also the regularity of the original variety (X, L). For the purpose of this article,

B. L. Tran (✉)
School of Mathematics, University of Edinburgh, James Clerk Maxwell Building, Peter Guthrie Tait Road, Edinburgh EH9 3FD, UK
e-mail: b.tran@sms.ed.ac.uk

A. M. Kasprzyk and B. Nill (eds.), *Interactions with Lattice Polytopes*, Springer Proceedings in Mathematics & Statistics 386,
https://doi.org/10.1007/978-3-030-98327-7_12

we will focus on the case that X is a fake weighted projective d-space and P_L a d-simplex.

For any fake weighted projective d-space X embedded in $\mathbb{P}^r$ via a very ample line bundle, Ogata [17] gives an upper bound for its k-normality:

$$k_X \leq \dim X + \left\lfloor \frac{\dim X}{2} \right\rfloor - 1.$$

In this article, we will improve Ogata's bound by giving a new upper bound for the k-normality of very ample lattice simplices and show that

$$\operatorname{reg}(X) \leq \deg(X) - \operatorname{codim}(X) + \left\lfloor \frac{\dim X}{2} \right\rfloor. \tag{12.1}$$

Recently, McCullough and Peeva showed some counterexamples to the Eisenbud–Goto conjecture and that the difference $\operatorname{reg}(X) - \deg(X) + \operatorname{codim}(X)$ can be arbitrary large [14, Counterexample 1.8]. However, for any fake weighted projective space X embedded in $\mathbb{P}^r$ via a very ample line bundle, it follows from (12.1) that $\operatorname{reg}(X) - \deg(X) + \operatorname{codim}(X)$ is bounded above by $\dim(X)/2$. Furthermore, we will show that the Eisenbud–Goto conjecture holds for any projective toric variety corresponding to a very ample Fano simplex.

12.2 Background Material

12.2.1 Toric Varieties and Lattice Simplices

We begin this section by recalling the definition of the Castelnuovo–Mumford regularity:

Definition 1 Let $X \subseteq \mathbb{P}^r$ be an irreducible projective variety and $\mathcal{F}$ a coherent sheaf over X. We say that $\mathcal{F}$ is k-regular if

$$\mathrm{H}^i(X, \mathcal{F}(k-i)) = 0$$

for all $i > 0$. The regularity of $\mathcal{F}$, denoted by $\operatorname{reg}(\mathcal{F})$, is the minimum number k such that $\mathcal{F}$ is k-regular. We say that X is k-regular if the ideal sheaf $\mathcal{I}_X$ of X is k-regular and use $\operatorname{reg}(X)$ to denote the regularity of X (or of $\mathcal{I}_X$).

As the main object of the article is to find an upper bound for k-normality of very ample lattice simplices, it is important for us to revisit the definition of k-normality of lattice polytopes.

Definition 2 A lattice polytope P is k-normal if the map

$$\underbrace{P \cap M + \cdots + P \cap M}_{k \text{ times}} \to kP \cap M$$

is surjective. The k-normality of P, denoted by k_P, is the smallest positive integer k_P such that P is k-normal for all $k \geq k_P$.

Suppose now that X is a fake weighted projective d-space embedded in $\mathbb{P}^r$ via a very ample line bundle. Then the polytope P corresponding to the embedding is a very ample lattice d-simplex. Furthermore, $\operatorname{codim}(X) = |P \cap M| - (d+1)$, where M is the ambient lattice, and $\deg(X) = \operatorname{Vol}(P)$, the normalized volume of P.

We have a combinatorial interpretation of $\operatorname{reg}(X)$ in terms of k_P and $\deg(P)$ [21, Proposition 4.1.5] as follows:

$$\operatorname{reg}(X) = \max\{k_P, \deg(P)\} + 1. \tag{12.2}$$

From this, we obtain a combinatorial form of the Eisenbud–Goto conjecture: for very ample lattice polytope $P \subset M_{\mathbb{R}}$, is it always true that

$$\max\{\deg(P), k_P\} \leq \operatorname{Vol}(P) - |P \cap M| + d + 1?$$

The first inequality was confirmed to be true recently [11, Proposition 2.2]; namely,

$$\deg(P) \leq \operatorname{Vol}(P) - |P \cap M| + d + 1. \tag{12.3}$$

Therefore, in order to verify the Eisenbud–Goto conjecture for the polarized toric variety (X, L), it suffices to check if

$$k_P \leq \operatorname{Vol}(P) - |P \cap M| + d + 1. \tag{12.4}$$

12.2.2 Ehrhart Theory

We now recall some basic facts about Ehrhart theory of polytopes and the definition of their degree.

Let P be a lattice polytope of dimension d. We define $\operatorname{ehr}_P(k) = |kP \cap M|$, the number of lattice points in kP. Then from Ehrhart's theory [6, 19],

$$\operatorname{Ehr}_P(t) = \sum_{k=0}^{\infty} \operatorname{ehr}_P(k) t^k = \frac{h_P^*(t)}{(1-t)^{d+1}}$$

for some polynomial $h_P^* \in \mathbb{Z}_{\geq 0}[t]$ of degree less than or equal to d. Let $h_P^*(t) = \sum_{i=0}^{d} h_i^* t^i$. We have

$$h_0^* = 1, \;\; h_1^* = |P \cap M| - d - 1, \;\; h_d^* = |P^0 \cap M|, \text{ and } \sum_{i=0}^{d} h_i^* = \mathrm{Vol}(P).$$

Definition 3 ([1, Remark 2.6]) Let P be a lattice polytope of dimension d. We define the degree of P, denoted by $\deg(P)$, to be the degree of $h_P^*(t)$. Equivalently,

$$\deg(P) = \begin{cases} d & \text{if } |P^0 \cap M| \neq 0, \\ \min\left\{i \in \mathbb{Z}_{\geq 0} | (kP)^0 \cap M = \emptyset \text{ for all } 1 \leq k \leq d-i\right\} & \text{otherwise.} \end{cases}$$

12.3 k-Normality of Very Ample Simplices

The following lemma by Ogata is crucial to the main result of this article:

Lemma 4 ([17, Lemma 2.1]) *Let $P = \mathrm{conv}(v_0, \ldots, v_d)$ be a very ample lattice n-simplex. Suppose that $k \geq 1$ is an integer and $x \in kP \cap M$. For any $i = 0, \ldots, d$, we have*

$$x + (k-1)v_i = \sum_{j=1}^{2k-1} u_j$$

for some $u_j \in P \cap M$.

Using the ideas in [17, Lemma 2.5], we generalize the above lemma as follows.

Lemma 5 *Suppose that $P = \mathrm{conv}(v_0, \ldots, v_d)$ is a very ample d-simplex. Let $k \in \mathbb{N}_{\geq 1}$. Then for any $x \in kP \cap M$, $a_0, \ldots, a_d \in \mathbb{Z}_{\geq 0}$ such that $\sum_{i=0}^{d} a_i = k-1$, we have*

$$\sum_{i=0}^{d} a_i v_i + x = \sum_{i=1}^{2k-1} u_i$$

for some $u_i \in P \cap M$.

Proof We will use induction in this proof. The case $k = 1$ is trivial. Suppose that the lemma holds for $k = s-1$. We will now show that it holds for $k = s$; i.e., for any $x \in sP \cap M$, $a_1, \ldots, a_d \in \mathbb{Z}_{\geq 0}$ such that $\sum_{i=0}^{d} a_i = s-1$, we have

$$\sum_{i=0}^{d} a_i v_i + x = \sum_{i=1}^{2s-1} u_i \tag{12.5}$$

for some $u_i \in P \cap M$. Without loss of generality, we can take a_0 to be positive and move v_0 to the origin. By Lemma 4,

$$(s-1)v_0 + x = \sum_{i=1}^{2s-1} w_i$$

for some $w_i \in P \cap M$. Since $v_0 = 0$, we can write $x = \sum_{i=1}^{2s-1} w_i$. If $w_i + w_j \in P \cap M$ for any $i \neq j$, then we can let $t_i = w_{2i-1} + w_{2i}$ for $i = 1, \ldots, s-1$ and have $x = t_1 + \cdots + t_{s-1} + w_{2s-1}$, which lies in $\sum_{i=1}^{s} P \cap M$. Therefore,

$$\sum_{i=0}^{d} a_i v_i + x = \sum_{i=0}^{d} a_i v_i + \sum_{i=1}^{s-1} t_i + w_{2s-1},$$

which satisfies (12.5). Conversely, without loss of generality, suppose that $w_1 + w_2 \notin P \cap M$. Then since $x = w_1 + w_2 + (w_3 + \cdots + w_{2s-1}) \in sP \cap M$, we have $y := w_3 + \cdots + w_{2s-1} \in (s-1)P \cap M$ and $v_0 + x = w_1 + w_2 + y$. Using the induction hypothesis,

$$\underbrace{(a_0 - 1)v_0 + \sum_{i=1}^{d} a_i v_i}_{a_0 - 1 + \sum_{i=1}^{d} a_i = s-2} + y = \sum_{i=1}^{2(s-1)-1} w'_i$$

for some $w'_i \in P \cap M$. It follows that

$$\begin{aligned}\sum_{i=0}^{d} a_i v_i + x &= v_0 + x + (a_0 - 1)v_0 + \sum_{i=1}^{d} a_i v_i \\ &= w_1 + w_2 + y + (a_0 - 1)v_0 + \sum_{i=1}^{d} a_i v_i \\ &= w_1 + w_2 + \sum_{i=0}^{2(s-1)-1} w'_i.\end{aligned}$$

The conclusion follows. □

Now define the invariants d_P and ν_P as in [21, Definition 2.2.8]:

Definition 6 Let P be a lattice polytope with the set of vertices $\mathcal{V} = \{v_0, \ldots, v_{n-1}\}$. We define d_P to be the smallest positive integer such that for every $k \geq d_P$,

$$(k+1)P \cap M = P \cap M + kP \cap M.$$

We also define ν_P to be the smallest positive integer such that for any $k \geq \nu_P$,

$$(k+1)P \cap M = \mathcal{V} + kP \cap M.$$

Notice that for P an n-simplex, $d_P \leq \nu_P \leq n-1$.

Proposition 7 *Let $P = \mathrm{conv}(v_0, \ldots, v_d)$ be a very ample d-simplex. Then*

$$k_P \leq \nu_P + d_P - 1.$$

Proof For any $k \geq d_P + \nu_P - 1$ and $p \in kP \cap M$, by the definition of d_P and ν_P, we have

$$p = x + \sum_{i=1}^{\nu_P - d_P} u_i + \sum_{i=0}^{d} a_i v_i \tag{12.6}$$

for some $x \in d_P P \cap M, u_i \in P \cap M, \sum_{i=0}^{d} a_i = k - \nu_P$. By assumption, $k - \nu_P \geq d_P - 1$, so it follows from Lemma 5 that

$$x + \sum_{i=0}^{d} a_i v_i = \sum_{i=1}^{d_P + k - \nu_P} u_i' \tag{12.7}$$

for some $u_i' \in P \cap M$. Substitute (12.7) into (12.6), we have

$$p = \sum_{i=1}^{\nu_P - d_P} u_i + \sum_{i=1}^{d_P + k - \nu_P} u_i'.$$

The conclusion follows. □

Remark 8 This bound is stronger than [17, Proposition 2.4] since $\nu_P \leq d$ [21, Proposition 2.2] and $d_P \leq d/2$ [17, Proposition 2.2].

12.4 Eisenbud–Goto-Type Upper Bound for Very Ample Simplices

Suppose that P is a very ample simplex. If P is unimodularly equivalent to the standard simplex $\Delta_d = \operatorname{conv}(0, e_1, \ldots, e_d)$ then (12.4) holds. Now consider the case P is not unimodularly equivalent to Δ_d.

The following lemma is a rewording of [9, Proposition IV.10] to fit our purpose. We provide a proof for the sake of completeness.

Lemma 9 *Let $\mathcal{V} = \{v_0, \ldots, v_d\}$ and suppose that $P = \operatorname{conv}(\mathcal{V})$ is a lattice simplex not unimodularly equivalent to Δ_d. Then $\deg(P) \geq \nu_P$.*

Proof Since $\nu_P \leq d$, it suffices to show that for any $d \geq k \geq \deg(P)$,

$$\mathcal{V} + kP \cap M \twoheadrightarrow (k+1)P \cap M.$$

Indeed, any $x \in (k+1)P \cap M$ can be written as $x = \sum_{i=0}^{d} a_i v_i$ such that $a_i \geq 0$ and $\sum_{i=0}^{d} a_i = k + 1$. If $a_i < 1$ for all i, then $d > k$ and the point $\sum_{i=0}^{d} (1 - a_i) v_i$ is an interior lattice point of $(d - k)P$, a contradiction since $d - k \leq d - \deg(P)$. Hence, $a_i \geq 1$ for some i, say $a_0 \geq 1$. Then

$$x = v_0 + (a_0 - 1)v_0 + \sum_{i=1}^{d} a_i v_i = v_0 + \left((a_0 - 1)v_0 + \sum_{i=1}^{d} a_i v_i \right) \in \mathcal{V} + kP \cap M.$$

Hence, $k \geq \nu_P$. The conclusion follows. □

Proposition 10 *Let* $P = \text{conv}(v_0, \ldots, v_d)$ *be a very ample simplex. Then*

$$k_P \leq \text{Vol}(P) - |P \cap M| + d + \left\lfloor \frac{d}{2} \right\rfloor.$$

Proof From Proposition 7, (12.3), and Lemma 9,

$$\begin{aligned} k_P &\leq d_P + \nu_P - 1 \leq d_P + \deg(P) - 1 \\ &\leq d_P + \text{Vol}(P) - |P \cap M| + d. \end{aligned}$$

By [17, Proposition 2.2], $d_P \leq \frac{d}{2}$. Therefore, since k_P, $\text{Vol}(P)$, and $|P \cap M|$ are all integers,

$$k_P \leq \text{Vol}(P) - |P \cap M| + d + \left\lfloor \frac{d}{2} \right\rfloor.$$

□

Remark 11 We show some cases that the result of Proposition 10 is stronger than [17, Proposition 2.4]:

1. $\text{Vol}(P) \leq |P \cap M| + 2$. In this case,

$$\text{Vol}(P) - |P \cap M| + d + \left\lfloor \frac{d}{2} \right\rfloor \leq d + \left\lfloor \frac{d}{2} \right\rfloor - 2.$$

Example 12 Let Δ_d be the standard d-simplex. Then

$$\text{Vol}(\Delta_d) - |\Delta_d \cap M| + d + \left\lfloor \frac{d}{2} \right\rfloor = 1 - (d+1) + d + \left\lfloor \frac{d}{2} \right\rfloor = \left\lfloor \frac{d}{2} \right\rfloor.$$

This is clearly a better bound compared to $d + \left\lfloor \frac{d}{2} \right\rfloor - 1$.

2. $P^0 \cap M = \emptyset$ or equivalently $\deg(P) \leq d - 1$. Indeed, in this case,

$$k_P \leq d_P + \deg(P) - 1 \leq \left\lfloor \frac{d}{2} \right\rfloor + d - 2.$$

We will show in next section that this is the only case that we still need to consider in order to verify the Eisenbud–Goto conjecture for very ample simplices.

Example 13 Consider $P = 2\Delta_d$ for $d \geq 4$, where Δ_d is the standard d-simplex. Then $\deg(P) = 2$ and by Proposition 7,

$$k_P \leq d_P + 1 \leq \left\lfloor \frac{d}{2} \right\rfloor + 1 < \left\lfloor \frac{d}{2} \right\rfloor + d - 1.$$

Theorem 14 *Suppose that X is a fake weighted projective space embedded in $\mathbb{P}^r$ via a very ample line bundle. Then*

$$\operatorname{reg}(X) \leq \deg(X) - \operatorname{codim}(X) + \left\lfloor \frac{\dim(X)}{2} \right\rfloor.$$

Proof Let P be the corresponding polytope of the embedding. From (12.2), (12.3), and Proposition 10, it follows that

$$\operatorname{reg}(X) \leq \operatorname{Vol}(P) - |P \cap M| + d + \left\lfloor \frac{d}{2} \right\rfloor + 1 = \deg(X) - \operatorname{codim}(X) + \left\lfloor \frac{d}{2} \right\rfloor.$$

12.5 Eisenbud–Goto Conjecture for Non-hollow Very Ample Simplices

In this section, we will improve the bound of k-normality for non-hollow very ample simplices.

Definition 15 A lattice polytope is hollow if it has no interior lattice points.

We now show that the inequality (12.4) holds for non-hollow very ample simplices.

Proposition 16 *Let $P \subseteq M_{\mathbb{R}}$ be a non-hollow very ample lattice d-simplex. Then*

$$k_P \leq \operatorname{Vol}(P) - |P \cap M| + d + 1.$$

Proof We will consider two cases, namely $|P \cap M| = d + 2$ and $|P \cap M| \geq d + 3$. For the first case, we have the following lemma:

Lemma 17 *Suppose that $P = \operatorname{conv}(v_0, \ldots, v_d)$ is a very ample lattice d-simplex with u is the only lattice point beside the vertices. Then P is normal.* □

Proof Assume that $d_P \geq 2$. Then there exists a point $p \in d_P P \cap M$ such that p cannot be written as $p = x + w$ for some $x \in (d_P - 1)P \cap M$ and $w \in P \cap M$. Since P is a simplex, u and p can be uniquely written as

$$p = \sum_{i=0}^{d} \lambda_i v_i, \quad \sum_{i=0}^{d} \lambda_i = d_P$$

and

$$u = \sum_{i=0}^{d} \lambda_i^* v_i, \quad \sum_{i=0}^{d} \lambda_i^* = 1,$$

respectively. It follows from the condition of p that $\lambda_i < 1$ for all $0 \le i \le d$ and there exists $0 \le i \le d$ such that $\lambda_i < \lambda_i^*$, say $i = 0$. By Lemma 4,

$$p + (d_P - 1)v_1 = \sum_{i=0}^{d} a_i v_i + bu$$

for some $a_i, b \in \mathbb{Z}_{\ge 0}$ such that $b + \sum_{i=0}^{d} a_i = 2d_P - 1$. Replacing p by $\sum_{i=0}^{d} \lambda_i v_i$ and u by $\sum_{i=0}^{d} \lambda_i^* v_i$ yields

$$\begin{aligned} \lambda_0 &= a_0 + b\lambda_0^* \\ \lambda_1 + d_P - 1 &= a_1 + b\lambda_1^* \\ \lambda_2 &= a_2 + b\lambda_2^* \\ &\vdots \\ \lambda_d &= a_d + b\lambda_d^*. \end{aligned}$$

Since $\lambda_0 < \lambda_0^*$ and $\lambda_i < 1$ for all $0 \le i \le d$, it follows that $a_0 = a_2 = \cdots = a_d = 0$ and $b = 0$. Then $p = d_P v_1$, a contradiction to the choice of p. Therefore, P is normal. □

As a consequence, $1 = k_P \le \text{Vol}(P) - |P \cap M| + d + 1 = \text{Vol}(P) - 1$. Now we consider the case $|P \cap M| \ge d + 3$. By the hypothesis, $|P \cap M| - (d + 1) \ge 2$. Consider the Ehrhart vector $h^* = (h_0^*, \cdots, h_d^*)$ of P. We have

$$\begin{aligned} h_0^* &= 1 \\ h_1^* &= |P \cap M| - d - 1 \ge 2 \\ h_d^* &= |P^0 \cap M| \ge 2. \end{aligned}$$

By [10, Theorem 1.1], $2 \le h_1^* \le h_i^*$ for all $1 \le i < d$. Therefore,

$$\text{Vol}(P) - |P \cap M| + d + 1 = h_0^* + h_2^* + \cdots + h_d^* \ge 1 + 2(d - 1) = 2d - 1.$$

By [17, Proposition 2.4],

$$k_P \le \left\lfloor \frac{d}{2} \right\rfloor + d - 1 \le 2d - 1 \le \text{Vol}(P) - |P \cap M| + d + 1$$

for all $d \ge 3$. The conclusion follows. □

Let us now recall the definition of Fano polytopes:

Definition 18 A Fano polytope is a convex lattice polytope $P \subseteq M_{\mathbb{R}}$ such that $P^0 \cap M = \{0\}$ and each vertex v of P is a primitive point of M.

From Proposition 16, we obtain the following corollary:

Corollary 19 *The Eisenbud–Goto conjecture holds for any projective toric variety corresponding to a very ample Fano simplex.*

12.6 Final Remarks

We start with a remark that Proposition 7 fails in general.

Example 20 ([3]) Consider the polytope P which is the convex hull of the vertices given by the columns of the following matrix

$$M = \begin{pmatrix} 0 & 1 & 0 & 0 & 1 & 0 & 1 & 1 \\ 0 & 0 & 1 & 0 & 0 & 1 & 1 & 1 \\ 0 & 0 & 0 & 1 & 1 & 1 & s & s+1 \end{pmatrix}$$

with $s \geq 4$. Then $d_P = \nu_P = 2$, and by [2, Theorem 3.3], $k_P = s - 1$. It is clear that $k_P > d_P + \nu_P - 1$ for all $s \geq 6$.

Furthermore, it can be shown that P cannot be covered by very ample simplicies [21, Proposition 4.3.3]; hence, it is very unlikely that we can apply Proposition 7 to find a bound of the k-normality of generic very ample polytopes.

12.6.1 *Hollow Very Ample Simplices*

Finally, we would love to see a classification of hollow very ample lattice simplices. For dimension 2, Rabinotwiz [18, Theorem 1] showed that any such simplex is unimodularly equivalent to either $T_{p,1} := \text{conv}(0, (p, 0), (0, 1))$ for some $p \in \mathbb{N}$ or $T_{2,2} = \text{conv}(0, (2, 0), (0, 2))$. Now we will show a way to obtain some hollow very ample simplices in any dimension with arbitrary volume.

We recall the definition of lattice pyramids as in [16]:

Definition 21 Let $B \subseteq \mathbb{R}^k$ be a lattice polytope with respect to $\mathbb{Z}^k$. Then $\text{conv}(0, B \times \{1\}) \subseteq \mathbb{R}^{k+1}$ is a lattice polytope with respect to $\mathbb{Z}^{k+1}$, called the (1-fold) standard pyramid over B. Recursively, we define for $l \in \mathbb{N}_{\geq 1}$ in this way the l-fold standard pyramid over B. As a convention, the 0-fold standard pyramid over B is B itself.

Proposition 22 *Let P be a lattice polytope. Then the 1-fold pyramid over P is very ample if and only if P is normal.*

Proof Let $Q = \text{conv}(0, P \times \{1\})$ be the 1-fold pyramid over P. Then it is easy to see that if P is normal then so is Q. Now suppose that Q is very ample. We have $k_Q \geq k_P$ [21, Lemma 4.2.2] and each lattice point of $k_Q Q \cap M$ sits in $(tP \cap M) \times \{t\}$ for some $0 \leq t \leq k_Q$. In particular, suppose that $(x, t) \in (tP \cap M) \times \{t\} \subseteq k_Q Q \cap M$. Then

$$(x, t) = \sum_{i=1}^{t} (u_i, 1) + (k_Q - t)0$$

for some $u_i \in P \cap M$. It follows that $x = \sum_{i=1}^{t} u_i$. Hence, P is t-normal for all $k_Q \geq t \geq 1$. Since $k_Q \geq k_P$, it follows that P is normal. The conclusion follows. □

From Proposition 22, if we take any $(d-2)$-fold pyramid over either $T_{p,1}$ with $p \in \mathbb{Z}_{\geq 1}$ or $T_{2,2}$, which are all normal, then we obtain a hollow normal (hence very ample) d-simplex with normalized volume p. The Eisenbud–Goto conjecture holds for these.

Example 23 We give here an example to demonstrate the case that if Q is very ample but not normal then the 1-fold pyramid over Q is not very ample. Let Q be the convex polytope given by taking $s = 4$ in Example 20. Then Q is very ample; however, the 1-fold pyramid of Q, which is given by the convex hull of

$$\begin{pmatrix} 0&0&1&0&0&1&0&1&1 \\ 0&0&0&1&0&0&1&1&1 \\ 0&0&0&0&1&1&1&4&5 \\ 0&1&1&1&1&1&1&1&1 \end{pmatrix},$$

is not very ample.

Acknowledgements We would like to thank Milena Hering for reading the drafts of this article and for some valuable suggestions.

References

1. Batyrev, V., Nill, B.: Multiples of lattice polytopes without interior lattice points. Mosc. Math. J. **7**(2), 195–207, 349 (2007)
2. Beck, M., Delgado, J., Gubeladze, J., Michał ek, M.: Very ample and Koszul segmental fibrations. J. Algebraic Combin. **42**(1), 165–182 (2015)
3. Bruns, W., Gubeladze, J.: Polytopes, Rings, and K-Theory. Springer Monographs in Mathematics. Springer, Dordrecht (2009)
4. Cox, D.A., Little, J.B., Schenck, H.K.: Toric Varieties. Graduate Studies in Mathematics, vol. 124. American Mathematical Society, Providence (2011)
5. Derksen, H., Sidman, J.: A sharp bound for the Castelnuovo-Mumford regularity of subspace arrangements. Adv. Math. **172**(2), 151–157 (2002)
6. Ehrhart, E.: Sur les polyèdres rationnels homothétiques à n dimensions. C. R. Acad. Sci. Paris **254**, 616–618 (1962)

7. Eisenbud, D., Goto, S.: Linear free resolutions and minimal multiplicity. J. Algebra **88**(1), 89–133 (1984)
8. Giaimo, D.: On the Castelnuovo-Mumford regularity of connected curves. Trans. Amer. Math. Soc. **358**(1), 267–284 (2006)
9. Hering, M.S.: Syzygies of toric varieties. ProQuest LLC, Ann Arbor, MI (2006). Thesis (Ph.D.)–University of Michigan
10. Hibi, T.: A lower bound theorem for Ehrhart polynomials of convex polytopes. Adv. Math. **105**(2), 162–165 (1994)
11. Hofscheier, J., Katthän, L., Nill, B.: Ehrhart theory of spanning lattice polytopes. Int. Math. Res. Not. IMRN **19**, 5947–5973 (2018)
12. Kwak, S.: Castelnuovo regularity for smooth subvarieties of dimensions 3 and 4. J. Algebraic Geom. **7**(1), 195–206 (1998)
13. Lazarsfeld, R.: A sharp Castelnuovo bound for smooth surfaces. Duke Math. J. **55**(2), 423–429 (1987)
14. McCullough, J., Peeva, I.: Counterexamples to the Eisenbud-Goto regularity conjecture. J. Amer. Math. Soc. **31**(2), 473–496 (2018)
15. Miyazaki, C.: Sharp bounds on Castelnuovo-Mumford regularity. Trans. Amer. Math. Soc. **352**(4), 1675–1686 (2000)
16. Nill, B.: Lattice polytopes having h^*-polynomials with given degree and linear coefficient. Eur. J. Combin. **29**(7), 1596–1602 (2008)
17. Ogata, S.: k-normality of weighted projective spaces. Kodai Math. J. **28**(3), 519–524 (2005)
18. Rabinowitz, S.: A census of convex lattice polygons with at most one interior lattice point. Ars Combin. **28**, 83–96 (1989)
19. Stanley, R.P.: Decompositions of rational convex polytopes. Ann. Discrete Math. **6**, 333–342 (1980)
20. Sturmfels, B.: Equations defining toric varieties. In: Algebraic Geometry—Santa Cruz 1995, Proceedings of Symposia in Pure Mathematics, vol. 62, pp. 437–449. American Mathematical Society, Providence (1997)
21. Tran, B.L.: On k-normality and regularity of normal projective toric varieties. Ph.D. thesis, University of Edinburgh (2018)

Chapter 13
Toric Degenerations in Symplectic Geometry

Milena Pabiniak

Abstract Toric degeneration in algebraic geometry is a process of degenerating a given projective variety into a toric one. Then one can obtain information about the original variety via analyzing the toric one, which is a much easier object to study. Harada and Kaveh described how one incorporates a symplectic structure into this process, providing a very useful tool for solving certain problems in symplectic geometry. Below we present two applications of this method: questions about the Gromov width, and cohomological rigidity problems.

Keywords Symplectic toric manifold · Bott manifold · Toric degeneration · Gromov width

13.1 Introduction

Manifolds and algebraic varieties equipped with a group action are usually better understood as a presence of an action is a sign of certain symmetries. In particular, *toric varieties* form a very well understood class of varieties. These are varieties which contain an algebraic torus $T_{\mathbb{C}}^n := (\mathbb{C}^*)^n$ as a dense open subset and are equipped with an action of $T_{\mathbb{C}}^n$ which extends the usual action of $T_{\mathbb{C}}^n$ on itself. For more about toric varieties see, for example, [5, 12]. To understand a given projective variety X one can try to "degenerate" it to a toric one, i.e., form a family of varieties with generic member X and one special member some toric variety X_0. The varieties X and X_0 are closely related and thus one can obtain information about X by studying X_0. Moreover, such a degeneration gives a map from X to X_0 which, in certain situations, preserves some special structures X and X_0 might be equipped with (for example: a symplectic structure).

One can use the method of toric degenerations to solve problems in symplectic geometry. In this work we discuss the following two applications:

M. Pabiniak (✉)
Mathematisches Institut, Universität zu Köln, Weyertal 86-90, 50931 Köln, Germany
e-mail: pabiniak@math.uni-koeln.de

A. M. Kasprzyk and B. Nill (eds.), *Interactions with Lattice Polytopes*,
Springer Proceedings in Mathematics & Statistics 386,
https://doi.org/10.1007/978-3-030-98327-7_13

1. calculating lower bounds for the Gromov width, i.e. trying to find the largest ball which can be symplectically embedded into a given symplectic manifold;
2. constructing symplectomorphisms needed for a cohomological rigidity problem for symplectic toric manifolds. This problem is about checking whether any two symplectic toric manifolds with isomorphic integral cohomology rings (via an isomorphism preserving the class of symplectic form) are symplectomorphic.

Recall that an $2n$-dimensional symplectic manifold (M, ω) equipped with an effective Hamiltonian action of an n-dimensional torus $T = (S^1)^n$ is called a *symplectic toric manifold*. The action being Hamiltonian means that there exists a moment map[1] $\mu \colon M \to \mathbb{R}^n$. Such a manifold can be given a complex structure interacting well with the symplectic one so that one calls ω a Kähler form and (M, ω) a Kähler manifold. In particular, symplectic toric manifolds are toric varieties in the sense of algebraic geometry. A theorem of Delzant states that we have a bijection[2]

{2n-dim compact symplectic toric manifolds} up to equivariant symplectomorphisms	$\Longleftrightarrow$	{rational and smooth polytopes in $\mathbb{R}^n$} up to translations and $\mathrm{GL}(n, \mathbb{Z})$ transformations.

In this bijection, a manifold corresponds to an image of its moment map, therefore the associated polytope is often called a moment polytope or a moment image. Not much is known about a classification of symplectic toric manifolds up to symplectomorphism. The cohomological rigidity mentioned in the second bullet above asks if such classification might be given by the integral cohomology rings.

In Sects. 13.3 and 13.4 respectively we describe the above problems in detail and explain how one can use toric degenerations to solve problems of this type. In particular we prove (rather, outline the proofs of) the following two results, obtained in projects joint with I. Halacheva, X. Fang, P. Littelmann, and S. Tolman. As to apply a toric degeneration to (M, ω) one needs ω to be an integral symplectic form, in both theorems we demand that the symplectic form is integral up to scaling, i.e. that $a[\omega] \in \mathrm{H}^*(M; \mathbb{Z})$ for some $a \in \mathbb{R}$. To simplify the exposition we slightly abuse the notation and given a map F defined on $\mathrm{H}^*(M; \mathbb{Z})$ we use F to also denote the map induced by F on $\mathrm{H}^*(M; \mathbb{Z}) \otimes_{\mathbb{Z}} \mathbb{R}$.

Theorem 1 ([11, 14]) *Let K be a compact connected simple Lie group. The Gromov width of a coadjoint orbit $\mathcal{O}_\lambda$ through a point λ, integral up to scaling, equipped with the Kostant–Kirillov–Souriau symplectic form, is at least*

$$\min\{\, \left|\langle \lambda, \alpha^\vee \rangle\right| \,;\ \alpha^\vee \text{ a coroot and } \langle \lambda, \alpha^\vee \rangle \neq 0\}. \tag{13.1}$$

[1] A moment map is a T-invariant map $\mu \colon M \to \mathrm{Lie}(T)^* \cong \mathbb{R}^n$ such that for every $X \in \mathrm{Lie}(T)$ it holds that $\iota_{X^\sharp}\omega = d\mu^X$ where $X^\sharp$ denotes the vector field on M induced by X and $\mu^X \colon M \to \mathbb{R}$ is defined by $\mu^X(p) = \langle \mu(p), X \rangle$..

[2] Recall that a polytope in $\mathbb{R}^n$ is called rational if the directions of its edges are in $\mathbb{Z}^n$. It is called smooth if for every vertex the primitive vectors in the directions of edges meeting at that vertex form a $\mathbb{Z}$-basis for $\mathbb{Z}^n$.

Theorem 2 ([27]) *Let* (M, ω_M) *and* (N, ω_N) *be Bott manifolds with symplectic forms integral up to scaling. Moreover, assume that* $\mathrm{H}^*(M; \mathbb{Q})$ *and* $\mathrm{H}^*(N; \mathbb{Q})$ *are isomorphic to* $\mathbb{Q}[x_1, \ldots, x_n]/\langle x_1^2, \ldots, x_n^2 \rangle$. *For any ring isomorphism* $F\colon \mathrm{H}^*(M; \mathbb{Z}) \to \mathrm{H}^*(N; \mathbb{Z})$ *with* $F([\omega_M]) = [\omega_N]$, *there exists a symplectomorphism* $f\colon (N, \omega_N) \to (M, \omega_M)$ *such that the map* $\mathrm{H}^*(f)\colon \mathrm{H}^*(M; \mathbb{Z}) \to \mathrm{H}^*(N; \mathbb{Z})$ *induced by* f *on integral cohomology rings is exactly* F.

There are other applications of toric degenerations in symplectic geometry. For example, one can obtain information about Ginzburg–Landau potential function on X from that of X_0 and thus detect some non-displaceable Lagrangians of X, see [25].

13.2 Toric Degenerations

A **toric degeneration** of a projective variety X is a flat family $\pi\colon \mathfrak{X} \to \mathbb{C}$ with generic fiber X and one special fiber $X_0 = \pi^{-1}(0)$, a (not necessarily normal) toric variety. A construction of such a degeneration of a projective variety X, equipped with a very ample line bundle satisfying certain conditions, can be found in Anderson [1, Theorem 2].

Example 3 Using the Plücker embedding,[3] view $X = Gr(2, \mathbb{C}^4)$, the Grassmannian of 2-planes in $\mathbb{C}^4$, as a subset of $\mathbb{CP}^5$ with coordinates $\{x_{ij};\ 1 \le i < j \le 4\}$, consisting of points satisfying

$$x_{12}x_{34} - x_{13}x_{24} + x_{14}x_{23} = 0.$$

Consider the subset $\mathfrak{X} \subset \mathbb{CP}^5 \times \mathbb{C}$ consisting of points satisfying

$$x_{12}x_{34} - x_{13}x_{24} + t x_{14}x_{23} = 0,$$

where t denotes the coordinate in $\mathbb{C}$. Let $\pi\colon \mathfrak{X} \to \mathbb{C}$ be the restriction to $\mathfrak{X}$ of the projection onto the second factor. This family constitutes a toric degeneration of $Gr(2, \mathbb{C}^4)$. In fact, $\{x_{ij}\}$ form a SAGBI basis of the homogeneous coordinate ring of X and this ensures the flatness [8, Theorem 15.17]. Clearly $\pi^{-1}(1)$ is $Gr(2, \mathbb{C}^4)$. Moreover, performing a change of coordinates, one can show that $\pi^{-1}(t)$ for $t \neq 0$ is also bihomolomorhpic to $Gr(2, \mathbb{C}^4)$. The central fiber, $\pi^{-1}(0)$, is described by the binomial ideal $\langle x_{12}x_{34} - x_{13}x_{24} \rangle$ and thus is a toric variety.

Harada and Kaveh [16] enriched the construction of Anderson by incorporating a symplectic structure. They start with a smooth projective variety X, of complex

---

[3] Recall that the Plücker embedding sends a Grassmannian spanned by vectors $v, w \in \mathbb{C}^4$ to a point $[x_{12} : \ldots : x_{34}] \in \mathbb{CP}^5$ with x_{ij} equal to the determinant of the 2×2 minor of $[v^T, w^T]$ spanned by rows i and j.

dimension n, equipped with a very ample line bundle $\mathcal{L}$, with some fixed Hermitian structure. Let $L := \mathrm{H}^0(X, \mathcal{L})$ denote the vector space of holomorphic sections, $\Phi_{\mathcal{L}}\colon X \to \mathbb{P}(L^*)$ the Kodaira embedding and $\omega = \Phi^*_{\mathcal{L}}(\omega_{FS})$ the pull back of the Fubini–Study form, i.e., of the standard symplectic structure on complex projective spaces. Then (X, ω) is a Kähler manifold. With this data they construct (under certain assumptions) not only a flat family $\pi : \mathfrak{X} \to \mathbb{C}$ but also a Kähler structure $\tilde{\omega}$ on (the smooth part of) $\mathfrak{X}$ so that $(\pi^{-1}(1), \tilde{\omega}_{|\pi^{-1}(1)})$ is symplectomorphic to (X, ω). Moreover, the special fiber $X_0 = \pi^{-1}(0)$ obtains a Kähler form, the restriction of $\tilde{\omega}$, defined on its smooth part $U_0 := (X_0)_{\text{smooth}}$, and thus it also obtains a divisor. If X_0 is normal, then the polytope associated to X_0 and this divisor by the usual procedure of toric algebraic geometry (see, for example, [5, Chap. 4]) is the closure of the moment image of the (non-compact) symplectic toric manifold $(U_0, \tilde{\omega}_{|U_0})$. As we will see, this polytope can be computed by analyzing the behaviour of the holomorphic sections of $\mathcal{L}$. Here are more details of this procedure.

Denote by L^m the image of span $\langle f_1 \cdot \ldots \cdot f_m\,;\ f_i \in L\rangle$ in $\mathrm{H}^0(X, \mathcal{L}^{\otimes m})$ and by $R = \mathbb{C}[X] = \oplus_{m \geq 0} L^m$ the homogeneous coordinate ring of X with respect to the embedding $\Phi_{\mathcal{L}}$. An important ingredient of the construction is a choice of a *valuation with one dimensional leaves*, $\nu\colon \mathbb{C}(X) \setminus \{0\} \to \mathbb{Z}^n$, from the ring $\mathbb{C}(X)$ of rational functions on X. A precise definition of a general valuation can be found, for example, in [16, Definition 3.1]. In this paper we only use valuations induced by a flag of subvarieties and a special case of these, called *lowest/highest term valuations associated to a coordinate system.*

Example 4 (*Lowest/highest term valuations* [16, Example 3.2]) Fix a smooth point $p \in X$ and let $(u_1, \ldots, u_n)$ be a system of coordinates in a neighborhood of p, meaning that $u_1, \ldots, u_n$ are regular functions at p, vanishing at p, and such that their differentials $du_1, \ldots, du_n$ are linearly independent at p. Then any regular function at p can be represented as a power series $\sum_{\alpha \in \mathbb{Z}^n_{\geq 0}} c_\alpha u^\alpha$. Here by u^α, with $\alpha = (\alpha_1, \ldots, \alpha_n) \in \mathbb{Z}^n_{\geq 0}$, we mean $u_1^{\alpha_1} \cdot \ldots \cdot u_n^{\alpha_n}$. Choose and fix a total order $>$ on $\mathbb{Z}^n$ respecting the addition, for example the lexicographic order. Define a map ν from the set of functions regular at p to $\mathbb{Z}^n$ by

$$\nu\Big(\sum_{\alpha \in \mathbb{Z}^n_{\geq 0}} c_\alpha u^\alpha\Big) = \min\{\alpha;\ c_\alpha \neq 0\},$$

and extend it to $\mathbb{C}(X) \setminus \{0\}$ by setting $\nu(f/g) = \nu(f) - \nu(g)$. Then ν is a valuation with one dimensional leaves, called a *lowest term valuation*. If one uses max instead of min in the definition of ν, one obtains a *highest term valuation*.

Example 5 (*Valuations induced by a flag of subvarieties* [16, Example 3.3]) Take a flag of normal subvarieties (called a Parshin point) of X

$$\{p\} = Y_n \subset \ldots \subset Y_0 = X,$$

with $\dim_{\mathbb{C}}(Y_k) = n - k$ and Y_k non-singular along Y_{k+1}. By the non-singularity assumption there exists a collection of rational functions $u_1, \ldots, u_n$ on X such that $u_{k|Y_{k-1}}$ is a rational function on Y_{k-1} which is not identically zero and which has a zero of first order on Y_k. Then the lowest term valuation with respect to the lexicographic order can alternatively be described in the following way: for any $f \in \mathbb{C}(X)$, $f \neq 0$, the valuation $\nu(f) = (k_1, \ldots, k_n)$ where k_1 is the order of vanishing of f on Y_1, k_2 is the order of vanishing of $f_1 := (u_1^{-k_1} f)_{|Y_1}$ on Y_2, etc.

Given such $X, \mathcal{L}$, and ν we form a semigroup $S = S(\nu, \mathcal{L})$, in the following way. Fix a non-zero element $h \in L$ and use it to identify L with a subspace of $\mathbb{C}(X)$ by mapping $f \in L$ to $f/h \in \mathbb{C}(X)$. Similarly identify L^m with a subspace of $\mathbb{C}(X)$ by sending $f \in L^m$ to $f/h^m \in \mathbb{C}(X)$. As any valuation satisfies that $\nu(fg) = \nu(f) + \nu(g)$, the set

$$S = S(\nu, \mathcal{L}) = \bigcup_{m \geq 0} \{(m, \nu(f/h^m)) \mid f \in L^m \setminus \{0\}\}$$

is a semigroup with identity (i.e. a monoid). If S is finitely generated, one can construct a toric degeneration whose special fiber is a toric variety $\mathrm{Proj}\, \mathbb{C}[S]$ (which is normal if S is saturated). Moreover we obtain an Okounkov body

$$\Delta = \Delta(S) = \overline{\mathrm{conv}\Big(\bigcup_{m>0} \{x/m \mid (m, x) \in S\}\Big)} \subset \mathbb{R}^n.$$

Note that if S is finitely generated, then Δ is a rational convex polytope. The toric variety corresponding to Δ is the normalization of $\mathrm{Proj}\, \mathbb{C}[S]$.[4]

In the following theorem we rephrase several results from [16].

Theorem 6 ([16]) *Let $\mathcal{L}$ be a very ample Hermitian line bundle on a smooth n-dimensional projective variety X and $\omega = \Phi_{\mathcal{L}}^*(\omega_{FS})$ the induced symplectic form. Let $\nu\colon \mathbb{C}(X) \setminus \{0\} \to \mathbb{Z}^n$ be a valuation with one dimensional leaves, and such that the associated semigroup S is finitely generated. Then*

- *There exists a toric degeneration $\pi : \mathfrak{X} \to \mathbb{C}$ with generic fiber X and special fiber $X_0 :=$ Proj $\mathbb{C}[S]$, and a Kähler structure $\tilde{\omega}$ on (the smooth part of) $\mathfrak{X}$ such that $(\pi^{-1}(1), \tilde{\omega}_{|\pi^{-1}(1)})$ is symplectomorphic to (X, ω) and the closure of the moment image of symplectic toric manifold $(U_0, \tilde{\omega}_{|U_0})$, where $U_0 := (X_0)_{smooth}$, is the Okounkov body $\Delta(S)$. The set U_0 contains the preimage of the interior of $\Delta(S)$.*
- *Moreover, there exists a surjective continuous map $\phi\colon X \to X_0$ that restricts to a symplectomorphism from $(\phi^{-1}(U_0), \omega)$ to $(U_0, \tilde{\omega}_{|U_0})$.*

In particular, if $X_0 =$ Proj $\mathbb{C}[S]$ built from S is smooth (thus also normal), then ϕ^{-1} $(U_0) = X$ and therefore ϕ provides a symplectomorphism between (X, ω) and the symplectic toric manifold $(X_{\Delta(S)}, \omega_{\Delta(S)})$ associated to $\Delta(S)$ via Delzant's construction.

---

[4] Recall that for a graded algebra $A = \oplus_{j=0}^{\infty} A_j$ the set Proj A is the set of homogeneous prime ideals in A that do not contain all of $A_+ := \oplus_{j=1}^{\infty} A_j$. The topology on Proj A is defined by setting the closed sets to be $V(I) := \{J;\ J \subset I$ is a homogeneous prime ideal of A not containing all of$A_+\}$, for some homogeneous ideal I of A. For more details see, for example [17, II.2], [9, III.2], and [5, Chap. 7].

Checking whether S is finitely generated is a very difficult problem. However, it was observed by Kaveh [20] that even if S is not finitely generated one can still form a (not flat) family with generic fiber X and special fiber $(\mathbb{C}^*)^n$. Even though such a construction provides much less information about X, it still suffices for the purpose of finding lower bounds on the Gromov width. We describe this idea in Sect. 13.3.

13.3 Gromov Width

The *Gromov width* of a $2n$-dimensional symplectic manifold (X, ω) is the supremum of the set of the positive real numbers a such that the ball of capacity a (radius $\sqrt{\frac{a}{\pi}}$),

$$B_a^{2n} = B^{2n}\Big(\sqrt{\frac{a}{\pi}}\Big) = \Big\{(x_1, y_1, \ldots, x_n, y_n) \in \mathbb{R}^{2n} \;\Big|\; \pi \sum_{i=1}^{n}(x_i^2 + y_i^2) < a\Big\} \subset (\mathbb{R}^{2n}, \omega_{st}),$$

can be symplectically embedded in (X, ω). Here $\omega_{st} = \sum_j dx_j \wedge dy_j$ denotes the standard symplectic form on $\mathbb{R}^{2n}$. This question was motivated by the Gromov non-squeezing theorem which states that a ball $B^{2n}(r) \subset (\mathbb{R}^{2n}, \omega_{st})$ cannot be symplectically embedded into $B^2(R) \times \mathbb{R}^{2n-2} \subset (\mathbb{R}^{2n}, \omega_{st})$ unless $r \leq R$.

J-holomorphic curves give obstructions to ball embeddings, while Hamiltonian torus actions can lead to constructions of such embeddings (by extending a Darboux chart using the flow of the vector field induced by the action).

This is why toric degenerations provide a useful tool for finding lower bounds on the Gromov width. Given a toric degeneration of (X, ω), as described in Theorem 6, one can use the toric action on X_0 to construct embeddings of balls into a smooth symplectic toric manifold $(U_0, \tilde{\omega}_{|U_0})$, where $U_0 = (X_0)_{\text{smooth}}$. Postcomposing such embedding with the symplectomorphism ϕ^{-1} produces a symplectic embedding into (X, ω).

Moreover, many embeddings of balls into symplectic toric manifolds can be read off from the associated (by the Delzant classification theorem) polytope. Identify the dual of the Lie algebra of the compact torus T with Euclidean space using the convention that $S^1 = \mathbb{R}/\mathbb{Z}$, i.e. the lattice of $\mathfrak{t}^*$ is mapped to $\mathbb{Z}^{\dim T} \subset \mathbb{R}^{\dim T}$. With this convention, the moment map for the standard $(S^1)^n$ action on $(\mathbb{R}^{2n}, \omega_{st})$ maps B_a^{2n} onto an n-dimensional simplex of size a, closed on n sides

$$\mathfrak{S}^n(a) := \left\{(x_1, \ldots, x_n) \in \mathbb{R}^n \,|\, 0 \leq x_j < a,\ \sum_{j=1}^{n} x_j < a\right\}.$$

Moreover, if the moment image contains an open simplex of size a, then for any $\varepsilon > 0$ a ball of capacity $a - \varepsilon$ can be embedded into the given symplectic toric manifold: see [28, Lemma 5.3.1] and, independently, [26, Propositions 2.1 and 2.4].

Proposition 7 ([24, Proposition 1.3] and [26, Proposition 2.5]) *For any connected, proper (not necessarily compact) symplectic toric manifold U of dimension 2n, with a momentum map μ, the Gromov width of U is at least*

$$\sup\{a > 0 \mid \exists\ \Psi \in \mathrm{GL}(n, \mathbb{Z}), x \in \mathbb{R}^n, \textit{ such that } \Psi(\mathrm{int}\,\mathfrak{S}^n(a)) + x \subset \mu(U)\}.$$

The appearance of Ψ and x comes from the facts that the identification $\mathfrak{t}^* \cong \mathbb{R}^{\dim\ T}$ depends on a splitting of T into (dim T) circles, and that a translation of a moment map also provides a moment map.

The above results lead to the following method for finding lower bounds on the Gromov width.

Corollary 8 *Let X be a smooth projective variety of complex dimension n, $\mathcal{L}$ an ample line bundle on X, and $\omega = \Phi^*_{\mathcal{L}}(\omega_{FS}) \in \mathrm{H}^2(X, \mathbb{Z})$ an integral Kähler form obtained using the Kodaira embedding $\Phi_{\mathcal{L}}\colon X \to \mathbb{P}(L^*)$. Suppose that there exists a valuation ν giving a finitely generated and saturated semigroup $S = S(\nu, \mathcal{L})$. Let Δ be the associated Okounkov body. The Gromov width of (X, ω) is at least*

$$\sup\{a > 0 \mid \exists\ \Psi \in \mathrm{GL}(n, \mathbb{Z}), x \in \mathbb{R}^n, \textit{ such that } \Psi(\mathrm{int}\,\mathfrak{S}^n(a)) + x \subset \Delta\}.$$

Proof By the result of [16] cited here as Theorem 6, there exists a toric degeneration of (X, ω) to a normal toric variety $X_0 = \mathrm{Proj}\,\mathbb{C}[S]$, and a surjective continuous map $\phi\colon X \to X_0$ whose appropriate restriction is a symplectomorphism. The subset $U := \phi^{-1}(U_0)$ of X inherits a toric action whose moment image contains int Δ, the interior of Δ (recall that a moment map sends singular points of a toric variety to the boundary of the moment polytope). The corollary follows from Proposition 7. □

In fact one does not need S to be saturated. The same corollary holds even if X_0 is not a normal toric variety. This is because a normalization map for X_0 induces a biholomorphism between $(X_0)_{\mathrm{smooth}}$ and an appropriate subset of the normalization of X_0.

It is, however, necessary that S is finitely generated for a toric degeneration to exist. Otherwise one can still form a family of manifolds, but one cannot guarantee that this family is flat, and thus X and X_0 are no longer so strongly related. As we already mentioned, Kaveh in [20] observed that such a (not necessarily flat) family, with $X_0 = (\mathbb{C}^*)^n$, still provides information about the Gromov width of (X, ω). To state this result we need additional notation. In the notation of Sect. 13.2, for any $m \in \mathbb{Z}_{>0}$ let

$$\mathcal{A}_m := \{\nu(f/h^m) \mid f \in L^m \setminus \{0\}\} \subset \mathbb{Z}^n, \quad \Delta_m = \frac{1}{m}\,\mathrm{conv}(\mathcal{A}_m).$$

Note that $\Delta = \overline{\bigcup_{m>0}\Delta_m}$. Fix m and let $r = r_m$ denote the number of elements in $\mathcal{A}_m = \{\beta_1, \ldots, \beta_r\}$. From these data we form a symplectic form, ω_m, on $(\mathbb{C}^*)^n$ using a standard procedure: ω_m is the pull back of the Fubini–Study form on $\mathbb{CP}^{r-1}$ via the map $\Psi_m\colon (\mathbb{C}^*)^n \to \mathbb{CP}^{r-1}, u \mapsto (u^{\beta_1}c_1, \ldots, u^{\beta_r}c_r)$, where $c = [(c_1, \ldots, c_r)]$ is some element in $\mathbb{CP}^{r-1}$ with all $c_i \neq 0$. (In [20] the elements c_i come from coefficients of leading terms of elements in appropriately chosen basis of L^m. One also needs that the differences of elements in $\mathcal{A}_m$ span $\mathbb{Z}^n$, which, by [20, Remark 5.6], is always true for lowest term valuations.)

Kaveh proved that:

1. for every $m > 0$ there exists an open subset $U \subset X$ such that (U, ω) is symplectomorphic to $((\mathbb{C}^*)^n, \frac{1}{m}\omega_m)$ [20, Theorem 10.5];
2. the Gromov width of $((\mathbb{C}^*)^n, \frac{1}{m}\omega_m)$ is at least R_m, where R_m is the size of the largest open simplex that fits in the interior of $\Delta_m = \frac{1}{m}\operatorname{conv}(\mathcal{A}_m)$ [20, Corollary 12.3].

This leads to the following corollary.

Corollary 9 *([20, Corollary 12.4]) Let X be a smooth projective variety of dimension n, $\mathcal{L}$ an ample line bundle on X, and $\omega = \Phi^*_{\mathcal{L}}(\omega_{FS}) \in H^2(X, \mathbb{Z})$ an integral Kähler form. Let ν be a lowest term valuation on $\mathbb{C}(X)$, with values in $\mathbb{Z}^n$, and Δ the associated Okounkov body. The Gromov width of (X, ω) is at least R, where R is the size of the largest open simplex that fits in the interior of Δ.*

13.3.1 Results About Coadjoint Orbits

The methods for finding the Gromov width described in Corollaries 8 and 9 have been used in [11, 14] for coadjoint orbits of compact Lie groups.

Recall that given a compact Lie group K each orbit $\mathcal{O} \subset \mathfrak{k}^* := (\text{Lie } K)^*$ of the coadjoint action of K on $\mathfrak{k}^*$ is naturally a symplectic manifold. Namely it can be equipped with the Kostant–Kirillov–Souriau symplectic form ω^{KKS} defined by:

$$\omega^{KKS}_{\xi}(X^{\#}, Y^{\#}) = \langle \xi, [X, Y] \rangle, \quad \xi \in \mathcal{O} \subset \mathfrak{k}^*, \; X, Y \in \mathfrak{k},$$

where $X^{\#}, Y^{\#}$ are the vector fields on $\mathfrak{k}^*$ induced by $X, Y \in \mathfrak{k}$ via the coadjoint action of K. For more details see, for example, [7, Sect. 21.5, Homeworks 16 and 17]. Coadjoint orbits are in bijection with points in a positive Weyl chamber as every coadjoint orbit intersects such a chamber in a single point. An orbit is called generic (resp. degenerate) if this intersection point is an interior point of the chamber (resp. a boundary point). For example, when $K = \mathrm{U}(n, \mathbb{C})$ is the unitary group, a coadjoint orbit can be identified with the set of Hermitian matrices with a fixed set of eigenvalues. The orbit is generic if all eigenvalues are different, and in this case it is diffeomorphic to the manifold of complete flags in $\mathbb{C}^n$.

It has been unofficially conjectured[5] that the Gromov width of $(\mathcal{O}_\lambda, \omega^{KKS})$ of K, through a point λ in a positive Weyl chamber should be given by the following neat

[5] During the work on the project [18], about complex Grassmannians, Karshon and Tolman looked at various examples of other coadjoint orbits and got the impression that the above value might be the Gromov width of all coadjoint orbits. They never formulated this expectation formally as their

formula, expressed entirely in the Lie-theoretical language

$$\min\{\,|\langle\lambda,\alpha^\vee\rangle|\,;\ \alpha^\vee \text{ a coroot and } \langle\lambda,\alpha^\vee\rangle\neq 0\}.$$

For example, as $\{e_{ii}-e_{jj};\ i\neq j\}$ forms a root system for the unitary group $\mathrm{U}(n,\mathbb{C})$, the Gromov width of its coadjoint orbit $\mathcal{O}_\lambda$ passing through a point

$$\lambda = \mathrm{diag}(\lambda_1,\ldots,\lambda_n)\in \mathfrak{u}(n)^*,$$

integral up to scaling, is equal to $\min\{|\lambda_i-\lambda_j|;\ i,j\in\{1,\ldots,n\},\ \lambda_i\neq\lambda_j\}$. Here we identified $\mathfrak{u}(n)$ and $\mathfrak{u}(n)^*$ with the set of $n\times n$ Hermitian matrices.

This conjecture was motivated by the computation of the Gromov width of complex Grassmannians, i.e. degenerate coadjoint orbits of $\mathrm{U}(n,\mathbb{C})$, done by Karshon and Tolman [18], and independently by Lu [23]. Later, using holomorphic techniques, Zoghi [29] showed that the above formula provides an upper bound for the Gromov width for generic indecomposable[6] orbits of $\mathrm{U}(n,\mathbb{C})$. This result was generalized to all coadjoint orbits by Caviedes [2]. The fact that this formula also provides a lower bound was proved using explicit Hamiltonian torus actions by several authors: [29] gives a proof for generic indecomposable orbits of $\mathrm{U}(n,\mathbb{C})$ using the standard action of the maximal torus, Lane [21] proves this lower bound for generic orbits of the exceptional group G_2, and [26] settled the case of $\mathrm{U}(n,\mathbb{C})$, $\mathrm{SO}(2n,\mathbb{C})$ and $\mathrm{SO}(2n+1,\mathbb{C})$ orbits[7] using the Gelfand–Tsetlin torus action.

13.3.2 A Sketch of the Proof of Theorem 1

The first usage of toric degenerations in Gromov width problems appeared in [14], where the generic orbits of the symplectic group $\mathrm{Sp}(n)=\mathrm{U}(n,\mathbb{H})$ are considered. Then it was used in [11] to prove that the formula (13.1) is a lower bound for the Gromov width of any coadjoint orbit of any compact connected simple Lie group K, passing through a point in the Weyl chamber, integral up to scaling, i.e. to prove Theorem 1.

The rationality assumption comes from the fact that the toric degeneration method can be applied only to the orbits passing through an integral point λ of a positive Weyl

conjecture, but they shared this idea with other mathematicians in private communications. This is how this value became to be known as the expected Gromov width for coadjoint orbits.

[6] A coadjoint orbit through a point λ in the interior of a chosen positive Weyl chamber is called indecomposable in [29] if there exists a simple positive root α such that for any positive root α' there exists a positive integer k such that $\langle\lambda,\alpha'\rangle = k\langle\lambda,\alpha\rangle$.

[7] The result about $\mathrm{SO}(2n+1,\mathbb{C})$ holds only for orbits satisfying a mild technical condition: the point λ of intersection of the orbit and a chosen positive Weyl chamber should not belong to a certain subset of one wall of the chamber; see [26] for more details. In particular, all generic orbits satisfy this condition.

chamber, i.e., in the language of representation theory language, through a dominant weight.

Let G be a simply connected simple complex algebraic group and $K \subset G$ be its maximal compact subgroup. With a dominant weight λ one can associate an irreducible representation $V(\lambda)$ of G of highest weight λ. Let $\mathbb{C}_{v_\lambda}$ be the highest weight line and $P = P_\lambda$ be the normalizer in G of this line. Then the coadjoint orbit $\mathcal{O}_\lambda$ of K is diffeomorphic to G/P (and to $K/K \cap P$) and there exists a very ample line bundle $\mathcal{L}_\lambda$ on G/P such that the pull back of the Fubini–Study form on the projective space $\mathbb{P}(\mathrm{H}^0(G/P, \mathcal{L}_\lambda)^*) = \mathbb{P}(V(\lambda))$ via the Kodaira embedding $G/P \to \mathbb{P}(\mathrm{H}^0(G/P, \mathcal{L}_\lambda)^*)$ is exactly the Kostant–Kirillov–Souriau symplectic form ω^{KKS} on $\mathcal{O}_\lambda$ (see for example [2, Remark 5.5]). Thus for integral λ's one can try to construct toric degenerations of projective variety G/P with line bundle $\mathcal{L}_\lambda$ and obtain some lower bounds for the Gromov width of the orbit $\mathcal{O}_\lambda$. Rescaling of symplectic forms allows to extend such a result to orbits $\mathcal{O}_{a\lambda}$, for any $a \in \mathbb{R}_{>0}$.

It remains to discuss how one can construct desired toric degenerations.

A great advantage of working with coadjoint orbits of a complex algebraic group G is that a lot of information can be obtained from studying representations of G. This leads to a beautiful interplay between symplectic geometry and representation theory. A reduced decomposition of the longest word in the Weyl group, $\underline{w_0} = s_{i_{\alpha_1}} \cdot \ldots \cdot s_{i_{\alpha_N}}$ provides the following items (defined below) related in an interesting way:

1. a valuation $\nu_{\underline{w_0}}$;
2. a string parameterization of a crystal basis of V_λ^*.

We continue to denote by λ a dominant weight (i.e. an integral element in a positive Weyl chamber of $\mathfrak{g}^*$) and by V_λ the finite dimensional irreducible representation of G with highest weight λ. Let V_λ^* denote the dual representation. One often seeks for a basis of V_λ^* consisting of elements which behave nicely under the action of Kashiwara operators. A crystal basis is a basis whose elements are permuted under the Kashiwara operators. Given a crystal basis one can form a crystal graph of a given representation: vertices are elements of the crystal basis and $\{0\}$, and edges are labelled by simple roots and correspond to the action of Kashiwara operators. There are (not canonical) ways of embedding such graph into $\mathbb{R}^N$, $N = \dim_{\mathbb{C}} G/P$. A reduced decomposition of the longest word in the Weyl group (into a composition of reflections with respect to simple roots), $\underline{w_0} = s_{\alpha_1} \cdot \ldots \cdot s_{\alpha_N}$, provides a way of assigning to each vertex of the crystal graph a string of N integers (string parametrization), and thus gives such an embedding. A convex hull of the image of string parametrization is called a string polytope. It depends on λ and also on the chosen decomposition $\underline{w_0}$. String polytopes have been extensively studied in representation theory.

Moreover, a reduced decomposition $\underline{w_0} = s_{i_{\alpha_1}} \cdot \ldots \cdot s_{i_{\alpha_N}}$ defines a sequence of Schubert subvarieties

$$[P] = Y_N \subset \ldots \subset Y_0 = G/P,$$

where Y_j denotes the Schubert variety corresponding to element $s_{i_{\alpha_{j+1}}} \cdot \ldots \cdot s_{i_{\alpha_N}}$ of the Weyl group. We denote by $\nu_{\underline{w_0}}$ the highest term valuation associated with this flag of subvarieties.

A theorem of Kaveh relates these two objects.

Theorem 10 ([19]) *The string parametrization for $V_\lambda^* = \mathrm{H}^0(G/P, \mathcal{L}_\lambda)$ obtained using the reduced decomposition $\underline{w_0}$ is the restriction of the valuation $\nu_{\underline{w_0}}$ and thus the corresponding string polytope is the Okounkov body $\Delta(\nu_{\underline{w_0}})$.*

Detailed computations for the case of $G = SL(3, \mathbb{C})$ and $\underline{w_0} = s_1 s_2 s_2$ are presented in [19, Sect. 5].

Explicit descriptions of string polytopes for classical Lie groups and some well-chosen reduced decompositions of the longest words were presented in the work of Littelmann [22]. With a bit of work one can show that the string polytope for V_λ^* with $G = \mathrm{Sp}(2n, \mathbb{C})$ the symplectic group (with maximal compact subgroup $K = \mathrm{Sp}(n) = \mathrm{U}(n, \mathbb{H})$), described in [22], contains a simplex of size prescribed by (13.1). Then, the result of Kaveh, [19], quoted above together with Corollary 8 prove that the Gromov width of $\mathrm{Sp}(n)$ coadjoint orbit $(\mathcal{O}_\lambda, \omega^{KKS})$ is at least equal to the value prescribed by (13.1), i.e. proves Theorem 1 for the case of the symplectic group. This is exactly the argument used in [14].

Similar methods could be applied for other classical Lie groups. However, one would need to consider each type separately, as the descriptions of string polytopes contained in [22] depend on reduced decompositions which are different for different group types.

To obtain a unified proof which works for all group types, in [11] we use lowest term valuations ν arising from a system of parameters induced by an enumeration $\{\beta_1, \ldots, \beta_N\}$ of certain positive roots, also in the cases where this enumeration does not come from a reduced decomposition of the longest word. In these cases it might be very difficult to show that the associated semigroup $S(\nu)$ is finitely generated (if it is) and to find an explicit description of the associated Okounkov body. Moreover, on the representation theory side, we do not have a natural way of obtaining a string parametrization of a crystal basis of V_λ^* from such enumerations. However, in [10] the authors managed to give a representation-theoretic description of the associated semigroup $S(\nu)$ in the case when the enumeration is a good ordering in the sense of [10]. Here is the main idea. Given such enumeration one can define for each $\alpha \in \mathbb{Z}^N_{\geq 0}$ subspaces $V(\lambda)_{\leq \alpha}$ and $V(\lambda)_{<\alpha}$ of $V(\lambda)$. An element $\alpha \in \mathbb{Z}^N_{\geq 0}$ is called essential for λ if $\dim V(\lambda)_{\leq \alpha}/V(\lambda)_{<\alpha} = 1$. It was proved in [10] that the set $\{(l, \alpha);\ l \in \mathbb{Z}_{\geq 0},\ \alpha \text{ essential for } l\lambda\}$ is a semigroup which coincides with $S(\nu)$. Moreover, building on other results from [10] concerning essential elements, one can show that the Okounkov body associated to $S(\nu)$ contains a simplex of size prescribed by (13.1). Then, using the result of [20] cited here as Corollary 9 (which does not require the semigroup to be finitely generated), one proves Theorem 1. The details of this argument are presented in [11].

13.4 Cohomological Rigidity

The following section is based on a project joint with Sue Tolman [27].

Cohomological rigidity problems are problems where one tries to determine whether the integral cohomology ring can distinguish between manifolds of certain family, and whether all isomorphisms between integral cohomology rings are induced by maps (homeomorphisms or diffeomorphisms, depending on the setting) between manifolds. The integral cohomology ring is too weak to distinguish a homeomorphism type of a manifold. However, by a result of Freedman, it classifies (up to a homeomorphism) all closed, smooth, simply connected 4-manifolds. Masuda and Suh posed a question of whether the cohomological rigidity holds for the family of symplectic toric manifolds. The question was studied by its authors, Choi, and Panov. No counterexample was found and partial positive results were proved. (Interested reader is encouraged to consult a nice survey [4] and references therein.) Due to the presence of symplectic structure, it seems natural to consider the following symplectic variant of the above question.

Question 11 (*Symplectic cohomological rigidity for symplectic toric manifolds*)

1. (weak) Are two symplectic toric manifolds (M, ω_M) and (N, ω_N) necessarily symplectomorphic whenever there exists an isomorphism $F\colon \mathrm{H}^*(M;\mathbb{Z}) \to \mathrm{H}^*(N;\mathbb{Z})$ sending the class $[\omega_M]$ to the class $[\omega_N]$?
2. (strong) Is any such isomorphism $F\colon \mathrm{H}^*(M;\mathbb{Z}) \to \mathrm{H}^*(N;\mathbb{Z})$ induced by a symplectomorphism?

In [27] it is shown that weak and strong symplectic cohomological rigidity hold for the family of Bott manifolds with rational cohomology ring isomorphic to that of a product of copies of $\mathbb{CP}^1$. Bott manifolds can be viewed as higher dimensional generalizations of Hirzebruch surfaces discussed in the example below. For the definition see Sect. 13.4.2.

Remark 12 Strong (not symplecic) cohomological rigidity, with diffeomorphisms, was already proved for this family by Choi and Masuda [3]. Their diffeomorphisms usually do not preserve the complex structure. If they had, then our result would be an immediate consequence of theirs. Indeed, if $f\colon N \to M$ is a biholomorphism inducing F, then ω_N and $f^*(\omega_M)$ are both Kähler forms on N, defining the same cohomology class in $\mathrm{H}^*(N;\mathbb{Z})$, and thus in this case (N, ω_N) and $(N, f^*(\omega_M))$ are symplectomorphic by the Moser's trick.

Example 13 (*Hirzebruch surfaces*) Hirzebruch surfaces are $\mathbb{CP}^1$ bundles over $\mathbb{CP}^1$. As complex manifolds they are classified by integers (encoding the twisting of the bundle): for each $A \in \mathbb{Z}$ we denote by $\mathcal{H}_{-A}$ the bundle $\mathbb{P}(\mathcal{O}(A) \oplus \mathcal{O}(0)) \to \mathbb{CP}^1$. In particular, $\mathcal{H}_0 = \mathbb{CP}^1 \times \mathbb{CP}^1$. They can be equipped with a symplectic (even Kähler) structure and a toric action. A polytope corresponding to $\mathcal{H}_{-A}$ in Delzant classification is (up to $\mathrm{GL}(2,\mathbb{Z})$ action) a trapezoid with outward normals $(-1, 0)$, $(0, -1)$, $(1, 0)$, $(A, 1)$. The lengths of the edges of this trapezoid depend on the chosen symplectic structure and can be encoded in $\lambda = (\lambda_1, \lambda_2) \in (\mathbb{R}_{>0})^2$. We denote by $(\mathcal{H}_{-A}, \omega_\lambda)$ the symplectic toric manifold corresponding to the trapezoid $\Delta(A, \lambda) := \mathrm{conv}((0,0), (\lambda_1, 0), (\lambda_1, \lambda_2 - A\lambda_1), (0, \lambda_2))$. For example, Fig. 13.1 presents $(\mathcal{H}_0, \omega_{(1,3)})$ and $(\mathcal{H}_{-2}, \omega_{(1,5)})$.

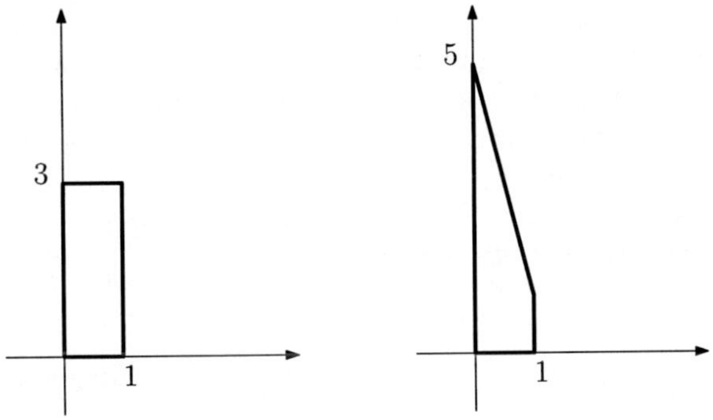

Fig. 13.1 Hirzebruch surfaces $(\mathcal{H}_0, \omega_{(1,3)})$ and $(\mathcal{H}_{-2}, \omega_{(1,5)})$

It was observed by Hirzebruch that $\mathcal{H}_{-A}$ and $\mathcal{H}_{-\widetilde{A}}$ are diffeomorphic if and only if $A \cong \widetilde{A} \mod 2$. Moreover, the symplectic toric manifolds $(\mathcal{H}_{-A}, \omega_\lambda)$ and $(\mathcal{H}_{-\widetilde{A}}, \omega_{\widetilde{\lambda}})$ are (not equivariantly) symplectomorphic if and only if $A \cong \widetilde{A} \mod 2$ and the widths and the areas of the associated polytopes agree, i.e. $\lambda_1 = \widetilde{\lambda}_1$ and $\lambda_2 - \frac{1}{2}A\lambda_1 = \widetilde{\lambda}_2 - \frac{1}{2}\widetilde{A}\widetilde{\lambda}_1$. For example, the manifolds presented on Fig. 13.1 are symplectomorphic. The cohomology ring can be presented as

$$\mathrm{H}^*(\mathcal{H}_{-A}; \mathbb{Z}) = \mathbb{Z}[x_1, x_2]/\langle x_2^2, x_1^2 + Ax_1x_2\rangle,$$

with $[\omega_\lambda] = \lambda_1 x_1 + \lambda_2 x_2$. If $A \cong \widetilde{A} \mod 2$, then the isomorphism $\mathbb{Z}[x_1, x_2] \to \mathbb{Z}[\widetilde{x}_1, \widetilde{x}_2]$ defined by $x_1 \mapsto \widetilde{x}_1 + \frac{1}{2}(\widetilde{A} - A)\widetilde{x}_2$, $x_2 \mapsto \widetilde{x}_2$ descends to an isomorphism between $\mathrm{H}^*(\mathcal{H}_{-A}; \mathbb{Z})$ and $\mathrm{H}^*(\mathcal{H}_{-\widetilde{A}}; \mathbb{Z})$. Note that this isomorphism sends $[\omega_\lambda] = \lambda_1 x_1 + \lambda_2 x_2$ to $\lambda_1 \widetilde{x}_1 + (\lambda_2 + \frac{\widetilde{A}-A}{2}\lambda_1)\,\widetilde{x}_2$ which is equal to $[\omega_{\widetilde{\lambda}}]$ if and only if $\lambda_1 = \widetilde{\lambda}_1$ and $\lambda_2 - \frac{A}{2}\lambda_1 = \widetilde{\lambda}_2 - \frac{\widetilde{A}}{2}\widetilde{\lambda}_1$. Therefore, for Hirzebruch surfaces (weak) symplectic cohomological rigidity holds.

To approach symplectic cohomological rigidity problem one needs a good method of constructing symplectomorphisms. Here is where toric degenerations come into play. By Theorem 6 a toric degeneration whose central fiber Proj $\mathbb{C}[S]$ is smooth produces a symplectomorphism between the symplectic manifold one started with and the central fiber. The main difficulty in this method of constructing symplectomorphisms lies in finding toric degenerations with smooth central fibers.

A great advantage of working with toric manifolds is that the sections of their line bundles are well understood and one can form very concrete constructions of toric degenerations.

13.4.1 Toric Degenerations for Symplectic Toric Manifolds

Let (X_P, ω_P) be a symplectic toric manifold with $\omega_P \in \mathrm{H}^2(M, \mathbb{Z})$, corresponding to a polytope $P \subset \mathbb{R}^n$ via Delzant construction. Then P is an integral polytope (i.e.

with vertices in $\mathbb{Z}^n$) and there exists a very ample line bundle $\mathcal{L}$ over X_P inducing ω_P. In this situation a basis of the space of holomorphic sections of $\mathcal{L}$ can be identified with the integral points of P, ([6], see also [15]). Without loss of generality we can assume that P in a neighborhood of some vertex looks like $(\mathbb{R}_{\geq 0})^n$ in a neighborhood of the origin in $\mathbb{R}^n$. Then we can identify $L = \mathrm{H}^0(X_P, \mathcal{L})$ with a subset of the ring of rational functions, $\mathbb{C}(X_P)$, as described on Sect. 13.2, using the section corresponding to the origin as the fixed element h:

$$f \mapsto \frac{f}{\text{section corresponding to the origin}}.$$

Note 14 For simplicity of notation, given a valuation ν we will write $\nu(L)$ to denote

$$\nu(L) := \{\nu(f/h);\ \ f \in L \setminus \{0\}\}.$$

Similarly, let $\nu(L^m) := \{\nu(f/h^m);\ \ f \in L^m \setminus \{0\}\}$ for any $m > 1$.

We denote by $f_j \in \mathbb{C}(X_P)$ the rational function coming from the section corresponding to the j-th basis vector, $j = 1, \ldots, n$. Note that $f_1, \ldots, f_n$ form a coordinate system around the fixed point of X_P corresponding to the origin via the moment map. To see this, one can, for example, use the description of X_P and f_j's from [15].

Choose and fix a non-negative integer c and two elements $k < l \in \{1, \ldots, n\}$. Then

$$\{u_1 = f_1, \ldots, u_{k-1} = f_{k-1}, u_k = f_k - f_l^c, u_{k+1} = f_{k+1}, \ldots, u_n = f_n\}$$

also gives a coordinate system. Let ν be the associated lowest term valuation (as in Example 4). The image $\nu(L)$ can be obtain by using a "sliding" operator $\mathcal{F}_{-e_k+ce_l}$, defined as follows. For each affine line ℓ in $\mathbb{R}^n$ in the direction of $-e_k + ce_l$, with $P \cap \ell \cap \mathbb{Z}^n \neq \emptyset$, translate the set $\{P \cap \ell \cap \mathbb{Z}^n\}$ by $a(-e_k + ce_l)$ with $a \geq 0$ maximal non-negative number for which $a(-e_k + ce_l) + \{P \cap \ell \cap \mathbb{Z}^n\} \subset (\mathbb{R}_{\geq 0})^n$.

Lemma 15 *One obtains $\nu(L)$ by sliding the integral points of P in the direction $-e_k + ce_l$, inside $(\mathbb{R}_{\geq 0})^n$, i.e.*

$$\nu(L) = \mathcal{F}_{-e_k+ce_l}(P \cap \mathbb{Z}^n).$$

Instead of the proof, which can be found in [27], we give the following example which ilustrates the main idea behind the proof.

Example 16 Let (X_P, ω_P) be the symplectic toric manifold corresponding to the polytope $P = \text{conv}\,\{(0,0), (1,0), (1,3), (0,3)\} \subset \mathbb{R}^2$. That is, X_P is diffeomorphic to $\mathbb{CP}^1 \times \mathbb{CP}^1$ with product symplectic structure (with different rescaling of the Fubini–Study symplectic form on each factor). Let ν be the lowest term valuation associated to the coordinate system

$$u_1 = f_1 - f_2^2,\ u_2 = f_2.$$

Line $\{(0,2)+t(1,-2);\ t\in\mathbb{R}\}$ intersects P in two integral points: $(1,0)$ and $(0,2)$. The corresponding functions are f_1 and f_2^2, and one can easily calculate that

$$\nu(f_1)=\nu(f_2^2)=(0,2) \quad \text{and} \quad \nu(f_1-f_2^2)=(1,0).$$

Similarly, using the integral points on the line $\{(0,3)+t(1,-2);\ t\in\mathbb{R}\}$ we obtain

$$\nu(f_1f_2)=\nu(f_2^3)=(0,3) \quad \text{and} \quad \nu(f_1f_2-f_2^3)=\nu((f_1-f_2^2)f_2)=(1,1).$$

More generally, if the integral points $(a,b),(a,b)+(1,-2),\dots,(a,b)+m(1,-2)$ are in P (implying that $b-2m>0$), then one can use the corresponding functions to construct functions with valuations $(0,b+2a),(0,b+2a)+(1,-2),\dots,(0,b+2a)+m(1,-2)=(m,2a+b-2m)$. Precisely, for any $l=0,\dots,m$

$$f_1^a f_2^{b-2l}(f_1-f_2^2)^l=\sum_{j=0}^{l}(-1)^{l-j}f_1^{a+j}f_2^{b-2j} \quad \text{and}$$
$$\nu(f_1^a f_2^{b-2l}(f_1-f_2^2)^l)=(l,2a+b-2l).$$

This proves that $\nu(L)\supset\mathcal{F}_{(-1,2)}(P\cap\mathbb{Z}^2)$. By [16, Proposition 3.4] the cardinality of $\nu(L)$ is the dimension of L, that is, the number of integral points in P. Therefore

$$\nu(L)=\mathcal{F}_{(-1,2)}(P\cap\mathbb{Z}^2).$$

The polytopes P and $\mathrm{conv}(\nu(L))$ are presented in Fig. 13.1.

Understanding $\nu(L)$ is not enough for constructing and understanding a toric degeneration. First of all, to construct a flat family with toric fiber $\pi^{-1}(0)$ one needs the associated semigroup $S=S(\nu)$ to be finitely generated. Additionally, this toric fiber $\pi^{-1}(0)=\mathrm{Proj}\,\mathbb{C}[S]$ is the toric variety associated to the Okounkov body Δ if $\mathrm{Proj}\,\mathbb{C}[S]$ is normal, that is, if S is saturated. Moreover, to describe the Okounkov body one also needs to find $\nu(L^m)$ for $m>1$. Note that in general L^m differs from $\mathrm{H}^0(X,\mathcal{L}^{\otimes m})$. The following proposition describes an especially nice situation where all these conditions simplify.

Proposition 17 *Let $(X,\omega=\Phi_{\mathcal{L}}^*(\omega_{FS}))$ be a $2n$ dimensional projective symplectic toric manifold associated to a smooth polytope P, with the projective embedding induced by a very ample line bundle $\mathcal{L}$. Let ν be a lowest term valuation associated to a coordinate system of the type presented on Sect. 13.4.1, and S the induced semigroup. Assume that there exists a smooth integral polytope $\Delta\subset\mathbb{R}^n$ such that*

$$S=(\mathrm{cone}\,(\{1\}\times\Delta))\cap(\mathbb{Z}\times\mathbb{Z}^n).$$

Then (X,ω) is symplectomorphic to the symplectic toric manifold (X_Δ,ω_Δ) associated to Δ via Delzant construction.

Here cone $(\{1\} \times \Delta)$ denotes the set $\{(t, tx);\, x \in \Delta, t \in \mathbb{R}_+\} \subset \mathbb{R}^{n+1}$.

***Proof** (sketch)* The assumptions imply that the semigroup S is saturated and (by Gordan's Lemma) finitely generated. Therefore there is a toric degeneration $(\mathfrak{X}, \tilde{\omega})$ with generic fiber (X, ω) and the special fiber $\pi^{-1}(0) = \text{Proj}\, \mathbb{C}[S]$ which is a normal toric variety. Moreover, the Okounkov body associated to the semigroup S is precisely Δ and therefore Proj $\mathbb{C}[S]$ equipped with the restriction of $\tilde{\omega}$, is the symplectic toric manifold $(X_\Delta, \omega_\Delta)$ associated to Δ via Delzant construction. □

Note that S=(cone $\Delta) \cap (\mathbb{Z} \times \mathbb{Z}^n)$ imply, in particular, that $\nu(L^m)$ contains "enough" of integral points, namely that

$$\forall\, m \geq 1 \;\; \nu(L^m) = m\,\Delta \cap \mathbb{Z}^n = \text{conv}(\nu(L^m)) \cap \mathbb{Z}^n.$$

To understand better the requirement conv $(\nu(L^m)) \cap \mathbb{Z}^n = \nu(L^m)$, consider the following example.

Example 18 (*"Enough" of integral points and saturation*) Let (X_P, ω_P) be the symplectic toric manifold corresponding to the polytope

$$P = \text{conv}\,\{(0,0), (2,0), (2,2), (0,2)\} \subset \mathbb{R}^2,$$

that is, X_P is diffeomorphic to $\mathbb{CP}^1 \times \mathbb{CP}^1$ as in the previous example, but the symplectic form is different. As before, let ν be the lowest term valuation associated to the coordinate system

$$u_1 = f_1 - f_2^2,\; u_2 = f_2.$$

Then

$$\begin{aligned}\nu(L) &= \mathcal{F}_{(-1,2)}(P \cap \mathbb{Z}^2)\\ &= \{(0, j);\, j = 0, \ldots, 6\} \cup \{(1,0), (1,2)\} \subsetneq \text{conv}\,(\nu(L)) \cap \mathbb{Z}^2.\end{aligned}$$

In fact conv $(\nu(L))$ is exactly the associated Okounkov body $\Delta(S(\nu))$. Indeed, $\Delta(S(\nu))$ must contain conv $(\nu(L))$. Moreover, $2!\,vol_2(\Delta(S(\nu)))$ is the degree of the Kodaira embedding $\Phi_{\mathcal{L}}\colon X_P \to \mathbb{P}(L^*)$ induced by the line bundle $\mathcal{L}$ corresponding to ω_P [16, Theorem 3.9]. Thus the area of $\Delta(S(\nu))$ must be equal to the area of P, which in this case is also the area of conv $(\nu(L))$. Therefore, in our example, $\nu(L)$ is "missing" the point $(1,1)$ in a sense that $\nu(L) = \Delta(S(\nu)) \cap \mathbb{Z}^2 \setminus \{(1,1)\}$, and thus $(1,1,1) \notin S(\nu)$. However, the line $\{t(1,1,1);\; t \in \mathbb{R}_+\}$ intersects $S(\nu)$: $(2,2,2) = \big(2, \nu(f_1(f_1 - f_2^2) \cdot (f_1 - f_2^2))\big) \in \{2\} \times \nu(L^2)$. Therefore the semigroup $S(\nu)$ is not saturated.

Let us analyse why in the above example the point $(1,1)$ is missing. Observe that the parallel lines $\ell_1 := \{(0,2) + t(-1,2);\; t \in \mathbb{R}\}$, $\ell_2 := \{(0,3) + t(-1,2);\; t \in \mathbb{R}\}$ and $\ell_3 := \{(0,4) + t(-1,2);\; t \in \mathbb{R}\}$ intersect P at intervals of the same length but with, respectively, 2, 1 and 2 integral points. Therefore the intersections of ℓ_1, ℓ_2

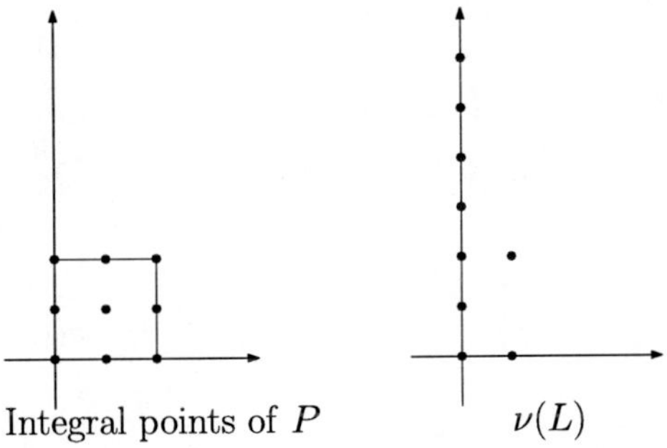

Fig. 13.2 Illustration of Example 18

and ℓ_3 with $\nu(L) = \mathcal{F}_{(-1,2)}(P \cap \mathbb{Z}^2)$ also contain, respectively, 2, 1 and 2 integral points. As a result, the points $(1, 0)$ and $(1, 2)$ are in $\nu(L)$, but $(1, 1)$ is not. The following condition is sufficient, though not necessary, to guarantee that we do not encounter that problem and have enough of integral points (Fig. 13.2).

Lemma 19 *Let* $\lambda_1, \lambda_2, c \in \mathbb{Z}_{>0}$ *and*

$$\Delta = \left\{p \in \mathbb{R}^2 \;\middle|\; 0 \le \langle p, e_1\rangle \le \lambda_1,\ 0 \le \langle p, e_2\rangle \text{ and } \langle p, e_2 + Ae_1\rangle \le \lambda_2\right\}.$$

Assume that

$$\lambda_2 - c\lambda_1 > 0.$$

Then the polytope conv $\mathcal{F}_{(-1,c)}(\Delta \cap \mathbb{Z}^2)$ *is a trapezoid of the same area as* Δ *and*

$$(\text{conv}\, \mathcal{F}_{(-1,c)}(\Delta \cap \mathbb{Z}^2)) \cap \mathbb{Z}^2 = \mathcal{F}_{(-1,c)}(\Delta \cap \mathbb{Z}^2).$$

If $c \le A$ then conv $\mathcal{F}_{(-1,c)}(\Delta \cap \mathbb{Z}^2)$ is simply Δ, and if $c > A$ then it is

$$\left\{p \in \mathbb{R}^2 \;\middle|\; 0 \le \langle p, e_1\rangle \le \lambda_1,\ 0 \le \langle p, e_2\rangle \text{ and } \langle p, e_2 + (2c - A)e_1\rangle \le \lambda_2 + (c - A)\lambda_1)\right\}.$$

An example is illustrated in Fig. 13.3.

13.4.2 Cohomological Rigidity for Bott Manifolds

A Bott manifold is a manifold obtained as a total space of a tower of iterated bundles with fiber $\mathbb{CP}^1$ and the first base space $\mathbb{CP}^1$. Such a manifold naturally carry an algebraic torus action, and can be viewed as a toric manifold. Note that 4-dimensional Bott manifolds are exactly the Hirzebruch surfaces discussed in Example 13. For more information about Bott manifolds see, for example, [13].

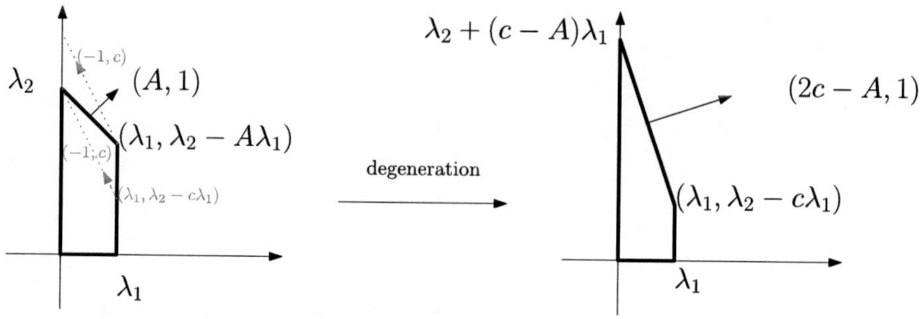

Fig. 13.3 Toric degeneration of a Hirzebruch surface

The simplest example of an $2n$-dimensional Bott manifold is the product of n copies of $\mathbb{CP}^1$'s. Equipped with a product symplectic structure $\omega = \pi_1^*(a_1\omega_{FS}) + \ldots + \pi_n^*(a_n\omega_{FS})$, for some $a_j \in \mathbb{R}_{>0}$, and the standard toric action[8] it becomes a symplectic toric manifold, whose Delzant polytope is a product of intervals, with lengths depending on a_j's. Here $\pi_j \colon \mathbb{CP}^1 \times \ldots \times \mathbb{CP}^1 \to \mathbb{CP}^1$ denotes the projection onto the j-th factor, and ω_{FS} stands for the Fubini–Study symplectic form. In particular, if all a_j's are equal, then the moment image is a hypercube.

A moment image for a general $2n$-dimensional Bott manifold is combinatorially an n-dimensional hypercube. By applying a translation and a $\mathrm{GL}(n, \mathbb{Z})$ transformation one can always arrange that the moment image is a polytope of the form

$$\Delta = \Delta(A, \lambda) = \Big\{p \in \mathbb{R}^n \mid \langle p, e_j\rangle \geq 0 \text{ and } \big\langle p, e_j + \sum_i A^i_j e_i\big\rangle \leq \lambda_j \;\forall\, 1 \leq j \leq n\Big\},$$

where $A \in M_n(\mathbb{Z})$ is an $n \times n$ strictly upper-triangular integral matrix, that is $A^i_j = 0$ unless $i < j$, and $\lambda \in (\mathbb{R}_{>0})^n$. Certain relation between A and λ must be satisfied in order for $\Delta(A, \lambda)$ to have 2^n facets and be combinatorially equivalent to a hypercube (see [27].) In that case we say that (A, λ) *defines a symplectic toric Bott manifold* (M_A, ω_λ) corresponding to the Delzant polytope $\Delta(A, \lambda)$. The matrix A encodes the twisting of consecutive $\mathbb{CP}^1$ bundles, and thus determines a diffeomorphism type of M_A, while λ determines the symplectic structure. By a classical result of Danilov [6]

$$\mathrm{H}^*(M_A; \mathbb{Z}) = \mathbb{Z}[x_1, \ldots, x_n]/\big(x_i^2 + \sum_j A^i_j x_j x_i\big), \tag{13.2}$$

with $[\omega_\lambda] = \sum_i \lambda_i x_i \in \mathrm{H}^*(M_A; \mathbb{Z}) \otimes_{\mathbb{Z}} \mathbb{R}$. If all coefficients λ_i are integral then $[\omega_\lambda]$ is an integral symplectic. Note that this particular presentation of $\mathrm{H}^*(M_A; \mathbb{Z})$ depends

[8] In the standard action of $(S^1)^n$ on $(\mathbb{CP}^1)^n$ each S^1 in $(S^1)^n$ acts on the respective copy of $\mathbb{CP}^1$ by $e^{it} \cdot [(z_0, z_1)] = [(z_0, e^{it}z_1)]$.

on A. (The element x_j is the Poincaré dual to the preimage of facet $\Delta(A,\lambda)\cap\{\langle p, e_j+\sum_i A^i_j e_i\rangle=\lambda_j\}$.)

We say that a Bott manifold is $\mathbb{Q}$-*trivial* if $\mathrm{H}^*(M;\mathbb{Q})\simeq \mathrm{H}^*((\mathbb{CP}^1)^n;\mathbb{Q})$. For example, observe that all Hirzebruch surfaces are $\mathbb{Q}$-trivial Bott manifolds.

Recall that we want to prove Theorem 2 which says that for $\mathbb{Q}$-trivial Bott manifolds (N,ω_N) and (M,ω_M), and any ring isomorphism $F\colon \mathrm{H}^*(M;\mathbb{Z})\to \mathrm{H}^*(N;\mathbb{Z})$, with $F([\omega_M])=[\omega_N]$, there exists a symplectomorphism $f\colon (N,\omega_N)\to(M,\omega_M)$ inducing F. The key ingredient of the proof of Theorem 2 is the following construction of symplectomorphisms, which uses toric degenerations.

Proposition 20 ([27]) *Let (M,ω) and $(\widetilde{M},\widetilde{\omega})$ be symplectic Bott manifolds associated to strictly upper triangular A and $\widetilde{A}$ in $M_n(\mathbb{Z})$ and λ and $\widetilde{\lambda}$ in $\mathbb{Z}^n$, respectively. Assume that there exist integers $1\le k<\ell\le n$ so that A^k_ℓ and $\widetilde{A}^k_\ell$ are of the same parity and the isomorphism from $\mathbb{Z}[x_1,\dots,x_n]$ to $\mathbb{Z}[\widetilde{x}_1,\dots,\widetilde{x}_n]$ that sends x_k to $\widetilde{x}_k+\frac{\widetilde{A}^k_\ell-A^k_\ell}{2}\widetilde{x}_\ell$ and x_i to $\widetilde{x}_i$ for all $i\neq k$ descends to an isomorphism from $\mathrm{H}^*(M;\mathbb{Z})$ to $\mathrm{H}^*(\widetilde{M};\mathbb{Z})$ and takes $\sum\lambda_i x_i$ to $\sum\widetilde{\lambda}_i\widetilde{x}_i$. If $A^k_\ell+\widetilde{A}^k_\ell\ge 0$, then M and $\widetilde{M}$ are symplectomorphic.*

Proof *(sketch)* Without loss of generality we can assume that the polytope $\Delta(A,\lambda)$ associated to (A,λ) is *normal*, that is, any integral point of $m\,\Delta(A,\lambda)$ can be expressed as a sum of m integral points of $\Delta(A,\lambda)$. Indeed, if $\Delta(A,\lambda)$ is not a normal polytope, replace (M,ω) and $(\widetilde{M},\widetilde{\omega})$ by $(M,(n-1)\,\omega)$ and $(\widetilde{M},(n-1)\,\widetilde{\omega})$. This dilates the corresponding polytopes by $(n-1)$. For any integral polytope $P\subset\mathbb{R}^n$ its dialate mP with $m\ge n-1$ is normal (see, for example, [5, Theorem 2.2.12]). Obviously if $(M,(n-1)\,\omega)$ and $(\widetilde{M},(n-1)\,\widetilde{\omega})$ are symplectomorphic, then so are (M,ω) and $(\widetilde{M},\widetilde{\omega})$. As usually, let $\mathcal{L}$ denote the very ample line bundle over M corresponding to ω and L the space of its holomorphic sections. Note that normality implies that L^m can be identified with $H^0(M,\mathcal{L}^{\otimes m})$ because a basis for both of these vector spaces is given by the integral points $m\,\Delta(A,\lambda)\cap\mathbb{Z}^n$.

Also without loss of generality we can assume that $\widetilde{A}^k_\ell\ge A^k_\ell$. Let $c=\frac{1}{2}(A^k_\ell+\widetilde{A}^k_\ell)\ge 0$. We will work with a lowest term valuation ν associated to the following coordinate system

$$\{u_1=f_1,\dots,u_{k-1}=f_{k-1},u_k=f_k-f_l^c,u_{k+1}=f_{k+1},\dots,u_n=f_n.\}$$

From Lemma 15 and the normality assumption, for all $m\ge 1$ we have that

$$\nu(L^m)=\mathcal{F}_{-e_k+ce_l}(m\,\Delta(A,\lambda)\cap\mathbb{Z}^n).$$

To understand $\mathcal{F}_{-e_k+ce_l}(m\,\Delta(A,\lambda)\cap\mathbb{Z}^n)$ consider the action of $\mathcal{F}_{-e_k+ce_l}$ on 2-dimensional "slices", that is, the intersections of $m\,\Delta(A,\lambda)$ with affine subspaces which are translations of (e_k,e_l)-planes. Such slices are either empty or are trapezoids like in Example 16 and Corollary 19, possibly with a cut. A bit tedious computation shows that

$$\mathcal{F}_{-e_k+ce_l}(m\,\Delta(A,\lambda)\cap\mathbb{Z}^n)=m\,\Delta(\widetilde{A},\widetilde{\lambda})\cap\mathbb{Z}^n.$$

For that computation one uses relations between A, λ, $\widetilde{A}$ and $\widetilde{\lambda}$ which are implied by the facts that $\Delta(A, \lambda)$ and $\Delta(\widetilde{A}, \widetilde{\lambda})$ are combinatorially hypercubes, and by the existence of the isomorphism described in the statement of the proposition. In particular, these relations also allow to generalize Corollary 19 (precisely: to show that the equivalent of condition $\lambda_2 - c\lambda_1 > 0$ holds). Therefore the semigroup S associated to the valuation ν of (M, ω) is exactly $S = (\text{cone } \Delta(\widetilde{A}, \widetilde{\lambda})) \cap (\mathbb{Z} \times \mathbb{Z}^n)$. Then the claim follows from Proposition 17. □

Using Proposition 20 we show below (Corollary 23) that each $\mathbb{Q}$-trivial Bott manifold is associated to a matrix A of a particularly easy form. To explain this idea we need few more definitions. Recall the presentation of the cohomology of symplectic Bott manifold M_A given in (13.2). We define the following special elements

$$\alpha_k = -\sum_j A^k_j x_j \in \mathrm{H}^*(M_A; \mathbb{Z}), \quad y_k = x_k - \frac{1}{2}\alpha_k \in \mathrm{H}^*(M_A; \mathbb{Q})$$

for all k. We say x_k is of *even (odd) exceptional type* if $\alpha_k = c y_l$ for some l, where c is an even (respectively, odd) integer. In "coordinates", this means that $A^k_j = 0$ for $j < l$ and $A^k_j = \frac{1}{2} A^k_l A^l_j$ for $j > l$. Note that if x_k is even (resp. odd) exceptional, say $\alpha_k = m y_l$, then one can construct an isomorphism of Proposition 20 from $\mathrm{H}^*(M_A; \mathbb{Z})$ to $\mathrm{H}^*(M_{\widetilde{A}}; \mathbb{Z})$ for some $\widetilde{A}$ with $\widetilde{A}^k_l$ equal to 0 (resp. -1). For example if x_k is of even exceptional type, i.e. $\alpha_k = 2m y_l$ for some m and l, implying that $A^k_l = -2m$ and $A^k_j = -m A^l_j$ for $j \neq l$, then one should put $\widetilde{A}^k_l = 0$, $\widetilde{A}^i_j = A^i_j$ for all i and all $j \neq l$, and $\widetilde{A}^i_l = A^i_l + m A^i_k$ for all $i \neq k$. Therefore, consecutive applications of the above proposition lead to simplifying the description of a given Bott manifold.

Corollary 21 *Any symplectic toric Bott manifold, with integral symplectic form is symplectomorphic to one for which $A^k_l = 0$ (resp. $A^k_l = -1$) whenever x_k has even (resp. odd) exceptional type and $\alpha_k = m y_l$.*

In the case of $\mathbb{Q}$-trivial Bott manifolds all x_i have exceptional type, [3, Proposition 3.1]. Therefore, any $\mathbb{Q}$-trivial symplectic toric Bott manifold with integral symplectic form must be a product of the following standard models of $\mathbb{Q}$-trivial Bott manifolds.

Example 22 ($\mathbb{Q}$-*trivial Bott manifold*) Take $n \in \mathbb{Z}_{>0}$. Let $A^i_n = -1$ for all $1 \leq i < n$, and $A^i_j = 0$ otherwise. For such upper triangular matrix $A = [A^i_j]$ and any $\lambda \in (\mathbb{R}_{>0})^n$, the polytope $\Delta(A, \lambda)$ is combinatorially a hypercube, thus it defines a symplectic toric Bott manifold, which we will denote by $\mathcal{H} = \mathcal{H}(\lambda_1, \dots, \lambda_n)$. Observe that

$$\mathrm{H}^*(\mathcal{H}; \mathbb{Z}) = \mathbb{Z}[x_1, \dots, x_n] / \big(x_1^2 - x_1 x_n, \dots, x_{n-1}^2 - x_{n-1} x_n, x_n^2\big).$$

Consider elements $y_i = x_i - \frac{1}{2} x_n \in \mathrm{H}^*(\mathcal{H}; \mathbb{Q})$ for all $i < n$, and $y_n = x_n$, and note that they form a basis for $\mathrm{H}^*(\mathcal{H}; \mathbb{Q})$. Moreover, as $y_i^2 = 0$ for all i, we get that $\mathrm{H}^*(\mathcal{H}; \mathbb{Q}) \simeq \mathbb{Q}[y_1, \dots, y_n] / \big(y_1^2, \dots, y_n^2\big)$, that is, $\mathcal{H}$ is $\mathbb{Q}$-trivial.

More generally, any partition of n, $\sum_{i=1}^m l_i = n$ together with $\lambda \in (\mathbb{R}_{>0})^n$, define a $\mathbb{Q}$-trivial Bott manifold

$$\mathcal{H}(\lambda_1, \ldots, \lambda_{l_1}) \times \ldots \times \mathcal{H}(\lambda_{n-l_m+1}, \ldots, \lambda_n).$$

Corollary 23 *Each $2n$-dimensional $\mathbb{Q}$-trivial Bott manifold M with integral symplectic form is symplectomorphic to*

$$\mathcal{H}(\lambda_1, \ldots, \lambda_{l_1}) \times \cdots \times \mathcal{H}(\lambda_{n-l_m+1}, \ldots, \lambda_n),$$

for some partition $n = \sum_{i=1}^m l_i$ of n and some $\lambda_1, \ldots, \lambda_n \in \mathbb{Z}_{>0}$.

The above standard model is easy enough, so that one can understand all possible ring isomorphisms between cohomology rings and prove that they are induced by maps on manifolds.

Lemma 24 *Fix $n \in \mathbb{Z}_{>0}$. Let $\sum_{i=1}^m l_i = \sum_{i=1}^{\widetilde{m}} \widetilde{l}_i = n$ be partitions of n, and let $\lambda, \widetilde{\lambda} \in (\mathbb{R}_{>0})^n$. Consider symplectic Bott manifolds*

$$(M, \omega) = \mathcal{H}(\lambda_1, \ldots, \lambda_{l_1}) \times \cdots \times \mathcal{H}(\lambda_{n-l_m+1}, \ldots, \lambda_n),$$
$$(\widetilde{M}, \widetilde{\omega}) = \mathcal{H}(\widetilde{\lambda}_1, \ldots, \widetilde{\lambda}_{\widetilde{l}_1}) \times \cdots \times \mathcal{H}(\widetilde{\lambda}_{n-\widetilde{l}_{\widetilde{m}}+1}, \ldots, \widetilde{\lambda}_n).$$

Given a ring isomorphism $F\colon \mathrm{H}^*(M; \mathbb{Z}) \to \mathrm{H}^*(\widetilde{M}; \mathbb{Z})$ such that $F[\omega] = [\widetilde{\omega}]$, there exists a symplectomorphism f from $(\widetilde{M}, \widetilde{\omega})$ to (M, ω) so that $\mathrm{H}^*(f) = F$.

Proof *(sketch)* First consider the situation when

$$(M, \omega) = \mathcal{H}(\lambda_1, \ldots, \lambda_n) \quad \text{and} \quad (\widetilde{M}, \widetilde{\omega}) = \mathcal{H}(\widetilde{\lambda}_1, \ldots, \widetilde{\lambda}_n).$$

The $\mathbb{Q}$-triviality assumption implies that there are exactly $2n$ primitive classes in $\mathrm{H}^2(M; \mathbb{Z})$ which square to 0. A short computation shows that these are $\pm z_1, \ldots, \pm z_n$, where $z_n = x_n$ and $z_i = 2x_i - x_n$ for all $i < n$. Similarly for $\widetilde{M}$. As the cohomology of a symplectic toric manifold is generated in degree 2, any ring isomorphism between $\mathrm{H}^*(M; \mathbb{Z})$ and $\mathrm{H}^*(\widetilde{M}; \mathbb{Z})$ restricts to a bijection on the set of such elements, that is, there exists $\epsilon = (\epsilon_1, \ldots, \epsilon_n) \in \{-1, 1\}^n$ and a permutation $\sigma \in \mathcal{S}_n$ such that $F(z_j) = \epsilon_j \widetilde{z}_{\sigma(j)}$. Moreover, presenting $[\omega]$ (resp. $[\widetilde{\omega}]$) in $\mathbb{R}$-basis $\{z_1, \ldots, z_n\}$ of $\mathrm{H}^*(M; \mathbb{Z}) \otimes_{\mathbb{Z}} \mathbb{R}$ (resp. $\{\widetilde{z}_1, \ldots, \widetilde{z}_n\}$) and recalling that the isomorphism F is to map $[\omega]$ to $[\widetilde{\omega}]$, one can deduce that F acts by a permutation: $F(z_j) = \widetilde{z}_{\sigma(j)}$ for some permutation $\sigma \in \mathcal{S}_n$ with $\sigma(n) = n$, and that $\lambda_j = \widetilde{\lambda}_{\sigma(j)}$. Moreover F takes x_i to $x_{\sigma(i)}$ and it holds that $A^i_j = \widetilde{A}^{\sigma(i)}_{\sigma(j)}$ for all i, j. If $\Lambda \in \mathrm{GL}(n, \mathbb{Z})$ denotes the unimodular matrix taking e_i to $e_{\sigma(i)}$, then $\Lambda^T(\Delta(\widetilde{A}, \widetilde{\lambda})) = \Delta(A, \lambda)$; Therefore, by the Delzant theorem, the manifolds (M, ω) and $(\widetilde{M}, \widetilde{\omega})$ are (equivariantly) symplectomorphic, by some symplectomorphism f. Moreover, as Λ^T maps the facet $\{\langle p, e_{\sigma(j)}\rangle = 0\} \cap \Delta(\widetilde{A}, \widetilde{\lambda})$ to the facet $\{\langle p, e_j\rangle = 0\} \cap \Delta(A, \lambda)$, and $\{\langle p, e_{\sigma(j)} + \sum_i \widetilde{A}^i_{\sigma(j)} e_i\rangle = \widetilde{\lambda}_{\sigma(j)}\} \cap \Delta(\widetilde{A}, \widetilde{\lambda})$ to $\{\langle p, e_j + \sum_i A^i_j e_i\rangle = \lambda_j\} \cap \Delta(A, \lambda)$, the map $\mathrm{H}^*(f)$ induced by f on cohomology must map the Poincaré duals of preimages of these facets accordingly. That is, $\mathrm{H}^*(f) = F$.

In a general case, denote by λ^{l_s} the l_s-tuple of numbers $(\lambda_{l_1+\dots+l_{s-1}+1}, \dots, \lambda_{l_1+\dots+l_s})$, and define $\widetilde{\lambda}^{l_s}$ similarly. Again, we look at primitive elements with trivial squares. In $\mathrm{H}^*(M; \mathbb{Z})$ these are precisely

$$\pm x_{l_s} \text{ and } \pm (2x_i - x_{l_s}) \text{ for } s = 1, \dots, m \text{ and } i_{s-1} < i < i_s.$$

Note that each such element is contained in some subring $\mathrm{H}^*(\mathcal{H}(\lambda^{l_s}); \mathbb{Z}) \subseteq \mathrm{H}^*(M; \mathbb{Z})$, and that all primitive square zero elements in $\mathrm{H}^*(\mathcal{H}(\lambda^{l_s}); \mathbb{Z})$ are equal modulo 2. Therefore F must restrict to an isomorphism from $\mathrm{H}^*(\mathcal{H}(\lambda^{l_s}); \mathbb{Z})$ to some $\mathrm{H}^*(\mathcal{H}(\widetilde{\lambda}^{\widetilde{l}_r}); \mathbb{Z})$ with $l_s = \widetilde{l}_r$. This implies that both partitions of n are equal, up to permutation of factors. Repeating the arguments of the previous paragraph one can construct a symplectomorphism inducing the ring isomorphism F. □

Proof (*Proof of Theorem* 2) Let (M, ω), $(\widetilde{M}, \widetilde{\omega})$ be two $\mathbb{Q}$-trivial Bott manifolds with symplectic forms integral up to scaling and let $F\colon \mathrm{H}^*(M; \mathbb{Z}) \to \mathrm{H}^*(\widetilde{M}; \mathbb{Z})$ be a ring isomorphism such that $F[\omega] = [\widetilde{\omega}]$. Rescaling the symplectic forms if necessary we can assume that both ω and $\widetilde{\omega}$ are integral. As the cohomology of a symplectic toric manifold is generated in degree 2, the isomorphism F must map $\mathrm{H}^2(M; \mathbb{Z})$ to $\mathrm{H}^2(\widetilde{M}; \mathbb{Z})$. Using (13.2) we see that $\dim \mathrm{H}^2(M; \mathbb{Z}) = \frac{1}{2} \dim M$, and similarly $\dim \mathrm{H}^2(\widetilde{M}; \mathbb{Z}) = \frac{1}{2} \dim \widetilde{M}$. Therefore $\dim M = \dim \widetilde{M}$. We will denote this dimension by $2n$. By Corollary 23 and the assumption that the symplectic forms are integral we have that

$$(M, \omega) = \mathcal{H}(\lambda_1, \dots, \lambda_{l_1}) \times \dots \times \mathcal{H}(\lambda_{n-l_m+1}, \dots, \lambda_n),$$
$$(\widetilde{M}, \widetilde{\omega}) = \mathcal{H}(\widetilde{\lambda}_1, \dots, \widetilde{\lambda}_{\widetilde{l}_1}) \times \dots \times \mathcal{H}(\widetilde{\lambda}_{n-\widetilde{l}_{\widetilde{m}}+1}, \dots, \widetilde{\lambda}_n).$$

for some $\sum_{i=1}^{m} l_i = \sum_{i=1}^{\widetilde{m}} \widetilde{l}_i = n$ partitions of n, and some $\lambda, \widetilde{\lambda} \in (\mathbb{Z}_{>0})^n$. Now Lemma 24 gives that there exist a symplectomorphism f from $(\widetilde{M}, \widetilde{\omega})$ to (M, ω) so that $\mathrm{H}^*(f) = F$. □

Acknowledgements First of all, the author would like to thank her collaborators: Xin Fang, Iva Halacheva, Peter Littelmann and Sue Tolman. Results contained in this manuscript were obtained in collaboration with the above mathematicians [11, 14, 27], and therefore all of them could also be considered as the authors of this paper. The author also thanks the organisers of the workshops "Interactions with Lattice Polytopes" for giving her the opportunity to participate and present her results at these workshops. The author is supported by the DFG (Die Deutsche Forschungsgemeinschaft) grant CRC/TRR 191 "Symplectic Structures in Geometry, Algebra and Dynamics".

References

1. Anderson, D.: Okounkov bodies and toric degenerations. Math. Ann. **356**(3), 1183–1202 (2013)
2. Caviedes Castro, A.: Upper bound for the Gromov width of coadjoint orbits of compact Lie groups. J. Lie Theory **26**(3), 821–860 (2016)
3. Choi, S., Masuda, M.: Classification of $\mathbb{Q}$-trivial Bott manifolds. J. Symplectic Geom. **10**(3), 447–461 (2012)
4. Choi, S., Masuda, M., Suh, D.Y.: Rigidity problems in toric topology: a survey. Tr. Mat. Inst. Steklova **275**(Klassicheskaya i Sovremennaya Matematika v Pole Deyatel'nosti Borisa Nikolaevicha Delone), 188–201 (2011)
5. Cox, D.A., Little, J.B., Schenck, H.K.: Toric varieties, Graduate Studies in Mathematics, vol. 124. American Mathematical Society, Providence, RI (2011)
6. Danilov, V.I.: The geometry of toric varieties. Uspekhi Mat. Nauk **33**(2(200)), 85–134, 247 (1978)
7. da Silva, A.C.: Lectures on symplectic geometry. Lecture Notes in Mathematics, vol. 1764. Springer, Berlin (2001)
8. Eisenbud, D.: Commutative algebra. Graduate Texts in Mathematics, vol. 150. Springer, New York (1995). With a view toward algebraic geometry
9. Eisenbud, D., Harris, J.: The geometry of schemes. In: Graduate Texts in Mathematics, vol. 197. Springer, New York (2000)
10. Fang, X., Fourier, G., Littelmann, P.: Essential bases and toric degenerations arising from birational sequences. Adv. Math. **312**, 107–149 (2017)
11. Fang, X., Littelmann, P., Pabiniak, M.: Simplices in Newton-Okounkov bodies and the Gromov width of coadjoint orbits. Bull. Lond. Math. Soc. **50**(2), 202–218 (2018)
12. Fulton, W.: Introduction to toric varieties. Annals of Mathematics Studies, vol. 131. Princeton University Press, Princeton, NJ (1993). The William H. Roever Lectures in Geometry
13. Grossberg, M., Karshon, Y.: Bott towers, complete integrability, and the extended character of representations. Duke Math. J. **76**(1), 23–58 (1994)
14. Halacheva, I., Pabiniak, M.: The Gromov width of coadjoint orbits of the symplectic group. Pacific J. Math. **295**(2), 403–420 (2018)
15. Hamilton, M.D.: The quantization of a toric manifold is given by the integer lattice points in the moment polytope. In: Toric topology, Contemporary Mathematics, vol. 460, pp. 131–140. American Mathematical Society, Providence, RI (2008)
16. Harada, M., Kaveh, K.: Integrable systems, toric degenerations and Okounkov bodies. Invent. Math. **202**(3), 927–985 (2015)
17. Hartshorne, R.: Algebraic Geometry. Springer, New York, Heidelberg (1977). Graduate Texts in Mathematics, No. 52
18. Karshon, Y., Tolman, S.: The Gromov width of complex Grassmannians. Algebr. Geom. Topol. **5**, 911–922 (2005)
19. Kaveh, K.: Crystal bases and Newton-Okounkov bodies. Duke Math. J. **164**(13), 2461–2506 (2015)
20. Kaveh, K.: Toric degenerations and symplectic geometry of smooth projective varieties. J. Lond. Math. Soc. (2) **99**(2), 377–402 (2019)
21. Lane, J.: A Completely Integrable System on g_2 Coadjoint Orbits (2016). arXiv:1605.01676
22. Littelmann, P.: Cones, crystals, and patterns. Transform. Groups **3**(2), 145–179 (1998)
23. Lu, G.: Gromov-Witten invariants and pseudo symplectic capacities. Israel J. Math. **156**, 1–63 (2006)
24. Lu, G.: Symplectic capacities of toric manifolds and related results. Nagoya Math. J. **181**, 149–184 (2006)
25. Nishinou, T., Nohara, Y., Ueda, K.: Toric degenerations of Gelfand-Cetlin systems and potential functions. Adv. Math. **224**(2), 648–706 (2010)
26. Pabiniak, M.: Gromov width of non-regular coadjoint orbits of $U(n)$, $SO(2n)$ and $SO(2n+1)$. Math. Res. Lett. **21**(1), 187–205 (2014)

27. Pabiniak, M., Tolman, S.: Symplectic cohomological rigidity via toric degnerations. In preparation
28. Schlenk, F.: Embedding problems in symplectic geometry. In: De Gruyter Expositions in Mathematics, vol. 40. Walter de Gruyter GmbH & Co. KG, Berlin (2005)
29. Zoghi, M.: The Gromov Width of Coadjoint Orbits of Compact Lie Groups. ProQuest LLC, Ann Arbor, MI (2010). Thesis (Ph.D.)–University of Toronto (Canada)

Chapter 14
On Deformations of Toric Fano Varieties

Andrea Petracci

Abstract In this note we collect some results on the deformation theory of toric Fano varieties.

Keywords Deformations · Fano varieties · Toric varieties · Reflexive polytopes

14.1 Introduction

A *Fano variety* is a normal projective variety X over $\mathbb{C}$ such that its anticanonical divisor $-K_X$ is $\mathbb{Q}$-Cartier and ample. Fano varieties constitute the basic building blocks of algebraic varieties, according to the Minimal Model Program. The geometry of Fano varieties is a well studied area. In particular, moduli (and consequently deformations) of Fano varieties constitute a very interesting and important topic in algebraic geometry, e.g. [21, 62, 69].

Here we will concentrate on deformations and smoothings of *toric* Fano varieties. These varieties occupy a prominent role in Mirror Symmetry, a large part of which is based on the phenomenon of toric degeneration as in [17, 18, 30, 43].

Toric Fano varieties correspond to certain polytopes which are called *Fano polytopes*. The goal of this note is to present some combinatorial criteria on Fano polytopes which can detect whether the corresponding toric Fano variety is smoothable, i.e. can be deformed to a smooth (Fano) variety.

Special attention is given to toric Fano threefolds with Gorenstein singularities. These varieties correspond to the 4319 reflexive polytopes of dimension 3, which were classified by Kreuzer and Skarke [66]. In this case, thanks to the use of the software MAGMA [22], we were able to produce a lot of examples for the combinatorial criteria discussed in this note.

A. Petracci (✉)
Dipartimento di Matematica, Università di Bologna, Piazza di Porta San Donato 5, 40126 Bologna, Italy
e-mail: a.petracci@unibo.it

A. M. Kasprzyk and B. Nill (eds.), *Interactions with Lattice Polytopes*, Springer Proceedings in Mathematics & Statistics 386,
https://doi.org/10.1007/978-3-030-98327-7_14

14.1.1 Outline

In Sect. 14.2.1 the very classical theory of infinitesimal deformations of algebraic varieties is recalled. In Sect. 14.2.2 we survey some properties of smoothings of algebraic varieties. In Sect. 14.2.3 two well-studied deformation invariants for Fano varieties are introduced.

In Sect. 14.3.1 we recall some results on the deformation theory of affine toric varieties. We provide an example in Sect. 14.3.2.

The core of this note is Sect. 14.4. We recall the definition of Fano polytopes in Sect. 14.4.1. In Sect. 14.4.2 we present a couple of sufficient conditions that ensure that a toric Fano variety is non-smoothable. The rigidity of toric Fano varieties is examined in Sect. 14.4.3. In Sects. 14.4.4 and 14.4.5 we study the smoothability of toric Fano surfaces and toric Fano threefolds with isolated singularities; an example is presented in Sect. 14.4.6. In Sect. 14.4.7 we present another sufficient condition that ensures that a toric Fano threefold is non-smoothable. In Sect. 14.4.8 we include more results on deformations of toric Fano varieties.

In Sect. 14.5 we write down the lists of the reflexive polytopes of dimension 3 which satisfy the several combinatorial conditions considered in Sect. 14.4.

14.1.2 Notation and Conventions

We work over $\mathbb{C}$, but everything will hold over a field of characteristic zero with appropriate modifications.

In Sects. 14.3 and 14.4 we assume that the reader is familiar with the basic notions of toric geometry, which can be found in [34, 41]. All toric varieties considered here are normal. A lattice is a finitely generated free abelian group. The letters $N, \overline{N}, \tilde{N}$ stand for lattices and $M, \overline{M}, \tilde{M}$ for their duals, e.g. $M = \mathrm{Hom}_{\mathbb{Z}}(N, \mathbb{Z})$; the duality pairing $M \times N \to \mathbb{Z}$ and its extension $M_{\mathbb{R}} \times N_{\mathbb{R}} \to \mathbb{R}$ are denoted by $\langle \cdot, \cdot \rangle$.

In a real vector space of finite dimension a polytope is the convex hull of finitely many points, or equivalently a compact subset which is the intersection of finitely many closed halfspaces. We refer the reader to the book [99] for the geometry of polytopes.

14.2 Deformations

14.2.1 Infinitesimal Deformations

Let (Comp) be the category of noetherian complete local $\mathbb{C}$-algebras with residue field $\mathbb{C}$. For every $R \in$ (Comp) we denote by $\mathfrak{m}_R$ the maximal ideal of R. Let (Art) be the subcategory of (Comp) whose objects are artinian, i.e. local finite $\mathbb{C}$-algebras.

A functor of Artin rings is a functor F from the category (Art) to the category of sets such that $F(\mathbb{C})$ is the set with one element. We will only consider functor of Artin rings which satisfy some additional properties: Schlessinger's axioms (H1) and (H2) [91] and Fantechi–Manetti condition (L) [37, (2.9)]. We will not specify these conditions here, but we refer the reader to [37, Sect. 2] for a quick introduction. Precise formulations and additional details about the notions we introduce below can be found in any reference about deformation theory, e.g. [13, 36, 50, 70, 91, 92, 95, 97].

A natural transformation (or briefly map) of functors $\phi\colon F \to G$ is called *smooth* if the lifting property in Grothendieck's definition of formally smooth morphisms holds, i.e. for every local surjection $A' \twoheadrightarrow A$ in (Art) the natural map $F(A') \to F(A) \times_{G(A)} G(A')$ is surjective; in particular, if ϕ is smooth then $\phi(A)\colon F(A) \to G(A)$ is surjective for all $A \in$ (Art). A functor F is called smooth if the map from F to the trivial functor is smooth.

For a functor F, the set $F(\mathbb{C}[t]/(t^2))$ has a natural structure of a $\mathbb{C}$-vector space, denoted by $\mathrm{T}F$ and called the *tangent space* of F. One can prove that F is the trivial functor if and only if $\mathrm{T}F = 0$. If $\phi\colon F \to G$ is a map, then the function $\phi(\mathbb{C}[t]/(t^2))$ is linear and denoted by $\mathrm{T}\phi\colon \mathrm{T}F \to \mathrm{T}G$.

If $R \in$ (Comp) one can consider the functor $h_R = \mathrm{Hom}(\cdot, R)$ prorepresented by R. A map $h_R \to F$ is equivalent to a pro-object of F on $R = \varprojlim R/\mathfrak{m}_R^{n+1}$, i.e. an element of the set $\varprojlim F(R/\mathfrak{m}_R^{n+1})$. A *hull* for a functor F is a ring $R \in$ (Comp) together with a smooth morphism $\phi\colon h_R \to F$ such that $\mathrm{T}\phi$ is bijective. A hull exists if and only if $\mathrm{T}F$ has finite dimension. If a hull exists, it is unique. Provided that $\mathrm{T}F$ has finite dimension r, then F is smooth if and only if the hull of F is isomorphic to $\mathbb{C}[\![t_1, \dots, t_r]\!]$.

For a functor F, consider the set $\mathcal{E}$ made up of pairs (π, ξ), where $\pi\colon A' \to A$ is a surjection in (Art) such that $\mathfrak{m}_{A'} \cdot (\ker \pi) = 0$ and $\xi \in F(A)$. A $\mathbb{C}$-vector space V is called an *obstruction space* for F if there exists a function $\omega\colon \mathcal{E} \to \coprod_{(\pi,\xi)\in\mathcal{E}} \ker \pi \otimes_{\mathbb{C}} V$ such that the two following conditions are satisfied:

1. for every $(\pi, \xi) \in \mathcal{E}$, $\omega(\pi, \xi) \in \ker \pi \otimes_{\mathbb{C}} V$;
2. for every $(\pi, \xi) \in \mathcal{E}$, we have that $\omega(\pi, \xi) = 0$ if and only if there exists $\xi' \in F(A')$ which maps to ξ.

There are infinitely many obstruction spaces for a functor F because any vector space containing an obstruction space is an obstruction space. A functor F is smooth if and only if 0 is an obstruction space for F; in this case we also say that F is unobstructed. There is a notion of compatible obstruction spaces for a map $\phi\colon F \to G$: this will be a linear map $\mathrm{o}\phi$ from an obstruction space of F to an obstruction space of G with some compatibility properties with respect to ϕ.

The following is an important smoothness criterion. Assume that $\phi\colon F \to G$ is a map with compatible obstruction map $\mathrm{o}\phi$ from an obstruction space of F to an obstruction space of G. If $\mathrm{T}\phi$ is surjective and $\mathrm{o}\phi$ is injective, then ϕ is smooth.

Let X be a scheme of finite type over $\mathbb{C}$. We denote by Def_X the functor of (infinitesimal) deformations of X. If $R \in$ (Comp), a pro-object of Def_X on R is called a *formal deformation* of X over R. If R is a hull for Def_X, then the corresponding formal deformation of X over R is called the *miniversal deformation* of X. We say that X is *rigid* if all deformations of X are trivial. If X is reduced, then the tangent space of Def_X is $\mathrm{Ext}^1(\Omega_X, \mathcal{O}_X)$; in this case X is rigid if and only if $\mathrm{Ext}^1(\Omega_X, \mathcal{O}_X) = 0$. If X is either normal or reduced and local complete intersection (l.c.i. for short), then $\mathrm{Ext}^2(\Omega_X, \mathcal{O}_X)$ is an obstruction space for Def_X. If X is smooth, then $\mathrm{H}^i(X, T_X) = \mathrm{Ext}^i(\Omega_X, \mathcal{O}_X)$ for all $i \geq 0$. In particular, if X is smooth and affine then it is rigid.

Proposition 1 *If X is a smooth Fano variety, then $\mathrm{H}^i(X, T_X) = 0$ for each $i \geq 2$. In particular, the infinitesimal deformations of X are unobstructed, i.e. Def_X is smooth.*

Proof Let n be the dimension of X. Since the anticanonical line bundle $\omega_X^\vee$ is ample, by Kodaira–Nakano vanishing we have $\mathrm{H}^i(X, \Omega_X^{n-1} \otimes \omega_X^\vee) = 0$ whenever $i + n - 1 > n$, i.e. $i \geq 2$. We conclude because the tangent sheaf T_X is isomorphic to $\Omega_X^{n-1} \otimes \omega_X^\vee$. □

Let X be a scheme of finite type over $\mathbb{C}$ and let Def_X^{lt} be the subfunctor of Def_X made up of the locally trivial deformations of X. The tangent space of Def_X^{lt} is $\mathrm{H}^1(X, T_X)$ and $\mathrm{H}^2(X, T_X)$ is an obstruction space for Def_X^{lt}.

Proposition 2 *Let X be a reduced scheme of finite type over $\mathbb{C}$ such that X is either l.c.i. or normal. If $\mathrm{H}^0(X, \mathcal{E}xt^1(\Omega_X, \mathcal{O}_X)) = 0$, then all deformations of X are locally trivial, i.e. $Def_X^{\mathrm{lt}} = Def_X$.*

Proof The local-to-global spectral sequence for Ext gives the following exact sequence.

$$0 \to \mathrm{H}^1(T_X) \to \mathrm{Ext}^1(\Omega_X, \mathcal{O}_X) \to \mathrm{H}^0(\mathcal{E}xt^1(\Omega_X, \mathcal{O}_X)) \\ \to \mathrm{H}^2(T_X) \to \mathrm{Ext}^2(\Omega_X, \mathcal{O}_X)$$

The vanishing of $\mathrm{H}^0(\mathcal{E}xt^1(\Omega_X, \mathcal{O}_X))$ implies that the inclusion $\phi \colon Def_X^{\mathrm{lt}} \hookrightarrow Def_X$ induces an isomorphism on tangent spaces and an injection on obstruction spaces. Therefore ϕ is smooth, and consequently surjective. □

In particular, all deformations of a smooth scheme are locally trivial.

Let X be a reduced scheme of finite type over $\mathbb{C}$ with isolated singularities. For each singular point $x \in X$, let U_x be an affine open neighbourhood of x such that $U_x \setminus \{x\}$ is smooth. Then define

$$Def_X^{\mathrm{loc}} := \prod_{x \in \mathrm{Sing}(X)} Def_{U_x}.$$

The tangent space of Def_X^{loc} is $\mathrm{H}^0(X, \mathcal{E}xt^1(\Omega_X, \mathcal{O}_X))$. If X is either l.c.i. or normal, then $\mathrm{H}^0(X, \mathcal{E}xt^2(\Omega_X, \mathcal{O}_X))$ is an obstruction space for Def_X^{loc}. There is an obvious map $Def_X \to Def_X^{\mathrm{loc}}$ which restricts a deformation of X to a deformation of U_x for each x.

Proposition 3 *Let X be a reduced scheme of finite type over $\mathbb{C}$ with isolated singularities. Assume that X is either l.c.i. or normal. If $\mathrm{H}^2(X, T_X) = 0$ then there are no local-to-global obstructions for the infinitesimal deformations of X, i.e. the map $Def_X \to Def_X^{\mathrm{loc}}$ is smooth.*

Proof We consider the local-to-global spectral sequence for $\mathrm{Ext}^\bullet(\Omega_X, \mathcal{O}_X)$. The second page is given by $E_2^{p,q} = \mathrm{H}^p(\mathcal{E}xt^q(\Omega_X, \mathcal{O}_X))$. Since X has isolated singularities, the sheaves $\mathcal{E}xt^q(\Omega_X, \mathcal{O}_X)$ are supported on isolated points for $q \geq 1$; in particular they do not have higher cohomology. This means that $E_2^{p,q}$ is supported on the lines $p = 0$ and $q = 0$. Therefore, in E_2 the only non-zero differential is

$$d_2 \colon \mathrm{H}^0(\mathcal{E}xt^1(\Omega_X, \mathcal{O}_X)) \longrightarrow \mathrm{H}^2(T_X).$$

We obtain that the bottom left corner of the third page E_3 is the following.

$$\begin{array}{llll} \mathrm{H}^3(T_X) & 0 & 0 & 0 \\ \mathrm{coker} d_2 & 0 & 0 & 0 \\ \mathrm{H}^1(T_X) & 0 & 0 & 0 \\ \mathrm{H}^0(T_X) & \ker d_2 & \mathrm{H}^0(\mathcal{E}xt^2(\Omega_X, \mathcal{O}_X)) & \mathrm{H}^0(\mathcal{E}xt^3(\Omega_X, \mathcal{O}_X)) \end{array}$$

In E_3 the only non-zero differential is

$$d_3 \colon \mathrm{H}^0(\mathcal{E}xt^2(\Omega_X, \mathcal{O}_X)) \longrightarrow \mathrm{H}^3(T_X).$$

The bottom left corner of the fourth page E_4 is the following.

$$\begin{array}{llll} \mathrm{coker} d_3 & 0 & 0 & 0 \\ \mathrm{coker} d_2 & 0 & 0 & 0 \\ \mathrm{H}^1(T_X) & 0 & 0 & 0 \\ \mathrm{H}^0(T_X) & \ker d_2 & \ker d_3 & \mathrm{H}^0(\mathcal{E}xt^3(\Omega_X, \mathcal{O}_X)) \end{array}$$

From the fourth page on, the pieces of total degree ≤ 3 do not change any more. Therefore we have two short exact sequences:

$$0 \longrightarrow \mathrm{H}^1(T_X) \longrightarrow \mathrm{Ext}^1(\Omega_X, \mathcal{O}_X) \longrightarrow \ker d_2 \longrightarrow 0,$$
$$0 \longrightarrow \mathrm{coker} d_2 \longrightarrow \mathrm{Ext}^2(\Omega_X, \mathcal{O}_X) \longrightarrow \ker d_3 \longrightarrow 0.$$

These can be joined to construct the following long exact sequence.

$$\begin{array}{llllllll} 0 \longrightarrow & \mathrm{H}^1(T_X) & \longrightarrow & \mathrm{Ext}^1(\Omega_X, \mathcal{O}_X) & \longrightarrow & \mathrm{H}^0(\mathcal{E}xt^1(\Omega_X, \mathcal{O}_X)) & \xrightarrow{d_2} \\ \xrightarrow{d_2} & \mathrm{H}^2(T_X) & \longrightarrow & \mathrm{Ext}^2(\Omega_X, \mathcal{O}_X) & \longrightarrow & \mathrm{H}^0(\mathcal{E}xt^2(\Omega_X, \mathcal{O}_X)) \end{array}$$

So far we did not use the assumption $H^2(T_X) = 0$. From this vanishing, via the long exact sequence above we deduce that the map $Def_X \to Def_X^{loc}$ induces a surjection on tangent spaces and an injection on obstruction spaces. □

14.2.2 Smoothings

Here we discuss smoothability conditions for schemes of finite type over $\mathbb{C}$. We will only consider the case of equidimensional schemes and we will refer the reader to [50, Sect. 29] for a more general treatment, which uses the Lichtenbaum–Schlessinger functors.

If X is a proper scheme over $\mathbb{C}$, a *smoothing* of X is a proper flat morphism $\mathcal{X} \to B$ such that B is an integral scheme of finite type over $\mathbb{C}$ of positive dimension and there exists a closed point $b_0 \in B$ such that the fibre over b_0 is X and all the other fibres are smooth. By restricting to a curve in B and normalising it, we may require that the base B is a smooth affine curve and that the maximal ideal corresponding to b_0 is principal. We say that X is *smoothable* if it admits a smoothing.

For every $n \geq 0$, set $S_n := \operatorname{Spec} \mathbb{C}[t]/(t^{n+1})$. If X is a scheme of finite type over $\mathbb{C}$ with pure dimension d, then a *formal smoothing* of X is a formal deformation $\{X_n \to S_n\}_n$ of X over $\mathbb{C}[\![t]\!]$ such that there exists m such that t^m is in the dth Fitting ideal of Ω_{X_m/S_m}. We refer the reader to [35, Sect. 20.2] for the definition and the properties of Fitting ideals. We say that X is *formally smoothable* if it admits a formal smoothing. It is clear that if X is formally smoothable, then every open subscheme of X is formally smoothable.

Remark 4 If $\{X_n \to S_n\}_n$ is a formal deformation of X over $\mathbb{C}[\![t]\!]$ and t^m is in the dth Fitting ideal of Ω_{X_m/S_m}, then for all $n \geq m$ we have that t^n is in the dth Fitting ideal of Ω_{X_n/S_n}.

The proof of this fact is as follows. We have $\mathcal{O}_{X_n} = \mathcal{O}_{X_{n+1}}/t^{n+1}\mathcal{O}_{X_{n+1}}$. Since the formation of Fitting ideals commutes with base change, we have the equality

$$\operatorname{Fitt}_d(\Omega_{X_n/S_n}) = (\operatorname{Fitt}_d(\Omega_{X_{n+1}/S_{n+1}}) + t^{n+1}\mathcal{O}_{X_{n+1}})/t^{n+1}\mathcal{O}_{X_{n+1}}.$$

Therefore if $t^n \in \operatorname{Fitt}_d(\Omega_{X_n/S_n})$ then $t^n \in \operatorname{Fitt}_d(\Omega_{X_{n+1}/S_{n+1}}) + t^{n+1}\mathcal{O}_{X_{n+1}}$, hence $t^{n+1} \in t\operatorname{Fitt}_d(\Omega_{X_{n+1}/S_{n+1}}) \subseteq \operatorname{Fitt}_d(\Omega_{X_{n+1}/S_{n+1}})$ as $t^{n+2} = 0$ in $\mathcal{O}_{X_{n+1}}$.

Lemma 5 *Let X be a Cohen–Macaulay proper scheme over $\mathbb{C}$ of pure dimension d. Let B be a smooth curve over $\mathbb{C}$, $b_0 \in B$ be a closed point, and $\pi : \mathcal{X} \to B$ be a proper flat morphism such that the fibre over b_0 is X. Let ξ be the formal $\mathfrak{m}_{b_0}$-adic completion of π at b_0, i.e. $\xi = \{\mathcal{X} \times_B \operatorname{Spec} \mathcal{O}_{B,b_0}/\mathfrak{m}_{b_0}^{n+1} \to \operatorname{Spec} \mathcal{O}_{B,b_0}/\mathfrak{m}_{b_0}^{n+1}\}_n$. Then:*

1. *if π is a smoothing of X, then ξ is a formal smoothing of X;*
2. *if ξ is a formal smoothing of X, then there exists an open neighbourhood B' of b_0 in B such that $\mathcal{X} \times_B B' \to B'$ is a smoothing of X.*

Proof This proof comes from [14, Sect. 0E7S].

Notice that ξ does not change if we restrict π to an open neighbourhood of b_0 in B. Therefore, in order to prove the statements (1) and (2) we can arbitrarily restrict to an open neighbourhood of b_0 in B. Hence we may assume that B is affine and the maximal ideal corresponding to the point b_0 is principal, generated by $t \in \mathcal{O}_B$.

We consider the set $W \subseteq \mathcal{X}$ made up of the points $x \in \mathcal{X}$ such that the local ring of the fibre $\mathcal{X}_{\pi(x)}$ at x is Cohen–Macaulay. By [44, 12.1.7], W is open in $\mathcal{X}$. As π is closed, $B \setminus \pi(\mathcal{X} \setminus W)$ is an open neighbourhood of b_0 in B. Therefore, if we restrict B to an open neighbourhood of b_0 in B, we may assume that all fibres of π are Cohen–Macaulay. By [14, Lemma 02NM], we may assume that π has relative dimension d.

Let $I \subseteq \mathcal{O}_\mathcal{X}$ be the dth Fitting ideal of $\Omega_{\mathcal{X}/B}$. For each n, set

$$S_n = \operatorname{Spec} \mathcal{O}_{B,b_0}/\mathfrak{m}_{b_0}^{n+1} = \operatorname{Spec} \mathcal{O}_B/t^{n+1}\mathcal{O}_B$$

and $X_n = \mathcal{X} \times_B S_n$; let $I_n \subseteq \mathcal{O}_{X_n}$ be the dth Fitting ideal of Ω_{X_n/S_n}. Since Fitting ideals commute with base change, we have $\mathcal{O}_{X_n} = \mathcal{O}_\mathcal{X}/t^{n+1}\mathcal{O}_\mathcal{X}$ and $I_n = I\mathcal{O}_{X_n} = (I + t^{n+1}\mathcal{O}_\mathcal{X})/t^{n+1}\mathcal{O}_\mathcal{X}$.

Since π is flat of relative dimension d, the zero locus of I is the singular locus of π. Moreover, the fibre over b_0 is the closed subset $\mathrm{V}(t)$. Therefore, the fibre of b_0 is the unique singular fibre if and only if $t \in \sqrt{I}$.

(1) If π is a smoothing, then there exists m such that $t^m \in I$. Since $I_m = (I + t^{m+1}\mathcal{O}_\mathcal{X})/t^{m+1}\mathcal{O}_\mathcal{X}$, this implies that $t^m \in I_m$. So ξ is a formal smoothing.

(2) If ξ is a formal smoothing, then $t^m \in I_m = (I + t^{m+1}\mathcal{O}_\mathcal{X})/t^{m+1}\mathcal{O}_\mathcal{X}$ for some m. So in $\mathcal{O}_\mathcal{X}$ we have the equality $t^m = p + t^{m+1}q$, for some $p \in I$ and $q \in \mathcal{O}_\mathcal{X}$. Writing $t^m(1 - tq) = p$ and noticing that the function $1 - tq$ does not vanish at the points of $X = \mathrm{V}(t)$, we deduce that t^m belongs to the stalk I_x of I at all points $x \in X$. This implies that t^m lies in I in an open neighbourhood U of X in $\mathcal{X}$. Since π is closed, by restricting B to $B \setminus \pi(\mathcal{X} \setminus U)$ we have $t^m \in I$. Therefore π is a smoothing. □

Proposition 6 *Let X be a Cohen–Macaulay scheme proper over $\mathbb{C}$.*

1. *If X is smoothable, then every open subscheme of X is formally smoothable.*
2. *Assume that X is projective and $\mathrm{H}^2(X, \mathcal{O}_X) = 0$; if X is formally smoothable, then X is smoothable.*

Proof We may assume that X is connected. Therefore X has pure dimension, say d.

(1) This follows immediately from Lemma 5 and from the fact that if X is formally smoothable then every open subscheme of X is formally smoothable.

(2) Set $d := \dim X$. Let $\xi = \{X_n \to S_n\}_n$ be a formal smoothing of X, where S_n is $\operatorname{Spec} \mathbb{C}[t]/(t^{n+1})$ as usual. Let m be such that t^m is in the dth Fitting ideal of Ω_{X_m/S_m}.

As X is proper over $\mathbb{C}$, the tangent space of Def_X has finite dimension, therefore Def_X has a hull $R \in (\mathrm{Comp})$. Let $\eta = \{\eta_n \colon Y_n \to \operatorname{Spec} R/\mathfrak{m}_R^{n+1}\}_n$ be the miniversal deformation of X. By [95, Proposition 6.51] or [92, Theorem 2.5.13], from $\mathrm{H}^2(\mathcal{O}_X) = 0$ we deduce that η is effective, i.e. there exists a projective flat morphism $\mathscr{X} \to \operatorname{Spec} R$ whose $\mathfrak{m}_R$-adic completion is η.

By a theorem of Artin [11, Theorem 1.6] (see also [50, Theorem 21.3]), the morphism $\mathscr{X} \to \operatorname{Spec} R$ is algebraizable in the following sense: there exist a scheme Z of finite type over $\mathbb{C}$, a closed point $z_0 \in Z$, and a proper flat morphism $\mathcal{X} \to Z$, with fibre X over z_0, such that R is the completion $\widehat{\mathcal{O}}_{Z,z_0}$ of the local ring of Z at z_0 and $\mathscr{X}$ is isomorphic, as R-schemes, to $\mathcal{X} \times_Z \operatorname{Spec} R$. In particular, the miniversal deformation η is the collection $\{\mathcal{X} \times_Z \operatorname{Spec} \mathcal{O}_{Z,z_0}/\mathfrak{m}_{z_0}^{n+1} \to \operatorname{Spec} \mathcal{O}_{Z,z_0}/\mathfrak{m}_{z_0}^{n+1}\}_n$. The situation is summarised in the following cartesian squares, for all n.

$$\begin{array}{ccccc} Y_n & \hookrightarrow & \mathscr{X} & \longrightarrow & \mathcal{X} \\ \downarrow{\scriptstyle \eta_n} & & \downarrow & & \downarrow \\ \operatorname{Spec} R/\mathfrak{m}_R^{n+1} & \hookrightarrow & \operatorname{Spec} R = \operatorname{Spec} \widehat{\mathcal{O}}_{Z,z_0} & \longrightarrow & Z \end{array}$$

As η is miniversal, there exists a local $\mathbb{C}$-algebra homomorphism

$$\varphi \colon \widehat{\mathcal{O}}_{Z,z_0} = R \longrightarrow \mathbb{C}[\![t]\!]$$

such that ξ is induced by η via φ, i.e. X_n is isomorphic to $Y_n \times_{\operatorname{Spec} R/\mathfrak{m}_R^{n+1}} S_n$ as S_n-schemes for every n. By another theorem of Artin [10, Corollary 2.5], the map φ has an algebraic approximation up to order m in the following sense: there exist a smooth affine curve B over $\mathbb{C}$ with a closed point $b_0 \in B$ and a $\mathbb{C}$-morphism $f \colon B \to Z$ such that $f(b_0) = z_0$ and the completion

$$\varphi' \colon \widehat{\mathcal{O}}_{Z,z_0} = R \longrightarrow \widehat{\mathcal{O}}_{B,b_0} = \mathbb{C}[\![t]\!]$$

of $f^\#_{b_0} \colon \mathcal{O}_{Z,z_0} \to \mathcal{O}_{B,b_0}$ satisfies the following property:

$$\varphi \equiv \varphi' \text{ modulo } t^{m+1}. \tag{14.1}$$

Let π be the base change $\mathcal{X} \times_Z B \to B$ along $f \colon B \to Z$. Let ξ' be the formal $\mathfrak{m}_{b_0}$-adic completion of π, i.e. $\xi' = \{\mathcal{X} \times_Z \operatorname{Spec} \mathcal{O}_{B,b_0}/\mathfrak{m}_{b_0}^{n+1} \to \operatorname{Spec} \mathcal{O}_{B,b_0}/\mathfrak{m}_{b_0}^{n+1}\}_n$. The two formal deformations ξ and ξ' of X over $\mathbb{C}[\![t]\!]$ are in general different, but they coincide up to order m because of (14.1). This implies that t^m is in the dth Fitting ideal of the sheaf of Kähler differentials of $\mathcal{X} \times_Z \operatorname{Spec} \mathcal{O}_{B,b_0}/\mathfrak{m}_{b_0}^{m+1} \to \operatorname{Spec} \mathcal{O}_{B,b_0}/\mathfrak{m}_{b_0}^{m+1}$. Therefore, ξ' is a formal smoothing. By Lemma 5, up to restrict B to an open neighbourhood of b_0 in B, we have that $\pi \colon \mathcal{X} \times_Z B \to B$ is a smoothing. □

The following theorem ensures that a projective scheme with formally smoothable isolated singularities is smoothable, provided that some local and cohomological conditions hold.

Theorem 7 *Let X be a projective scheme over $\mathbb{C}$ such that:*

1. *X is reduced and Cohen–Macaulay;*
2. *X is either l.c.i. or normal;*

3. $\mathrm{H}^2(X, T_X) = 0$ *and* $\mathrm{H}^2(X, \mathcal{O}_X) = 0$*;*
4. *X has isolated singularities and for each singular point $x \in X$ there exists an open affine neighbourhood of x which is formally smoothable.*

Then X is smoothable.

Proof Set $d = \dim X$. Let $x_1, \ldots, x_r$ be the singular points of X. Let U_i be an affine open neighbourhood of x_i in X which is formally smoothable and such that $U_i \setminus \{x_i\}$ is smooth. Let $\xi_i = \{U_{i,n} \to S_n\}_n$ be a formal smoothing of U_i, where S_n is $\operatorname{Spec} \mathbb{C}[t]/(t^{n+1})$ as usual.

By Proposition 3, from $\mathrm{H}^2(T_X) = 0$ we deduce that the map $\mathit{Def}_X \to \mathit{Def}_X^{\mathrm{loc}} = \prod_{i=1}^r \mathit{Def}_{U_i}$ is smooth. Therefore there exists a formal deformation $\xi = \{X_n \to S_n\}_n$ of X over $\mathbb{C}[\![t]\!]$ such that for each i the restriction of ξ to U_i is ξ_i, i.e. for all n the restriction of X_n to U_i is $U_{i,n}$. By Remark 4 we have that ξ is a formal smoothing of X. We conclude by Proposition 6. □

We now see some conditions that imply that a scheme is not smoothable.

Proposition 8 *Let X be a singular scheme of finite type over $\mathbb{C}$ of pure dimension. Assume that at least one of the following conditions holds:*

1. *every infinitesimal deformation of X is locally trivial;*
2. *the functor Def_X has an artinian hull.*

Then X is not formally smoothable.

Proof Set $d = \dim X$.

(1) Let U be a singular affine open subscheme of X. Let $\{X_n \to S_n\}_n$ be a formal deformation of X over $\mathbb{C}[\![t]\!]$. Let U_n be the restriction of X_n to U. By (1) we get that U_n is isomorphic, as S_n-scheme, to the trivial deformation $U \times_{\operatorname{Spec} \mathbb{C}} S_n$. Therefore $\operatorname{Fitt}_d(\Omega_{U_n/S_n}) = \operatorname{Fitt}_d(\Omega_{U/\mathbb{C}})\mathcal{O}_{U_n}$. As U is singular, $\operatorname{Fitt}_d(\Omega_{U/\mathbb{C}}) \subsetneq \mathcal{O}_U$. This implies that $t^n \notin \operatorname{Fitt}_d(\Omega_{U_n/S_n})$.

(2) Let R be the hull of Def_X. Every formal deformation of X over $\mathbb{C}[\![t]\!]$ is induced by the miniversal one via a local $\mathbb{C}$-algebra homomorphism $f \colon R \to \mathbb{C}[\![t]\!]$. As every element in $\mathfrak{m}_R$ is nilpotent and $\mathbb{C}[\![t]\!]$ is a domain, the homomorphism f factors as $R \twoheadrightarrow R/\mathfrak{m}_R = \mathbb{C} \hookrightarrow \mathbb{C}[\![t]\!]$. This implies that every formal deformation of X over $\mathbb{C}[\![t]\!]$ is trivial. Using a similar argument as in (1), we can prove that X cannot have a formal smoothing. □

The following corollary, which is a direct consequence of Propositions 2, 6 and 8, gives some obstructions to the smoothability of a Cohen–Macaulay proper scheme.

Corollary 9 *Let X be a Cohen–Macaulay scheme proper over $\mathbb{C}$. Let $U \subseteq X$ be an open subscheme of X such that U is singular, reduced, and either l.c.i. or normal. If* $\mathrm{H}^0(U, \mathcal{E}xt^1(\Omega_U, \mathcal{O}_U)) = 0$ *or Def_U has an artinian hull (e.g. if* $\operatorname{Ext}^1(\Omega_U, \mathcal{O}_U) = 0$*), then X is not smoothable.*

14.2.3 Invariants

Here we introduce a couple of invariants for Fano varieties.

The *Hilbert series* of a Fano variety X is the power series defined by its anti-plurigenera:

$$\mathrm{Hilb}(X, -K_X) := \sum_{m\geq 0} h^0(X, -mK_X)t^m \in \mathbb{Z}[\![t]\!].$$

The *(anticanonical) degree* of a Fano variety X is the positive rational number $(-K_X)^n$, where $n = \dim X$. If X is Gorenstein, i.e. K_X is Cartier, then the degree is an integer. The degree can be recovered from the Hilbert series because, up to a constant which depends on the dimension of X, it is the leading term of the Hilbert polynomial of $-K_X$.

The following proposition shows that the Hilbert series and the anticanonical degree are deformation invariants for Fano varieties with Gorenstein log terminal singularities.

Proposition 10 *Let S be a noetherian scheme over $\mathbb{Q}$ and let $\pi : X \to S$ be a proper flat morphism whose geometric fibres are Fano varieties with Gorenstein log terminal singularities. Then the Hilbert series and the degree of the fibres are locally constant on S.*

Proof The morphism π is a relatively Gorenstein. Therefore, by [47, V.9.7], the dualising sheaf ω_π is a line bundle on X and its restriction to each fibre X_s is $\mathcal{O}_{X_s}(K_{X_s})$.

By Serre duality and Kawamata–Viehweg vanishing [63, Theorem 2.70], we get $\mathrm{H}^1(X_s, \mathcal{O}_{X_s}(-mK_{X_s})) = 0$ for all $m \geq 0$ and $s \in S$. By cohomology and base change [48, Theorem III.12.11], for all $m \geq 0$, we get that the sheaf $\pi_*\omega_\pi^{\otimes -m}$ is locally free and has rank $h^0(X_s, \mathcal{O}_{X_s}(-mK_{X_s}))$ at the point $s \in S$. This implies that the Hilbert series of the fibres is locally constant on S. □

14.3 Deformations of Affine Toric Varieties

14.3.1 Toric Singularities

In this section we will consider deformations of toric singularities, that is affine toric varieties. We refer the reader to [34, 41] for an introduction to toric geometry.

If X is an affine toric variety of dimension 2, then X is a cyclic quotient surface singularity. There is extensive literature about deformations of this kind of singularities, e.g. [19, 27, 64, 88, 93, 94]. In particular, it is known that every affine toric variety of dimension 2 is smoothable [12].

The study of the deformation theory of affine toric varieties of dimension at least 3 has been initiated by Altmann [4–8]. For example, he computed the tangent space of the deformation functor of an affine toric variety. We will not write down the explicit description of $\mathrm{Ext}^1(\Omega_X, \mathcal{O}_X)$ when X is an affine toric variety, but we will mention a consequence.

Proposition 11 (Altmann [5, Corollary 6.5.1]) *If X is a $\mathbb{Q}$-Gorenstein affine toric variety which is smooth in codimension 2 and $\mathbb{Q}$-factorial in codimension 3, then X is rigid.*

Corollary 12 *Every isolated $\mathbb{Q}$-Gorenstein toric singularity of dimension ≥ 4 is rigid.*

Now we need to do a brief detour on Minkowski sums. If $F_0, F_1, \dots, F_r$ are polytopes in a real vector space, their *Minkowski sum* is the polytope

$$F_0 + F_1 + \cdots + F_r := \{v_0 + v_1 + \cdots + v_r \mid v_0 \in F_0, v_1 \in F_1, \dots, v_r \in F_r\}.$$

When we have $F = F_0 + F_1 + \cdots + F_r$, we say that we have a *Minkowski decomposition* of the polytope F. We consider Minkowski decompositions up to translation: for instance, we consider the Minkowski decomposition $F = (v + F_0) + (-v + F_1)$ to be equivalent to $F = F_0 + F_1$ for every vector v. Moreover, in what follows we require that the summands F_j are lattice polytopes, i.e. their vertices belong to a fixed lattice.

Altmann [5] has noticed that certain Minkowski decompositions induce deformations of affine toric varieties. In Sect. 14.3.2 we will see an example of this fact. For the proof we refer the reader to the original reference [5] and to [71, 81].

Now let us concentrate on Gorenstein toric singularities. They are associated to lattice polytopes of dimension one less than the dimension of the singularity. More precisely, let F be a lattice polytope of dimension $n-1$ in a lattice $\overline{N}$ of rank $n-1$ and let U_F be the affine toric variety associated to the cone $\sigma_F = \mathbb{R}_{\geq 0}(F \times \{1\})$ in the lattice $N := \overline{N} \oplus \mathbb{Z}$, i.e. $U_F = \operatorname{Spec} \mathbb{C}[\sigma_F^\vee \cap M]$, where $M = \overline{M} \oplus \mathbb{Z}$ is the dual of N and $\sigma_F^\vee$ is the dual cone of σ_F. We have that U_F has dimension n and is Gorenstein. All Gorenstein affine toric varieties without torus factors arise in this way from a lattice polytope. The isomorphism class of U_F does not change if we change F via an affine transformation in $\overline{N} \rtimes \mathrm{GL}(\overline{N}, \mathbb{Z})$.

As usual in toric geometry, the geometric properties of U_F can be deduced from the combinatorial properties of F. For instance:

1. U_F is smooth in codimension k if and only if all faces of F with dimension $< k$ are standard simplices;
2. U_F is $\mathbb{Q}$-factorial in codimension k if and only if all faces of F with dimension $< k$ are simplices.

It is always the case that U_F is smooth in codimension 1 and $\mathbb{Q}$-factorial in codimension 2.

If F is a segment of lattice length $m+1$, then U_F is the A_m surface singularity $\operatorname{Spec} \mathbb{C}[x, y, z]/(xy - z^{m+1})$. This is an isolated hypersurface singularity, therefore it is very easy to write down the miniversal deformation: $xy = z^{m+1} + t_m z^{m-1} + \cdots + t_1$ over $\mathbb{C}[\![t_1, \dots, t_m]\!]$. It is clear that this singularity is smoothable.

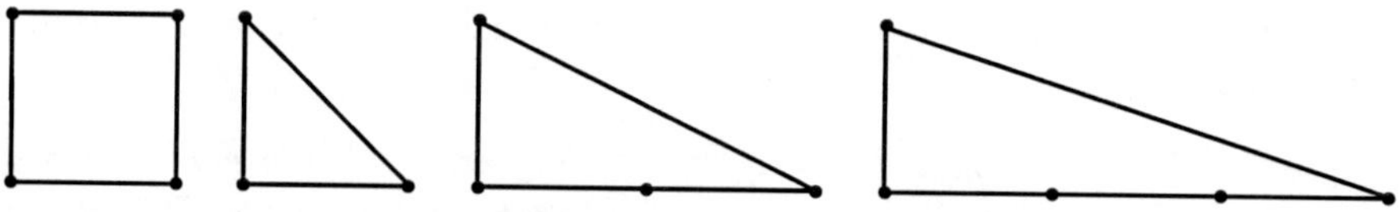

Fig. 14.1 A standard square, a standard triangle, an A_1-triangle and an A_2-triangle

If F is a lattice polygon, then the affine toric threefold U_F has the following properties:

1. U_F has, at most, an isolated singularity if and only if the edges of F are unitary, i.e. have lattice length 1;
2. U_F is $\mathbb{Q}$-factorial if and only if F is a triangle.

Now we provide some examples of lattice polygons and their corresponding toric Gorenstein affine threefolds.

Example 13 A lattice polygon F is called a *standard square* (Fig. 14.1) if it is a quadrilateral such that all its lattice points are vertices, or equivalently if it is $\mathbb{Z}^2 \rtimes \mathrm{GL}_2(\mathbb{Z})$-equivalent to $\mathrm{conv}\{(0,0),(1,0),(1,1),(0,1)\} \subseteq \mathbb{R}^2$. If F is a standard square, then U_F is the ordinary double point (i.e. node) $\mathrm{Spec}\,\mathbb{C}[x,y,z,w]/(xy - zw)$. This singularity is clearly smoothable as it is a hypersurface singularity. Its miniversal deformation is given by $xy - zw = t$ over $\mathbb{C}[\![t]\!]$.

A lattice polygon F is called a *standard triangle* if it is a triangle such that all its lattice points are vertices, or equivalently if it is $\mathbb{Z}^2 \rtimes \mathrm{GL}_2(\mathbb{Z})$-equivalent to $\mathrm{conv}\{(0,0),(1,0),(0,1)\} \subseteq \mathbb{R}^2$. F is a standard triangle if and only if U_F is isomorphic to $\mathbb{A}^3$.

If $m \geq 1$, then a lattice polygon F is called an *A_m-triangle* if it is a triangle such that there are no interior lattice points and the edges have lattice lengths $1, 1, m+1$, respectively. Equivalently, a polygon is an A_m-triangle if and only if it is $\mathbb{Z}^2 \rtimes \mathrm{GL}_2(\mathbb{Z})$-equivalent to $\mathrm{conv}\{(0,0),(m+1,0),(0,1)\} \subseteq \mathbb{R}^2$. If F is an A_m-triangle, then U_F is the cA_m-singularity $\mathrm{Spec}\,\mathbb{C}[x,y,z,w]/(xy - z^{m+1})$. This singularity is clearly smoothable as it is a hypersurface singularity.

Altmann [7] explicitly constructed the miniversal deformation of an isolated Gorenstein toric singularity of dimension 3. (By Corollary 12 it is trivial to construct the miniversal deformation of an isolated Gorenstein toric singularity of dimension ≥ 4.) A consequence of his construction is the following description of the irreducible components of the base of the miniversal deformation.

Theorem 14 (Altmann [7]) *Let F be a lattice polygon with unitary edges and let U_F be the corresponding isolated Gorenstein toric singularity of dimension* 3. *Let R be the hull of Def_{U_F}. Then there exists a one-to-one correspondence between minimal primes of R and maximal Minkowski decompositions of F. Moreover, if a minimal prime $\mathfrak{p} \subset R$ corresponds to the maximal Minkowski decomposition $F = F_0 + F_1 + \cdots + F_r$, then $r = \dim R/\mathfrak{p}$.*

Corollary 15 *Let F be a lattice polygon with unitary edges and let U_F be the associated isolated Gorenstein toric singularity of dimension 3. Then Def_{U_F} has an artinian hull if and only if F is Minkowski indecomposable.*

14.3.2 The Affine Cone over the Del Pezzo Surface of Degree 7

Here we study an explicit example of what has been considered in Sect. 14.3.1. In the lattice $\overline{N} = \mathbb{Z}^2$ consider the pentagon

$$F = \operatorname{conv}\left\{\begin{pmatrix}1\\0\end{pmatrix}, \begin{pmatrix}1\\1\end{pmatrix}, \begin{pmatrix}0\\1\end{pmatrix}, \begin{pmatrix}-1\\0\end{pmatrix}, \begin{pmatrix}0\\-1\end{pmatrix}\right\} \subseteq \overline{N}_{\mathbb{R}}, \tag{14.2}$$

which is depicted on the left of Fig. 14.2. The toric variety associated to the face fan of F is the smooth del Pezzo surface of degree 7, which is denoted by dP_7 and is the blow up of $\mathbb{P}^2$ in 2 distinct points. The anticanonical map of dP_7 is a closed embedding into $\mathbb{P}^7$.

Now we put the pentagon F at height 1 in the lattice $N = \overline{N} \oplus \mathbb{Z}$ and we consider the cone over it:

$$\sigma_F = \operatorname{cone}\left\{\begin{pmatrix}1\\0\\1\end{pmatrix}, \begin{pmatrix}1\\1\\1\end{pmatrix}, \begin{pmatrix}0\\1\\1\end{pmatrix}, \begin{pmatrix}-1\\0\\1\end{pmatrix}, \begin{pmatrix}0\\-1\\1\end{pmatrix}\right\} \subseteq \overline{N}_{\mathbb{R}} \oplus \mathbb{R}.$$

The affine toric variety $U_F = \operatorname{Spec} \mathbb{C}[\sigma_F^\vee \cap (\overline{M} \oplus \mathbb{Z})]$ is the affine cone over the anticanonical embedding of dP_7 and has an isolated Gorenstein canonical non-terminal singularity at the vertex of the cone.

Altmann [7, (9.1)] shows that the hull of Def_{U_F} is $\mathbb{C}[\![t_1, t_2]\!]/(t_1^2, t_1 t_2)$, which is a line with an embedded point. The reduction of the miniversal deformation, i.e. the base change to the reduction of the hull, is induced by the unique maximal Minkowski decomposition of the pentagon F in the following way.

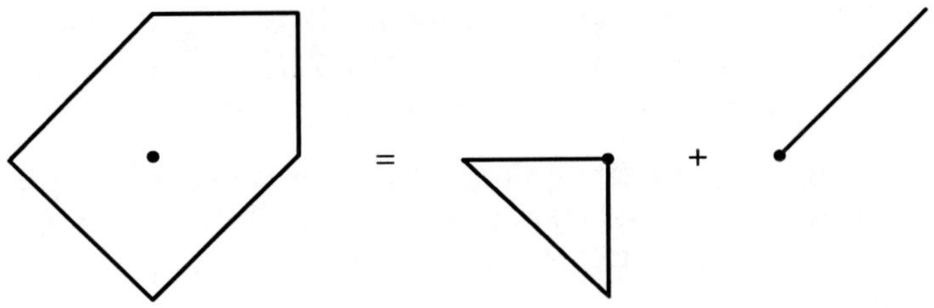

Fig. 14.2 The Minkowski decomposition (14.3) of the pentagon F in (14.2)

In the lattice $\overline{N}$ we have the Minkowski decomposition

$$F = \operatorname{conv}\left\{\begin{pmatrix}0\\0\end{pmatrix}, \begin{pmatrix}-1\\0\end{pmatrix}, \begin{pmatrix}0\\-1\end{pmatrix}\right\} + \operatorname{conv}\left\{\begin{pmatrix}0\\0\end{pmatrix}, \begin{pmatrix}1\\1\end{pmatrix}\right\}, \tag{14.3}$$

which is illustrated in Fig. 14.2. Following [5, (3.4)], in the lattice $\tilde{N} = \overline{N} \oplus \mathbb{Z}e_1 \oplus \mathbb{Z}e_2$ we construct the cone

$$\tilde{\sigma} = \operatorname{cone}\left\{\begin{pmatrix}0\\0\\1\\0\end{pmatrix}, \begin{pmatrix}-1\\0\\1\\0\end{pmatrix}, \begin{pmatrix}0\\-1\\1\\0\end{pmatrix}, \begin{pmatrix}0\\0\\0\\1\end{pmatrix}, \begin{pmatrix}1\\1\\0\\1\end{pmatrix}\right\} \subseteq \tilde{N}_{\mathbb{R}}.$$

Notice that the the first three rays of $\tilde{\sigma}$ come from the vertices of the first summand of F in (14.3), whereas the last two rays of $\tilde{\sigma}$ come from the vertices of the second summand of F in (14.3). Let $\tilde{U} = \operatorname{Spec} \mathbb{C}[\tilde{\sigma}^\vee \cap \tilde{M}]$ be the affine toric variety associated to the cone $\tilde{\sigma}$, where $\tilde{M}$ denotes the dual of $\tilde{N}$. One can prove that $\tilde{U}$ has only an isolated terminal Gorenstein singularity. Let f_1 and f_2 be the regular functions on $\tilde{U}$ associated to the characters $(0, 0, 1, 0) \in \tilde{M}$ and $(0, 0, 0, 1) \in \tilde{M}$, respectively. The variety U_F is the zero locus of the function $f_1 - f_2$, i.e. we have a cartesian diagram

$$\begin{array}{ccc} U_F & \longrightarrow & \tilde{U} \\ \downarrow & & \downarrow \pi \\ \operatorname{Spec} \mathbb{C} & \longrightarrow & \mathbb{A}^1_{\mathbb{C}} \end{array} \tag{14.4}$$

where π is given by the function $f_1 - f_2$ and the bottom morphism is given by the origin of $\mathbb{A}^1_{\mathbb{C}}$. Since $f_1 - f_2$ is not constant and $\mathbb{A}^1_{\mathbb{C}}$ is regular of dimension 1, the morphism π is flat. The reduction of the miniversal deformation of U_F is the formal deformation of U_F over $\mathbb{C}[\![t]\!]$ obtained from the square (14.4) by base change via $\operatorname{Spec} \mathbb{C}[t]/(t^{n+1}) \hookrightarrow \operatorname{Spec} \mathbb{C}[t] = \mathbb{A}^1_{\mathbb{C}}$ for all n. The following proposition shows that this is a formal smoothing.

Proposition 16 *Let F be the pentagon defined in (14.2) and let U_F be the corresponding Gorenstein toric threefold singularity. Then the collection of the base change of π in (14.4) via* $\operatorname{Spec} \mathbb{C}[t]/(t^{n+1}) \to \operatorname{Spec} \mathbb{C}[t] = \mathbb{A}^1_{\mathbb{C}}$ *for all n is a formal smoothing of U_F. In particular, U_F is formally smoothable.*

Proof We want to study the closed fibres of π. The fibre over the origin of $\mathbb{A}^1_{\mathbb{C}}$ is U_F. Let us fix $\lambda \in \mathbb{C} \setminus \{0\}$ and we consider the fibre $\pi^{-1}(\lambda)$ of π over the closed point $(t - \lambda)$ of $\mathbb{A}^1_{\mathbb{C}}$ corresponding to λ. We consider the subcone τ_1 (resp. τ_2) of $\tilde{\sigma}$ that is generated by the first three (resp. last two) rays of $\tilde{\sigma}$. We consider the affine toric variety $W_j = \operatorname{Spec} \mathbb{C}[\tau_j^\vee \cap \tilde{M}]$, for $j = 1, 2$. We have that W_1 and W_2 are open subschemes of $\tilde{U}$.

We have that W_1 is the open subset of $\tilde{U}$ where the function f_2 does not vanish, i.e. $W_1 = \{f_2 \neq 0\} \subseteq \tilde{U}$, and analogously $W_2 = \{f_1 \neq 0\} \subseteq \tilde{U}$. It is clear that there is an isomorphism $W_1 \simeq \mathbb{A}^3 \times \mathbb{G}_m$ with respect to which the function $f_1|_{W_1}$ becomes a projection onto a $\mathbb{A}^1$-factor in $\mathbb{A}^3$ and the function $f_2|_{W_1}$ becomes the projection onto the $\mathbb{G}_m$-factor. There is also an isomorphism $W_2 \simeq \mathbb{A}^2 \times \mathbb{G}_m^2$ with respect to which the function $f_1|_{W_2}$ becomes a projection onto a $\mathbb{G}_m$-factor and the function $f_2|_{W_2}$ becomes a projection onto an $\mathbb{A}^1$-factor.

Now $\pi^{-1}(\lambda) \cap W_1 = \{f_1 = f_2 + \lambda\} \cap W_1$ is isomorphic to $\mathbb{A}^2 \times \mathbb{G}_m$ and $\pi^{-1}(\lambda) \cap W_2 = \{f_2 = f_1 - \lambda\} \cap W_2$ is isomorphic to $\mathbb{A}^1 \times \mathbb{G}_m^2$. Since $\lambda \neq 0$, it is clear that $\pi^{-1}(\lambda) \subseteq W_1 \cup W_2$. Therefore we have proved that $\pi^{-1}(\lambda)$ is smooth.

One can also show that the generic fibre of π is smooth over the generic point of $\mathbb{A}^1_{\mathbb{C}}$. In order to prove this, it is enough to base change to the spectrum of the field $\mathbb{C}(t)$ of rational function of $\mathbb{A}^1_{\mathbb{C}}$ and pursue a similar argument, which deals with toric varieties over the field $\mathbb{C}(t)$.

In particular, π is flat of relative dimension 3 and has Cohen–Macaulay fibres. As in the proof of Lemma 5, from the fact that all non-special fibres of π are smooth we can deduce that π induces a formal smoothing of U_F. □

14.4 Deformations of Toric Fano Varieties

14.4.1 Fano Polytopes

Fano polytopes are the combinatorial-polyhedral avatars of toric Fano varieties.

Definition 17 A polytope P in a lattice N of rank n is called *Fano* if:

1. P has dimension n;
2. the origin 0 lies in the interior of P;
3. the vertices of P are primitive lattice elements of N.

If P is a Fano polytope, we denote by X_P the complete toric variety associated to the *spanning fan* (also called the *face fan*) of P.

If P is a Fano polytope, then X_P is a Fano variety. All toric Fano varieties arise in this way from a Fano polytope [34, Sect. 8.3]. The variety X_P is Gorenstein, i.e. its (anti)canonical divisor is Cartier, if and only if P is reflexive, i.e. the facets of P lie on hyperplanes with height 1 with respect to the origin. The maximal toric affine charts of X_P (or equivalently the torus-fixed points of X_P) are in one-to-one correspondence with the facets of P. If n is the dimension of P, for every $0 \leq k \leq n$ there is a one-to-one correspondence between the k-dimensional torus-orbits of X_P and the $(n-k-1)$-dimensional faces of P.

Fano polytopes of small dimension with specific properties have been classified [15, 16, 57–59, 65–67, 77, 78, 89, 90, 98]. We refer the reader to [60] for a survey on the classification of Fano polytopes.

14.4.2 Two Sufficient Conditions for Non-smoothability

It is an open problem to understand whether an arbitrary toric Fano variety is smoothable. Here we state a couple of conditions that forbid the smoothability. Both conditions on a toric Fano variety X are based on the existence of an open affine singular subscheme U such that U is not formally smoothable.

Theorem 18 *Let N be a lattice, let P be a Fano polytope in N, and let X be the toric Fano variety associated to the spanning fan of P. Assume that there exists a face F of P which satisfies the following conditions:*

1. *for each 1-face F' of F, there exists a basis of N which contains the two vertices of F';*
2. *each 2-face of F is a triangle;*
3. *there exists no basis of N which contains all the vertices of F.*

Then X is not smoothable.

Proof Let U be the affine toric open subscheme of X associated to the cone spanned by the face F. The condition (1) means that U is smooth in codimension 2. The condition (2) means that U is $\mathbb{Q}$-factorial in codimension 3. Therefore U is rigid by Proposition 11. The condition (3) implies that U is singular. Therefore, by Corollary 9, X is not smoothable. □

If P is a reflexive polytope of dimension 3, then the theorem above applies if there exists a triangular facet F with unitary edges and such that it is not a standard triangle. Below we see that we can relax the condition of F being triangular to F being Minkowski-indecomposable.

Proposition 19 *Let P be a reflexive polytope of dimension 3 and let X be the toric Fano threefold associated to the spanning fan of P. Assume that there exists a facet F of P such that:*

1. *F has unitary edges;*
2. *F is Minkowski-indecomposable;*
3. *F is not a standard triangle (i.e. the vertices of F do not form a basis of the lattice).*

Then X is not smoothable.

Proof The proof is very similar to the proof of Theorem 18. Let U be the affine toric open subscheme of X associated to the cone spanned by F. The conditions (1) and (3) means that U has an isolated singularity. Since P is reflexive, U is Gorenstein. By Corollary 15, from (2) we deduce that Def_U has an artinian hull. Therefore, by Proposition 8, U is not formally smoothable. By Corollary 9, X is not smoothable. □

14.4.3 Rigidity

Here we will see that if a toric Fano variety has very mild singularities then it is rigid.

Lemma 20 *Let X be a toric Fano variety. Then* $\mathrm{H}^i(X, T_X) = 0$ *for each* $i \geq 1$. *In particular, all locally trivial deformations of X are trivial.*

Proof Set $n = \dim X$. Consider the smooth locus $j : U \hookrightarrow X$. Let D be the toric boundary of X. The sheaves T_X and $(j_* \Omega_U^{n-1} \otimes \mathcal{O}_X(D))^{\vee\vee}$ are reflexive on X and their restrictions to U coincide, because U is smooth and T_U is isomorphic to $\Omega_U^{n-1} \otimes \omega_U^{\vee}$. Therefore, since the complement of U has codimension at least 2, by [49, Proposition 1.6] we have that T_X is isomorphic to $(j_* \Omega_U^{n-1} \otimes \mathcal{O}_X(D))^{\vee\vee}$. Since D is ample, we conclude by Bott–Steenbrink–Danilov vanishing [34, Theorem 9.3.1] (see also [24, 40, 75]). □

An immediate consequence of the lemma above is the following result.

Proposition 21 *Every smooth toric Fano variety is rigid.*

This result was originally proved by Bien and Brion [20]. Later de Fernex and Hacon [38] proved the rigidity of $\mathbb{Q}$-factorial terminal toric Fano varieties. The following theorem, due to Totaro, is the most general rigidity theorem for toric Fano varieties of which we are aware.

Theorem 22 (Totaro [96, Theorem 5.1]) *A Fano toric variety which is smooth in codimension* 2 *and* $\mathbb{Q}$*-factorial in codimension* 3 *is rigid.*

Proof By Lemma 20, $\mathrm{H}^1(T_X) = 0$. By Proposition 11, the sheaf $\mathcal{E}xt^1(\Omega_X, \mathcal{O}_X)$ is zero. From the five term exact sequence of Ext, which is written in the proof of Proposition 2, we deduce that $\mathrm{Ext}^1(\Omega_X, \mathcal{O}_X)$ is zero. □

If P is a Fano polytope, then X_P satisfies the hypotheses of this theorem if and only if all 2-faces of P are triangles and each edge, i.e. 1-face, of P has lattice length 1 and is contained in some hyperplane which has height 1 with respect to the origin.

Corollary 23 *Let X be a toric Fano variety of dimension* ≥ 4. *If X has isolated singularities, then X is rigid.*

In Sects. 14.4.4 and 14.4.5 we will study deformations of toric Fanos with isolated singularities and of dimension 2 or 3.

14.4.4 Toric del Pezzo Surfaces

A *del Pezzo surface* is a Fano variety of dimension 2. A toric del Pezzo surface is associated to a Fano polygon, which is a Fano polytope of dimension 2.

Theorem 24 *Every toric del Pezzo surface is smoothable.*

Proof Let X be an arbitrary toric del Pezzo surface. It is well known that X is a normal Cohen–Macaulay projective variety. By Demazure vanishing [34, Theorem 2.9.3], $\mathrm{H}^2(\mathcal{O}_X) = 0$. By Lemma 20, $\mathrm{H}^2(T_X) = 0$. Since X is normal and of dimension 2, X has isolated singularities. By Theorem 7 it is enough to check that the singularities of X are formally smoothable.

The singularities of X are cyclic quotient surface singularities. This kind of singularities is always smoothable; indeed, it is enough to pick the Artin component of the base of the miniversal deformation [12]. □

Remark 25 When the canonical divisor of a normal variety X is not Cartier, flat deformations of X are too wild for hoping to study moduli of varieties. For a normal $\mathbb{Q}$-Gorenstein non-Gorenstein variety X one should consider a subfunctor of Def_X which is made up of the deformations of X in which the canonical divisor deforms well. This is the theory of $\mathbb{Q}$-Gorenstein deformations, developed by Kollár–Shepherd-Barron [64] (see also [1, 9, 45, 68]).

In the context of $\mathbb{Q}$-Gorenstein deformations the analogous statement of Theorem 24 is false: there exist non-Gorenstein toric del Pezzo surfaces which cannot be deformed via $\mathbb{Q}$-Gorenstein deformations to a smooth del Pezzo surface, e.g. the weighted projective space $\mathbb{P}(1, 1, 3)$. Nonetheless, it is true that for $\mathbb{Q}$-Gorenstein deformations of del Pezzo surfaces there are no local-to-global obstructions [2, Lemma 6]. Therefore, a del Pezzo surface is $\mathbb{Q}$-Gorenstein smoothable if and only if its singularities are $\mathbb{Q}$-Gorenstein smoothable.

Since the main focus of this note is the study of deformations of Gorenstein toric Fano threefolds, we will omit to discuss the theory of $\mathbb{Q}$-Gorenstein deformations. We refer the reader to [46, 82] for the study of toric del Pezzo surfaces which have $\mathbb{Q}$-Gorenstein smoothings.

14.4.5 Toric Fano Threefolds with Isolated Singularities

Theorem 26 *Let X be a toric Fano variety of dimension* 3 *with isolated singularities. Then X is smoothable if and only if its singularities are formally smoothable.*

Proof By Proposition 6, if X is smoothable then its singularities are formally smoothable. Conversely, suppose that the singularities of X are formally smoothable. Then we argue as in the proof of Theorem 24: X is a normal Cohen–Macaulay projective variety with $\mathrm{H}^2(\mathcal{O}_X) = 0$, by [34, Theorem 2.9.3], and $\mathrm{H}^2(T_X) = 0$, by Lemma 20. By Theorem 7, X is smoothable. □

Corollary 27 *Let P be a reflexive polytope of dimension 3 and let X be the toric Fano threefold associated to the spanning fan of P. If each facet of P is either a standard triangle or a standard square (see the definition in Example 13), then X is smoothable.*

Proof By Example 13 we have that the singularities of X are at most ordinary double points (i.e. nodes). These singularities are formally smoothable. By Theorem 26 we conclude. □

The proof of this corollary is essentially a specific case of [39, Sect. 4.a]. The corollary could have been deduced also from a more general result by Namikawa according to which every Fano threefold with Gorenstein terminal singularities is smoothable [76]. The smooth Fano threefolds which are the smoothings of the toric Fano threefold appearing in Corollary 27 have been studied by Galkin [42].

For $d \in \{6, 7\}$, let dP_d be the smooth del Pezzo surface of degree d; it is toric. The complete anticanonical linear system on dP_d induces a closed embedding $\mathrm{dP}_d \hookrightarrow \mathbb{P}^d$. We consider the projective cone $C(\mathrm{dP}_d) \subseteq \mathbb{P}^{d+1}$ over this embedding; we have that $C(\mathrm{dP}_d)$ is a toric Fano threefold with a Gorenstein canonical non-terminal isolated singularity. In Sect. 14.4.6 we will see that $C(\mathrm{dP}_7)$ is smoothable. In [80] it is shown that $C(\mathrm{dP}_6)$ has two smoothings (see also [56, Example 3.3]).

14.4.6 The Projective Cone over the Del Pezzo Surface of Degree 7

Here we study the deformations of an explicit toric Fano threefold with an isolated Gorenstein non-terminal singularity.

Fix the lattice $\overline{N} = \mathbb{Z}^2$. Consider the pentagon $F \subseteq \overline{N}_{\mathbb{R}}$ defined in (14.2), imagine to put it into the plane $\overline{N}_{\mathbb{R}} \times \{1\}$ in $\overline{N}_{\mathbb{R}} \oplus \mathbb{R} \simeq \mathbb{R}^3$, and create the pyramid over it with apex at the point $(0, 0, -1)$: this is the polytope

$$P = \mathrm{conv}\left\{ \begin{pmatrix}1\\0\\1\end{pmatrix}, \begin{pmatrix}1\\1\\1\end{pmatrix}, \begin{pmatrix}0\\1\\1\end{pmatrix}, \begin{pmatrix}-1\\0\\1\end{pmatrix}, \begin{pmatrix}0\\-1\\1\end{pmatrix}, \begin{pmatrix}0\\0\\-1\end{pmatrix} \right\} \tag{14.5}$$

in the lattice $\overline{N} \oplus \mathbb{Z}$ and is depicted in Fig. 14.3. It is clear that P is a Fano polytope.

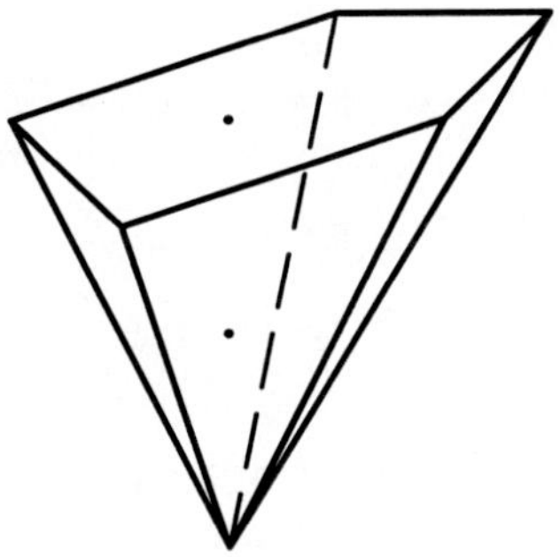

Fig. 14.3 The 3-dimensional lattice polytope P defined in (14.5) and associated to the projective cone over the del Pezzo surface of degree 7

Let X be the toric variety associated to the spanning fan of P. Then X is the projective cone over the anticanonical embedding of the smooth del Pezzo surface of degree 7. The affine toric variety U_F considered in Sect. 14.3.2 is the affine open toric subscheme of X associated to the pentagonal facet F of P. We have that X is a Fano threefold with an isolated non-terminal canonical Gorenstein singularity at the vertex of the cone.

Proposition 28 *Let X be the toric Fano threefold associated to the polytope P in (14.5), i.e. X is the projective cone over the anticanonical embedding of the smooth del Pezzo surface of degree 7. Then X is smoothable and can be deformed to the smooth Fano threefold $\mathbb{P}(\mathcal{O}_{\mathbb{P}^2} \oplus \mathcal{O}_{\mathbb{P}^2}(1))$.*

Proof By Proposition 16, X has an isolated singularity which is formally smoothable. By Theorem 26 we know that X is smoothable. We need to know to which smooth Fano threefold X can be deformed.

From toric geometry [34, Theorem 13.4.3], we have that the anticanonical degree $(-K_X)^3$ is the normalised volume of the polar polytope of P, which is 56 in this case. Since X has Gorenstein canonical singularities, by Proposition 10 we have that the anticanonical degree is preserved in the smoothing. By inspecting the list of smooth Fano threefolds (see [53, 54, 72–74] or [55, Sect. 12]), there is a unique smooth Fano threefold of anticanonical degree 56, namely $\mathbb{P}(\mathcal{O}_{\mathbb{P}^2} \oplus \mathcal{O}_{\mathbb{P}^2}(1))$. □

14.4.7 Another Sufficient Condition for Non-smoothability

In addition to the result of Proposition 19, here we present another obstruction for the smoothability of a toric Fano threefold with Gorenstein singularities.

Theorem 29 ([79]) *Let N be a lattice of rank 3, let $M = \mathrm{Hom}_{\mathbb{Z}}(N, \mathbb{Z})$, let $\langle \cdot, \cdot \rangle \colon M \times N \to \mathbb{Z}$ be the duality pairing, let P be a reflexive polytope in N, and let X be the toric Fano threefold associated to the spanning fan of P. Assume that there are two adjacent facets F_0 and F_1 of P such that:*

1. *both F_0 and F_1 are A_n-triangles for some integer $n \geq 1$ (see the definition in Example 13);*
2. *$F_0 \cap F_1$ is a segment with $n+2$ lattice points;*
3. *$\langle w_1, v_0 \rangle = 0$, where $w_1 \in M$ is such that $F_1 \subseteq \{v \in N_{\mathbb{R}} \mid \langle w_1, v \rangle = 1\}$ and $v_0 \in N$ is the vertex of F_0 which does not lie on the segment $F_0 \cap F_1$.*

Then X is not smoothable.

With the terminology of [79], the two triangles F_0 and F_1 are called "two adjacent almost-flat A_n-triangles".

Proof (Sketch of the proof of Theorem 29) We refer the reader to [79] for all the details missing here. Let U_i be the toric open affine subscheme of X associated to the facet F_i, for each $i = 0, 1$. Set $U = U_0 \cup U_1$.

One can show that U admits an A_n-bundle structure over $\mathbb{P}^1$. More precisely, one can construct a toric morphism $\pi : U \to \mathbb{P}^1$ such that, for each $i = 0, 1$, if V_i denotes the ith standard affine chart of $\mathbb{P}^1$ then $\pi^{-1}(V_i) = U_i$ and the restriction $\pi|_{U_i} : U_i \to V_i$ is the projection $\operatorname{Spec} \mathbb{C}[x, y, z, w]/(xy - z^{n+1}) \to \operatorname{Spec} \mathbb{C}[w]$. This A_n-bundle may be non trivial, depending on the relative position of the two triangles F_0 and F_1. Set $d = \langle w_1, v_0 \rangle$. By [79, Proposition 3.5] there exists an isomorphism of coherent sheaves on $\mathbb{P}^1$:

$$\pi_* \mathcal{E}xt^1_{\mathcal{O}_U}(\Omega_U, \mathcal{O}_U) \simeq \bigoplus_{2 \le j \le n+1} \mathcal{O}_{\mathbb{P}^1}(-jd - j).$$

Since $d = 0$, the sheaf on the right is a direct sum of negative line bundles on $\mathbb{P}^1$, hence we have $\mathrm{H}^0(U, \mathcal{E}xt^1_{\mathcal{O}_U}(\Omega_U, \mathcal{O}_U)) = 0$. By Corollary 9, X is not smoothable. □

With the same technique of the theorem above one can also construct some rigid toric Fano threefolds with only cA_1-singularities (see [79, Theorem 1.2]). This refutes a conjecture of Prokhorov [85] according to which every Fano threefold with compound Du Val singularities is smoothable.

14.4.8 Other Methods

Here we briefly collect some other results on deformations and smoothings of toric Fano varieties. Most of these results have been motivated by Mirror Symmetry for Fano varieties (see [2, 3, 26, 30, 31, 61, 86, 87]).

By analysing cluster transformations of tori, Akhtar–Coates–Galkin–Kasprzyk [3] have introduced the notion of *mutation* of Fano polytopes. A mutation is a combinatorial procedure that, under certain conditions, transforms a Fano polytope P into another Fano polytope P'. Ilten [51] has proved that mutations of Fano polytopes induce deformations of the corresponding toric Fano varieties; more precisely, if P and P' are related via a mutation, then he has constructed a flat family over $\mathbb{P}^1$ such that the fibre over 0 is X_P and the fibre over ∞ is $X_{P'}$.

Ilten, Lewis and Przyjalkowski [52] have constructed toric degenerations of smooth Fano threefolds with Picard rank 1.

Christophersen and Ilten [29] have constructed degenerations of smooth Fano threefolds of low degree to certain unobstructed Fano Stanley–Reisner schemes. Since these unobstructed Fano Stanley–Reisner schemes are also degenerations of singular toric Fano varieties, this implies the following result.

Theorem 30 (Christophersen–Ilten [28, Proposition 4.2, Theorems 5.1, 7.1]) *Let X be a toric Fano threefold with Gorenstein singularities. If $(-K_X)^3 \in \{4, 6, 8, 10, 12\}$, then X is smoothable.*

Coates–Kasprzyk–Prince [32] have introduced a combinatorial gadget, called *scaffolding*, on a Fano polytope P which induces a closed embedding of the toric Fano variety X_P into a bigger toric variety Y. Often X_P is a complete intersection in the Cox coordinates of Y, therefore it is easy to construct embedded deformations of X_P in Y. In many cases this produces smoothings of X_P. For instance, Cavey and Prince [25] have successfully applied the scaffolding method to construct deformations of toric del Pezzo surfaces to del Pezzo surfaces with a single $\frac{1}{k}(1, 1)$ singularity.

Moreover, Prince [83] has found necessary and sufficient conditions in order to have that the ambient toric variety Y is smooth: this is the notion of *cracked polytope*. He has also found a sufficient condition for a smoothing of X_P to exist inside Y. Via the scaffolding method and cracked polytopes, in [84] he constructs a degeneration of each smooth Fano threefold with very ample anticanonical bundle and Picard rank ≥ 2 to a Gorenstein toric Fano threefold.

14.5 Lists of Reflexive Polytopes of Dimension 3

Below we write lists of reflexive polytopes of dimension 3 which satisfy specific properties. There are exactly 4319 reflexive polytopes of dimension 3: the classification is due to Kreuzer and Skarke [66]. The IDs we use are numbers between 1 and 4319 and come from the Graded Ring Database [23]. All polytopes we consider below are reflexive of dimension 3. They correspond to toric Fano threefolds with Gorenstein singularities. We denote by X_P the toric Fano threefold associated to the spanning fan of P.

Let $\mathcal{S}_{\text{smoothable}}$ be the set of polytopes P such that the corresponding toric Fano threefold X_P is smoothable. It is an open question to explicitly compute $\mathcal{S}_{\text{smoothable}}$.

Let $\mathcal{S}_{\text{smooth}}$ be the set of polytopes which have only standard triangles as facets. These 18 polytopes correspond to the smooth toric Fano threefolds.

Let $\mathcal{S}_{\text{isol}}$ be the set of polytopes with unitary edges such that at least one facet is not a standard triangle. These 137 polytopes correspond to the singular toric Fano threefolds with isolated Gorenstein singularities.

Let $\mathcal{S}_{\text{nodes}}$ be the set of polytopes such that all facets are either standard triangles or standard squares and there is at least a square facet. These 82 polytopes correspond to the singular toric Fano threefolds with at most ordinary double points, or equivalently to the singular toric Fano threefolds with Gorenstein terminal singularities. By Corollary 27 these varieties are smoothable.

Let $\mathcal{S}_{\text{low}}$ be the set of polytopes P such that the normalised volume of the polar P^* of P belongs to $\{4, 6, 8, 10, 12\}$. These 220 polytopes correspond to the toric Gorenstein Fano threefolds X such that $(-K_X)^3 \in \{4, 6, 8, 10, 12\}$.

Let $\mathcal{S}_{\text{indec}}$ be the set of polytopes which contain a facet F which has unitary edges, is Minkowski indecomposable and is not a standard triangle. By Proposition 19 the corresponding toric Fano threefolds are not smoothable.

Let $\mathcal{S}_{\text{aft}}$ be the set of polytopes which contain a pair of adjacent almost-flat A_n-triangles, for some $n \geq 1$. In other words, the set $\mathcal{S}_{\text{aft}}$ contains exactly all polytopes P to which Theorem 29 applies. Therefore, the corresponding toric Fano threefolds are not smoothable.

Let $\mathcal{S}$ denote the set of all reflexive polytopes of dimension 3, i.e. the set of positive integers not greater than 4319. We have:

$$\mathcal{S}_{\text{nodes}} \subseteq \mathcal{S}_{\text{isol}} \subseteq \mathcal{S} \setminus \mathcal{S}_{\text{smooth}},$$
$$\mathcal{S}_{\text{smooth}} \cup \mathcal{S}_{\text{nodes}} \cup \mathcal{S}_{\text{low}} \subseteq \mathcal{S}_{\text{smoothable}},$$
$$\mathcal{S}_{\text{indec}} \cup \mathcal{S}_{\text{aft}} \subseteq \mathcal{S} \setminus \mathcal{S}_{\text{smoothable}}.$$

Below we write down the elements of most of the sets mentioned above.

$\mathcal{S}_{\text{smooth}}$ = {1, 5, 6, 7, 8, 25, 26, 27, 28, 29, 30, 31, 82, 83, 84, 85, 219, 220}

$\mathcal{S}_{\text{isol}}$ = {3, 4, 11, 12, 17, 21, 22, 23, 24, 42, 48, 49, 50, 51, 54, 68, 69, 70, 71, 72, 73, 74, 75, 76, 77, 78, 79, 80, 81, 155, 156, 158, 159, 160, 167, 168, 170, 177, 187, 188, 198, 199, 200, 201, 202, 203, 204, 205, 206, 207, 208, 209, 210, 211, 212, 213, 214, 215, 216, 217, 218, 360, 363, 364, 365, 366, 376, 377, 378, 380, 385, 403, 410, 411, 412, 413, 414, 415, 416, 417, 418, 419, 420, 421, 422, 423, 424, 425, 426, 427, 686, 688, 689, 692, 693, 694, 695, 696, 707, 710, 725, 729, 730, 731, 732, 733, 734, 735, 736, 737, 738, 739, 740, 741, 1085, 1086, 1087, 1091, 1092, 1093, 1109, 1110, 1111, 1112, 1113, 1114, 1517, 1518, 1519, 1524, 1528, 1529, 1530, 1941, 1943, 2355, 2356}

$\mathcal{S}_{\text{nodes}}$ = {4, 21, 22, 23, 24, 68, 69, 70, 71, 72, 73, 74, 75, 76, 77, 78, 79, 80, 81, 198, 199, 200, 201, 202, 203, 204, 205, 206, 207, 208, 209, 210, 211, 212, 213, 214, 215, 216, 217, 218, 410, 411, 412, 413, 414, 415, 416, 417, 418, 419, 420, 421, 422, 423, 424, 425, 426, 427, 729, 730, 731, 732, 733, 734, 735, 736, 737, 738, 739, 740, 741, 1109, 1110, 1111, 1112, 1113, 1114, 1528, 1529, 1530, 1943, 2356}

$\mathcal{S}_{\text{low}}$ = {1946, 2711, 2756, 2817, 3043, 3051, 3053, 3079, 3314, 3319, 3329, 3331, 3349, 3350, 3390, 3393, 3406, 3416, 3447, 3452, 3453, 3505, 3573, 3620, 3625, 3626, 3667, 3683, 3702, 3727, 3728, 3731, 3733, 3735, 3736, 3738, 3739, 3740, 3756, 3760, 3762, 3777, 3790, 3791, 3792, 3795, 3796, 3844, 3845, 3846, 3848, 3853, 3857, 3868, 3869, 3874, 3875, 3879, 3901, 3903, 3922, 3923, 3927, 3928, 3933, 3936, 3937, 3938, 3946, 3962, 3964, 3965, 3966, 3967, 3981, 3983, 3984, 3985, 3991, 3995, 4003, 4004, 4005, 4006, 4007, 4022, 4023, 4024, 4027, 4031, 4032, 4041, 4042, 4043, 4044, 4056, 4058, 4059, 4060, 4070, 4074, 4075, 4076, 4080, 4088, 4092, 4094, 4095, 4102, 4104, 4117, 4118, 4119, 4122, 4124, 4131, 4132, 4133, 4134, 4135, 4143, 4144, 4145, 4149, 4159, 4160, 4161, 4167, 4168, 4169, 4170, 4179, 4180, 4181, 4182, 4183, 4184, 4186, 4190, 4191, 4194, 4200, 4202, 4203, 4205, 4206, 4214, 4215, 4216, 4217, 4218, 4219, 4220, 4225, 4228, 4229, 4231, 4232, 4233, 4235, 4236, 4238, 4239, 4241, 4244, 4245, 4246, 4247, 4249, 4250, 4251, 4252, 4254, 4255, 4256, 4258, 4260, 4261, 4263, 4267, 4268, 4269, 4270, 4272, 4273, 4275, 4278, 4280, 4281, 4282, 4284, 4285, 4286, 4287, 4288, 4290, 4291, 4292, 4293, 4294, 4295, 4297, 4298, 4299, 4300, 4301, 4303, 4304, 4307, 4308, 4309, 4310, 4311, 4312, 4313, 4314, 4315, 4317, 4318, 4319}

$\mathcal{S}_{\text{indec}}$ = {3, 12, 17, 32, 38, 48, 49, 51, 54, 88, 91, 94, 98, 99, 100, 101, 102, 103, 105, 115, 119, 121, 134, 137, 138, 141, 142, 155, 158, 159, 170, 188, 228, 235, 239, 242, 243, 247, 248, 252, 254, 256, 260, 262, 265, 271, 278, 293, 294, 298, 299, 301, 317, 318, 330, 351, 353, 360, 378, 380, 438, 439, 440, 443, 445, 455, 468, 480, 491,

492, 493, 497, 501, 502, 515, 525, 526, 529, 530, 532, 539, 541, 543, 546, 550, 553, 562, 570, 575, 604, 608, 609, 614, 620, 645, 650, 660, 663, 688, 744, 752, 753, 754, 756, 760, 774, 775, 776, 780, 784, 790, 791, 792, 800, 834, 841, 844, 845, 852, 856, 859, 864, 866, 887, 900, 908, 912, 914, 923, 935, 963, 979, 990, 991, 1012, 1019, 1020, 1130, 1151, 1154, 1183, 1199, 1204, 1205, 1208, 1215, 1218, 1220, 1261, 1275, 1277, 1283, 1299, 1302, 1309, 1311, 1352, 1370, 1384, 1397, 1547, 1585, 1598, 1631, 1636, 1638, 1679, 1683, 1687, 1693, 1728, 1750, 1751, 1777, 1791, 1992, 2014, 2046, 2047, 2050, 2051, 2080, 2081, 2084, 2096, 2124, 2129, 2379, 2404, 2425, 2427, 2455, 2456, 2716, 2750, 2751, 2755}

$\mathcal{S}_{\text{aft}} = \{15, 16, 36, 41, 45, 53, 58, 59, 61, 65, 66, 102, 105, 110, 111, 112, 113, 116, 117, 124, 125, 128, 135, 141, 142, 144, 146, 147, 148, 149, 152, 162, 172, 179, 183, 189, 192, 193, 197, 230, 236, 244, 248, 261, 268, 271, 272, 277, 278, 279, 280, 281, 282, 286, 288, 290, 292, 302, 310, 324, 325, 327, 331, 332, 333, 334, 335, 337, 340, 343, 347, 349, 351, 355, 356, 358, 361, 362, 386, 399, 400, 407, 443, 445, 448, 452, 453, 456, 457, 463, 467, 487, 490, 496, 497, 499, 501, 502, 505, 507, 508, 509, 511, 512, 516, 523, 540, 545, 550, 563, 569, 577, 579, 581, 582, 583, 594, 599, 600, 601, 605, 606, 617, 629, 633, 658, 670, 671, 672, 674, 679, 682, 687, 705, 760, 764, 770, 771, 780, 781, 786, 787, 792, 797, 799, 809, 811, 812, 815, 816, 824, 859, 865, 868, 873, 875, 878, 883, 884, 889, 891, 892, 893, 894, 895, 902, 905, 929, 956, 960, 965, 987, 1003, 1004, 1006, 1011, 1021, 1038, 1045, 1051, 1156, 1160, 1168, 1175, 1177, 1199, 1203, 1209, 1216, 1217, 1225, 1232, 1234, 1251, 1252, 1253, 1255, 1256, 1260, 1262, 1265, 1275, 1286, 1287, 1293, 1300, 1305, 1308, 1324, 1327, 1351, 1371, 1383, 1398, 1533, 1545, 1550, 1551, 1554, 1561, 1579, 1589, 1613, 1614, 1615, 1620, 1637, 1638, 1656, 1665, 1666, 1671, 1686, 1690, 1693, 1697, 1711, 1747, 1748, 1760, 1763, 1989, 2000, 2001, 2027, 2045, 2051, 2052, 2068, 2071, 2072, 2076, 2084, 2096, 2098, 2102, 2379, 2380, 2385, 2403, 2405, 2423, 2424, 2425, 2427, 2738, 2777, 2778, 2792, 3047, 3057, 3063, 3064\}$

We have $|\mathcal{S}_{\text{indec}} \cup \mathcal{S}_{\text{aft}}| = 442$. Therefore there exist at least 442 non-smoothable toric Fano threefolds with Gorenstein singularities.

Acknowledgements I am indebted to Alessio Corti, Paul Hacking, Alexander Kasprzyk, and Thomas Prince for many fruitful conversations about the topics of this survey. I am very grateful to Tom Coates and Alexander Kasprzyk for having provided me with access to their MAGMA machine and to their database of Fano polytopes [33]. Finally, I would like to thank the anonymous referee for having done a thorough job.

References

1. Abramovich, D., Hassett, B.: Stable varieties with a twist. In: Classification of algebraic varieties, EMS Series of Congress Reports, pp. 1–38. European Mathematical Society, Zürich (2011)
2. Akhtar, M., Coates, T., Corti, A., Heuberger, L., Kasprzyk, A.M., Oneto, A., Petracci, A., Prince, T., Tveiten, K.: Mirror symmetry and the classification of orbifold del Pezzo surfaces. Proc. Amer. Math. Soc. **144**(2), 513–527 (2016)

3. Akhtar, M., Coates, T., Galkin, S., Kasprzyk, A.M.: Minkowski polynomials and mutations. SIGMA Symmetry Integr. Geom. Methods Appl. **8**, Paper 094, 17 (2012)
4. Altmann, K.: Computation of the vector space T^1 for affine toric varieties. J. Pure Appl. Algebra **95**(3), 239–259 (1994)
5. Altmann, K.: Minkowski sums and homogeneous deformations of toric varieties. Tohoku Math. J. (2) **47**(2), 151–184 (1995)
6. Altmann, K.: Infinitesimal deformations and obstructions for toric singularities. J. Pure Appl. Algebra **119**(3), 211–235 (1997)
7. Altmann, K.: The versal deformation of an isolated toric Gorenstein singularity. Invent. Math. **128**(3), 443–479 (1997)
8. Altmann, K.: One parameter families containing three-dimensional toric-Gorenstein singularities. In: Explicit birational geometry of 3-folds. London Mathematical Society Lecture Note series, vol. 281, pp. 21–50. Cambridge University Press, Cambridge (2000)
9. Altmann, K., Kollár, J.: The dualizing sheaf on first-order deformations of toric surface singularities. J. Reine Angew. Math. **753**, 137–158 (2019)
10. Artin, M.: Algebraic approximation of structures over complete local rings. Inst. Hautes Études Sci. Publ. Math. **36**, 23–58 (1969)
11. Artin, M.: Algebraization of formal moduli. I. In: Global Analysis (Papers in Honor of K. Kodaira), pp. 21–71. University Tokyo Press, Tokyo (1969)
12. Artin, M.: Algebraic construction of Brieskorn's resolutions. J. Algebra **29**, 330–348 (1974)
13. Artin, M.: Lectures on Deformations of Singularities. Tata Institute of Fundamental Research, Bombay (1976)
14. The Stacks project (2018). https://stacks.math.columbia.edu
15. Batyrev, V.V.: Toric Fano threefolds. Izv. Akad. Nauk SSSR Ser. Mat. **45**(4), 704–717, 927 (1981)
16. Batyrev, V.V.: On the classification of toric Fano 4-folds. pp. 1021–1050 (1999). Algebraic geometry, 9
17. Batyrev, V.V.: Toric degenerations of Fano varieties and constructing mirror manifolds. In: The Fano Conference, pp. 109–122. University Torino, Turin (2004)
18. Batyrev, V.V., Ciocan-Fontanine, I., Kim, B., van Straten, D.: Mirror symmetry and toric degenerations of partial flag manifolds. Acta Math. **184**(1), 1–39 (2000)
19. Behnke, K., Riemenschneider, O.: Quotient surface singularities and their deformations. In: Singularity Theory (Trieste, 1991), pp. 1–54. World Scientific Publishing, River Edge (1995)
20. Bien, F., Brion, M.: Automorphisms and local rigidity of regular varieties. Compositio Math. **104**(1), 1–26 (1996)
21. Birkar, C.: Singularities of linear systems and boundedness of Fano varieties. Ann. of Math. (2) **193**(2), 347–405 (2021)
22. Bosma, W., Cannon, J., Playoust, C.: The Magma algebra system. I. The user language. pp. 235–265 (1997). Computational algebra and number theory (London, 1993)
23. Brown, G., Kasprzyk, A.M.: Graded Ring Database. http://www.grdb.co.uk
24. Buch, A., Thomsen, J.F., Lauritzen, N., Mehta, V.: The Frobenius morphism on a toric variety. Tohoku Math. J. (2) **49**(3), 355–366 (1997)
25. Cavey, D., Prince, T.: Del Pezzo surfaces with a single $1/k(1, 1)$ singularity. J. Math. Soc. Japan **72**(2), 465–505 (2020)
26. Cheltsov, I., Katzarkov, L., Przyjalkowski, V.: Birational geometry via moduli spaces. In: Birational Geometry, Rational Curves, and Arithmetic, Simons Symposia, pp. 93–132. Springer, Cham (2013)
27. Christophersen, J.A.: On the components and discriminant of the versal base space of cyclic quotient singularities. In: Singularity theory and its applications, Part I (Coventry, 1988/1989). Lecture Notes in Math., vol. 1462, pp. 81–92. Springer, Berlin (1991)
28. Christophersen, J.A., Ilten, N.: Hilbert schemes and toric degenerations for low degree Fano threefolds. J. Reine Angew. Math. **717**, 77–100 (2016)
29. Christophersen, J.A., Ilten, N.O.: Degenerations to unobstructed Fano Stanley-Reisner schemes. Math. Z. **278**(1–2), 131–148 (2014)

30. Coates, T., Corti, A., Galkin, S., Golyshev, V., Kasprzyk, A.M.: Mirror symmetry and Fano manifolds. In: European Congress of Mathematics, pp. 285–300. European Mathematical Society, Zürich (2013)
31. Coates, T., Corti, A., Galkin, S., Kasprzyk, A.M.: Quantum periods for 3-dimensional Fano manifolds. Geom. Topol. **20**(1), 103–256 (2016)
32. Coates, T., Kasprzyk, A., Prince, T.: Laurent inversion. Pure Appl. Math. Q. **15**(4), 1135–1179 (2019)
33. Coates, T., Kasprzyk, A.M.: Code repository (2019). https://bitbucket.org/fanosearch/magma-core
34. Cox, D.A., Little, J.B., Schenck, H.K.: Toric varieties. Graduate Studies in Mathematics, vol. 124. American Mathematical Society, Providence (2011)
35. Eisenbud, D.: Commutative algebra. Graduate Texts in Mathematics, vol. 150. Springer, New York (1995). With a view toward algebraic geometry
36. Fantechi, B., Göttsche, L.: Local properties and Hilbert schemes of points. In: Fundamental Algebraic Geometry. Mathematical Surveys and Monographs, vol. 123, pp. 139–178. American Mathematical Society, Providence (2005)
37. Fantechi, B., Manetti, M.: Obstruction calculus for functors of Artin rings. I. J. Algebra **202**(2), 541–576 (1998)
38. de Fernex, T., Hacon, C.D.: Deformations of canonical pairs and Fano varieties. J. Reine Angew. Math. **651**, 97–126 (2011)
39. Friedman, R.: Simultaneous resolution of threefold double points. Math. Ann. **274**(4), 671–689 (1986)
40. Fujino, O.: Multiplication maps and vanishing theorems for toric varieties. Math. Z. **257**(3), 631–641 (2007)
41. Fulton, W.: Introduction to toric varieties. Annals of Mathematics Studies, vol. 131. Princeton University Press, Princeton (1993). The William H. Roever Lectures in Geometry
42. Galkin, S.: Small toric degenerations of Fano threefolds (2018). arXiv:1809.02705 [math.AG]
43. Gross, M., Siebert, B.: From real affine geometry to complex geometry. Ann. of Math. (2) **174**(3), 1301–1428 (2011)
44. Grothendieck, A.: Éléments de géométrie algébrique. IV. Étude locale des schémas et des morphismes de schémas. III. Inst. Hautes Études Sci. Publ. Math. (28), 255 (1966)
45. Hacking, P.: Compact moduli of plane curves. Duke Math. J. **124**(2), 213–257 (2004)
46. Hacking, P., Prokhorov, Y.: Smoothable del Pezzo surfaces with quotient singularities. Compos. Math. **146**(1), 169–192 (2010)
47. Hartshorne, R.: Residues and duality. Lecture notes of a seminar on the work of A. Grothendieck, given at Harvard 1963/64. With an appendix by P. Deligne. Lecture Notes in Mathematics, No. 20. Springer, Berlin (1966)
48. Hartshorne, R.: Algebraic Geometry. Springer, New York (1977). Graduate Texts in Mathematics, No. 52
49. Hartshorne, R.: Stable reflexive sheaves. Math. Ann. **254**(2), 121–176 (1980)
50. Hartshorne, R.: Deformation theory. Graduate Texts in Mathematics, vol. 257. Springer, New York (2010)
51. Ilten, N.O.: Mutations of Laurent polynomials and flat families with toric fibers. SIGMA Symmetry Integr. Geom. Methods Appl. **8**, Paper 047, 7 (2012)
52. Ilten, N.O., Lewis, J., Przyjalkowski, V.: Toric degenerations of Fano threefolds giving weak Landau-Ginzburg models. J. Algebra **374**, 104–121 (2013)
53. Iskovskih, V.A.: Fano threefolds. I. Izv. Akad. Nauk SSSR Ser. Mat. **41**(3), 516–562, 717 (1977)
54. Iskovskih, V.A.: Fano threefolds. II. Izv. Akad. Nauk SSSR Ser. Mat. **42**(3), 506–549 (1978)
55. Iskovskikh, V.A., Prokhorov, Y.G.: Fano varieties. In: Algebraic Geometry, V, Encylopaedia of Mathematical Sciences, vol. 47, pp. 1–247. Springer, Berlin (1999)
56. Jahnke, P., Radloff, I.: Terminal Fano threefolds and their smoothings. Math. Z. **269**(3–4), 1129–1136 (2011)

57. Kasprzyk, A.M.: Toric Fano three-folds with terminal singularities. Tohoku Math. J. (2) **58**(1), 101–121 (2006)
58. Kasprzyk, A.M.: Canonical toric Fano threefolds. Canad. J. Math. **62**(6), 1293–1309 (2010)
59. Kasprzyk, A.M., Kreuzer, M., Nill, B.: On the combinatorial classification of toric log del Pezzo surfaces. LMS J. Comput. Math. **13**, 33–46 (2010)
60. Kasprzyk, A.M., Nill, B.: Fano polytopes. In: Strings. Gauge Fields, and the Geometry Behind, pp. 349–364. World Scientific Publishing, Hackensack (2013)
61. Katzarkov, L., Przyjalkowski, V.: Landau-Ginzburg models—old and new. In: Proceedings of the Gökova Geometry-Topology Conference 2011, pp. 97–124. International Press, Somerville, MA (2012)
62. Kollár, J., Miyaoka, Y., Mori, S.: Rational connectedness and boundedness of Fano manifolds. J. Differ. Geom. **36**(3), 765–779 (1992)
63. Kollár, J., Mori, S.: Birational geometry of algebraic varieties. Cambridge Tracts in Mathematics, vol. 134. Cambridge University Press, Cambridge (1998). With the collaboration of C. H. Clemens and A. Corti, Translated from the 1998 Japanese original
64. Kollár, J., Shepherd-Barron, N.I.: Threefolds and deformations of surface singularities. Invent. Math. **91**(2), 299–338 (1988)
65. Kreuzer, M., Nill, B.: Classification of toric Fano 5-folds. Adv. Geom. **9**(1), 85–97 (2009)
66. Kreuzer, M., Skarke, H.: Classification of reflexive polyhedra in three dimensions. Adv. Theor. Math. Phys. **2**(4), 853–871 (1998)
67. Kreuzer, M., Skarke, H.: Complete classification of reflexive polyhedra in four dimensions. Adv. Theor. Math. Phys. **4**(6), 1209–1230 (2000)
68. Lee, Y., Nakayama, N.: Grothendieck duality and $\mathbb{Q}$-Gorenstein morphisms. Publ. Res. Inst. Math. Sci. **54**(3), 517–648 (2018)
69. Li, C., Wang, X., Xu, C.: On the proper moduli spaces of smoothable Kähler-Einstein Fano varieties. Duke Math. J. **168**(8), 1387–1459 (2019)
70. Manetti, M.: Differential graded Lie algebras and formal deformation theory. In: Algebraic Geometry—Seattle 2005. Part 2. Proceedings of Symposia in Pure Mathematics, vol. 80, pp. 785–810. American Mathematical Society, Providence (2009)
71. Mavlyutov, A.R.: Deformations of toric varieties via Minkowski sum decompositions of polyhedral complexes (2009). arXiv:0902.0967 [math.AG]
72. Mori, S., Mukai, S.: Classification of Fano 3-folds with $B_2 \geq 2$. Manuscripta Math. **36**(2), 147–162 (1981/82)
73. Mori, S., Mukai, S.: On Fano 3-folds with $B_2 \geq 2$. In: Algebraic Varieties and Analytic Varieties (Tokyo, 1981). Advanced Studies in Pure Mathematics, vol. 1, pp. 101–129. North-Holland, Amsterdam (1983)
74. Mori, S., Mukai, S.: Erratum: "Classification of Fano 3-folds with $B_2 \geq 2$" [Manuscripta Math. **36**(2), 147–162 (1981/82); MR0641971 (83f:14032)]. Manuscripta Math. **110**(3), 407 (2003)
75. Mustaţă, M.: Vanishing theorems on toric varieties. Tohoku Math. J. (2) **54**(3), 451–470 (2002)
76. Namikawa, Y.: Smoothing Fano 3-folds. J. Algebraic Geom. **6**(2), 307–324 (1997)
77. Nill, B., Øbro, M.: $\mathbb{Q}$-factorial Gorenstein toric Fano varieties with large Picard number. Tohoku Math. J. (2) **62**(1), 1–15 (2010)
78. Øbro, M.: An algorithm for the classification of smooth Fano polytopes (2007). arXiv:0704.0049 [math.CO]
79. Petracci, A.: Some examples of non-smoothable Gorenstein Fano toric threefolds. Math. Z. **295**(1–2), 751–760 (2020)
80. Petracci, A.: An example of mirror symmetry for Fano threefolds. In: Birational geometry and moduli spaces. Springer INdAM Serirs, vol. 39, pp. 173–188. Springer, Cham ([2020] © 2020)
81. Petracci, A.: Homogeneous deformations of toric pairs. Manuscripta Math. **166**(1–2), 37–72 (2021)
82. Prince, T.: Smoothing toric Fano surfaces using the Gross-Siebert algorithm. Proc. Lond. Math. Soc. (3) **117**(3), 617–660 (2018)
83. Prince, T.: Cracked polytopes and Fano toric complete intersections. Manuscripta Math. **163**(1–2), 165–183 (2020)

84. Prince, T.: From cracked polytopes to Fano threefolds. Manuscripta Math. **164**(1–2), 267–320 (2021)
85. Prokhorov, Y.G.: The degree of Fano threefolds with canonical Gorenstein singularities. Mat. Sb. **196**(1), 81–122 (2005)
86. Przhiyalkovskiĭ, V.V.: Weak Landau-Ginzburg models of smooth Fano threefolds. Izv. Ross. Akad. Nauk Ser. Mat. **77**(4), 135–160 (2013)
87. Przyjalkowski, V.: On Landau-Ginzburg models for Fano varieties. Commun. Number Theory Phys. **1**(4), 713–728 (2007)
88. Riemenschneider, O.: Deformationen von Quotientensingularitäten (nach zyklischen Gruppen). Math. Ann. **209**, 211–248 (1974)
89. Sato, H.: Toward the classification of higher-dimensional toric Fano varieties. Tohoku Math. J. (2) **52**(3), 383–413 (2000)
90. Sato, H.: Smooth toric Fano five-folds of index two. Proc. Japan Acad. Ser. A Math. Sci. **82**(7), 106–110 (2006)
91. Schlessinger, M.: Functors of Artin rings. Trans. Amer. Math. Soc. **130**, 208–222 (1968)
92. Sernesi, E.: Deformations of algebraic schemes, Grundlehren der Mathematischen Wissenschaften [Fundamental Principles of Mathematical Sciences], vol. 334. Springer, Berlin (2006)
93. Stevens, J.: On the versal deformation of cyclic quotient singularities. In: Singularity Theory and Its Applications, Part I (Coventry, 1988/1989). Lecture Notes in Mathematics, vol. 1462, pp. 302–319. Springer, Berlin (1991)
94. Stevens, J.: The versal deformation of cyclic quotient singularities. In: Deformations of Surface Singularities. Bolyai Society Mathematical Studies, vol. 23, pp. 163–201. János Bolyai Mathematical Society, Budapest (2013)
95. Talpo, M., Vistoli, A.: Deformation theory from the point of view of fibered categories. In: Handbook of Moduli. Vol. III, Advanced Lectures in Mathematics (ALM), vol. 26, pp. 281–397. International Press, Somerville (2013)
96. Totaro, B.: Jumping of the nef cone for Fano varieties. J. Algebraic Geom. **21**(2), 375–396 (2012)
97. Vistoli, A.: The deformation theory of local complete intersections (1997). arXiv:alg-geom/9703008
98. Watanabe, K., Watanabe, M.: The classification of Fano 3-folds with torus embeddings. Tokyo J. Math. **5**(1), 37–48 (1982)
99. Ziegler, G.M.: Lectures on Polytopes. Graduate Texts in Mathematics, vol. 152. Springer, New York (1995)

Chapter 15
Polygons of Finite Mutation Type

Thomas Prince

Abstract We classify Fano polygons with finite mutation class. This classification exploits a correspondence between Fano polygons and cluster algebras, refining the notion of singularity content due to Akhtar and Kasprzyk. We also introduce examples of cluster algebras associated to Fano polytopes in dimensions greater than two.

Keywords Polytopes · Toric Varieties · Mirror Symmetry

15.1 Introduction

The notion of combinatorial, or polytope, mutation was introduced by Akhtar–Coates–Galkin–Kasprzyk [3] to describe mirror partners to Fano manifolds. Following Givental [17–19], Kontsevich [28], and Hori–Vafa [25], the mirror partner to a Fano manifold consists of a complex manifold together with a holomorphic function, the *superpotential*. If this mirror manifold contains a complex torus we can write down a collection of volume preserving birational maps of this complex torus which preserve the regularity of the superpotential. We call these rational maps (algebraic) mutations, following [3] and work of Galkin–Usnich [16]. Combinatorial mutation is the operation induced on the Newton polyhedra of the restriction of the superpotential to such tori.

All the polytopes we consider are *Fano*, that is, polytopes which contain the origin in the interior and such that its vertices are primitive lattice vectors. In joint work [27] with Kasprzyk and Nill we showed that, in dimension two, the notion of polytope mutation is compatible with the construction of a quiver and cluster algebras one can associate to each Fano polygon.

The idea of associating a polygon with a quiver—or toric diagram—has a reasonably long history, particularly in the physics literature. In that setting the polygon describes a toric Calabi–Yau singularity and the quiver is used to describe the matter

T. Prince (✉)
Mathematical Institute, University of Oxford, Woodstock Road, Oxford OX2 6GG, UK
e-mail: thomas.prince@magd.ox.ac.uk

A. M. Kasprzyk and B. Nill (eds.), *Interactions with Lattice Polytopes*,
Springer Proceedings in Mathematics & Statistics 386,
https://doi.org/10.1007/978-3-030-98327-7_15

content of a gauge theory arising on a stack of D3-branes probing the toric Calabi–Yau singularity, (see for example [1, 5, 10, 15, 21, 22, 29] for a selection of the literature on this subject). The construction of a quiver (and cluster algebra) from a polygon has also been used by Gross–Hacking–Keel [20] in the study of associated log Calabi–Yau varieties, and to study the derived category of the toric variety, or the associated local toric Calabi–Yau as pursued, for example, in [7, 23, 24, 31, 32]. In each setting the basic construction is the same, and we recall the version relevant to our applications in Sect. 15.3.

Our main result, Theorem 30, is a classification of the mutation classes of polygons which contain only finitely many polygons. This parallels a finite type result of Mandel [30], for rank two cluster varieties. In particular we see that finite mutation classes of polygons fall into four types A_1^n, for $n \in \mathbb{Z}_{\geq 0}$, A_2, A_3, and D_4.

There is a close connection between mutation classes of Fano polygons and $\mathbb{Q}$-Gorenstein deformations of the corresponding toric varieties which is described in detail in [2]. Following these ideas we predict the existence of a finite type parameter space for these deformations, together with a boundary stratification such that each 0-stratum corresponds to a polygon in the given mutation class, and the 1-strata corresponds to the mutation families constructed by Ilten [26].

While our main result applies in dimension two, we note that polytope mutation is defined in all dimensions, and the construction of a quiver and cluster algebra we provide applies to 'compatible collection' of mutations in any dimension, see Definition 16. This definition is, unfortunately, less well behaved in dimensions greater than two, but we provide an example indicating that polytope mutation can detect known examples of cluster structures appearing on linear sections of Grassmannians of planes. We expect this to extend to a wide variety of other cluster structures found in Fano manifolds and their mirror manifolds.

15.2 Quivers and Cluster Algebras

We devote this section to fixing the various conventions and notation, as well as recalling the basic definitions. We recall the definition of cluster algebra, and in order to address both geometric and combinatorial applications we shall adapt our treatment from the work of Fomin–Zelevinsky [13], and the work of Fock–Goncharov [11] and Gross–Hacking–Keel [20]. We first fix the following data:

1. N, a fixed lattice with skew-symmetric form $\{-,-\}\colon N \times N \to \mathbb{Z}$;
2. a saturated sublattice $N_{uf} \subseteq N$, the *unfrozen* sublattice;
3. an index set I, $|I| = \mathrm{rk}(N)$ together with a subset $I_{uf} \subseteq I$ such that $|I_{uf}| = \mathrm{rk}(N_{uf})$. For later convenience we set $n := |I_{uf}|$.

Remark 1 The requirement that the form is integral is not necessary, but is sufficiently general for our applications and simplifies the exposition considerably.

Definition 2 A *(labelled) seed* is a pair $\mathbf{s} = (\mathcal{E}, C)$, where:

1. $\mathcal{E}$ is a basis of N indexed by I, such that the subset indexed by I_{uf} is a basis of N_{uf};

2. C is a transcendence basis of $\mathcal{F}$, the field of rational functions in n independent variables over $\mathbb{Q}(x_i : i \in I \backslash I_{uf})$, referred to as a *cluster*. Elements of C are also referred to as *cluster variables*.

Remark 3 The basis $\mathcal{E}$ is referred to as *seed data* in [11, 20]. Since we have fixed the lattice N and skew-symmetric form $\{-,-\}$ the elements of C can be identified with coordinate functions on the *seed torus* T_N.

Definition 4 Fix a seed $\mathbf{s} = (\mathcal{E}, C)$, where $\mathcal{E} = \{e_i : i \in I\}$ and $C = \{x_i : i \in I_{uf}\}$. Fixing an element $j \in I_{uf}$, the *jth mutation* of $(\mathcal{E}, C)$ is the seed $(\mathcal{E}', C')$, where $\mathcal{E}' = \{e'_i : i \in I\}$ is defined by setting

$$e'_k = \begin{cases} -e_j & \text{if } k = j, \\ e_k + \max\big(b_{kj}, 0\big)e_j & \text{otherwise,} \end{cases}$$

where $b_{kl} = \{e_k, e_l\}$. While the cluster $C' = \{x'_i : i \in I_{uf}\}$ is defined by setting,

$$x'_k = x_k, \text{ if } k \neq j, \qquad \text{and} \qquad x_j x'_j = \prod_{\substack{k \text{ such that} \\ b_{jk}>0}} x_k^{b_{jk}} + \prod_{\substack{l \text{ such that} \\ b_{jl}<0}} x_l^{b_{lj}}. \tag{15.1}$$

Recall that the matrix $B := (b_{kl})_{k,l \in I_{uf}}$ is typically referred to as the *exchange matrix* of the seed.

Definition 5 A *cluster algebra* is the subalgebra of $\mathcal{F}$ generated by the union of all clusters obtained by mutation from a given seed.

Any skew-symmetric $n \times n$ matrix B determines a skew-symmetric form on a (based) lattice $\mathbb{Z}^n$. Set $N = \mathbb{Z}^n$, $I = I_{uf} = \{1, \dots, n\}$, $\mathcal{E}$ to be the standard basis on $\mathbb{Z}^n$, and let $C = \{x_1, \dots, x_n\}$. We let $\mathcal{A}(B)$ denote the cluster algebra associated to the seed $(\mathcal{E}, C)$.

Definition 5 is really a special case of the definition of a cluster algebra, a class referred to as the *skew-symmetric cluster algebras of geometric type*. In the general case the form $\{-,-\}$ need only be *skew-symmetrizable*. One consequence of the skew-symmetry of the form $\{-,-\}$ is the identification of each exchange matrix with an (unfrozen) *quiver*. One may assign this quiver in the obvious way, assigning a vertex v_i to each element $i \in I_{uf}$, and b_{ij} arrows $v_i \to v_j$, oriented according to the sign of b_{ij}. We may also add 'frozen' vertices v_i for each element of $i \in I \backslash I_{uf}$, with arrows introduced between frozen and unfrozen vertices similarly. Equivalently we may consider the quiver associated to the *extended exchange matrix*, but we do not make further use of this terminology. There is a well-known notion of quiver mutation, going back to Fomin–Zelevinsky [13], generalising the reflection functors of Bernstein–Gelfand–Ponomarev [6]. Mutating a seed in a skew-symmetric cluster algebra induces a corresponding mutation of the associated quiver.

Definition 6 Given a quiver Q and an element $i \in I_{uf}$, the *mutation of Q at v_i* is the quiver $\mathrm{mut}(Q, v)$ obtained from Q by:

1. adding, for each subquiver $v_1 \to v \to v_2$, an arrow from v_1 to v_2;
2. deleting a maximal set of disjoint two-cycles;
3. reversing all arrows incident to v.

The resulting quiver is well-defined up to isomorphism, regardless of the choice of two-cycles in (2).

Since we shall refer to quivers frequently we shall make the following conventions. Given a quiver Q, we define:

1. Q_0 to be the set of vertices of Q;
2. $\mathrm{Arr}(v_i, v_j)$ to be the set of arrows from $v_i \in Q_0$ to $v_j \in Q_0$;
3. b_{ij} to be the cardinality of $\mathrm{Arr}(v_i, v_j)$, with sign indicating orientation.

We shall always assume Q has no vertex-loops or 2-cycles.

Given a seed $\mathbf{s}$ we shall also fix notation for the dual basis $\mathcal{E}^\star$ of $M := \hom(N, \mathbb{Z})$ and for each $i \in I$, set $v_i := \{e_i, -\} \in M$. We now define the $\mathcal{A}$ and $\mathcal{X}$ cluster *varieties* defined by Fock–Goncharov [11]. Toward this, observe to a seed $\mathbf{s}$ we can associate a pair of tori:

$$\mathcal{X}_{\mathbf{s}} = T_M \qquad \mathcal{A}_{\mathbf{s}} = T_N.$$

The dual pair of bases for the respective lattices define identifications of these tori with split tori,

$$\mathcal{X}_{\mathbf{s}} \longrightarrow \mathbb{G}_m^{|I|}, \qquad \mathcal{A}_{\mathbf{s}} \longrightarrow \mathbb{G}_m^{|I|}.$$

Letting $\mathbf{s}'$ denote the kth mutation of $\mathbf{s}$, we associate birational maps $\mu_k \colon \mathcal{X}_{\mathbf{s}} \dashrightarrow \mathcal{X}_{\mathbf{s}'}$ and $\mu_k \colon \mathcal{A}_{\mathbf{s}} \dashrightarrow \mathcal{A}_{\mathbf{s}'}$ to each seed, defined by setting

$$\mu_k^\star z^n = z^n(1 + z^{e_k})^{-\{n, e_k\}} \qquad \mu_k^\star z^m = z^m(1 + z^{v_k})^{\langle e_k, m\rangle},$$

where $n \in N$, $m \in M$, z^m (resp. z^n) denotes the function on $T_N := \mathrm{Spec}(\mathbb{C}[M])$ corresponding to m (resp. the function on T_M corresponding to n), and $\langle -, - \rangle$ denotes the canonical pairing between M and N. Note that the tori $\mathcal{X}_{\mathbf{s}}$ and $\mathcal{X}_{\mathbf{s}'}$ are canonically identified with T_N, but the maps μ_k between them are not isomorphisms.

Pulling these birational maps back along the identifications with the split torus given by the seed, the birational map $\mu_k \colon \mathcal{A}_{\mathbf{s}} \dashrightarrow \mathcal{A}_{\mathbf{s}'}$ is given by the exchange relation (15.1). That is, this birational map is the coordinate-free expression of the exchange relation once we identify the standard coordinates on T_N with the cluster variables $x_i \in C$ (including the frozen variables $x_{n+1} \dots, x_m$). We obtain schemes $\mathcal{X}$ and $\mathcal{A}$ by gluing the seed tori $\mathcal{A}_s$ and $\mathcal{X}_s$ along the birational maps defined by the mutations μ_k. For more details and related results we refer to [11, 20].

We conclude this section by recalling the classifications of cluster algebras of *finite type* and *finite mutation type*.

Definition 7 A cluster algebra is said to be of *finite type* if it contains finitely many clusters.

Given an undirected graph G we say that a quiver Q is an *orientation* of G if it has the same set of vertices and for each edge of G there is precisely one arrow between the respective vertices. Given a simply-laced Dynkin diagram D we say that Q is of *type* D if it is an orientation of the underlying graph of D.

Theorem 8 ([14]) *There is a canonical bijection between the Cartan matrices of finite type and cluster algebras (without frozen variables) of finite type. Under this bijection, a Cartan matrix* A *of finite type corresponds to the cluster algebra* $\mathcal{A}(B)$*, where* B *is an arbitrary skew-symmetrizable matrix with Cartan companion equal to* A.

Theorem 8 describes skew-symmetric cluster algebras with finitely many clusters. We can ask instead for the weaker condition that only finitely many *quivers* appear associated to seeds of the cluster algebra. This is the notion of *finite mutation type* cluster algebra, for which a classification is also known.

Theorem 9 ([9, Theorem 6.1]) *Given a quiver* Q *with finite mutation class, its adjacency matrix* b_{ij} *is the adjacency matrix of a triangulation of a bordered surface or is mutation equivalent to one of eleven exceptional types.*

The class of quivers coming from triangulations of surfaces is well-studied and we make use of a combinatorial characterisation of this class of quivers via *block decomposition*. A quiver Q is said to admit a *block decomposition* if it may be assembled from the six *blocks* shown in Fig. 15.5 by identifying the vertices of quivers shown with unfilled circles, the *outlets*. More precisely, we choose an injection from a subset of the combined set of outlets O into O such that no outlet is mapped to a vertex of the same block, including itself. We form Q by gluing the quiver along these vertices and cancelling any two cycles formed by this process. See [12, Definition 13.1] for further discussion and examples of this definition.

Theorem 10 ([12, Theorem 13.3]) *A quiver* Q *given by the adjacency matrix of a triangulation of a surface is mutation equivalent to a quiver which admits a* block decomposition.

15.3 Mutations of Polytopes

In two dimensions all combinatorial mutations are 'tropicalisations' of cluster mutations. While this ceases to be true in higher dimensions there is a natural class of combinatorial mutations, the *edge mutations* which do appear in this way. In terms of the definition of combinatorial mutation given in [3], edge mutations are those which have one-dimensional *factor*. In particular each edge mutation is obtained by studying the effect of the following birational maps—an *algebraic mutation* [3]—on the Newton polyhedra of certain Laurent polynomials. Throughout this section N denotes an n-dimensional lattice (not necessarily related to the definition of a cluster algebra). We recall that, working over $\mathbb{C}$, if M is the lattice dual to N, the torus T_M is defined to be $\mathrm{Spec}(\mathbb{C}[N])$.

Definition 11 Given an element $w \in M$, the *weight vector*, and $f \in \mathrm{Ann}(w)$, the *factor*, define a birational map $\phi_{w,f} \colon T_M \dashrightarrow T_M$ sending

$$z^n \mapsto z^n(1+z^f)^{\langle w,n\rangle}.$$

Given a Laurent polynomial $W \in \mathbb{C}[N]$ such that $\phi^{\star}_{w,f}(W) \in \mathbb{C}[N]$ say that W is mutable with weight vector w and factor f.

Definition 12 (*Cf.* [3, pg. 12]) Fix a Fano polytope $P \subset N_{\mathbb{Q}}$ and its dual $P^{\star} \subset M_{\mathbb{Q}}$, a weight vector $w \in M$, and factor $f \in \mathrm{Ann}(w)$. Define a piecewise linear map $T_{w,f} \colon M_{\mathbb{Q}} \to M_{\mathbb{Q}}$ by setting

$$T_{w,f} \colon m \mapsto m + \max(0, \langle m, f\rangle)w.$$

If $T_{w,f}(P^{\star})$ is a convex polytope then we say P admits the mutation (w, f) and that P mutates to $(T_{w,f}(P^{\star}))^{\star}$.

Remark 13 This definition of mutation is really a 'dual characterisation' of [3, Definition 5], which encodes how the Newton polytope of a Laurent polynomial changes under algebraic mutation.

Remark 14 In [3] the authors show that the result of applying a mutation to a Fano polytope produces another Fano polytope, so the last dualization in Definition 12 is well-defined.

Proposition 15 *Given* $w \in M$, $f \in \mathrm{Ann}(w)$ *and a mutable Laurent polynomial* $W \in \mathbb{C}[N]$ *we have the following identity;*

$$\mathrm{Newt}\left(\phi^{\star}_{w,f}W\right)^{\star} = T_{w,f}\left(\mathrm{Newt}(W)^{\star}\right).$$

Proof The notion of combinatorial mutation is compatible with the mutation W by construction. The interpretation of a combinatorial mutation as a piecewise linear map is made in the proof of Proposition 4 in [3]. □

Definition 16 We define *mutation data* to be elements $(w, f) \in M \oplus N$ such that w and f are primitive, and $f \in \mathrm{Ann}(w)$. A set of mutation data $\{(w_i, f_i) \in M \oplus N : i \in I\}$, for a finite index set I, is called *compatible* if

$$\langle w_i, f_j\rangle = -\langle w_j, f_i\rangle \qquad \text{for all } i, j \in I.$$

Remark 17 If $\dim N = 2$ mutation data is automatically compatible; indeed $\langle w_i, f_j\rangle$ can be identified with $w_i \wedge w_j$ for a suitable orientation of M.

Definition 18 To a compatible collection of mutation data $\mathcal{E}$ we define a quiver $Q_{\mathcal{E}}$ as follows:

1. the vertex set of $Q_{\mathcal{E}}$ is $\mathcal{E}$;
2. between two vertices (w_i, f_i) and (w_j, f_j) there are $\langle w_i, f_j\rangle$ arrows, with sign indicating orientation.

Observe that, as $\langle w_i, f_j\rangle$ is skew-symmetric, the quiver $Q_{\mathcal{E}}$ contains no loops or two cycles. Note that we can use this definition to assign a *cluster algebra* to a compatible collection of mutation data. We define a rule governing how compatible collections of mutations themselves mutate.

Definition 19 Given a compatible collection of mutation data $\mathcal{E}$, let L be the sublattice of $M \oplus N$ generated by the elements of $\mathcal{E}$, and let $\{(w_i, f_i), (w_j, f_j)\} := \langle w_i, f_j\rangle$ define a skew-symmetric form on L. Fixing a pair $E_k = (w_k, f_k) \in \mathcal{E}$ we *mutate* $\mathcal{E}$ to a new collection $\mathcal{E}_k$ as follows:

1. $E_k \mapsto -E_k$;
2. $E_i \mapsto E_i - \max(\{E_i, E_k\}, 0)E_k$, if $i \neq k$.

This formula is identical to the mutation of seed data given in [11]; a connection we now make precise. Fix a compatible collection of mutations $\mathcal{E}$ and define a skew-symmetric form $[-,-]$ on $\mathbb{Z}^{|\mathcal{E}|}$ defined by setting $[e_i, e_j] := \{\theta(e_i), \theta(e_j)\}$, where $\theta\colon \mathbb{Z}^{|\mathcal{E}|} \to M \oplus N$ is defined by sending $e_i \mapsto (w_i, f_i)$. The following Lemma follows immediately by comparison of the formulae for mutating seed data in a cluster algebra with Definition 19.

Lemma 20 *The operations of mutation given in Definition 19, and of mutation of the seeds defined above, are intertwined by θ.*

Example 21 In dimensions higher than two a compatible collection of mutation data which defines a set of combinatorial mutations of a given polytope can transform by mutation to a compatible collection of mutation data which does not define a set of combinatorial mutations of the transformed polytope. In particular the piecewise linear maps may fail to preserve convexity. This appears to be a important obstruction to generalising the two-dimensional theory of mutations to higher dimensional polytopes. For example, consider the polytope

$$P := \operatorname{conv}\left(\begin{pmatrix}1\\0\\0\end{pmatrix}, \begin{pmatrix}0\\1\\0\end{pmatrix}, \begin{pmatrix}0\\0\\1\end{pmatrix}, \begin{pmatrix}0\\0\\-1\end{pmatrix}, \begin{pmatrix}-1\\-1\\0\end{pmatrix}, \begin{pmatrix}-1\\-1\\-1\end{pmatrix}\right).$$

Consider mutation data $(w_1, f_1) := (e_1^\star, e_3)$ and $(w_2, f_2) = (e_2^\star, e_3)$. Since $f_1 = f_2$, we have that $\langle w_i, f_j\rangle = 0$ for all $i, j \in \{1, 2\}$; hence these mutations are compatible. However, while P admits both these mutations, the composition of these two (in either order) is the mutation corresponding to the pair $(w, f) := (e_1^\star + e_2^\star, e_3)$, which does *not* define a mutation of P.

Proposition 22 *Given seed data $\mathcal{E}$ such that $Q_{\mathcal{E}}$ is a directed simply-laced Dynkin diagram the number of polytopes obtained by successive edge mutation is bounded by the numbers of seeds in the cluster algebra determined by $Q_{\mathcal{E}}$. If $Q_{\mathcal{E}}$ is of type A_n this bound is the Catalan number C_{n+1}.*

In fact, compatible collections of mutations appear whenever we have a cluster algebra with skew-symmetric exchange matrix.

Proposition 23 *Every compatible collection of mutations determines and is determined by a skew-symmetric cluster algebra without frozen variables, together with a subspace V of the kernel of the skew-symmetric form $\{-,-\}$ defined by the exchange matrix.*

Proof Fix a skew-symmetric cluster algebra without frozen variables and a nominated subspace $V \subset \ker\{-,-\}$. Recall that a seed defines a basis e_i of a lattice, which we denote $\widetilde{N}$. Define $M := \widetilde{N}/V$ and let $p\colon \widetilde{N} \to M$ be the canonical projection. The map $\theta\colon \widetilde{N} \to M \oplus \hom(M,\mathbb{Z})$ defined by $\theta\colon n \mapsto (p(n), \{n,-\})$ defines a compatible collection of mutation data with weight vectors in the lattice M. □

Note that N and M play dual roles to those in [11], and we insist throughout that $P \subset N_{\mathbb{Q}}$. This exchange of roles explains the odd definition of M in the proof of Proposition 23. To compare the birational maps associated to the two notions of mutations let $\mathbf{s}$ be a seed of the cluster algebra determined by a compatible collection of mutation data, and let $\mathcal{E}$ be the compatible collection corresponding to $\mathbf{s}$. Fix an element $E_k = (w_k, f_k) \in \mathcal{E}$ and consider the following diagram,

$$\begin{array}{ccc} \mathcal{A}_{\mathbf{s}} & \overset{\mu_k}{\dashrightarrow} & \mathcal{A}_{\mu_k(\mathbf{s})} \\ \downarrow p & & \downarrow p \\ T_M & \underset{\phi_{(w_k,f_k)}}{\dashrightarrow} & T_M, \end{array} \tag{15.2}$$

Proposition 24 *The diagram shown in* (15.2) *commutes.*

Proof This is an exercise in writing out the definitions of the respective mutations: see [27, Sect. 3]. □

Example 25 The del Pezzo surface of degree 5 admits a toric degeneration to a toric surface Z with a pair of A_1 singularities. Given a three-dimensional linear section X of the Grassmannian $\mathrm{Gr}(2,5)$ X admits a toric degeneration to the projective cone over Z. The fan determined by this toric threefold is formed by taking the cones over the faces of the reflexive polytope with PALP id 245.

In Fig. 15.1 we show a pentagon of polytopes obtained by successively mutating the polytope shown in the top-right with respect to the mutation data

$$\mathcal{E} := \{(w_1, f_1), (w_2, f_2)\}$$

where,

$$\begin{aligned} w_1 &:= (-1,0,0), & f_1 &:= (0,1,1)^T, \\ w_2 &:= (0,0,-1), & f_2 &:= (-1,0,0)^T. \end{aligned}$$

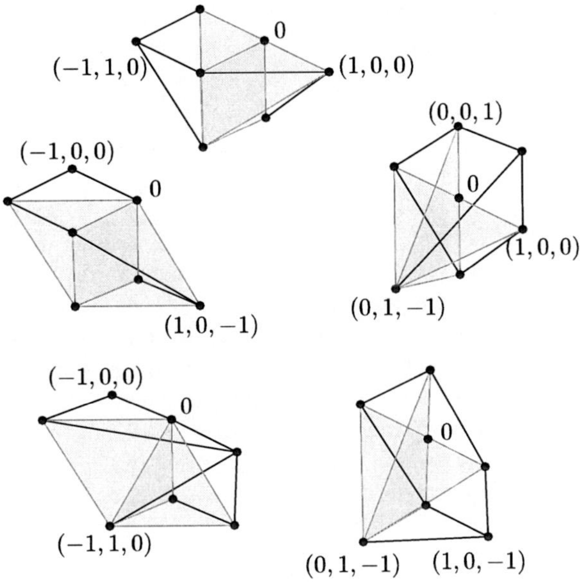

Fig. 15.1 Pentagon of edge mutations among toric degenerations of B_5

We recall that there is an A_2 cluster structure on the co-ordinate ring of the Grassmannian, and a toric degeneration of Gr(2, 5) for each cluster chart in the dual Grassmannian [33]. We expect that cluster structures in the mirror to a Fano variety to be detected by such compatible collections of mutations.

Note that the polytopes we show in Fig. 15.1 are not dual to Fano polytopes. However, recalling that B_5 has Fano index 2, we can obtain a reflexive polytope by dilating each of the polytopes shown in Fig. 15.1 by a factor of two, and translating.

In the two dimensional case, we can canonically define a maximal set of compatible mutations, making use of the notion of singularity content [4].

Definition 26 (*Cf.* [27, Sect. 1.2]) Given a Fano polygon $P \subset N_{\mathbb{Q}}$ with singularity content $(n, \mathcal{B})$ and $m := |\mathcal{B}| + n$, we define:

1. an index set I of size m containing a subset I_{uf} of size n, together with functions

$$\phi_{uf}\colon I_{uf} \to \{\text{edges of } P\} \qquad \phi_f\colon I\backslash I_{uf} \to \mathcal{B}$$

   such that fibres $\phi_{uf}^{-1}(E)$ contain $m_E := \lfloor \ell(E)/r_E \rfloor$ elements, where $\ell(E)$ is the lattice length of the edge E, and r_E is the Gorenstein (or *local*) index of the cone over E, while the map ϕ_f is a bijection;
2. a lattice map $\rho\colon \mathbb{Z}^m \to M$ sending each basis element to the primitive, inward-pointing normal to the edge of P defined by the cone given by the specified functions ϕ_{uf} and ϕ;
3. a form $\{e_i, e_j\} := \rho(e_i) \wedge \rho(e_j)$. Note that this is an integral skew-symmetric form.

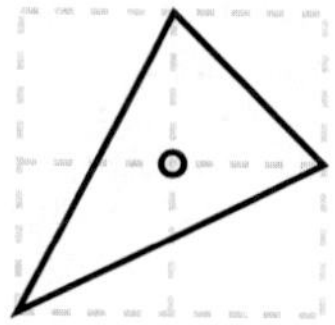

Fig. 15.2 The Fano polygon for $\mathbb{P}^2$

The value m_E appears in the definition of the *singularity content* of a two dimensional cone; and is equal to the maximal number of T-cones of Gorenstein index r_E which fit inside the cone on E.

By [27, Proposition 3.17] the construction of a quiver from mutation data provided by Definition 26 intertwines polygon and quiver mutations. We let $(\mathcal{E}_P, C_P)$ denote the seed associated to a Fano polygon, where $\mathcal{E}_P$ is the standard basis e_i of $\mathbb{Z}^m$, and C_P is the standard transcendence basis of the field of rational functions in n variables over $\mathbb{Q}(x_i : i \in I\backslash I_{uf})$. We let Q_P denote the *unfrozen* quiver associated to $(\mathcal{E}_P, C_P)$. We say a Fano polygon is of *finite mutation type* if it is mutation equivalent to only finitely many Fano polygons.

Conjecture 27 The cluster algebra C_P associated to a Fano polygon P, together with a bijection between the set of frozen variables and $\mathcal{B}$, is a complete mutation invariant of the Fano polygon P.

Example 28 Consider the Fano polygon P for $\mathbb{P}^2$ (this is depicted in Fig. 15.2). Computing the determinant of the inward-pointing normals we obtain the quiver Q_P

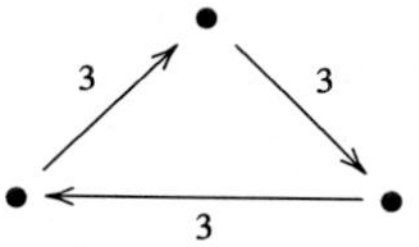

The mutations of this quiver are well-known, and the triple $(3a, 3b, 3c)$ of non-zero entries of the exchange matrix satisfy the Markov equation $a^2 + b^2 + c^2 = 3abc$. Indeed, as the polygon P is mutated the corresponding toric surfaces are $\mathbb{P}(a^2, b^2, c^2)$ for the same triples (a, b, c). We see that in this case the mutations of the quivers exactly capture the mutations of the polygon.

Example 29 Consider the toric surface (using the notation for these surfaces appearing in [8]), $X_{5,5/3}$ associated with the Fano polygon shown in Fig. 15.3. The unfrozen quiver associated to this surface is simply the A_2 quiver:

This example is important, both in this section, because it is an example of a *finite type* polygon, and since a smoothing of this surface is given by 5 Pfaffian equations, see [8, Sect. 3.3], a fact closely connected to the A_2 quiver we construct here.

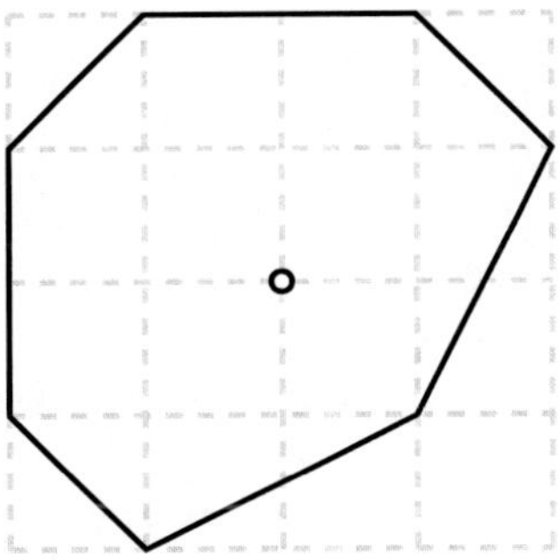

Fig. 15.3 The Fano polygon for $X_{5,5/3}$

15.4 Finite Type Classification

We now make use of the classification of finite type and finite mutation type cluster algebras to establish the following result.

Theorem 30 *P is of finite mutation type if and only if* Q_P *is mutation equivalent to a quiver of type* $(A_1)^n$, A_2, A_3, *or* D_4.

Remark 31 The types referred to in Theorem 30 may also be referred to as type I_n, II, III, and IV respectively; in analogy with Kodaira's monodromy matrices. The relationship between these matrices, log Calabi–Yau manifolds, and monodromy in certain integral affine manifolds is explored by Mandel in [30].

Remark 32 We remark that all the cases which appear in Theorem 30 do occur as (unfrozen) quivers associated to polygons. Several examples can be found in [8, p. 42] and are tabulated below.

Quiver Q_P	Polygon	Surface
$\varnothing = A_1^0$	9	$X_{6,2}$
A_1^3	11	$X_{4,7/3}$
A_1^6	12	$B_{2,8/3}$
A_2	7	$X_{5,5/3}$
A_3	17	$X_{3,4}$
D_4	5	$X_{4,4/3}$

Examples of polygons P with $Q_P = A_1^{2k+2}$ and A_1^{2k+1} for $k \geq 5$ are given by the quadrilaterals

$$\operatorname{conv}\left(\begin{pmatrix}-1\\-3\end{pmatrix}, \begin{pmatrix}1\\-3\end{pmatrix}, \begin{pmatrix}-1\\k-2\end{pmatrix}, \begin{pmatrix}1\\k-2\end{pmatrix}\right), \text{ and}$$

$$\operatorname{conv}\left(\begin{pmatrix}-1\\-3\end{pmatrix}, \begin{pmatrix}1\\-2\end{pmatrix}, \begin{pmatrix}-1\\k-2\end{pmatrix}, \begin{pmatrix}1\\k-2\end{pmatrix}\right).$$

We first make two straightforward observations. First we note that the cluster algebra C_P induces a sequence of surjections:

$$\begin{array}{c}\{\text{Clusters of } C_P\} \\ \downarrow \\ \{\text{Polygons mutation equivalent to } P\} \\ \downarrow \\ \{\text{Quivers mutation equivalent to } Q_P\}.\end{array} \tag{15.3}$$

The first vertical arrow follows from the fact that algebraic mutations determine combinatorial mutations, the second from Lemma 20. For example, using this tower of surjections in the case of a type A_2 cluster algebra, we can immediately state the following result.

Proposition 33 *If a Fano polygon P has singularity content $(2, \mathcal{B})$ and the primitive inward-pointing normal vectors of the two edges corresponding to the unfrozen variables of C_P form a basis of the lattice M, then the mutation-equivalence class of P has at most five members.*

Proof The quiver associated to P is precisely an orientation of the A_2 quiver. The cluster algebra C_P is well-known and its cluster exchange graph forms a pentagon. Note however that the *quiver* mutation graph is trivial, as the A_2 quiver mutates only to itself. □

Proposition 24 implies that the mutation class of P has at most five elements. Note that we do not have a non-trivial lower bound: there is only one polygon in mutation equivalent to the polygon described in Example 29 up to $\mathrm{GL}(2, \mathbb{Z})$ equivalence. Next observe that the sequence of surjections shown in (15.3) immediately implies that

$$C_P \text{ finite type} \Rightarrow P \text{ finite mutation type} \Rightarrow C_P \text{ finite mutation type.}$$

Lemma 34 *Given a Fano polygon P of finite mutation type, Q_P does not contain a Kronecker subquiver*

$$Q_k := \{\, v_1 \xrightarrow{\;k\;} v_2 \,\},$$

where $k > 1$ is the number of arrows from v_1 to v_2.

Remark 35 This result is expected from results on the corresponding cluster algebra. The Kronecker quiver defines a rank 2 cluster algebra which is known not to be of finite type when $k > 1$. Given that P is the Newton polygon of a superpotential which is itself a combination of cluster monomials, we expect the polygon P to grow as we mutate.

Proof (***Proof of Lemma*** 34) Assume there is a Q_k subquiver of Q_P, with vertices v_1, v_2 corresponding to edges E_1 E_2 of P. We define $\rho\colon \mathbb{Z}^2 \to M$ by mapping the standard basis to the primitive inward normal vectors w_i to E_i for $i \in \{1, 2\}$.

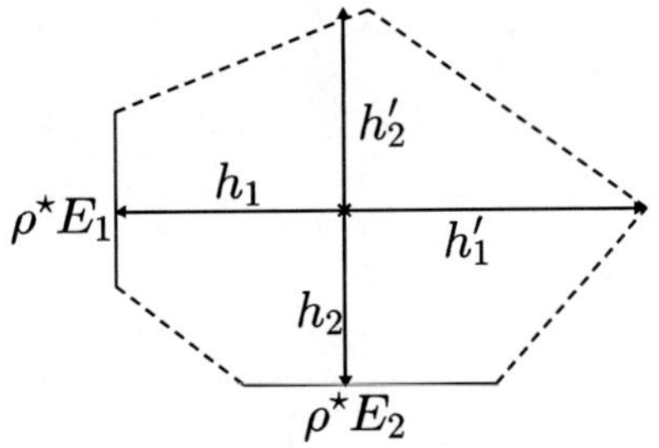

Fig. 15.4 Schematic diagram of a polygon in standard form

Let $P' \subset \mathbb{Q}^2$ be the image of P under $\rho^\star$. The resulting polygon in $\mathbb{Q}^2$ is shown schematically in Fig. 15.4.

The *local index* of each cone in P is the integral height of the edge from the origin. Let h_i denote the local indices of E_i for $i \in \{1, 2\}$. Note that, as $h_i = \langle e_i, \rho^\star u\rangle$ for any $u \in E_i$, h_i is also the local index of $\rho^\star(E_i)$ in P'. Mutating at v_1 and v_2 we denote the new local indices,

$$(h_1, h_2') \longleftarrow (h_1, h_2) \longrightarrow (h_1', h_2).$$

We first show that $\rho^\star$ increases the lattice lengths of E_i by a factor of $k := |w_1 \wedge w_2|$ for each $i \in \{1, 2\}$. Let u_1^i and u_2^i denote the vertices of E_i, and $f_i := (u_1^i - u_2^i)/\ell(E_i)$; where $\ell(E_i)$ denotes the lattice length of E_i. Note that $\langle w_1, f_2\rangle = \langle w_2, f_1\rangle = w_1 \wedge w_2$ for a suitable choice of orientation of M. Moreover, since – for each $i \in \{1, 2\}$ – $\langle \rho^\star(u), e_i\rangle = h_i$ is constant as u varies in E_i, the edge $\rho^\star(E_i)$ has direction vector $e^\star_{3-i}$; where $\{e_1^\star, e_2^\star\}$ is the dual basis to $\{e_1, e_2\}$. In other words,

$$\rho^\star(u_1^i) - \rho^\star(u_2^i) = \ell(\rho^\star(E_i))e^\star_{3-i},$$

and hence we have that

$$\begin{aligned}
\ell(\rho^\star(E_i)) &= \langle \ell(\rho^\star(E_i))e^\star_{3-i}, e_{3-i}\rangle \\
&= \langle \rho^\star(u_1^i) - \rho^\star(u_2^i), e_{3-i}\rangle \\
&= \langle u_1^i - u_2^i, \rho(e_{3-i})\rangle \\
&= \ell(E_i)\langle f_i, w_{3-i}\rangle \\
&= \pm\ell(E_i)(w_1 \wedge w_2),
\end{aligned}$$

where signs and orientations are chosen such that $\ell(E)$ is always positive. Studying Fig. 15.4 note that $h_1 + h_1' \geq \ell(\rho^\star(E_2))$, however—by the calculation above—$\ell(\rho^\star(E_2)) = k\ell(E_2)$. Moreover, we have that $\ell(E_2) \geq h_2$, since the Fano polygon P admits a mutation along this edge. Hence we observe that

$$h_1' \geq kh_2 - h_1 \qquad h_2' \geq kh_1 - h_2.$$

Consider the case $k \geq 3$, and assume without loss of generality that $h_2 \geq h_1$. We have that $h_1' \geq 3h_2 - h_1 \geq 2h_2 \geq 2h_1$. Thus in this case the values in the pair (h_1, h_2) grow (at least) exponentially with mutation, and in particular take infinitely many values.

Next consider the case $k = 2$. The inequalities above become,

$$h_1' \geq 2h_2 - h_1 \qquad h_2' \geq 2h_1 - h_2,$$

and we are again free to assume that $h_2 \geq h_1$. Thus $h_1' \geq 2h_2 - h_1 \geq h_1$, and if $h_2 > h_1$, $h_1' \geq 2h_2 - h_1 > h_2$. Thus, assuming $h_1 \neq h_2$, one can generate an infinite increasing sequence of local indices. The only remaining case is if $h := h_1 = h_2 = h_1' = h_2'$. To eliminate this possibility observe that, since $k = 2$, the edges $\rho^\star(E_1)$, $\rho^\star(E_2)$ must meet in a vertex with coordinates $(-h, -h)$ (indeed, assuming this does not hold, a mutation returns us to the previous case and one of the above inequalities is strict). Note that the sublattice $\rho^\star(N)$ is determined by the fact that $\rho^\star$ doubles the edge lengths of E_1 and E_2. The lattice vectors (a, a) are in this sublattice for all $a \in \mathbb{Z}$. Thus, by primitivity of the vertices in P, $h = 1$. Since the origin is in the interior of P, mutating in one of v_1 or v_2 returns us to the previous case. □

Remark 36 Proposition 34 implies all the quivers that we consider from now on are directed graphs. Hence we refer to vertices as *adjacent* if they are adjacent in the underlying graph.

As well as the non-existence of Kronecker quivers in Q_P for finite mutation type polygons P, we use heavy use of a connectedness result for quivers Q_P which follows immediately from the definition of Q_P via determinants in the plane; or equivalently from the fact the exchange matrix has rank 2 (Fig. 15.5).

Lemma 37 *Given a Fano polygon P and vertices v_1, v_2, v_3 of Q_P such that v_i and v_{i+1} are not adjacent for $i = 1, 2$, then v_1 and v_3 are not adjacent.*

Proof (*Proof of Theorem* 30) By Lemma 37, if Q_P is not connected, $Q_P \cong A_1^n$ for some n. Similarly, if Q_P is of type A or D, then it must be one of A_2, A_3 or D_4. Thus we only need to show that there is no Fano polygon P of finite mutation type such that C_P is not of finite-type. However C_P is of finite mutation type, and we use the classification described in Theorems 9 and 10, following [9, 12]. In fact, using Lemma 37, none of the eleven exceptional types can occur as Q_P for a Fano polygon P. Hence we can restrict to quivers which admit a *block decomposition* and work case-by-case.

We claim that every quiver Q_P associated to a Fano polygon P which admits a block decomposition is either mutation equivalent to an orientation of a simply-laced Dynkin diagram or to a quiver which contains a subquiver Q_k for $k > 1$. We assume for contradiction that Q_P is the quiver associated to a Fano polygon P of finite-type which is not mutation equivalent to a simply laced Dynkin diagram.

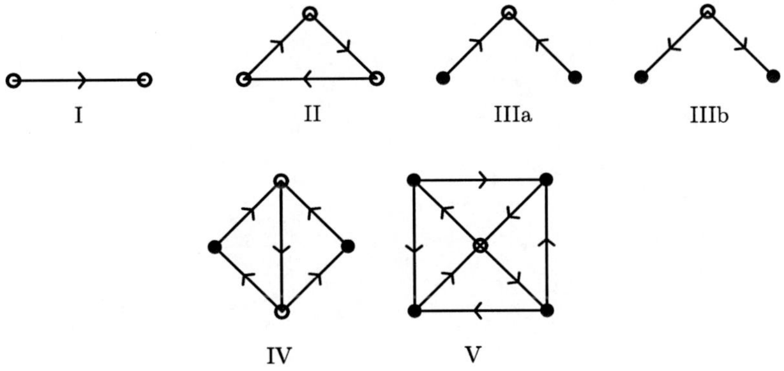

Fig. 15.5 The blocks of a *block decomposition*

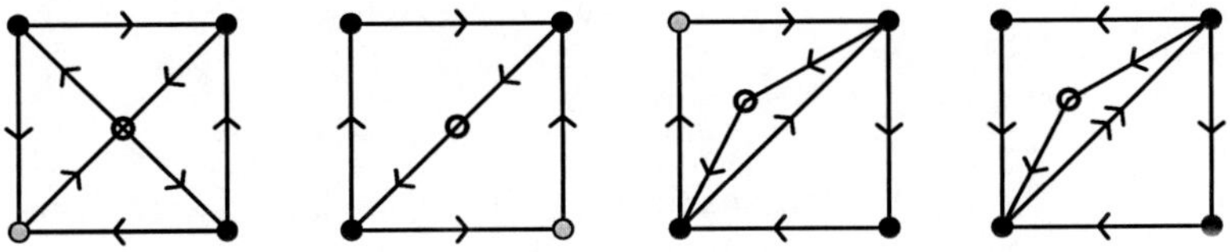

Fig. 15.6 Mutations of block V

Fig. 15.7 Quiver V′

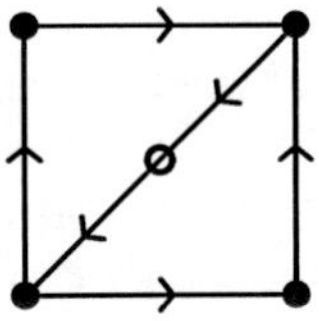

Block V: First observe that, since only one vertex of the block V is an outlet, the V block quiver is a subquiver of any quiver which contains the V block in its decomposition. However this quiver mutates to a quiver with a Q_2 subquiver as shown in Fig. 15.6. Therefore block V never appears in a decomposition of a quiver Q_P. For later use we shall fix the following intermediate quiver, V′, shown in Fig. 15.7.

Blocks IIIa and IIIb: Assume there is a type III block (a or b) connected to a quiver Q' at a vertex v. If there is a vertex v' of Q' such that v and v' are not adjacent, the quiver violates Lemma 37. In particular the vertex set of Q' must be the vertex set of a single block. In particular, using the previous part, Q' has at most four vertices. Case by case study shows that only the A_3 and D_4 types appear.

Block IV: Consider the case of a decomposition only using type IV blocks. Note that the type IV block is itself of type D4. Consider attaching two type IV blocks. If the blocks are attached at a single outlet the resulting quiver contradicts Lemma 37. In fact it is easy to see that it is impossible to add additional type IV blocks to meet this condition. If both pairs of outlets are matched there are two possible quivers depending on the relative orientations of the arrow between the outlets, one orienta-

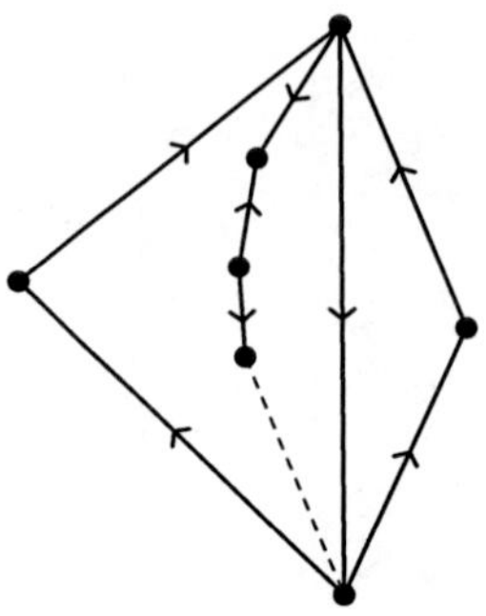

(a) Attaching I blocks to a IV block

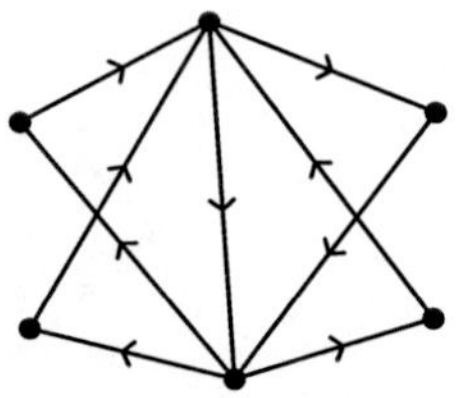

(b) Attaching II blocks to a IV block

Fig. 15.8 A type IV block

tion produces a Q_2 subquiver automatically, the other produces a quiver containing the quiver V$'$ as a subquiver. Thus, for a type IV block to appear in a decomposition of Q_P it must include a type I or II block.

Now consider decompositions using type I and II blocks as well as type IV blocks. First note there must be exactly one IV block (assuming there is at least one). Indeed, if type IV blocks are not connected using both vertices, a non-outlet vertex of a IV block is not adjacent to some outlet, and some non-outlet vertex of a (different) IV block. However outlets and non-outlets of a type IV block are always adjacent, violating Lemma 37.

Thus we must attach I and II blocks to a single type IV block. By Lemma 37 the vertex set of the final quiver must be equal to the vertex set obtained by attaching a single block to each outlet of the IV block. Considering these cases in turn, we note first that attaching a type I block to cancel the arrow between the two outlets produces a quiver mutation equivalent to D_4 and therefore eliminated. For chains type I blocks of length two, if a 3-cycle is produced, a mutation in the vertex between the type I blocks produces the V$'$ quiver. If not, the same mutation produces a Q_2 subquiver.

Attaching a type II block along two outlets of the type IV block recovers the V$'$ or Q_2 subquiver cases we have already seen. Attaching type II blocks to a single outlet each we observe that every new vertex must be adjacent to both outlets of the IV block. Hence the only case without a Q_2 subquiver is shown on the right of Fig. 15.8, however this quiver mutates to one with a Q_2 subquiver. Attaching further type II blocks any quiver we obtain must contradict Lemma 37.

Blocks I and II: From what we have shown above, the block decomposition of Q_P consists only of type I and type II blocks. Any connected quiver with a block decomposition into type I blocks is a path (with possibly changing orientations), which possibly closes up into a cycle. The only cases not violating Lemma 37 are mutation equivalent to orientations of simply laced Dynkin diagrams.

For decompositions of Q_P with type I and II blocks we divide the proof into cases indexed by the number of type II blocks. For a single type II block, we can attach a type I block to two outlets and in this way reduce to the type III case. Attaching each type I block to a type II block in at most one outlet, we use the fact that every new vertex must be adjacent to at least two of the vertices of the type II block. Thus we can obtain only two undirected graphs—the underlying graph of a type IV block or an orientation of a tetrahedron, these cases can easily be eliminated. For example, there is no orientation of the tetrahedron making every cycle oriented; hence after a single mutation we obtain a quiver violating Lemma 34.

Consider the case of a pair of type II blocks. If these have disjoint vertex sets, each outlet of a type II block cannot be adjacent to *two* of the outlets of the other type II block. Thus we must cancel the arrow between these two outlets with a type I block. However this creates a pair of 1-valent non-outlet vertices which can be eliminated similarly to the type III case. At the other extreme, if we attach along all three outlets, we produce two easy cases. Attaching along a pair of outlets we generate either a Q_2 subquiver or a 4-cycle. Considering the 4-cycle with two outlets v_1 and v_2 (on non-adjacent corners) to meet the conditions of Lemma 37 any vertex adjacent to one of v_1 or v_2 must be adjacent to the other. Moreover, if the resulting quiver contains an arrow between v_1 and v_2, a mutation at one of the non-outlet vertices gives a Q_2 subquiver. Given a vertex v adjacent to v_1 and v_2, if this defines a path between them, mutating at this node and a non-outlet in the four cycle produces a Q_2 subquiver. If v does not lie on a path between v_1 and v_2 then mutating at both outlets produces a Q_2 subquiver.

Attaching the type II blocks at a single outlet, the four arrows incident to this vertex are now fixed, so any new vertex must be adjacent to each of the remaining four outlets by Lemma 37. However this cannot be achieved with type I blocks.

Attaching more than two type II blocks together, we can eliminate the case where two are connected to form a 4-cycle as above. Since we can easily eliminate the case that two type II blocks meet in three outlets, we assume that each type II block meets every other in at most one outlet. Some pair of type II blocks must be attached in an outlet (otherwise we can argue as in the case of type II block separated by type I blocks). Thus, since every new vertex must be adjacent to all four outlets formed by attaching two type II blocks, all possible quivers can be represented as an octahedron with some orientation, see Fig. 15.9.

Considering an orientation of the octahedron; if any triangular face does not form a cycle we can mutate to form a Q_2 subquiver. Assuming every triangle is a cycle, and possibly mutating, the vertices adjacent to the 'top' of the octahedron form a type V block subquiver. Following the same reasoning as for the type V block case (although note that the type V block is not part of a block decomposition here) these cases can be eliminated. □

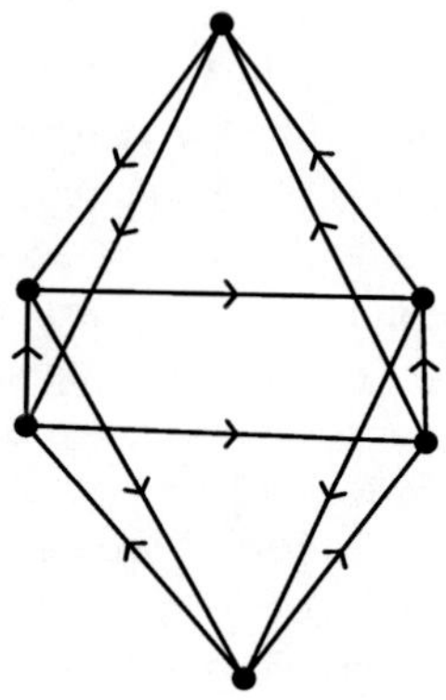

Fig. 15.9 Octahedron of type II blocks

Acknowledgements We thank Alexander Kasprzyk for his insights on polytope mutation, and our many conversations about quivers. The author is supported by a Fellowship by Examination at Magdalen College, Oxford. This work was undertaken while the author was a graduate student at Imperial College London.

References

1. Aharony, O., Hanany, A.: Branes, superpotentials and superconformal fixed points. Nucl. Phys. B **504**(1–2), 239–271 (1997)
2. Akhtar, M., Coates, T., Corti, A., Heuberger, L., Kasprzyk, A.M., Oneto, A., Petracci, A., Prince, T., Tveiten, K.: Mirror symmetry and the classification of orbifold del Pezzo surfaces. Proc. Am. Math. Soc. **144**(2), 513–527 (2016)
3. Akhtar, M., Coates, T., Galkin, S., Kasprzyk, A.M.: Minkowski polynomials and mutations. SIGMA Symmetry Integrability Geom. Methods Appl. **8**, 094, 17 (2012)
4. Akhtar, M., Kasprzyk, A.M.: Singularity Content (2014). arXiv:1401.5458
5. Bergman, A., Proudfoot, N.: Moduli spaces for D-branes at the tip of a cone. J. High Energy Phys. (3), 073, 9 (2006)
6. Bernšteĭn, I.N., Gel/ fand, I.M., Ponomarev, V.A.: Coxeter functors, and Gabriel's theorem. Uspehi Mat. Nauk **28**(2(170)), 19–33 (1973)
7. Bridgeland, T., Stern, D.: Helices on del Pezzo surfaces and tilting Calabi-Yau algebras. Adv. Math. **224**(4), 1672–1716 (2010)
8. Corti, A., Heuberger, L.: Del Pezzo surfaces with $\frac{1}{3}(1, 1)$ points. Manuscripta Math. **153**(1–2), 71–118 (2017)
9. Felikson, A., Shapiro, M., Tumarkin, P.: Skew-symmetric cluster algebras of finite mutation type. J. Eur. Math. Soc. (JEMS) **14**(4), 1135–1180 (2012)
10. Feng, B., Hanany, A., He, Y.H.: Phase structure of D-brane gauge theories and toric duality. J. High Energy Phys. (8), 40, 25 (2001)
11. Fock, V.V., Goncharov, A.B.: Cluster ensembles, quantization and the dilogarithm. II. The intertwiner. In: Algebra, arithmetic, and geometry: in honor of Yu. I. Manin, vol. I, Progress in Mathematics, vol. 269, pp. 655–673. Birkhäuser Boston, Inc., Boston, MA (2009)
12. Fomin, S., Shapiro, M., Thurston, D.: Cluster algebras and triangulated surfaces. I. Cluster complexes. Acta Math. **201**(1), 83–146 (2008)
13. Fomin, S., Zelevinsky, A.: Cluster algebras. I. Foundations. J. Am. Math. Soc. **15**(2), 497–529 (2002)

14. Fomin, S., Zelevinsky, A.: Cluster algebras. II. Finite type classification. Invent. Math. **154**(1), 63–121 (2003)
15. Franco, S., Hanany, A., Martelli, D., Sparks, J., Vegh, D., Wecht, B.: Gauge theories from toric geometry and brane tilings. J. High Energy Phys. (1), 128, 40 (2006)
16. Galkin, S., Usnich, A.: Mutations of Potentials (2010). Preprint IPMU 10-0100
17. Givental, A.: A mirror theorem for toric complete intersections. In: Topological Field Theory, Primitive Forms and Related Topics (Kyoto, 1996), Progress in Mathematics, vol. 160, pp. 141–175. Birkhäuser Boston, Boston, MA (1998)
18. Givental, A.B.: Homological geometry and mirror symmetry. In: Proceedings of the International Congress of Mathematicians, vol. 1, 2 (Zürich, 1994), pp. 472–480. Birkhäuser, Basel (1995)
19. Givental, A.B.: Equivariant Gromov-Witten invariants. Internat. Math. Res. Not. **13**, 613–663 (1996)
20. Gross, M., Hacking, P., Keel, S.: Birational geometry of cluster algebras. Algebr. Geom. **2**(2), 137–175 (2015)
21. Hanany, A., Kazakopoulos, P., Wecht, B.: A new infinite class of quiver gauge theories. J. High Energy Phys. (8), 054, 30 (2005)
22. Hanany, A., Vegh, D.: Quivers, tilings, branes and rhombi. J. High Energy Phys. (10), 029, 35 (2007)
23. Herzog, C.P.: Seiberg duality is an exceptional mutation. J. High Energy Phys. (8), 064, 31 (2004)
24. Hille, L., Perling, M.: Exceptional sequences of invertible sheaves on rational surfaces. Compos. Math. **147**(4), 1230–1280 (2011)
25. Hori, K., Katz, S., Klemm, A., Pandharipande, R., Thomas, R., Vafa, C., Vakil, R., Zaslow, E.: Mirror symmetry, *Clay Mathematics Monographs*, vol. 1. American Mathematical Society, Providence, RI; Clay Mathematics Institute, Cambridge, MA (2003). With a preface by Vafa
26. Ilten, N.O.: Mutations of Laurent polynomials and flat families with toric fibers. SIGMA Symmetry Integrability Geom. Methods Appl. **8**, 047, 7 (2012)
27. Kasprzyk, A.M., Nill, B., Prince, T.: Minimality and mutation-equivalence of polygons. Forum Math. Sigma **5**, e18, 48 (2017)
28. Kontsevich, M.: Lectures at ENS Paris (1998). Set of notes taken by J. Bellaiche, J.-F. Dat, I. Martin, G. Rachinet and H. Randriambololona
29. Leung, N.C., Vafa, C.: Branes and toric geometry. Adv. Theor. Math. Phys. **2**(1), 91–118 (1998)
30. Mandel, T.: Classification of rank 2 cluster varieties. SIGMA Symmetry Integrability Geom. Methods Appl. **15**, 042, 32 (2019)
31. Mukhopadhyay, S., Ray, K.: Seiberg duality as derived equivalence for some quiver gauge theories. J. High Energy Phys. (2), 070, 22 (2004)
32. Perling, M.: Examples for exceptional sequences of invertible sheaves on rational surfaces. In: Geometric methods in representation theory. II, Sémin. Congr., vol. 24, pp. 371–392. Society Mathematics France, Paris (2012)
33. Rietsch, K., Williams, L.: Newton-Okounkov bodies, cluster duality, and mirror symmetry for Grassmannians. Duke Math. J. **168**(18), 3437–3527 (2019)

Chapter 16
Orbit Spaces of Maximal Torus Actions on Oriented Grassmannians of Planes

Hendrik Süß

Abstract Motivated by Buchstaber's and Terzić's work on the complex Grassmannians $G_{\mathbb{C}}(2, 4)$ and $G_{\mathbb{C}}(2, 5)$ we describe the moment map and the orbit space of the oriented Grassmannians $G^+_{\mathbb{R}}(2, n)$ under the action of a maximal compact torus. Our main tool is the realisation of these oriented Grassmannians as smooth complex quadric hypersurfaces and the relatively simple Geometric Invariant Theory of the corresponding algebraic torus action.

Keywords Oriented Grassmannian · Complexity-one T-variety · GIT quotients

16.1 Introduction

We denote an algebraic torus $(\mathbb{C}^*)^k$ by $T_{\mathbb{C}}$ and the corresponding compact torus $(S^1)^k \subset T_{\mathbb{C}}$ by T. A complex algebraic variety with a $T_{\mathbb{C}}$-action is called a $T_{\mathbb{C}}$-variety. The *complexity* of a $T_{\mathbb{C}}$-variety is the minimal (complex) codimension of an orbit. In this paper we study the T-orbit spaces of projective $T_{\mathbb{C}}$-varieties and apply our findings to the case of oriented Grassmannians of planes and that of smooth $T_{\mathbb{C}}$-varieties of complexity 1. Our main goal is to determine the corresponding T-orbit spaces up to homeomorphism.

We consider the Grassmannian $G^+_{\mathbb{R}}(2, n)$ parametrising oriented planes in $\mathbb{R}^2$ with the natural action of a maximal torus in SO_n. Our main result determines the orbit space of this action.

Theorem 1 *The orbit space $G^+_{\mathbb{R}}(2, n)/T$ is homeomorphic to the join*

$$S^{\lfloor n/2 \rfloor - 1} * \mathbb{P}_{\mathbb{C}}^{\lceil n/2 \rceil - 2}.$$

H. Süß (✉)
Friedrich-Schiller-Universität Jena, Institut für Mathematik, Ernst-Abbe-Platz 2, 07743 Jena, Germany
e-mail: hendrik.suess@uni-jena.de

A. M. Kasprzyk and B. Nill (eds.), *Interactions with Lattice Polytopes*, Springer Proceedings in Mathematics & Statistics 386,
https://doi.org/10.1007/978-3-030-98327-7_16

For smooth varieties with a torus action of complexity 1 we derive the following general results on the structure of their orbit spaces.

Theorem 2 *Consider a smooth projective $T_{\mathbb{C}}$-variety X of complexity 1. Then the corresponding orbit space X/T is a topological manifold with boundary.*

Theorem 3 *Consider a smooth projective $T_{\mathbb{C}}$-variety X of complexity 1 with only finitely many lower dimensional $T_{\mathbb{C}}$-orbits. Then the orbit space X/T is homeomorphic to a sphere.*

Note that results comparable to Theorem 2 have been proved by Ayzenberg [3] and Cherepanov [8], Theorem 3 has been proved independently, but using similar methods, by Karshon and Tolman [15]. Their work covers the more general setting of symplectic manifolds with Hamiltonian torus actions. In their paper they also prove Theorem 1 for the cases of complexity 1, i.e. for $n = 5, 6$.

Our main tool is Geometric Invariant Theory (GIT) and its symplectic counterpart in combination with the Kempf–Ness Theorem. This approach suggest to stratify the manifold and eventually the orbit space via a polyhedral subdivision of the momentum polytope, which encodes the variation of GIT quotients. In general these stratifications can become arbitrarily complicated. However, in the cases considered in this paper they turn out to be almost trivial allowing us to derive concrete results about the orbits spaces.

In Sect. 16.2 we fix our setting for compact torus actions induced by algebraic torus actions on complex varieties and recall crucial results from Geometric Invariant Theory. Moreover, we derive first results on the structure of orbits spaces in suitable situations. We then apply these to the special cases of oriented Grassmannians of planes in Sect. 16.3 and $T_{\mathbb{C}}$-varieties of complexity 1 in Sect. 16.4.

In order to distinguish between the algebraic and the topological category, we are going to denote isomorphism of algebraic varieties by $\cong$ and homeomorphisms of topological spaces by $\approx$.

16.2 $T_{\mathbb{C}}$-Varieties and Their T-Orbit Spaces

Fix a linearised action of an algebraic torus $T_{\mathbb{C}} = (\mathbb{C}^*)^k$ on $\mathbb{P}^N_{\mathbb{C}}$ with weights $u_0, \ldots, u_N \in \mathbb{Z}^k$, i.e. for $t = (t_1, \ldots, t_k) \in T_{\mathbb{C}}$ we have

$$t.(z_0 : \ldots : z_N) = (t^{u_0} z_1 : \ldots : t^{u_N} z_N),$$

where $t^{u_j} := t_1^{(u_j)_1} \cdots t_k^{(u_j)_k}$. Then a moment map of this action is given by

$$\nu : \mathbb{P}^N_{\mathbb{C}} \to \mathbb{R}^k; \quad (z_0 : \ldots z_N) \mapsto \frac{\sum_j |z_j|^2 u_j}{\sum_j |z_j|^2}.$$

For an embedded projective variety $X \subset \mathbb{P}^N_{\mathbb{C}}$, which is invariant under under this torus action, a moment map of the induced torus action on X is given by the restriction $\mu = \nu|_X$. The moment image $P = \mu(X)$ is known to be a convex polytope [2, 11]. We start with some notions known as variation of GIT quotients with [9, 12, 16] being the most relevant references. For a point $x \in X$ the moment image $\Delta(x) = \mu(\overline{T_{\mathbb{C}}.x}) \subset P$ of its orbit closure is again a polytope and the orbit $T_{\mathbb{C}}.x$ is mapped to the relative interior $\Delta^\circ(x) \subset \Delta(x)$.

For a point $u \in P$ we define

$$X^{ss}(u) = \{x \in X \mid u \in \Delta(x)\}, \quad X^{ps}(u) = \{x \in X \mid u \in \Delta^\circ(x)\}.$$

Hence, $X^{ss}(u)$ consists of those points in X whose orbit closures intersect $\mu^{-1}(u)$ and $X^{ps}(u)$ consists of those points whose orbits intersect $\mu^{-1}(u)$. Equivalenty $X^{ps}(u)$ is the union of closed $T_{\mathbb{C}}$-orbits in $X^{ss}(u)$.

Now for every $u \in P$ we may consider

$$\lambda(u) = \bigcap_{x, u \in \Delta(x)} \Delta(x); \quad \lambda^\circ(u) = \bigcap_{x, u \in \Delta^\circ(x)} \Delta^\circ(x)$$

Since only finitely many polytopes occur as moment images of orbit closures their intersections are again polytopes. We denote the set of all these polytopes $\lambda(u)$ by Λ. This set is partially ordered by the face relation $\prec$. The polytopes $\lambda \in \Lambda$ form a polyhedral subdivision of P and one obtains a stratification of P via their relative interiors.

$$P = \bigsqcup_{\lambda \in \Lambda} \lambda^\circ.$$

For $u \in P$ let us denote by $\lambda(u)$ the unique element of $\lambda \in \Lambda$ such that $u \in \lambda^\circ$.

From the definitions above it follows that $X^{ss}(u) = X^{ss}(v)$ if and only if $\lambda(u) = \lambda(v)$, i.e. u and v are contained in the relative interior of the same element of Λ. In this case also $X^{ps}(u) = X^{ps}(v)$ holds. Hence, we may define $X^{ss}_\lambda = X^{ss}(u)$ and $X^{ps}_\lambda = X^{ps}(u)$ for $u \in \lambda^\circ$.

Example 4 We consider the linear $T = \mathbb{C}^*$-action on $\mathbb{P}^2_{\mathbb{C}}$ given by $t.(x : y : z) = (tx : t^{-1}y : z)$. Then the moment map is given by $\mu(x : y : z) = \frac{|x|^2 - |y|^2}{|x|^2 + |y|^2 + |z|^2}$. We get $P = \mu(\mathbb{P}^2) = [-1, 1] \subset \mathbb{R}$. The orbits can be described as follows. We have the fixed points $(1 : 0 : 0)$, $(0 : 1 : 0)$ and $(0 : 0 : 1)$ with moment images 1, -1 and 0, respectively. The moment images of the other $T_{\mathbb{C}}$-orbits are

$$\begin{aligned}
\mu(T.(1:0:1)) &= (0, 1) \subset \mathbb{R}, \\
\mu(T.(0:1:1)) &= (-1, 0) \subset \mathbb{R}, \\
\mu(T.(1:1:0)) &= (-1, 1) \subset \mathbb{R}, \\
\mu(T.(\alpha:1:1)) &= (-1, 1) \subset \mathbb{R}, \ \alpha \in \mathbb{C}^*.
\end{aligned}$$

Hence, in this case Λ is obtained by subdividing the interval $[-1, 1]$ at the point 0, or more formally $\Lambda = \{[-1, 0], [0, 1], \{-1\}, \{0\}, \{1\}\}$.

By the Kempf–Ness Theorem for rational values of u the definition of $X^{ss}(u)$ coincides with the semi-stable locus of Mumford's Geometric Invariant Theory. Hence, there exists a categorical quotient morphism $q_\lambda : X^{ss}_\lambda \to Y_\lambda = X^{ss}_\lambda /\!\!/ T_{\mathbb{C}}$ where Y_λ is an orbit space for the $T_{\mathbb{C}}$-action on X^{ps}_λ and the corresponding quotient map is given by the restriction of q_λ to X^{ps}_λ. The occurring quotients Y_u have the expected dimension for $u \in P^\circ$, but can be lower-dimensional for elements $u \in \partial P$. By [17, Lemma 7.2] for $u \in \lambda^\circ$ every $T_{\mathbb{C}}$-orbit in X^{ps}_λ intersects $\mu^{-1}(u)$ in exactly one T-orbit. Hence, the restriction of q_λ to $\mu^{-1}(u)$ induces a homeomorphisms between $Y_\lambda = X^{ss}_\lambda /\!\!/ T_{\mathbb{C}}$ and the topological orbit space $\mu^{-1}(u)/T$. Moreover the inclusion $X^\lambda \subset X^\gamma$ for $\gamma \prec \lambda$ induces contraction morphisms on the level of quotients $p_{\gamma\lambda} : Y_\lambda \to Y_\gamma$ forming an inverse system.

Example 5 When $\dim X = \dim T_{\mathbb{C}}$, i.e. if the variety is toric, then the moment image P completely determines the variety. The moment images $\Delta(x)$ of $T_{\mathbb{C}}$-orbit closures are just the faces of the polytope P and the stratification of P is the decomposition of P into the relative interiors of its faces. The preimage $\mu^{-1}(u)$ consists of exactly on T-orbit with dimension equal to dimension of the face containing u in its interior. Consequently $X/T \approx P$ with μ coinciding with the quotient map. Alternatively, we may apply Proposition 12 below and obtain $X/T \approx S^{k-1} * \{\text{pt}\} \approx D^k \approx P$.

Example 6 Consider a projective toric variety X corresponding to a polytope $Q \subset \mathbb{R}^d$. Then the inclusion of a k-dimensional subtorus $T_{\mathbb{C}}' \subset T_{\mathbb{C}}$ induces a surjection $F : \mathbb{R}^d \to \mathbb{R}^k$. Given a moment map μ for the $T_{\mathbb{C}}$-action a corresponding moment map $\mu' : X \to \mathbb{R}^k$ is given by $\mu' = F \circ \mu$. Hence, the moment image for the $T_{\mathbb{C}}'$-action is $P := F(Q)$ and the stratification of P is induced by the images of the faces of Q. More precisely, the stratification consists of the relative interiors of the polytopes

$$\lambda(u) = \bigcap_{\tau \prec Q,\ u \in F(\tau)} F(\tau).$$

Moreover, the GIT quotients $X^{ss}(u) /\!\!/ T_{\mathbb{C}}'$ are again toric varieties corresponding to the polytope $F^{-1}(u) \cap Q$, see [13, Proposition 3.5].

Already in [10] it has been observed that orbit space of the T-action on X can be constructed out of the inverse system of GIT quotients.

Theorem 7 ([10, Sect. 5]) *We have*

$$X/T \approx \left(\bigsqcup_{\lambda \in \Lambda} \lambda \times Y_\lambda \right) /\!\sim,$$

where $(u, y) \sim (u, y')$ *if* $(u, y) \in \gamma \times Y_\gamma$, $(u, y') \in \lambda \times Y_\lambda$ *with* $\gamma \prec \lambda$ *and* $p_{\gamma\lambda}(y') = y$.

We easily derive the following result, which turns out to be a little bit handier in some situations.

Corollary 8 *Assume that we have a compact topological space Y and with proper surjective maps $r_\lambda \colon Y \to Y_\lambda$ being compatible with the inverse system above. Then we have the following homeomorphism.*

$$X/T \approx (P \times Y)/\sim_r.$$

Here, the equivalence relation is generated by

$$(u, y) \sim_r (u, y') \Leftrightarrow r_{\lambda(u)}(y) = r_{\lambda(u)}(y')$$

for $u \in \lambda^\circ$.

Proof There is a canonical map

$$P \times Y \to \left(\bigsqcup_{\lambda \in \Lambda} \lambda \times Y_\lambda \right) /\sim; \quad (u, y) \mapsto [(u, r_{\lambda(u)}(y))]$$

This map is surjective and continuous and identifies exactly those pairs which are equivalent under $\sim_r$. The quotient $(P \times Y)/\sim_r$ is compact as $(P \times Y)$ is and by Theorem 7 the codomain of the map is homeomorphic to X/T, which is a Hausdorff space. Hence, the induced continuous bijection $(P \times Y)/\sim_r \to X/T$ is a homeomorphism. □

Remark 9 In [6, 7] such a Y is called a *universal parameter space* for the $T_\mathbb{C}$-obits. In algebraic geometry a natural choice for such a dominating algebraic object Y would be the inverse limit of the Y_λ or the Chow quotient of X by $T_\mathbb{C}$, which can be identified with a distinguished irreducible component of this inverse limit.

If the structure of the inverse system $\{Y_\lambda\}_{\lambda \in \Lambda}$ of GIT quotients is complicated Corollary 8 might not give much concrete information about the orbit space X/T. However, in certain situations this structure turns out to be almost trivial allowing us to effectively calculate the orbit space.

Definition 10 We say the $T_\mathbb{C}$-action on $X \subset \mathbb{P}^N_\mathbb{C}$ has an *almost trivial variation of GIT* if for $\lambda \not\subset \partial P$ the quotients Y_λ are all isomorphic to some Y.

Example 11 If the torus action has complexity one than the quotients Y_λ are smooth algebraic curves or just a point, where the latter happens at most over the boundary of P. The only contraction morphisms here are isomorphisms or the contraction of a curve to a point. Hence, the definition is automatically fulfilled.

Proposition 12 *Consider a $T_\mathbb{C}$-action on X with almost trivial variation of GIT and only finitely many lower-dimensional $T_\mathbb{C}$-orbits. Then X/T is homeomorphic to the topological join $S^{k-1} * Y$.*

Proof Having an almost trivial variation of GIT means that $Y_\lambda \cong Y$ for $\lambda \not\subset \partial P$. On the other hand, having only finitely many lower-dimensional $T_\mathbb{C}$-orbits implies that the moment fibre of a boundary point $u \in \partial P$ consists of exactly one (lower-dimensional) orbit and therefore $Y_{\lambda(u)}$ is just a point. Hence, by Corollary 8 we have

$$X/T \approx (P \times Y)/\sim_\partial,$$

where the equivalence relation $\sim_\partial$ is generated by $(u, y) \sim_\partial (u, y')$ for $u \in \partial P$. Now, the claim follows from Lemma 13 below. □

Lemma 13 *Consider the closed unit disc D^k and the unit sphere S^{k-1}. Then for any compact topological manifold Y we have*

$$S^{k-1} * Y \approx (D^k \times Y)/\sim_\partial,$$

where the equivalence relation $\sim_\partial$ is generated by $(u, y) \sim_\partial (u, y')$ for $u \in \partial D^k$.

Proof Recall that the join $S^{k-1} * Y$ is defined as $(S^{k-1} \times Y \times [0, 1])/\sim$, with the equivalence relation being generated by $(s, y, 0) \sim (s', y, 0)$ and $(s, y, 1) \sim (s, y', 1)$. Now the homeomorphism is given by

$$(S^{k-1} \times Y \times [0, 1])/\sim \longrightarrow (D^k \times Y)/\sim_\partial, \quad [(u, y, t)] \mapsto [(tu, y)].$$

As a special case of Lemma 13 we may consider the situation when $Y \approx S^m$. Then Lemma 13 implies $(D^k \times S^m)/\sim_\partial \approx S^{k-1} * S^m$, which is known to be homeomorphic to S^{m+k}. The lemma below gives a slightly more general statement.

Lemma 14 *For any closed $H \subset \mathbb{R}^k$ we have*

$$((D^k \cap H) \times S^m)/\sim_\partial \ \approx \ S^{k+m} \cap (H \times \mathbb{R}^m)$$

where the equivalence relation $\sim_\partial$ is generated by $(u, y) \sim_\partial (u, y')$ for $u \in \partial D^k$.

Proof We can state the homeomorphism explicitly

$$((D^k \cap H) \times S^m)/\sim_\partial \to S^{k+m} \cap H \times \mathbb{R}^m, \quad [(u, y)] \mapsto (u, \sqrt{1 - |u|^2} \cdot y).$$

For every $(u, y) \in (\partial D^k \cap H) \times S^m$ we have $(u, \sqrt{1 - |u|^2} \cdot y) = (u, 0)$. Hence, the map is a well-defined continuous bijection from a compact space to a Hausdorff space and, therefore a homeomorphism. □

16.3 Oriented Grassmanians of Planes as $T_{\mathbb{C}}$-Varieties

We consider the smooth manifold $G^+_{\mathbb{R}}(2,n)$ parametrising oriented planes in $\mathbb{R}^n$. An oriented plane is given by an orthonormal basis (v_1, v_2). Another orthonormal pair (v_1', v_2') gives rise to the same oriented plane if and only if $(v_1', v_2') = (v_1, v_2)Q_\phi$ with

$$Q_\phi = \begin{pmatrix} \cos\phi & -\sin\phi \\ \sin\phi & \cos\phi \end{pmatrix} \in \mathrm{SO}_2 .$$

Hence,

$$G^+_{\mathbb{R}}(2,n) = \{(v_1, v_2) \in \mathbb{R}^{n\times 2} \mid \langle v_i, v_j\rangle = \delta_{ij}\}/\,\mathrm{SO}_2 .$$

A (compact) torus action on $G^+_{\mathbb{R}}(2,n)$ is induced by the choice of a maximal torus in SO_n via its action on the pair (v_1, v_2). A maximal torus T is given by block diagonal matrices of the form $\mathrm{diag}(Q_{\phi_1}, \ldots, Q_{\phi_k})$ in the case $n = 2k$ or $\mathrm{diag}(Q_{\phi_1}, \ldots, Q_{\phi_k}, 1)$ in the case $n = 2k+1$. In the even-dimensional case the induced action on the oriented planes is not effective as $-I$ acts trivially. To obtain an effective action one has to pass to the quotient $T/\langle \pm I\rangle$. However, this does not effect the orbit structure of the action.

It is well-known that the oriented Grassmannian of planes can be identified with the underlying smooth manifold of the complex smooth quadric Q^{n-2} in $\mathbb{P}^{n-1}_{\mathbb{C}}$, see e.g. [18, p. 280]. Indeed, the map

$$\Psi\colon \mathbb{R}^{n\times 2} \to \mathbb{C}^n; \quad (v_1, v_2) \mapsto w = v_1 + i\cdot v_2$$

induces an embedding $\bar\Psi\colon G^+_{\mathbb{R}}(2,n) \hookrightarrow \mathbb{P}^n_{\mathbb{C}}$. This is well-defined as $\Psi((v_1, v_2)Q_\phi) = e^{i\phi}\cdot \Psi(v_1, v_2)$. Moreover the condition $\langle v_1, v_2\rangle = 0$ is equivalent to $\Im(\sum_j w_j^2) = 0$ and $|v_1|^2/|v_2|^2 = 1$ is equivalent to $\Re(\sum_j w_j^2) = 0$. Hence, the image of the embedding in $\mathbb{P}^n_{\mathbb{C}}$ is cut out by the equation $\sum_j w_j^2 = 0$. A change of coordinates

$$z_{2j-1} = w_{2j-1} + i\cdot w_{2j}; \quad z_{2j} = w_{2j-1} - i\cdot w_{2j} \quad \text{for } j = 1, \ldots, k$$

in the case $n = 2k$ and additionally $z_n = w_n$ in the case $n = 2k+1$ leads to the equivalent equation

$$\sum_{j=1}^{k} z_{2j-1}z_{2j} = 0 \tag{16.1}$$

or

$$z_n^2 + \sum_{j=1}^{k} z_{2j-1}z_{2j} = 0, \tag{16.2}$$

respectively. Now in these coordinates one easily checks that for an oriented plane $E \in G^+_{\mathbb{R}}(2,n)$ with

$$\bar{\Psi}(E) = (z_1 : \ldots : z_n)$$

we have

$$\bar{\Psi}(\mathrm{diag}(Q_{\phi_1}, \ldots, Q_{\phi_k})E) = \left(e^{i\phi_1} z_1 : e^{-i\phi_1} z_2 : \ldots : e^{i\phi_k} z_{2k-1} : e^{-i\phi_1} z_{2k}\right)$$

in the case $n = 2k$ and similarly

$$\bar{\Psi}(\mathrm{diag}(Q_{\phi_1}, \ldots, Q_{\phi_k}, 1)E) = \left(e^{i\phi_1} z_1 : e^{-i\phi_1} z_2 : \ldots : e^{i\phi_k} z_{2k-1} : e^{-i\phi_1} z_{2k} : z_{2k+1}\right)$$

in the case $n = 2k + 1$. Let e_j denote the jth canonical basis vector of $\mathbb{Z}^k$. Then the action of $T = (S^1)^k$ above is induced by an algebraic torus action of $T_{\mathbb{C}} = (\mathbb{C}^*)^k$ with weights

$$\deg(z_{2j-1}) = e_j,\ \deg(z_{2j}) = -e_j; \quad j = 1, \ldots, k.$$

and $\deg(z_{2k+1}) = 0$ in the case $n = 2k + 1$.

We are now going to describe the GIT quotients in order to eventually construct the orbit space using Corollary 8. We only describe the case of even $n = 2k$ in detail. The situation for n odd is very similar.

The moment map is given by

$$\mu(z_1 : \ldots : z_n) = \frac{1}{\sum_{i=1}^n |z_i|^2} \sum_{j=1}^k (|z_{2j-1}|^2 - |z_{2j}|^2) e_j. \tag{16.3}$$

The moment image of X is the cross-polytope given as the convex hull of the weights $P = \beta_k = \mathrm{conv}(\pm e_1, \ldots, \pm e_k)$. The fixed point $(1 : 0 : \ldots : 0)$ is mapped to e_1, $(0 : 1 : \ldots : 0)$ to $-e_1$ and similarly for the other coordinates.

Remark 15 The proper faces of the cross-polytope P are exactly the convex hulls of subsets of $\{\pm e_1, \ldots, \pm e_k\}$ where for every j at most one of e_j and $-e_j$ is contained.

Lemma 16 *The moment preimage of a boundary point $u \in \partial P$ consists of exactly one T-orbit. Hence, the quotient $Y_{\lambda(u)}$ is just a single point.*

Proof It follows from Remark 15 that a point $(z_1 : \ldots : z_n) \in X$ is mapped to the boundary of P if and only if all the products $z_{2j-1}z_{2j}$ vanish. Indeed, assume $\mu(z)$ lies in the convex hull $\mathrm{conv}(\sigma_1 e_1, \ldots, \sigma_n e_n)$, where $\sigma_i \in \{-1, 1\}$ for $i = 1, \ldots, n$. Then the coefficients of $\mu(z)$ in the corresponding barycentric coordinates are up to sign the same coefficients as in (16.3). Hence, we must have that

$$\frac{\sum_{j=1}^k \left||z_{2j-1}|^2 - |z_{2j}|^2\right|}{\sum_{i=1}^n |z_i|^2} = 1$$

or equivalently

$$\sum_{j=1}^{k} \left||z_{2j-1}|^2 - |z_{2j}|^2\right| = \sum_{j=1}^{k} \left(|z_{2j-1}|^2 + |z_{2j}|^2\right).$$

This implies that for every $j = 1, \ldots, k$ either $z_{2j-1} = 0$ or $z_{2j} = 0$.

Now, assume $z = (z_1 : \ldots : z_n)$ and $z' = (z'_1 : \ldots : z'_n)$ have the same moment image and that $z_{2j-1}z_{2j} = z'_{2j-1}z'_{2j} = 0$, for $j = 1, \ldots k$. By choosing a suitable representative of the homogeneous coordinates we may assume that $\sum_j |z_j|^2 = \sum_j |z'_j|^2 = 1$. With these choice of homogenous coordinates $|z_{2j-1}|^2 - |z_{2j}|^2 = |z_{2j-1}|^2 - |z_{2j}|^2$ holds for $j = 1, \ldots k$, since $\mu(z) = \mu(z')$. For sign reasons we have either $z_{2j-1} = z'_{2j-1} = 0$ or $z_{2j} = z'_{2j} = 0$. This implies $|z_{2j}| = |z'_{2j}|$ or $|z_{2j-1}| = |z'_{2j-1}|$, respectively. In either case we have $(z'_{2j-1}, z'_{2j}) = (s_j \cdot z_{2j-1}, s_j^{-1} \cdot z_{2j})$ for some element $s_j \in S^1 \subset \mathbb{C}^*$. Hence, z and z' lie in the same T-orbit. □

We consider the rational map

$$q\colon \mathbb{P}^{n-1}_{\mathbb{C}} \dashrightarrow \mathbb{P}^{k-2}_{\mathbb{C}}, \quad (z_1 : \ldots : z_n) \mapsto (z_3 z_4 : \ldots : z_{n-1} z_n).$$

This map is easily seen to be invariant under the $T_{\mathbb{C}}$-action. It is well-defined on the locus of points where at least one of the products $z_{2j-1}z_{2j}$ for $j = 2, \ldots, k$ does not vanish. For a point $z \in X$ this is equivalent to the fact that $\mu(z) \in P^\circ$.

Lemma 17 *For $u \in P^\circ$ the map $q|_{\mu^{-1}(u)}\colon \mu^{-1}(u) \to \mathbb{P}^{k-2}_{\mathbb{C}}$ is a quotient map to the T-orbit space of the fibre.*

Proof Consider $z, z' \in X$ with $\mu(z) = \mu(z') \in P^\circ$ and $q(z) = q(z')$. By choosing a suitable representative of the homogeneous coordinates for z and z' we may assume that $z_{2j-1}z_{2j} = z'_{2j-1}z'_{2j}$ for $j = 2, \ldots, k-1$. Then the defining equation of $X \subset \mathbb{P}^{n-1}_{\mathbb{C}}$ implies $\sum_j z_{2j-1}z_{2j} = \sum_j z'_{2j-1}z'_{2j} = 0$. Hence, also $z_1 z_2 = z'_1 z'_2$ must hold. Let us set $N = \sum_i |z_i|^2$ and $N' = \sum_i |z'_i|^2$. Assume $z_{2j-1}z_{2j} = 0$ then $\mu(z) = \mu(z')$ implies

$$\frac{|z_{2j-1}|^2 - |z_{2j}|^2}{N} = \frac{|z_{2j-1}|^2 - |z_{2j}|^2}{N'}.$$

Hence, for sign reasons we have $z_{2j-1} = z'_{2j-1} = 0$ or $z_{2j} = z'_{2j} = 0$ in this case. Now, for each $j = 1, \ldots k$ we set $s_j = z'_{2j-1}/z_{2j-1}$ or $s_j = z_{2j}/z'_{2j}$ whichever is defined. If they are both defined they have to coincide, since $z_{2j-1}z_{2j} = z'_{2j-1}z'_{2j}$. If $z_{2j-1} = z'_{2j-1} = z_{2j} = z'_{2j} = 0$, then we set $s_j = 1$. By these choices we have $s.z = z'$ with $s = (s_1, \ldots, s_k) \in T_{\mathbb{C}}$. It remains to show, that $s \in T \subset T_{\mathbb{C}}$.

W.l.o.g we may assume that

$$|s_1| = \max\{|s_1|, |s_1^{-1}|, \ldots, |s_k|, |s_k^{-1}|\}. \tag{16.4}$$

The condition $\mu(z) = \mu(z')$ implies

$$\frac{|z_1| - |z_2|}{N} = \frac{|z_1'| - |z_2'|}{N'},$$

Note, that by (16.4) we have $|s_1|^{-2} \leq \frac{N'}{N} \leq |s_1|^2$. Now, from $(z_1', z_2') = (s_1 z_1, s_1^{-1} z_2)$ we obtain

$$\frac{|z_1|^2 - |z_2|^2}{N} = \frac{|s_1|^2|z_1|^2 - |s_1|^{-2}|z_2|^2}{N'},$$

which implies $(N'/N - |s_1|^2)|z_1|^2 = (N'/N - |s_1|^{-2})|z_2|^2$. For sign reasons this is only possible if $|s_1|^2 = N'/N = 1$. Now, it follows from (16.4) that $|s_1| = \cdots = |s_k| = 1$ and $s \in T$. □

Since $X^{ss}(u)$ consists exactly of the orbits whose closures intersect $\mu^{-1}(u)$ it follows also that $q|_{X^{ss}(u)}$ is a good quotient in the sense of Geometric Invariant Theory. Hence, it coincides with the GIT quotient.

Proof (Proof of Theorem 1) We use Corollary 8. Here, Y is just given as $Y = \mathbb{P}^{\lceil n/2 \rceil - 2}$ and by Lemma 16 we have $r_{\lambda(u)} : Y \to \{\mathrm{pt}\}$ for $u \in \partial P$ and by Lemma 17 $r_{\lambda(u)} = \mathrm{id}_Y$. In particular, the equivalent relation on $P \times Y$ is just given by $(u, y) \sim_r (u, y')$ for $u \in \partial P$. Now, the claim follows from Lemma 13. □

We conclude this section by studying the moment images $\Delta(x)$ of $T_{\mathbb{C}}$-orbit closures and the induced subdivision of P from Sect. 16.2. For $n = 2k + 1$ the convex hull of every subset of vertices of P occurs as a moment image. For $n = 2k$ such convex hulls are moment images if an only if they are faces of P or contain at least two pairs of opposite vertices $\{e_i, -e_i\}$, $\{e_j, -e_j\}$. Indeed, for every such polytope Δ a corresponding $T_{\mathbb{C}}$-orbit is given by $T_{\mathbb{C}} \cdot (z_1 : z_2 : \ldots : z_n)$ with $z_{2j-1} \neq 0 \Leftrightarrow e_j \in \Delta$ and $z_{2j} \neq 0 \Leftrightarrow -e_i \in \Delta$ for $j = 1, \ldots, k$. Note, that for $n = 2k + 1$ such $(z_1 : z_2 : \ldots : z_n)$ fulfilling (16.2) always exist, but for $n = 2k$ there is obviously no non-trivial solution of (16.1) where all but one monomial vanish. In both cases, with the exception of $k = 2$, the induced subdivision of P is the same and coincides with the stellar subdivision of P obtained by starring in the origin.

Remark 18 In [7] Buchstaber and Terzić introduced the notion of $(2n, k)$-manifold. It's relatively straightforward to check that for $n = 2k$, the axioms of this notion are indeed fulfilled for the associated effective torus action by $T_{\mathbb{C}}/\langle \pm 1 \rangle$. However, in the odd case on a generic point of the hyperplane section $[z_n = 0]$ we have finite stabilisers of order 2, which violates the conditions for a $(2n, k)$-manifold.

Remark 19 Note that $G_{\mathbb{R}}^+(2, 6)$ can be identified with $G_{\mathbb{C}}(2, 4)$ as both are given by the smooth quadric hypersurface in $\mathbb{P}_{\mathbb{C}}^5$. Hence, for this case we just rediscover the results of [5]. Combinatorially this fact is reflected by coincidence of the moment polytopes, i.e. the cross-polytope β_3 and the hypersimplex $\Delta_{4,2}$.

16.4 Complexity-One $T_{\mathbb{C}}$-Varieties

If the complexity of the torus action is 1 the possible GIT quotients Y_λ are either single points or isomorphic to a fixed algebraic curve Y. Hence, when applying Corollary 8 to this situation the maps $p_\lambda \colon Y \to Y_\lambda$ are either isomorphisms or contractions to a point. Our main aim in this section is to prove that in this situation the resulting orbits spaces are topological manifolds with boundary and even spheres if the number of lower dimensional $T_{\mathbb{C}}$-orbits is finite.

Remark 20 In the toric case the orbit space can be identified with the moment polytope. In particular, it is also a topological manifold with boundary. Hence, Theorem 2 can be seen as generalisation of this fact. On the other hand, this phenomenon is very special to complexity 0 and 1. In higher dimensions this will almost never be the case. For example for a smooth projective variety Y the join $S^n * Y$, which occurs as an orbit space in the situation of Proposition 12, is a topological manifold if and only if $Y \cong \mathbb{P}^1_{\mathbb{C}}$.

Proposition 21 *Consider the projective d-space with a $(d-1)$-torus $T_{\mathbb{C}}$ acting effectively by weights $u_0 = 0, u_1, \dots, u_d \in \mathbb{Z}^{d-1}$ on the coordinates $z_0, \dots, z_d$. Then the orbit space $\mathbb{P}^d_{\mathbb{C}}/T$ is homeormorphic to either a disc or a sphere. In particular, it is a topological manifold with boundary.*

Proof The moment image of $\mathbb{P}^d_{\mathbb{C}}$ is given by the convex hull of the weights $u_0, \dots, u_d$. The weights $u_0, \dots, u_d$ are necessarily affinely dependent in $\mathbb{R}^{d-1}$. On the other hand they span $\mathbb{R}^{d-1}$ as an affine space due to the effectiveness of the torus action. Hence, there is a non-trivial choice of $\alpha_j \in \mathbb{Z}$, such that $0 = \sum_{i=0}^d \alpha_i u_i$ and $0 = \sum_{i=0}^d \alpha_i$ and the coefficients are unique up to simultaneous scaling.

Set $K = \{i \in \{0, \dots, d\} \mid \alpha_i \neq 0\}$. Then P is obtained as the join $Q * \Delta$ of the lower-dimensional polytopes $Q = \operatorname{conv}\{u_i\}_{i \in K}$ and $\Delta = \operatorname{conv}\{u_i\}_{i \notin K}$ of dimensions $m := (\#K - 2)$ and $n := (d - \#K)$, respectively. Here, we allow that $\Delta = \varnothing$ and use the non-standard convention $Q * \varnothing := Q$. Note, that Δ is a simplex (or empty). Hence, a $u \in \Delta$ has a unique representation as $u = \sum_{j \in K} \lambda_j u_j$ with $\lambda_j \geq 0$ and $\sum_j \lambda_j = 1$. For $u \in Q$ such a representation $u = \sum_{j \notin K} \lambda_j u_j$ is unique if and only if $u \in \partial Q$. It follows that $u \in P = Q * \Delta$ has a unique such representation if and only if $u \in \partial Q * \Delta$.

Now, $Y_{\lambda(u)} = \mu^{-1}(u)/T$ is a point whenever $u \in P$ has a unique representation as $u = \sum_j \lambda_j u_j$ and $\mu^{-1}(u)/T \approx \mathbb{P}^1_{\mathbb{C}}$ otherwise. This is just a special case of Example 6, when $F : \mathbb{R}^d \to \mathbb{R}^k$ is given by $F(e_i) = u_i$ for $i = 1, \dots, d$. Then the intersection of $F^{-1}(u)$ and the standard simplex consists of all linear combinations $\sum_{i=1}^n \lambda_i e_i$ with non-negative coefficients, such that $u = \sum_{i=1}^n \lambda_i u_i$ and $\sum \lambda_i = 1$. The result is a point if the linear combination is unique or a line segment if not. The corresponding toric varieties are a single point and $\mathbb{P}^1_{\mathbb{C}}$, respectively. Alternatively, it not hard to show that the non-trivial quotient morphisms $X^{\mathrm{ss}}_\lambda \to Y_\lambda = \mathbb{P}^1$ are all restrictions of the rational map

$$\mathbb{P}^d \dashrightarrow \mathbb{P}^1, \qquad (z_0 : \ldots : z_d) \mapsto \left(z_0^{\sum_i \alpha_i} : \prod_i z_i^{\alpha_i} \right).$$

From Corollary 8 we obtain that

$$\mathbb{P}^d_{\mathbb{C}}/T \approx \left(P \times \mathbb{P}^1_{\mathbb{C}}\right)/\sim,$$

where the equivalence relation $\sim$ is generated by $(u, y) \sim (u, y')$ with $y, y' \in \mathbb{P}^1_{\mathbb{C}}$ and $u \in \partial Q * \Delta \subset \partial P$. Topologically ∂Q can be identified with a sphere S^{m-1} and Q with the cone $S^{m-1} * \{\text{pt}\}$. Similarly we have a homeomorphism $\Delta \approx S^{n-1} * \{\text{pt}\}$ for $\Delta \neq \varnothing$. For the pair $(P, \partial Q * \Delta)$ we obtain

$$\begin{aligned}(P, \partial Q * \Delta) &\approx (S^{m-1} * \text{pt} * S^{n-1} * \{\text{pt}\},\ S^{m-1} * (S^{n-1} * \{\text{pt}\}))\\ &\approx (S^{m+n-1} * \{\text{pt}\} * \{\text{pt}\},\ S^{m+n-1} * \{\text{pt}\})\\ &\approx (D^{m+n} * \{\text{pt}\},\ D^{m+n}).\end{aligned}$$

Note, that the disc D^{m+n} can be identified with the hemisphere via projection and $D^{m+n} * \{\text{pt}\}$ with the corresponding halfdisc. Now, by choosing H to be an arbitrary halfspace it follows from Lemma 14 that the orbit space is a hemisphere.

For $\Delta = \varnothing$ and $P = Q$ we see directly $(P, \partial Q * \Delta) = (Q, \partial Q) \approx (D^{d-1}, S^{d-2})$ and we obtain $\mathbb{P}^d_{\mathbb{C}}/T \approx S^{d+1}$ from invoking Lemma 14 again, this time with $H = \mathbb{R}^{d-1}$. □

Proof (Proof of Theorem 2) We first consider the situation of a complexity-one torus action on the affine space $\mathbb{C}^d$. Such an action is linearisable by [4]. We may equivariantly compactify the $T_{\mathbb{C}}$-action on $\mathbb{C}^d$ to a $T_{\mathbb{C}}$-action on $\mathbb{P}^d_{\mathbb{C}}$. Then $\mathbb{C}^d/T$ is an open subset of $\mathbb{P}^d_{\mathbb{C}}/T$. Hence, the claim follows from the observation in Proposition 21 that the orbit space $\mathbb{P}^d_{\mathbb{C}}/T$ is a manifold with boundary.

To deduce the general case we consider the two situations from Lemma 23. If contractions to a point do not occur the equivalence relation $\sim_r$ is trivial and by Corollary 8 we have $X/T \approx P \times Y$ which is a product of topological manifolds with boundary.

In the second situation, we have $Y \cong \mathbb{P}^1$. Then by [1, Theorem 5] we have an equivariant open cover of X by copies of $\mathbb{C}^n$ and the result follows directly from the consideration above. □

Remark 22 To deduce the general case from the case $X = \mathbb{C}^d$ in the proof of Theorem 2 we may alternatively consider the induced $T_{\mathbb{C}}$-action on the tangent space at a fixed point and reduce everything to this situation by applying Luna's Slice Theorem.

Lemma 23 ([14, Lemma 5.7]) *In the situation of a complexity-one $T_{\mathbb{C}}$-action on a smooth projective variety either*

1. *the GIT quotients are all isomorphic to Y, or*
2. *$Y \cong \mathbb{P}^1_{\mathbb{C}}$.*

Consider the following definition.

Definition 24 A k-holed n-sphere is defined as the intersection of $S^n \subset \mathbb{R}^{n+1}$ with k affine closed halfspaces, such that S^n is contained in none of them and their boundary hyperplanes intersect only outside the sphere.

In other words, we remove the interiors of k disjoint closed discs from S^n.

Proposition 25 *Consider a smooth $T_{\mathbb{C}}$-variety X of complexity 1 with moment polytope P and stratification Λ. Let Λ' be the set of $\lambda \in \Lambda$, such that $\lambda \subset \partial P$ and $Y_\lambda \neq \{\text{pt}\}$. Assume that Λ' consists of k disjoint polytopes. Then X/T is a k-holed sphere.*

Proof In our situation the equivalence relation $\sim_r$ from Corollary 8 is generated by $(u, y) \sim_r (u, y')$ for $u \in \partial P \setminus \bigcup_{\lambda \in \Lambda'} \lambda^\circ$. Note that by the preconditions $\partial P \setminus \bigcup_{\lambda \in \Lambda'} \lambda^\circ$ is a k-holed sphere. Hence, by Lemma 14 the orbit space $X/T_{\mathbb{C}} \approx (P \times \mathbb{P}^1)/\sim_r$ is a k-holed sphere as well. □

This statement is useful to determine the orbit space in concrete situation. To demonstrate this we look at the classification of Fano threefolds from [19]. In [20, 21] the ones within the classification which admit a $T_{\mathbb{C}}$-action of complexity 1 where identified. In the following we determine the orbit spaces for all of them.

Theorem 26 *Using the notation of the Mori-Mukai classification [19] for the Fano threefolds with $(\mathbb{C}^*)^2$-action we obtain the orbits spaces being k-holed spheres with $k = 0$ for Q, 2.24, 2.29, 2.32, 3.10 and 3.20 with $k = 1$ for 2.30, 2.31, 3.18, 3.21, 3.23 and 3.24, with $k = 2$ for 3.19, 3.22, 4.4, 4.5, 4.7 and 4.8.*

Proof For every of the above Fano varieties the combinatorial data provided in [21, Sect. 5] consists of the moment polytope P and a piecewise linear map

$$\Psi \colon P \to \operatorname{Div}_{\mathbb{R}} \mathbb{P}^1_{\mathbb{C}}$$

from the polytope to the vector space of $\mathbb{R}$-divisors on $\mathbb{P}^1$. By [21, Sect. 3.1] we have $Y_{\lambda(u)} = \{\text{pt}\}$ if and only if $\deg \Psi(u) = 0$. Now, one checks that in every case the subset $\{u \in \partial P \mid \deg \Psi(u) > 0\} \subset \partial P$ consists of the interior of k disjoint facts of P. Applying Proposition 25 gives the desired result. □

Remark 27 Note, that Q is the smooth quadric and by Sect. 16.3 coincides with $G^+_{\mathbb{R}}(2, 5)$. Moreover, 2.32 is the variety of complete flags in $\mathbb{C}^3$. Hence, we recover results of [7, Proposition 8] and [15].

Proof (Proof of Theorem 3) To have finitely many lower dimensional $T_{\mathbb{C}}$-orbits implies that the GIT quotients Y_u for $u \in \partial P$ are just points. Then by Lemma 23 we conclude that $Y_u = \mathbb{P}^1_{\mathbb{C}}$ for every $u \in P^\circ$. By applying Proposition 12 we obtain

$$X/T \approx \mathbb{P}^1 * S^{k-1} \approx S^2 * S^{k-1} \approx S^{k+2}.$$

Acknowledgements This research was supported by the program *Interdisciplinary Research* conducted jointly by the Skolkovo Institute of Science and Technology and the Interdisciplinary Scientific Center J.-V. Poncelet. In particular, I am grateful for the warm hospitality offered by Center Poncelet. This work was also partially supported by the grant 346300 for IMPAN from the Simons Foundation and the matching 2015–2019 Polish MNiSW fund. I would like to thank Anton Ayzenberg, Victor Buchstaber, Alexander Perepechko and Nigel Ray for stimulating discussions on the subject of this paper. Finally, I want to thank Yael Karshon and Susan Tolman for making their draft [15] available to me.

References

1. Arzhantsev, I., Perepechko, A., Süß, H.: Infinite transitivity on universal torsors. J. Lond. Math. Soc. (2) **89**(3), 762–778 (2014)
2. Atiyah, M.F.: Convexity and commuting Hamiltonians. Bull. Lond. Math. Soc. **14**(1), 1–15 (1982)
3. Aĭzenberg, A.A.: Torus actions of complexity 1 and their local properties. Tr. Mat. Inst. Steklova **302**(Topologiya i Fizika), 23–40 (2018). English version published in Proc. Steklov Inst. Math. 302(1), 16–32 (2018)
4. Biał ynicki Birula, A.: Remarks on the action of an algebraic torus on k^n. Bull. Acad. Polon. Sci. Sér. Sci. Math. Astronom. Phys. **14**, 177–181 (1966)
5. Buchstaber, V.M., Terzić, S.: Topology and geometry of the canonical action of T^4 on the complex Grassmannian $G_{4,2}$ and the complex projective space $\mathbb{CP}^5$. Mosc. Math. J. **16**(2), 237–273 (2016)
6. Buchstaber, V.M., Terzić, S.: Toric topology of the complex Grassmann manifolds. Mosc. Math. J. **19**(3), 397–463 (2019)
7. Bukhshtaber, V.M., Terzich, S.: The foundations of $(2n, k)$-manifolds. Mat. Sb. **210**(4), 41–86 (2019)
8. Cherepanov, V.V.: Orbit spaces of torus actions on Hessenberg varieties (2019). arXiv:1905.02294 [math.AT]
9. Dolgachev, I.V., Hu, Y.: Variation of geometric invariant theory quotients. Inst. Hautes Études Sci. Publ. Math. (87), 5–56 (1998). With an appendix by Nicolas Ressayre
10. Goresky, M., MacPherson, R.: On the topology of algebraic torus actions. In: Algebraic groups Utrecht 1986, Lecture Notes in Mathematics, vol. 1271, pp. 73–90. Springer, Berlin (1987)
11. Guillemin, V., Sternberg, S.: Convexity properties of the moment mapping. Invent. Math. **67**(3), 491–513 (1982)
12. Hu, Y.: The geometry and topology of quotient varieties of torus actions. Duke Math. J. **68**(1), 151–184 (1992)
13. Kapranov, M.M., Sturmfels, B., Zelevinsky, A.V.: Quotients of toric varieties. Math. Ann. **290**(4), 643–655 (1991)
14. Karshon, Y., Tolman, S.: Centered complexity one Hamiltonian torus actions. Trans. Amer. Math. Soc. **353**(12), 4831–4861 (2001)
15. Karshon, Y., Tolman, S.: Topology of complexity one quotients. Pacific J. Math. **308**(2), 333–346 (2020)
16. Kirwan, F.: Momentum maps and reduction in algebraic geometry. pp. 135–171 (1998). Symplectic geometry
17. Kirwan, F.C.: Cohomology of quotients in symplectic and algebraic geometry, Mathematical Notes, vol. 31. Princeton University Press, Princeton (1984)
18. Kobayashi, S., Nomizu, K.: Foundations of differential geometry. Vol. II. Interscience Tracts in Pure and Applied Mathematics, No. 15 Vol. II. Interscience Publishers John Wiley & Sons, Inc., New York (1969)

19. Mori, S., Mukai, S.: Classification of Fano 3-folds with $B_2 \geq 2$. Manuscripta Math. **36**(2), 147–162 (1981/82)
20. Przhiyalkovskiĭ, V.V., Chelʹtsov, I.A., Shramov, K.A.: Fano threefolds with infinite automorphism groups. Izv. Ross. Akad. Nauk Ser. Mat. **83**(4), 226–280 (2019)
21. Süss, H.: Fano threefolds with 2-torus action: a picture book. Doc. Math. **19**, 905–940 (2014)

Chapter 17
The Reflexive Dimension of (0, 1)-Polytopes

Akiyoshi Tsuchiya

Abstract Haase and Melnikov showed that every lattice polytope is unimodularly equivalent to a face of some reflexive polytope. The reflexive dimension of a lattice polytope P is the minimal d so that P is unimodularly equivalent to a face of some d-dimensional reflexive polytope. Computing the reflexive dimension of a lattice polytope is a hard problem in general. In this survey, we discuss the reflexive dimension of a (0, 1)-polytope. In particular, virtue of the algebraic technique on Gröbner bases and a linear algebraic technique, many families of reflexive polytopes arising from several classes of (0, 1)-polytopes are presented, and we see that the (0, 1)-polytopes are unimodularly equivalent to facets of some reflexive polytopes.

Keywords Reflexive polytope · Reflexive dimension · Order polytope · Stable set polytope · Edge polytope · Gröbner basis

17.1 Introduction

The reflexive polytope is one of the keywords belonging to the current trends in the research of convex polytopes. In fact, many authors have studied reflexive polytopes from the viewpoints of combinatorics, commutative algebra and algebraic geometry. Hence, finding new classes of reflexive polytopes is an important problem.

A *lattice polytope* is a convex polytope all of whose vertices have integer coordinates. Two lattice polytopes $\mathcal{P} \subseteq \mathbb{R}^d$ and $\mathcal{P}' \subseteq \mathbb{R}^{d'}$ are said to be *unimodularly equivalent* if there exists an affine map from the affine span $\mathrm{aff}(\mathcal{P})$ of $\mathcal{P}$ to the affine span $\mathrm{aff}(\mathcal{P}')$ of $\mathcal{P}'$ that maps $\mathbb{Z}^d \cap \mathrm{aff}(\mathcal{P})$ bijectively onto $\mathbb{Z}^{d'} \cap \mathrm{aff}(\mathcal{P}')$ and that maps $\mathcal{P}$ to $\mathcal{P}'$. Note that every lattice polytope is unimodularly equivalent to a full-dimensional one. A lattice polytope $\mathcal{P} \subset \mathbb{R}^d$ of dimension d is called *reflexive* if the origin of $\mathbb{R}^d$ is the unique lattice point belonging to the interior of $\mathcal{P}$ and its

---

A. Tsuchiya (✉)
Graduate School of Mathematical Sciences, University of Tokyo, 3-8-1 Komaba, Meguro-ku, Tokyo 153-8914, Japan
e-mail: akiyoshi@ms.u-tokyo.ac.jp

A. M. Kasprzyk and B. Nill (eds.), *Interactions with Lattice Polytopes*, Springer Proceedings in Mathematics & Statistics 386,
https://doi.org/10.1007/978-3-030-98327-7_17

dual polytope

$$\mathcal{P}^\vee := \{\mathbf{y} \in \mathbb{R}^d \mid \langle \mathbf{x}, \mathbf{y} \rangle \leq 1 \text{ for all } \mathbf{x} \in \mathcal{P}\}$$

is also a lattice polytope, where $\langle \mathbf{x}, \mathbf{y} \rangle$ is the usual inner product of $\mathbb{R}^d$. It is known that reflexive polytopes correspond to Gorenstein toric Fano varieties, and they are related to mirror symmetry, e.g., [1, 3]. In each dimension, there exist only finitely many reflexive polytopes up to unimodular equivalence [15]. Moreover, Haase and Melnikov [4] showed that every lattice polytope is unimodularly equivalent to a face of some reflexive polytope. From this result they defined the reflexive dimension of a lattice polytope.

Definition 1 Let $\mathcal{P}$ be a lattice polytope. The reflexive dimension of $\mathcal{P}$, denoted by $\text{rdim}(\mathcal{P})$, is the minimal integer d such that $\mathcal{P}$ is unimodularly equivalent to a face of some d-dimensional reflexive polytope.

We immediately know the following proposition from the fact that there are only finitely many reflexive polytopes up to unimodular equivalence.

Proposition 2 *Given a positive integer d, there exist up to unimodular equivalence only finitely many lattice polytopes whose reflexive dimensions are equal to d.*

Our interest is to classify lattice polytopes whose reflexive dimensions are a given integer. We remark that classifying lattice polytopes whose reflexive dimensions are equal to their dimensions is equivalent to classifying reflexive polytopes. In particular, it is known that there is one reflexive polytope in dimension one, there are 16 in dimension two, 4,319 in dimension three and 473,800,776 in dimension four according to computations by Kreuzer and Skarke [14]. As a next step, we focus on lattice polytopes whose reflexive dimensions are equal to their dimensions plus one. Namely, we consider the following question.

Question 3 For which lattice polytope $\mathcal{P}$, does it follow that $\text{rdim}(\mathcal{P}) = \dim(\mathcal{P}) + 1$?

We note that for a lattice polytope $\mathcal{P}$, $\text{rdim}(\mathcal{P}) = \dim(\mathcal{P}) + 1$ if and only if $\mathcal{P}$ is not reflexive and $\mathcal{P}$ is a facet of some reflexive polytope. For example, the reflexive dimensions of a d-dimensional unit cube and a d-dimensional unit simplex are $d + 1$. A stronger question is the following.

Question 4 For every $(0, 1)$-polytope $\mathcal{P}$, does it follow that $\text{rdim}(\mathcal{P}) = \dim(\mathcal{P}) + 1$? Equivalently, is every $(0, 1)$-polytope a facet of some reflexive polytope?

In order to show that this second Question 4 has a positive answer for some class of $(0, 1)$-polytopes, we give higher-dimensional construction of lattice polytopes. Given two lattice polytopes $\mathcal{P} \subset \mathbb{R}^d$ and $\mathcal{Q} \subset \mathbb{R}^d$, we set the lattice polytope $\Omega(\mathcal{P}, \mathcal{Q}) \subset \mathbb{R}^{d+1}$ with

$$\Omega(\mathcal{P}, \mathcal{Q}) := \text{conv}\{(\mathcal{P} \times \{1\}) \cup (-\mathcal{Q} \times \{-1\})\}.$$

If $\mathcal{P} = \mathcal{Q}$, then we will write $\Omega(\mathcal{P}) := \Omega(\mathcal{P}, \mathcal{P})$. We remark that the origin of $\mathbb{R}^{d+1}$ is always a relative interior lattice point of $\Omega(\mathcal{P})$. Assume that $\mathcal{P}$ is full-dimensional. Then $\Omega(\mathcal{P})$ is also full-dimensional. In particular, $\mathcal{P} \times \{1\}$ is a facet of $\Omega(\mathcal{P})$. Hence for a lattice polytope $\mathcal{P} \subset \mathbb{R}^d$ of dimension d, if $\Omega(\mathcal{P})$ is reflexive, then we know that $\mathcal{P}$ is unimodularly equivalent to a facet of some reflexive polytope.

In this survey, by using this construction, we will show that for several classes of (0, 1)-polytopes, Question 4 has a positive answer. In particular, we will present many large families of reflexive polytopes arising from well-known classes of (0, 1)-polytopes.

This survey is organized as follows: In Sect. 17.2, we will introduce an algebraic technique to show that a given lattice polytope is reflexive. In particular, we will recall basic materials and notation on toric ideals. In Sect. 17.3, we will give two families of reflexive polytopes arising from the order polytopes and the chain polytopes of finite partially ordered sets. In Sect. 17.4, we will give a family of reflexive polytopes arising from the stable set polytopes of perfect graphs. Finally, in Sect. 17.5, we will introduce a linear algebraic technique to show that a lattice polytope is reflexive, and we will give a family of reflexive polytopes arising from the edge polytopes of finite simple graphs.

17.2 Toric Ideals and Reflexive Polytopes

In this section, we introduce an algebraic technique to show that a lattice polytope is reflexive. First, we recall basic materials and notation on toric ideals. Let $K[\mathbf{t}^{\pm 1}, s] := K[t_1^{\pm 1}, \ldots, t_d^{\pm 1}, s]$ be the Laurent polynomial ring in $d + 1$ variables over a field K. If $\mathbf{a} = (a_1, \ldots, a_d) \in \mathbb{Z}^d$, then $\mathbf{t}^{\mathbf{a}}s$ is the Laurent monomial $t_1^{a_1} \cdots t_d^{a_d}s \in K[\mathbf{t}^{\pm 1}, s]$. Let $\mathcal{P} \subset \mathbb{R}^d$ be a lattice polytope of dimension d and $\mathcal{P} \cap \mathbb{Z}^d = \{\mathbf{a}_1, \ldots, \mathbf{a}_n\}$. Then, the *toric ring* of $\mathcal{P}$ is the subalgebra $K[\mathcal{P}]$ of $K[\mathbf{t}^{\pm 1}, s]$ generated by $\{\mathbf{t}^{\mathbf{a}_1}s, \ldots, \mathbf{t}^{\mathbf{a}_n}s\}$ over K. We regard $K[\mathcal{P}]$ as a homogeneous algebra by setting each $\deg \mathbf{t}^{\mathbf{a}_i}s = 1$. Let $K[\mathbf{x}] := K[x_1, \ldots, x_n]$ denote the polynomial ring in n variables over K. The *toric ideal* $I_{\mathcal{P}}$ of $\mathcal{P}$ is the kernel of the surjective homomorphism $\pi : K[\mathbf{x}] \to K[\mathcal{P}]$ defined by $\pi(x_i) = \mathbf{t}^{\mathbf{a}_i}s$ for $1 \leq i \leq n$. It is known that $I_{\mathcal{P}}$ is generated by homogeneous binomials, see, e.g., [27]. Let $<$ be a monomial order on $K[\mathbf{x}]$ and $\mathrm{in}_<(I_{\mathcal{P}})$ the initial ideal of $I_{\mathcal{P}}$ with respect to $<$. The initial ideal $\mathrm{in}_<(I_{\mathcal{P}})$ is called *squarefree* if $\mathrm{in}_<(I_{\mathcal{P}})$ is generated by squarefree monomials. The *reverse lexicographic order* on $K[\mathbf{x}]$ induced by the ordering $x_n <_{\mathrm{rev}} \cdots <_{\mathrm{rev}} x_1$ is the total order $<_{\mathrm{rev}}$ on the set of monomials in the variables $x_1, x_2, \ldots, x_n$ by setting $x_1^{a_1}x_2^{a_2} \cdots x_n^{a_n} <_{\mathrm{rev}} x_1^{b_1}x_2^{b_2} \cdots x_n^{b_n}$ if either (i) $\sum_{i=1}^n a_i < \sum_{i=1}^n b_i$, or (ii) $\sum_{i=1}^n a_i = \sum_{i=1}^n b_i$ and the rightmost nonzero component of the vector $(b_1 - a_1, b_2 - a_2, \ldots, b_n - a_n)$ is negative. A reverse lexicographic order is also called a *graded reverse lexicographic order*. Please refer to [5, Chaps. 1 and 5] for more details on Gröbner bases and toric ideals.

Next, we recall important classes of lattice polytopes. We say that a lattice polytope $\mathcal{P} \subset \mathbb{R}^d$ possesses the *integer decomposition property* if, for each integer $k \geq 1$ and for each $\mathbf{a} \in k\mathcal{P} \cap \mathbb{Z}^N$, there exist $\mathbf{a}_1, \ldots, \mathbf{a}_k$ belonging to $\mathcal{P} \cap \mathbb{Z}^N$ with $\mathbf{a} = \mathbf{a}_1 + \cdots + \mathbf{a}_k$, where $k\mathcal{P} := \{k\mathbf{x} : \mathbf{x} \in \mathcal{P}\}$ is the kth dilated polytope of $\mathcal{P}$. The integer decomposition property is particularly important in the theory and application of integer programming [25, Sect. 22.10]. Moreover, a lattice polytope which possesses the integer decomposition property is normal and very ample. Hence these properties play important roles in algebraic geometry. A lattice polytope $\mathcal{P}$ is called *compressed* [28] if all its "pulling triangulations" are unimodular. In particular, if any lattice point in $\mathbb{Z}^{d+1}$ is a linear integer combination of the lattice points in $\mathcal{P} \times \{1\}$, then $\mathcal{P}$ is compressed if and only if every reverse lexicographic initial ideal of $I_{\mathcal{P}}$ is squarefree [27]. We note that a lattice polytope with a unimodular triangulation possesses the integer decomposition property, hence, all compressed polytopes possess the integer decomposition property.

Now, we introduce an algebraic technique to show that a lattice polytope is reflexive.

Lemma 5 ([8, Lemma 1.1]) *Let $\mathcal{P} \subset \mathbb{R}^d$ be a lattice polytope of dimension d such that the origin $\mathbf{0}$ of $\mathbb{R}^d$ is contained in its interior and $\mathcal{P} \cap \mathbb{Z}^d = \{\mathbf{a}_1, \ldots, \mathbf{a}_n\}$. Suppose that any lattice point in $\mathbb{Z}^{d+1}$ is a linear integer combination of the lattice points in $\mathcal{P} \times \{1\}$ and there exists an ordering of the variables $x_{i_1} <_{\text{rev}} \cdots <_{\text{rev}} x_{i_n}$ for which $\mathbf{a}_{i_1} = \mathbf{0}$ such that the initial ideal* $\text{in}_{<_{\text{rev}}}(I_{\mathcal{P}})$ *of the toric ideal $I_{\mathcal{P}}$ with respect to the reverse lexicographic order $<_{\text{rev}}$ on the polynomial ring $K[\mathbf{x}]$ induced by the ordering is squarefree. Then $\mathcal{P}$ is a reflexive polytope with a regular unimodular triangulation.*

Example 6 Set $\mathbf{a}_i = \mathbf{e}_i$ for $i = 1, \ldots, d$, $\mathbf{a}_{d+1} = -\mathbf{e}_1 - \cdots - \mathbf{e}_d$ and $\mathbf{a}_{d+2} = \mathbf{0}$, where $\mathbf{e}_1, \ldots, \mathbf{e}_d$ are the standard coordinate unit vectors of $\mathbb{R}^d$. Let $\mathcal{P} \subset \mathbb{R}^d$ be the lattice polytope with $\mathcal{P} \cap \mathbb{Z}^d = \{\mathbf{a}_1, \ldots, \mathbf{a}_{d+2}\}$. Then one has

$$I_{\mathcal{P}} = (x_1 \cdots x_{d+1} - x_{d+2}^{d+1}).$$

Hence the initial ideal $\text{in}_{<_{\text{rev}}}(I_{\mathcal{P}})$ of the toric ideal $I_{\mathcal{P}}$ with respect to the reverse lexicographic order $<_{\text{rev}}$ on the polynomial ring $K[\mathbf{x}]$ induced by the ordering $x_{d+2} <_{\text{rev}} x_{d+1} <_{\text{rev}} \cdots <_{\text{rev}} x_1$ is squarefree. Therefore, by Lemma 5, $\mathcal{P}$ is a reflexive polytope with a regular unimodular triangulation.

Recently, several families of reflexive polytopes with regular unimodular triangulations have been given by using this technique [7–13, 21–24].

17.3 Reflexive Polytopes Arising from Order Polytopes and Chain Polytopes

In this section, we give two families of reflexive polytopes with regular unimodular triangulations arising from order polytopes and chain polytopes of finite partially ordered sets. In particular, we show that Question 4 has a positive answer for order and for chain polytopes.

First, we recall some terminologies of finite partially ordered sets and introduce two lattice polytopes arising from finite partially ordered sets. Let P denote a finite partially ordered set (*poset*, for short) on the ground set $[d] := \{1, \ldots, d\}$. A subset I of $[d]$ is called a *poset ideal* of P if $i \in I$ and $j \in P$ together with $j \leq i$ in P, then $j \in I$. Note that the empty set $\varnothing$ and $[d]$ are poset ideals of P. Let $\mathcal{J}(P)$ denote the set of poset ideals of P. A subset A of $[d]$ is called an *antichain* of P if i and j belonging to A with $i \neq j$ are incomparable. In particular, the empty set $\varnothing$ and each 1-element subsets $\{j\}$ are antichains of P. Let $\mathcal{A}(P)$ denote the set of antichains of P. For a poset ideal I of P, we write $\max(I)$ for the set of maximal elements of I. In particular, $\max(I)$ is an antichain. A *linear extension* of P is a permutation $\sigma = i_1 i_2 \cdots i_d$ of $[d]$ which satisfies $a < b$ if $i_a < i_b$ in P.

Stanley [26] introduced two classes of lattice polytopes arising from finite posets, which are called order polytopes and chain polytopes. The *order polytope* $\mathcal{O}_P$ of P is defined to be the convex polytope consisting of those $(x_1, \ldots, x_d) \in \mathbb{R}^d$ such that:

1. $0 \leq x_i \leq 1$ for $1 \leq i \leq d$;
2. $x_i \geq x_j$ if $i \leq j$ in P.

The *chain polytope* $\mathcal{C}_P$ is defined to be the convex polytope consisting of those $(x_1, \ldots, x_d) \in \mathbb{R}^d$ such that:

1. $x_i \geq 0$ for $1 \leq i \leq d$;
2. $x_{i_1} + \cdots + x_{i_k} \leq 1$ for every maximal chain $i_1 < \cdots < i_k$ of P.

For each subset $I \subset [d]$, we define the (0, 1)-vectors $\rho(I) := \sum_{i \in I} \mathbf{e}_i$. In particular $\rho(\varnothing)$ is the origin $\mathbf{0}$ of $\mathbb{R}^d$. Both order polytopes and chain polytopes are (0, 1)-polytopes of dimension d. In fact, in [26, Corollary 1.3 and Theorem 2.2], it is shown that

$$\{\text{the set of vertices of } \mathcal{O}_P\} = \{\rho(I) : I \in \mathcal{J}(P)\},$$
$$\{\text{the set of vertices of } \mathcal{C}_P\} = \{\rho(A) : A \in \mathcal{A}(P)\}.$$

Moreover, both order polytopes and chain polytopes are compressed, hence, they possess the integer decomposition property. However, the class of order polytopes is different from the class of chain polytopes [6, Example 3.5 and Corollary 3.9].

Now, we consider the two lattice polytopes $\Omega(\mathcal{O}_P)$ and $\Omega(\mathcal{C}_P)$ for a finite poset P on $[d]$. In particular, we will see that the toric ideals of these lattice polytopes are squarefree with respect to some monomial orders on $K[\mathbf{x}]$. Let

$$K[O] := K[\{x_I, y_I\}_{I \in \mathcal{J}(P)} \cup \{z\}]$$
$$K[C] := K[\{x_{\max(I)}, y_{\max(I)}\}_{I \in \mathcal{J}(P)} \cup \{z\}]$$

denote the polynomial rings over K, and define the surjective ring homomorphisms π_O and π_C by the following:

1. $\pi_O : K[O] \to K[\Omega(O_P)]$ by setting

$$\pi_O(x_I) = \mathbf{t}^{\rho(I \cup \{d+1\})} s,$$
$$\pi_O(y_J) = \mathbf{t}^{-\rho(J \cup \{d+1\})} s, \text{ and}$$
$$\pi_O(z) = s;$$

2. $\pi_C : K[C] \to K[\Omega(C_P)]$ by setting

$$\pi_C(x_{\max(I)}) = \mathbf{t}^{\rho(\max(I) \cup \{d+1\})} s,$$
$$\pi_C(y_{\max(J)}) = \mathbf{t}^{-\rho(\max(J) \cup \{d+1\})} s, \text{ and}$$
$$\pi_C(z) = s.$$

Here $I, J \in \mathcal{J}(P)$. Then the toric ideal $I_{\Omega(O_P)}$ (resp. $I_{\Omega(C_P)}$) is the kernel of π_O (resp. π_C).

Next, we introduce monomial orders $<_O$ and $<_C$, and $\mathcal{G}_O$ and $\mathcal{G}_C$ which are the set of binomials. Let $<_O$ denote a reverse lexicographic order on $K[O]$ satisfying:

1. $z <_O y_J <_O x_I$;
2. $x_{I'} <_O x_I$ if $I' \subset I$;
3. $y_{J'} <_O y_J$ if $J' \subset J$;

and $\mathcal{G}_O \subset K[O]$ the set of binomials:

$$x_I x_{I'} - x_{I \cup I'} x_{I \cap I'},$$
$$y_J y_{J'} - y_{J \cup J'} y_{J \cap J},$$
$$x_I y_J - x_{I \setminus \{i\}} y_{J \setminus \{i\}},$$
$$x_\varnothing y_\varnothing - z^2.$$

Let $<_C$ denote a reverse lexicographic order on $K[C]$ satisfying:

1. $z <_C y_{\max(J)} <_C x_{\max(I)}$;
2. $x_{\max(I')} <_C x_{\max(I)}$ if $I' \subset I$;
3. $y_{\max(J')} <_C y_{\max(J)}$ if $J' \subset J$;

and $\mathcal{G}_C \subset K[C]$ the set of binomials:

$$x_{\max(I)} x_{\max(I')} - y_{\max(I \cup I')} y_{\max(I * I')},$$
$$y_{\max(J)} y_{\max(J')} - y_{\max(J \cup J')} y_{\max(J * J')},$$
$$x_{\max(I)} y_{\max(J)} - x_{\max(I) \setminus \{i\}} y_{\max(J) \setminus \{i\}},$$
$$x_\varnothing y_\varnothing - z^2.$$

Here

1. I and I' (resp. J and J') are poset ideals of P which are incomparable in $\mathcal{J}(P)$;

2. $I * I'$ (resp. $J * J'$) is the poset ideal of P generated by $\max(I \cap I') \cap (\max(I) \cup \max(I'))$ (resp. $\max(J \cap J') \cap (\max(J) \cup \max(J'))$);
3. i is a maximal element of both I and J.

Proposition 7 ([11, Propositions 2.2 and 2.4]) *Let the notation be as above. Then $\mathcal{G}_O$ (resp. $\mathcal{G}_C$) is a Gröbner basis of $I_{\Omega(O_P)}$ (resp. $I_{\Omega(C_P)}$) with respect to $<_O$ (resp. $<_C$).*

By combining Lemma 5 and Proposition 7 we obtain the following theorem.

Theorem 8 ([11, Theorem 1.3]) *Let P be a finite poset on $[d]$. Then each of $\Omega(O_P)$ and $\Omega(C_P)$ is a reflexive polytope with a regular unimodular triangulation.*

Hence we have:

Corollary 9 *Let P be a finite poset on $[d]$. Then one has*

$$\text{rdim}(O_P) = \text{rdim}(C_P) = d + 1.$$

Remark 10 In [11], larger families of reflexive polytopes are given. In fact, for two finite posets P and Q with $|P| = |Q| = d$, the three lattice polytopes $\Omega(O_P, O_Q)$, $\Omega(O_P, C_Q)$ and $\Omega(C_P, C_Q)$ are studied. By the same technique, we know that $\Omega(O_P, O_Q)$ is a reflexive polytope (with a regular unimodular triangulation) if and only if P and Q have a common linear extension, and $\Omega(O_P, C_Q)$ and $\Omega(C_P, C_Q)$ are always reflexive polytopes with regular unimodular triangulations [11, Theorem 1.3].

Remark 11 In [7, 9, 10], by using other constructions of lattice polytopes, large families of reflexive polytopes are presented. Given two lattice polytopes $\mathcal{P} \subset \mathbb{R}^d$ and $\mathcal{Q} \subset \mathbb{R}^d$, we set the lattice polytope $\Gamma(\mathcal{P}, \mathcal{Q}) \subset \mathbb{R}^d$ with

$$\Gamma(\mathcal{P}, \mathcal{Q}) := \text{conv}\{\mathcal{P} \cup (-\mathcal{Q})\}.$$

If $\mathcal{P} = \mathcal{Q}$, then we will write $\Gamma(\mathcal{P}) := \Gamma(\mathcal{P}, \mathcal{P})$. The two lattice polytopes $\Omega(\mathcal{P}, \mathcal{Q})$ and $\Gamma(\mathcal{P}, \mathcal{Q})$ often have the same properties. In fact, $\Gamma(O_P, O_Q)$ is a reflexive polytope (with a regular unimodular triangulation) if and only if P and Q have a common linear extension [7, Corollary 2.2]. Moreover, $\Gamma(O_P, C_Q)$ and $\Gamma(C_P, C_Q)$ are always reflexive polytopes with regular unimodular triangulations; see [10, Corollary 1.2] and [9, Corollary 1.3]. In [9, 29], combinatorial properties of these polytopes, for example, their volumes, are studied.

17.4 Reflexive Polytopes Arising from the Stable Set Polytopes of Perfect Graphs

In this section, we give a family of reflexive polytopes with regular unimodular triangulations arising from the stable set polytopes of perfect graphs. In particular, we show that Question 4 has a positive answer for the stable set polytopes of perfect graphs.

First, we recall what perfect graphs are and introduce the stable set polytopes of finite simple graphs. Let G be a finite simple graph on the vertex set $[d]$ and $E(G)$ the set of edges of G. (A finite graph G is called *simple* if G possesses no loop and no multiple edge.) A subset $W \subset [d]$ is called *stable* if, for all i and j belonging to W with $i \neq j$, one has $\{i, j\} \notin E(G)$. We remark that a stable set is often called an *independent set*. A *clique* of G is a subset $W \subset [d]$ which is a stable set of the complementary graph $\overline{G}$ of G. The *clique number* $\omega(G)$ of G is the maximal cardinality of a clique of G. The *chromatic number* $\chi(G)$ of G is the smallest integer $t \geq 1$ for which there exist stable set $W_1, \ldots, W_t$ of G with $[d] = W_1 \cup \cdots \cup W_t$. In general, it follows that $\omega(G) \leq \chi(G)$. A finite simple graph G is said to be *perfect* [2] if, for any induced subgraph H of G including G itself, one has $\omega(H) = \chi(G)$. Perfect graphs include many important classes of graphs, for example, chordal graphs and comparability graphs. Moreover, it is known that the complementary graph of a perfect graph is perfect. This characterization of perfect graphs is called the perfect graph theorem. Recently, a stronger characterization of perfect graphs, which is called the strong perfect graph theorem, is known. An *odd hole* is an induced odd cycle of length ≥ 5 and an *odd antihole* is the complementary graph of an odd hole.

Proposition 12 ([2, Strong Perfect Graph Theorem]) *A finite simple graph G is perfect if and only if G has no odd hole and no odd antihole as induced subgraph.*

Next, we introduce the stable set polytopes of finite simple graphs. Let $S(G)$ denote the set of stable sets of G. One has $\varnothing \in S(G)$ and $\{i\} \in S(G)$ for each $i \in [d]$. The *stable set polytope* $Q_G \subset \mathbb{R}^d$ of G is the (0, 1)-polytope which is the convex hull of $\{\rho(W) : W \in S(G)\}$ in $\mathbb{R}^d$. Then the dimension Q_G is equal to d. It is known that every chain polytope is a stable set polytope. In fact, let P be a finite poset on $[d]$. Its *comparability graph* G_P is the finite simple graph on $[d]$ such that $\{i, j\} \in E(G_P)$ if and only if $i < j$ or $j < i$ in P. Then a stable set of G_P corresponds to an antichain of P. Moreover, one has $C_P = Q_{G_P}$. Since every comparability graph is perfect, the class of the chain polytopes is contained in the class of the stable set polytopes of perfect graphs. We see a characterization of perfect graphs in terms of the stable set polytopes.

Proposition 13 ([19, Example 1.3 (c)]) *Let G be a finite simple graph on $[d]$. Then G is perfect if and only if Q_G is compressed.*

Now, we consider the lattice polytope $\Omega(Q_G)$ for a perfect graph G on $[d]$. In particular, we see that the toric ideal of this lattice polytope is squarefree with respect to some monomial order on $K[\mathbf{x}]$. Let

$$K[Q] := K[\{x_S, y_S\}_{S \in S(G)} \cup \{z\}]$$

denote the polynomial ring over K and define the surjective ring homomorphism $\pi_Q : K[Q] \to K[\Omega(Q_G)]$ by:

$$\pi_Q(x_S) = \mathbf{t}^{\rho(S\cup\{d+1\})}s,$$
$$\pi_Q(y_T) = \mathbf{t}^{-\rho(T\cup\{d+1\})}s, \text{ and}$$
$$\pi_Q(z) = s.$$

Here $S, T \in S(G)$. Then the toric ideal $I_{\Omega(Q_G)}$ is the kernel of π_Q.

Next, we introduce monomial order $<_Q$ and $\mathcal{M}_Q$ which is the set of monomials. Let $<_Q$ denote a reverse lexicographic order on $K[Q]$ satisfying:

1. $z <_Q y_T <_O x_S$;
2. $x_{S'} <_Q x_S$ if $S' \subset S$;
3. $y_{T'} <_Q y_T$ if $T' \subset T$;

and set

$$\mathcal{M} := \{x_\varnothing y_\varnothing\} \cup \{x_S y_T : S, T \in S(G), S \cap T \neq \varnothing\} \cup \mathcal{M}_G \cup \mathcal{M}'_G,$$

where $\mathcal{M}_G$ (resp. $\mathcal{M}'_G$) is the minimal set of squarefree monomial generators of $\text{in}_{<_Q}(I_{Q_G})$ (resp. $\text{in}_{<_Q}(I_{(-Q_G)})$). Let $\mathcal{G}_Q$ be a finite set of binomials belonging to $I_{\Omega(Q_G)}$ with $\mathcal{M} = \{\text{in}_{<_Q}(f) : f \in \mathcal{G}_Q\}$.

Proposition 14 ([12, Theorem 2.1]) *Let the notation be as above. Then $\mathcal{G}_Q$ is a Gröbner basis of $I_{\Omega(Q_G)}$ with respect to $<_Q$.*

By combining Lemma 5 and Proposition 14, we can show that $\Omega(Q_G)$ is a reflexive polytope with a regular unimodular triangulation. Moreover, we can give a polyhedral characterization of perfect graphs.

Theorem 15 ([12, Theorem 1.1 (b)]) *Let G be a finite simple graph on $[d]$. Then the following arguments are equivalent:*

1. *G is perfect;*
2. *$\Omega(Q_G)$ is a reflexive polytope with a regular unimodular triangulation;*
3. *$\Omega(Q_G)$ has a regular unimodular triangulation.*

Hence we can obtain the following corollary.

Corollary 16 *Let G be a perfect graph on $[d]$. Then one has* $\text{rdim}(Q_G) = d + 1$.

Remark 17 In [12], a lager family of reflexive polytopes is given. In fact, for two finite simple graphs G_1 and G_2 on $[d]$, the lattice polytope $\Omega(Q_{G_1}, Q_{G_2})$ is studied. By the same technique, it follows from [12, Theorem 1.1 (b)] that the following arguments are equivalent:

1. G_1 and G_2 are perfect;
2. $\Omega(Q_{G_1}, Q_{G_2})$ is a reflexive polytope with a regular unimodular triangulation;
3. $\Omega(Q_{G_1}, Q_{G_2})$ has a regular unimodular triangulation.

Remark 18 In [21], the lattice polytope $\Gamma(Q_{G_1}, Q_{G_2})$ for two finite simple graphs G_1, G_2 on $[d]$ is studied. The two lattice polytopes $\Omega(Q_{G_1}, Q_{G_2})$ and $\Gamma(Q_{G_1}, Q_{G_2})$ have a very similar but different property. In fact, it follows from [21, Theorem 2.8] that the following arguments are equivalent:

1. G_1 and G_2 are perfect;
2. $\Gamma(Q_{G_1}, Q_{G_2})$ is a reflexive polytope with a regular unimodular triangulation;
3. $\Gamma(Q_{G_1}, Q_{G_2})$ is a reflexive polytope.

Remark 19 In [13], the lattice polytopes $\Omega(Q_G, O_P)$ and $\Gamma(Q_G, O_P)$ for a finite simple graph G on $[d]$ and a finite poset P on $[d]$ are studied. In fact, it follows from [13, Theorem 1.2] that the following arguments are equivalent:

1. G is perfect;
2. $\Omega(Q_G, O_P)$ has a regular unimodular triangulation;
3. $\Gamma(Q_G, O_P)$ is a reflexive polytope.

In particular, if G is perfect, then each of $\Omega(Q_G, O_P)$ and $\Gamma(Q_G, O_P)$ is a reflexive polytope with a regular unimodular triangulation.

17.5 Reflexive Polytopes Arising from Edge Polytopes

When we show that $\Omega(\mathcal{P})$ is reflexive for a lattice polytope $\mathcal{P} \subset \mathbb{Z}^d$ of dimension d by using Lemma 5, we need to assume that $\mathcal{P}$ possesses the integer decomposition property. In fact, order polytopes, chain polytopes and the stable set polytopes of perfect graphs are compressed, hence they possess the integer decomposition property. In order to show that Question 4 has a positive answer for a class of (0, 1)-polytopes which do not necessarily possess the integer decomposition property, we should consider other approaches. In this section, we introduce a linear algebraic technique to show that a lattice polytope which does not necessarily possess the integer decomposition property is reflexive, and we give one family of reflexive polytopes arising from the edge polytopes of finite simple graphs.

For two $d \times d$ integer matrices A, B, we write $A \sim B$ if B can be obtained from A by some row and column operations over $\mathbb{Z}$. Let $X = \{\mathbf{x}_1, \ldots, \mathbf{x}_n\}$ be a set of lattice points in $\mathbb{R}^d$. Given a subset $I = \{i_1, \ldots, i_{d+1}\} \subset [n]$, let X_I be the $(d+1) \times (d+1)$ matrix whose ith row vector is $(\mathbf{x}_i, 1)$.

Lemma 20 ([16, Proof of Theorem 2.1]) *Let $X = \{\mathbf{x}_1, \ldots, \mathbf{x}_n\}$ be a set of lattice points in $\mathbb{R}^d$ and $\mathcal{P} \subset \mathbb{R}^d$ a lattice polytope of dimension d all of whose vertices belong to X. Assume that for any subset $I = \{i_1, \ldots, i_{d+1}\} \subset [n]$ with $\det(X_I) \neq 0$, it follows that for some integer $0 \leq s \leq d$,*

$$X_I \sim \begin{pmatrix} 1 & & & & \\ & \ddots & & 0 & \\ & & 1 & & \\ & & & 2 & \\ & 0 & & & \ddots & \\ & & & & & 2 \end{pmatrix} \begin{matrix} \left.\vphantom{\begin{matrix} 1 \\ \ddots \\ 1 \end{matrix}}\right\} s \\ \\ \\ \end{matrix}.$$

Then $\Omega(\mathcal{P})$ is reflexive.

In order to give a family of reflexive polytopes, to find X which satisfies the condition of Lemma 20 is an interesting problem. In fact, by using this lemma, we can get a large family of reflexive polytopes.

Proposition 21 ([16, Proposition 2.3]) *Set*

$$X = \{\mathbf{0}\} \cup \{\mathbf{e}_i : 1 \leq i \leq d\} \cup \{\mathbf{e}_i + \mathbf{e}_j : 1 \leq i \leq j \leq d\} \subset \mathbb{R}^d.$$

Then for any subset $I = \{i_1, \ldots, i_{d+1}\} \subset [n]$ with $\det(X_I) \neq 0$, it follows that for some integer $0 \leq s \leq d$,

$$X_I \sim \begin{pmatrix} 1 & & & & \\ & \ddots & & 0 & \\ & & 1 & & \\ & & & 2 & \\ & 0 & & & \ddots & \\ & & & & & 2 \end{pmatrix} \begin{matrix} \left.\vphantom{\begin{matrix} 1 \\ \ddots \\ 1 \end{matrix}}\right\} s \\ \\ \\ \end{matrix}.$$

By combining Lemma 20 and Proposition 21, we can obtain the following theorem.

Theorem 22 ([16, Theorem 2.1]) *Let $\mathcal{P} \subset \mathbb{R}^d$ be a full-dimensional lattice polytope all of whose vertices belong to*

$$\{\mathbf{0}\} \cup \{\mathbf{e}_i : 1 \leq i \leq d\} \cup \{\mathbf{e}_i + \mathbf{e}_j : 1 \leq i \leq j \leq d\}.$$

Then $\Omega(\mathcal{P})$ is reflexive.

The class of lattice polytopes which satisfy the condition of Theorem 22 contains a well-known large family of (0, 1)-polytopes, namely the edge polytopes of finite simple graphs. Let G be a simple graph on $[d]$. The *edge polytope* $\mathcal{P}_G \subset \mathbb{R}^d$ of G is the convex hull of all vectors $\mathbf{e}_i + \mathbf{e}_j$ such that $\{i, j\} \in E(G)$. This means that the edge polytope of $\mathcal{P}_G$ of G is the convex hull of all row vectors of the *incidence matrix* A_G of G, where A_G is the matrix in $\{0, 1\}^{E(G)\times[d]}$ with

$$a_{e,v} = \begin{cases} 1 & \text{if } v \in e, \\ 0 & \text{otherwise.} \end{cases}$$

Moreover, the dimension of P_G equals $\text{rank}(A_G) - 1$. Hence edge polytopes are not full-dimensional. However, given an edge polytope $\mathcal{P}_G$, one can easily get a full-dimensional unimodularly equivalent copy $\mathcal{P}$ of $\mathcal{P}_G$ by considering the lattice polytope defined as the convex hull of the row vectors of A_G with some columns deleted. This implies that every edge polytope $\mathcal{P}_G$ is unimodularly equivalent to a lattice polytope which satisfies the condition of Theorem 22. In particular, $\Omega(\mathcal{P}_G)$ is unimodularly equivalent to $\Omega(\mathcal{P})$. Hence we can get the following corollary.

Corollary 23 *Let G be a finite simple graph on $[d]$. Then one has* $\text{rdim}(\mathcal{P}_G) = \dim(\mathcal{P}_G) + 1$.

In general, an edge polytope does not possess the integer decomposition property. In fact, it is known when the edge polytope of a connected finite simple graph possesses the integer decomposition property.

Proposition 24 ([18, Corollary 2.3]) *Let G be a connected finite simple graph on $[d]$. Then $\mathcal{P}_G$ possesses the integer decomposition property if and only if for any two odd cycles C and C' of G having no common vertex, there exists an edge of G joining a vertex of C with a vertex of C'.*

However, even if the edge polytope $\mathcal{P}_G$ of a connected finite simple graph G possesses the integer decomposition property, the reflexive polytope $\Omega(\mathcal{P}_G)$ does not always possess the integer decomposition property.

Example 25 Let G be the connected simple graph as follows:

G:

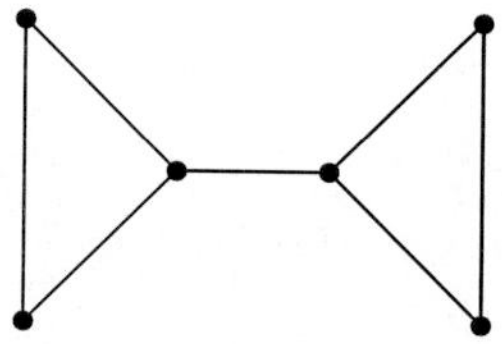

Then G satisfies the condition of Proposition 24. Hence $\mathcal{P}_G$ possesses the integer decomposition property. However, $\Omega(\mathcal{P}_G)$ does not possess the integer decomposition property.

Finally, we characterize when $\Omega(\mathcal{P}_G)$ possesses the integer decomposition property for a connected finite simple graph G.

Theorem 26 ([16, Theorem 3.2]) *Let G be a connected finite simple graph on $[d]$. Then $\Omega(\mathcal{P}_G)$ possesses the integer decomposition property if and only if G does not contain two disjoint odd cycles.*

Remark 27 We remark that the condition of Theorem 26 characterizes some class of edge polytopes. In fact, a connected finite simple graph G does not contain two disjoint odd cycles if and only if $\mathcal{P}_G$ is unimodular, that is, all its triangulations are unimodular [17, Example 3.6 b].

Remark 28 In [20], the lattice polytope $\Gamma(\mathcal{P}_G)$ for a connected finite simple graph G is studied. In fact, if G does not contain two disjoint odd cycles, then $\Gamma(\mathcal{P}_G)$ is a reflexive polytope which possesses the integer decomposition property. However, in general, $\Gamma(\mathcal{P}_G)$ is not always reflexive. For instance, if G is the connected finite simple graph which appears in Example 25, then $\Gamma(\mathcal{P}_G)$ is not reflexive and does not possess the integer decomposition property.

Acknowledgements The author is very grateful to the anonymous referee for his or her insightful report that led to significant improvements of the form of the paper. This manuscript was prepared as a contribution to the conference proceedings of the *Interactions with Lattice Polytopes* at Otto-von-Guericke-Universität Magdeburg on September 14th – 16th, 2017. The author thanks the organizers, Christopher Borger, Alexander Kasprzyk, Benjamin Nill and Johannes Hofscheier, for their support during this conference. The author was partially supported by Grant-in-Aid for JSPS Fellows 16J01549.

References

1. Batyrev, V.V.: Dual polyhedra and mirror symmetry for Calabi-Yau hypersurfaces in toric varieties. J. Algebraic Geom. **3**(3), 493–535 (1994)
2. Chudnovsky, M., Robertson, N., Seymour, P., Thomas, R.: The strong perfect graph theorem. Ann. of Math. (2) **164**(1), 51–229 (2006)
3. Cox, D.A., Little, J.B., Schenck, H.K.: Toric varieties, Graduate Studies in Mathematics, vol. 124. American Mathematical Society, Providence (2011)
4. Haase, C., Melnikov, I.V.: The reflexive dimension of a lattice polytope. Ann. Comb. **10**(2), 211–217 (2006)
5. Hibi, T.: A quick introduction to Gröbner bases. In: Gröbner bases, pp. 1–54. Springer, Tokyo (2013)
6. Hibi, T., Li, N., Li, T.X., Mu, L.L., Tsuchiya, A.: Order-chain polytopes. Ars Math. Contemp. **16**(2), 299–317 (2019)
7. Hibi, T., Matsuda, K.: Quadratic Gröbner bases of twinned order polytopes. Eur. J. Combin. **54**, 187–192 (2016)
8. Hibi, T., Matsuda, K., Ohsugi, H., Shibata, K.: Centrally symmetric configurations of order polytopes. J. Algebra **443**, 469–478 (2015)
9. Hibi, T., Matsuda, K., Tsuchiya, A.: Gorenstein Fano polytopes arising from order polytopes and chain polytopes (2015). arXiv:1507.03221 [math.CO]
10. Hibi, T., Matsuda, K., Tsuchiya, A.: Quadratic Gröbner bases arising from partially ordered sets. Math. Scand. **121**(1), 19–25 (2017)
11. Hibi, T., Tsuchiya, A.: Facets and volume of Gorenstein Fano polytopes. Math. Nachr. **290**(16), 2619–2628 (2017)
12. Hibi, T., Tsuchiya, A.: Reflexive polytopes arising from perfect graphs. J. Combin. Theory Ser. A **157**, 233–246 (2018)
13. Hibi, T., Tsuchiya, A.: Reflexive polytopes arising from partially ordered sets and perfect graphs. J. Algebraic Combin. **49**(1), 69–81 (2019)

14. Kreuzer, M., Skarke, H.: Complete classification of reflexive polyhedra in four dimensions. Adv. Theor. Math. Phys. **4**(6), 1209–1230 (2000)
15. Lagarias, J.C., Ziegler, G.M.: Bounds for lattice polytopes containing a fixed number of interior points in a sublattice. Canad. J. Math. **43**(5), 1022–1035 (1991)
16. Nagaoka, T., Tsuchiya, A.: Reflexive polytopes arising from edge polytopes. Linear Algebra Appl. **557**, 438–454 (2018)
17. Ohsugi, H., Herzog, J., Hibi, T.: Combinatorial pure subrings. Osaka J. Math. **37**(3), 745–757 (2000)
18. Ohsugi, H., Hibi, T.: Normal polytopes arising from finite graphs. J. Algebra **207**(2), 409–426 (1998)
19. Ohsugi, H., Hibi, T.: Convex polytopes all of whose reverse lexicographic initial ideals are squarefree. Proc. Amer. Math. Soc. **129**(9), 2541–2546 (2001)
20. Ohsugi, H., Hibi, T.: Centrally symmetric configurations of integer matrices. Nagoya Math. J. **216**, 153–170 (2014)
21. Ohsugi, H., Hibi, T.: Reverse lexicographic squarefree initial ideals and Gorenstein Fano polytopes. J. Commut. Algebra **10**(2), 171–186 (2018)
22. Ohsugi, H., Tsuchiya, A.: Enriched chain polytopes. Israel J. Math. **237**(1), 485–500 (2020)
23. Ohsugi, H., Tsuchiya, A.: Reflexive polytopes arising from bipartite graphs with γ-positivity associated to interior polynomials. Selecta Math. (N.S.) **26**(4), Paper No. 59, 22 (2020)
24. Ohsugi, H., Tsuchiya, A.: Enriched order polytopes and enriched Hibi rings. Eur. J. Math. **7**(1), 48–68 (2021)
25. Schrijver, A.: Theory of linear and integer programming. Wiley-Interscience Series in Discrete Mathematics. Wiley, Chichester (1986). A Wiley-Interscience Publication
26. Stanley, R.P.: Two poset polytopes. Discrete Comput. Geom. **1**(1), 9–23 (1986)
27. Sturmfels, B.: Gröbner bases and convex polytopes. University Lecture Series, vol. 8. American Mathematical Society, Providence (1996)
28. Sullivant, S.: Compressed polytopes and statistical disclosure limitation. Tohoku Math. J. (2) **58**(3), 433–445 (2006)
29. Tsuchiya, A.: Volume, facets and dual polytopes of twinned chain polytopes. Ann. Comb. **22**(4), 875–884 (2018)